COMPOSITE SATELLITE AND CABLE TELEVISION

COMPOSITE SATELLITE AND CABLE TELEVISION

R.R.GULATI

NEW AGE INTERNATIONAL (P) LIMITED, PUBLISHERS
(formerly Wiley Eastern Limited)
New Delhi • Bangalore • Chennai • Cochin • Guwahati • Hyderabad
Jalandhar • Kolkata • Lucknow • Mumbai • Ranchi
Visit us at www.newagepublishers.com

Published by New Age International (P) Ltd., Publishers
First Edition : 2000
Revised Second Edition : 2005
Reprint : 2006

ISBN : 81-224-1605-5

Rs. 195.00

C-06-07-894

2 3 4 5 6 7 8 9 10

Printed in India at Nisha Enterprises, Delhi.
Typesetter at Pagetek Graphics, Delhi.

PUBLISHING FOR ONE WORLD

NEW AGE INTERNATIONAL (P) LIMITED, PUBLISHERS
(formerly Wiley Eastern Limited)
4835/24, Ansari Road, Daryaganj, New Delhi - 110002
Visit us at **www.newagepublishers.com**

This book is dedicated to
Respected Dr. A. K. Chatterjee
M. S. (Cornell), Ph.D. (Illinois)—FIE (India)
Scientific Research Adviser
Birla Institute of Scientific Research Calcutta;
Ex, Vice-Principal,
Birla Institute of Technology and Science (BITS),
Pilani (Rajasthan) and
Principal, Birla Institute of Technology,
Mesra-Ranchi (Bihar) (Retd.)

PREFACE TO THE SECOND EDITION

The present edition of "Composite Satellite and Cable Television" offers a well planned course material, beginning with basics, to cover all aspects of satellite television transmission and reception. However, over the years, digital technology has made rapid strides in all fields of electronics. Its impact on television has been almost a complete transition from analog to digital form of signal encoding, transmission and reception. Such a transformation has resulted in improved picture reproduction and better sound output. Development work on "Extended Definition Television" has brought in Direct Satellite Broadcasting (DSB) where the customer receives satellite signals directly on a dish instead of getting its analog version through the cable network. Similarly, High Definition Television (HDTV) is now accessible where wider pictures with 16:9 aspect ratio can be viewed on TV receiver screens. These and other developments motivated the author to update the present edition of this book.

For this, the last chapter which describes briefly latest developments in digital technology has been split into seven chapters to provide sufficient coverage of each topic. Chapter 20, after describing basic tools of digital signal processing, as in the present edition, now concludes with a section to introduce the basics of digital transmission and reception. Chapter 21 explains video and audio signal encoding and compression methcds of digital data stream to conserve channel width which otherwise is large for digital transmission.

Chapter 22 gives a complete overview of digital satellite transmission and reception of TV signals. Chapter 23 is devoted to conditional access (CAS) techniques which restrict viewing of pay channels to only those customers who pay for accessing them.

Chapter 24 explains technical details with necessary circuit blocks of the Direct Satellite Broadcast system, also called Direct-To-Home service. In continuation, chapter 25 introduces HDTV, the transmission of which is now available on limited channels.

The last i.e. 26th chapter explores a new concept of watching television by setting an entertainment theatre at home. This is gaining popularity and as such all aspects of its implementation are fully explained. By installing top of the line TV receiver system and surround sound speakers, movies on large screens can be enjoyed as in PVRs.

In addition, Review Questions to all the chapters have been included after Glossary for the benefit of students and convenience of staff teaching this course. It is thus hoped that this revised edition will meet the requirements of a full fledged course on modern digital television and allied topics and also be useful to practising engineers and technicians.

R.R. GULATI

PREFACE TO THE FIRST EDITION

The operation of any cable television system encompasses a very wide frequency range from audio frequencies for sound signals to microwaves for satellite signals. As such, special equipments are needed and have been designed to receive and down-convert microwave satellite signals to the VHF range for processing and distribution over cable networks. Each equipment is unique for specified functions and employs special active devices and circuit formulations. This book has therefore been written to explain fully this somewhat complex system that enables reception of TV signals from satellites and their subsequent processing and distribution to customers.

The contents of the book are based on a well planned course material and cover various aspects of satellite signal transmission, reception, processing and distribution. There are separate chapters on cable system installation, signal measurements, cable plant maintenance and fault finding. The text has thus a full course material on satellite and cable television for electronics engineering students interested in this area of electrical engineering applications. Its contents also meet the needs of cable system engineers and cable station operators and technicians.

ORGANIZATION: The book is broadly divided into two parts, where the first half is devoted to the basics of signal reception techniques and its processing while the second half describes cable plant installation, testing and maintenance. After introducing the essentials of cable television in chapter 1, the following two chapters *i.e.* 2 and 3, describe generation and propagation of RF waves and antennas to receive them. Microwave techniques, devices, special components and dish antennas form the subject matter of chapter 4. Chapters 5 through chapter 10 provide necessary coverage on HF amplifiers, oscillators, frequency control, modulation, frequency translation and voltage regulation. Each chapter begins with the basics and develops it to the level necessary for use in later chapters that describe cable systems.

In the application section of the book, chapter 11 is devoted to terrestrial television to explain TV signal formation, its transmission, reception and circuitry used in TV receivers. Similarly, chapter 12 describes various aspects of geo-stationary satellites and reception of signals from communication satellites at ground stations. Chapters 13 to 17 cover cable systems, processing of CATV signals and their distribution to subscribers. In continuation, chapter 18 is devoted to cable signal measurements and chapter 19 to cable plant maintenance and fault localisation. The concluding 20th chapter outlines additional cable TV features which include two-way cable system, pay television and direct reception techniques like Direct-To-Home (DTH) for individual reception of satellite signals.

PRESENTATION: The book will serve as an excellent two semester course material for senior electronics stream students of polytechnics and Technical Training Schools. It should normally be supplemented in the laboratory with CATV equipment familiarisation exercises and a short duration training at a cable station. The text is equally suitable for a one semester optional course in engineering colleges for the final year students of electronics and electrical engineering.

The matter has been so presented that readers with a basic knowledge of electronics should have no difficulty in following contents of all the chapters. As such, cable operators interested in knowing more about modern cable plant practice will find the book very useful. Similarly technicians employed at CATV stations will find the chapters devoted to equipment installation, maintenance and servicing of special interest.

CREDITS: I am highly indebted to Mr. Sanjeev Kanth, a CATV professional for his useful suggestions and help in revising the manuscript of chapters which describe cable TV systems and associated equipments. I am also grateful to Shri S.K. Marwha, a systems engineer with a cable TV Company in USA and earlier my colleague at BITS, Pilani, for making available useful material on satellite and cable television. I also wish to record with great appreciation, the contribution of my son, Ravi K. Gulati, a CAD engineer with a multinational company in USA, for his efforts in collecting and sending me books and articles devoted to latest developments in the field of satellite communication, which in particular enabled me to upgrade the chapter on Digital Satellite Television and Interactive Cable Systems.

The photographs and technical data used in the book has been made available by many companies engaged in the manufacture of CATV equipments. This courtesy is greatly acknowledged. In particular, I am thankful to M/s. Shyam Communication Systems and M/s M. C. Electronics, both located at Delhi for their kind permission to use their technical leaflets for illustrations in various chapters.

I also express my sincere thanks to Miss Gupta for preparing the artwork for the book by computer graphics technique and Mr. C. P. Dargan for skilfully typing the manuscript.

R. R. GULATI

CONTENTS

1

ELEMENTS OF CABLE TELEVISION

Introduction

Cable television, from its inception, developed in western countries into two separate systems called—Master Antenna Television (MATV) and Community Cable Television (CATV). The purpose of MATV was to deliver strong RF signals from one or more suitably located antennas to every TV receiver in complexes like hotels and apartment buildings, while that of CATV was to receive, combine and distribute signals of terrestrial (OFF-THE-AIR) television broadcasts through a network of cables to subscribers residing in remote towns and villages beyond the service area of local TV stations.

However, with the advent of Satellite Television (SAT-TV), reception of signals from a large number of TV stations located far-away became possible. With this, MATV lost its significance and CATV developed into a complex multi-channel system and expanded in a big way both in big cities and small towns. This large expansion became possible, because cable TV does not have the restriction of channel allocations as is necessary in terrestrial television and can thus offer a variety of programmes on a large number of channels.

Modern Cable Systems carry and deliver signals of local and nearby TV stations that are available directly and of stations located far-away in and outside the country which are received through satellites. Besides, programmes earlier recorded on video tapes and discs are also distributed thereby offering a wide choice to CATV subscribers.

Cable systems meant for small towns, hotels and apartment buildings are less complex and offer six to eleven channels whereas CATV systems developed for large community areas and big hotels are quite complex and distribute television signals on 16 or 31 channels with provision to extend up to 50 or more channels.

From the technical point of view, a cable TV system can be divided into three main sections. These are:

(1) Signal Reception (Outdoor Equipment)
(2) Signal Processing (Indoor *i.e.* Control Room Equipment)
(3) Signal Distribution (Cable Network).

As an introduction to cable television, each section is briefly described.

1.1 Signal Reception

The purpose of this section is to collect TV signals from different sources and on conversion as necessary, deliver these through coaxial cables to corresponding signal processing units located in the control room. The reception of direct VHF and UHF broadcasts is done by installing conventional TV antennas and signals thus collected are sent directly through flat twin or coaxial cables to the signal processing section. However, collection of signals coming from satellites is quite complex, because the arriving signal is extremely weak and its frequency is very high, being in the super UHF *i.e.* microwave spectrum. Therefore, a high gain antenna becomes necessary. A horn type parabolic dish antenna of large diameter meets this requirement. This, when correctly oriented towards the satellite, collects signals arriving from it and reflects them to a common point called 'focal point' which is located in front and above the centre of antenna dish as shown in Fig. 1.1. A feed-horn is actually a small wave-guide section which is mounted at the focal point and its function is to receive signals reflected

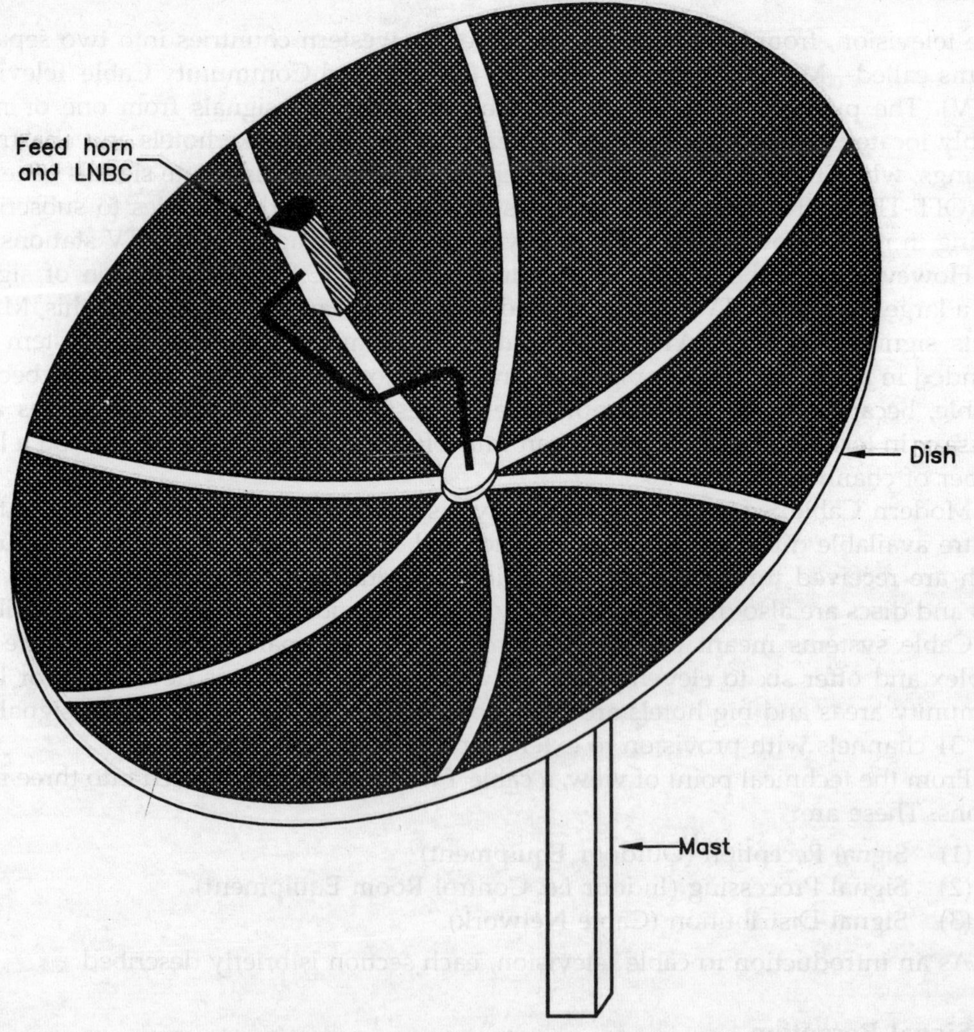

Fig. 1.1. A typical dish antenna assembly with feed-horn and LNBC mounted side-by-side at its focal point.

towards it by the dish and deliver these to the close-by located low noise block converter (LNBC) as shown in the figure.

The down-link signals from most communication satellites are in C-band of frequency spectrum in the range of 3.7 to 4.2 Giga-Hertz (GHz) *i.e.* 3700 to 4200 MHz. In order to minimise losses in coaxial cables that carry dish antenna signals to control room, the collected signals are first translated to a lower frequency range. This is done by a low-noise block converter (LNBC), the building blocks of which are shown in Fig. 1.2. The composite signal collected by the feed horn is fed to a low noise amplifier (LNA) which is specially designed to provide enough gain while maintaining maximum possible signal-to-noise ratio. The LNA output is fed to a converter (mixer) which translates the incoming microwave signals to a lower frequency range of 950 to 1450 MHz. This is achieved by fixing local oscillator (LO) frequency of the converter at 5150 MHz and selecting only the difference products from its output. The difference products will thus have the desired range of 950 MHz (5150 MHz - 4200 MHz) to 1450 MHz (5150 MHz - 3700 MHz). A band-pass filter (BPF) at the output of mixer separates the wanted IF signals from other signals. It is amplified by a multistage IF amplifier and then sent through a high grade coaxial cable to the CATV station. If necessary, a low noise amplifier (LNA) is provided in the middle of coaxial cable-run to make-up for any losses in it [see (Fig. 1.3)]. This LNA is often called 'Bullet Amplifier'.

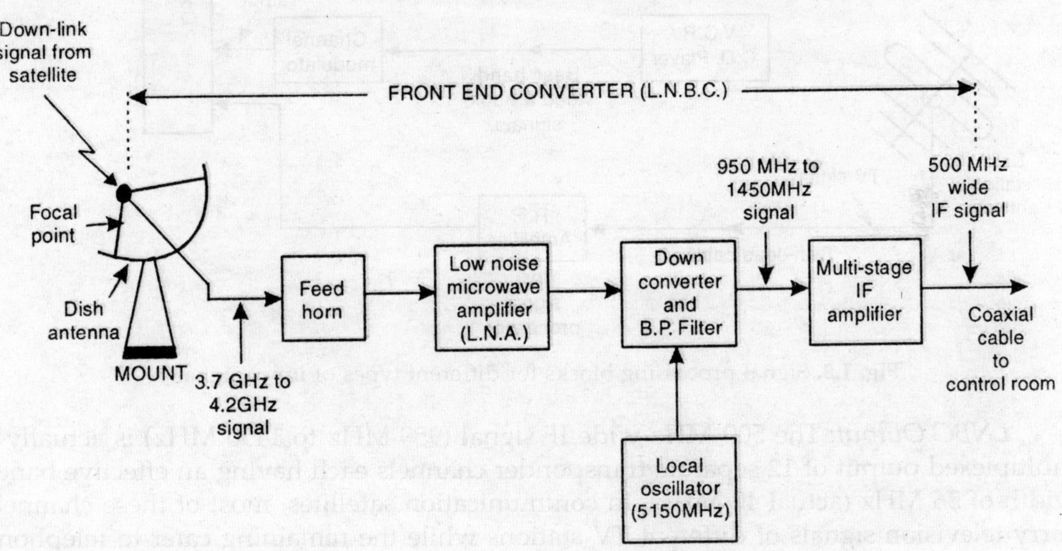

Fig. 1.2. Block diagram of an LNBC also called Front End Converter (FEC).

The feed horn and LNBC unit [see Fig. 1.3] is called Front-End-Converter (FEC) and also referred to as 'Outdoor Equipment' because it is located in the open and close to where the dish antenna is mounted. It is often necessary to install 2, 4, 6 or even 8 dish antenna units with associated feed horn and LNBCs to collect signals from different satellites. Television programmes like movies, plays and songs recorded on video tapes and discs are also distributed on one or two channels of the CATV network. For this, video cassette recorders (VCRs) and C.D. players are provided in the control room.

1.2 Signal Processing

The signal processing unit, also called 'Head-End Equipment' consists of power dividers, satellite receivers, channel modulators, signal processors/amplifiers, VCRs, C.D. players and a combining network. Fig. 1.3 shows necessary details of processing different types of input signals which are briefly described.

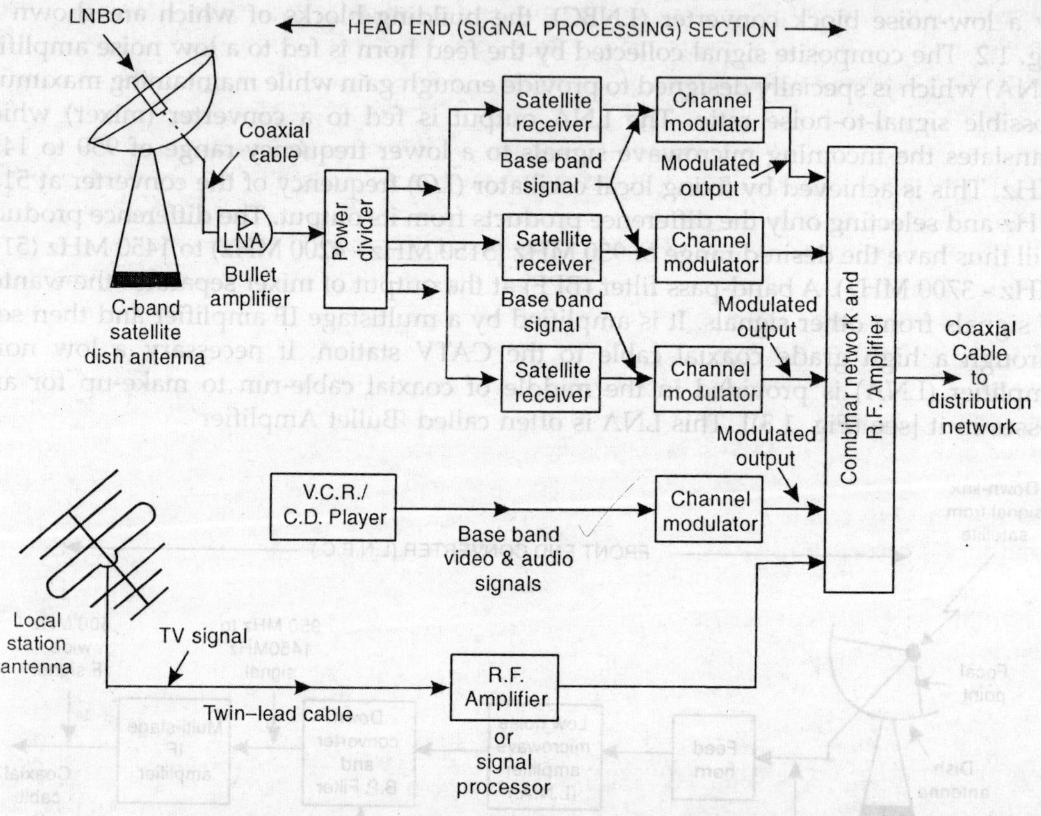

Fig. 1.3. Signal processing blocks for different types of input signals.

LNBC Output: The 500 MHz wide IF signal (950 MHz to 1450 MHz) is actually a multiplexed output of 12 separate transponder channels each having an effective bandwidth of 36 MHz (actual 40 MHz). In communication satellites, most of these channels carry television signals of different TV stations while the remaining cater to telephone and various data transmission services.

Power Divider: The IF signal from the LNBC is delivered to a signal splitter which is actually a multicoupler that divides the signal into independent paths. The signal splitter is commonly called a 'Power Divider' because it enables equal division of signal power at its output ports.

Satellite Receiver: The satellite receiver accepts inputs from one of the output ports on the power divider. It is designed to select manually or by remote control any one of the channels present in the 500 MHz IF signal. The receiver tuner can thus be set to select any desired channel out of many available. On selecting the wanted channel, the receiver functions like a TV receiver and demodulates the received frequency modulated

(FM) signal of the selected channel to provide base-band signal. The term baseband signal is used to indicate the combined video (0-5 MHz) and SIF (FM) signal around 5.5 MHz. The sound IF (SIF) signal is further demodulated as in a TV receiver to obtain audio signal. It is either done in the satellite receiver or at the input of channel modulator.

Channel Modulator: A channel modulator functions as a mini TV transmitter to provide output in the same format as produced at any TV station. Each channel modulator receives video and audio signals from one satellite receiver and modulates these on carrier frequencies of the assigned cable channel. It then combines these outputs to form the composite output signal. The output of each channel modulator feeds into the combiner unit.

Terrestrial TV Signal: As explained earlier, signals of local or nearby TV stations are picked up by multichannel antennas installed at suitable locations. Each antenna output is fed to the combiner after necessary amplification. However, if it is to be shifted to another VHF channel, it is processed by a unit called 'signal processor' to obtain desired output before combining it with other modulator outputs. As necessary, the signal processor demodulates video and audio signals and then modulates these on carrier frequencies of the assigned cable channel.

VCR/C.D. Player: All VCRs/C.D. players enable two types of outputs. One is in the form of video (V) and audio (A) signals and the other in the standard TV format either on VHF channel 3/4 or any UHF channel between 30 to 39. In Fig. 1.3, baseband signal has been chosen as the output and fed to a channel modulator for further processing. The resulting output is then combined with other similar RF signals in the combiner (mixer) unit.

Combiner: This is a network with provision to accept outputs of different channel modulators and combine them to form a single wideband RF signal which is then distributed to subscribers over the cable network.

1.3 Signal Distribution

A representative signal distribution plan is shown in Fig. 1.4. The multiplexed output from the combiner is amplified before taking it to the distribution amplifier which is located at the geographical centre of the area to be served. From there, trunk lines carry the combined RF signal to distant locations. Trunk amplifiers are located on such trunk routes to compensate for signal loss in cables. Signal splitters also called direction-couplers are used to distribute cable signal from trunk lines to branch routes as shown in the figure. Passive matching networks, usually called 'TAP-OFFS', are provided along branch cable runs from where individual cable runs take the signal to TV receiver locations.

The signal level at the input of each TV receiver should be of the order of 1.5 mV (P-P) which is enough to generate good quality pictures. Since improperly terminated lines develop standing waves *i.e.* signal reflections, which result in disturbance and signal loss, the end of each distribution cable is terminated in a resistance (usually 75 ohms) which is equal to the characteristic impedance of the line. Such a termination provides correct impedance match to the line and hence, reflections are prevented.

Set-Top Converter: Drops *i.e.* connections to subscriber TV receivers are made either directly or through a VCR if in use. If the number of channels is not more than

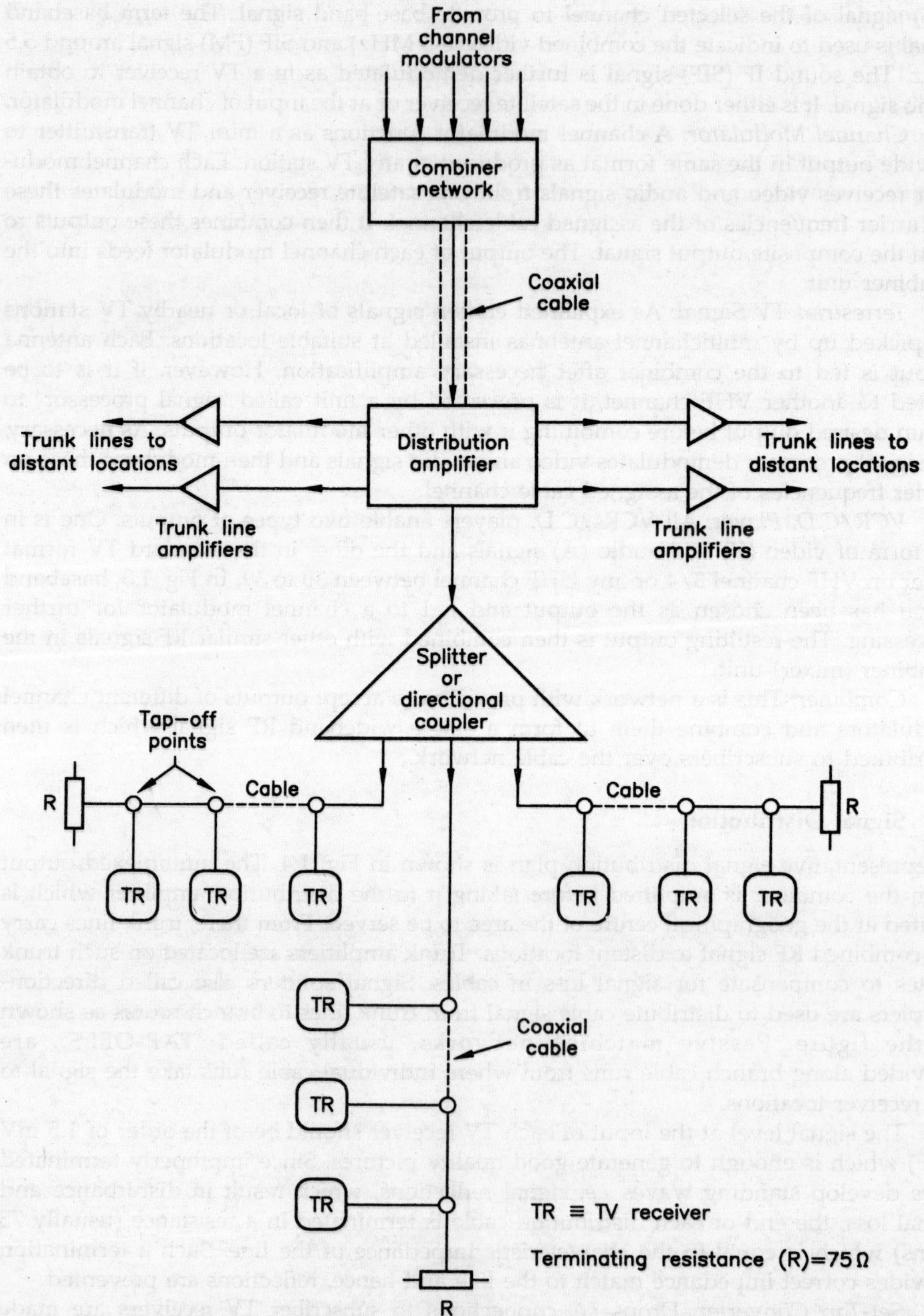

Fig. 1.4. A representative signal distribution plan of a Cable TV system.

eleven, channel modulators are allotted the usual TV channels that lie in VHF I and III bands.

Depending on the mode of connection, the corresponding tuner of the TV receiver or VCR, is used to select incoming programmes on various channels. However, if the number of cable channels is large, it becomes necessary to look for an alternative to accommodate additional channels. For this, recent models of TV receivers have modified tuners that extend selection in the mid band, super band and hyper band *i.e.* higher frequency range than available on earlier receivers. Such TV receivers are called 'CABLE READY' and can also receive 'S' band channels.

However, with earlier TV receivers which do not have additional band provision, it becomes necessary to use an external converter which translates all incoming cable channels to a common VHF channel which is either channel No. 3/4 or a UHF channel. Such a converter is called 'SET-TOP CONVERTER' because it is usually put on top or along-side the television receiver.

The channel selection is done at the converter either manually or by remote control depending on its design. The cable coming from tap-off is connected at the input socket of the converter and its output goes to antenna input point of the TV receiver. The receiver is tuned to the chosen common channel and left there to receive all the incoming cable channels.

2

RF WAVES—GENERATION AND PROPAGATION

Introduction

The functioning of cable television depends on reception of television signals by receiving antennas either directly or via satellites from transmitting ground stations. The carrier frequencies of such signals are high frequency radio waves which are sent out or propagated in free space from various transmitting sources. While CTV, as such, is not directly linked to generation and propagation of radio waves, yet it is desirable to be familiar with such techniques and processes. This chapter, is therefore devoted to basics of radio waves, their generation and various modes of propagation.

2.1 Radio Waves

Radio Frequency (RF) waves are in the form of electric energy that escapes into free space from a transmitting source. These are actually electro-magnetic (EM) radiations like that from the sun, though at much lesser frequencies as compared to sun rays. However, both travel freely in straight lines in empty space or vacuum and thus need no medium for their propagation. The velocity of propagation of radio waves in free space is the same as the velocity of light waves which is equal to a constant 'C' = 300,000 km/sec. Hence, the frequency 'f' of any radio wave = $\dfrac{C\,(\text{meters/sec})}{\lambda\,(\text{meters})}$ where λ is the distance equal to one wave-length of the radio wave. Thus, the time 't' taken in seconds by an RF wave to travel any distance $d = \dfrac{d\,(\text{km})}{C\,(\text{km/sec.})}$. Since the velocity of propagation is extremely high, radio wave travel-time is very small. For example, it is only 1.25 seconds for a distance close to 3,68,000 km between moon and earth. Thus, the journey of RF waves even across a vast country appears almost instantaneous.

The similarity between sun rays and EM radio waves does not end in that they travel in straight lines in free space and take extremely small time to travel, but both also obey the light laws of Reflection and Refraction. For example, appearance of a ghost image on the TV screen is caused by reflection of television signals from any obstruction or nearby buildings. The signal after reflection arrives a small fraction of a second late at the receiving antenna than the strong direct signal, because of additional distance travelled by it along the reflected path. This time delayed signal also produces a weak image on the receiver screen slightly displaced from the main image towards its right side which is referred to as the unwanted or ghost image.

Similarly, it is the refraction of radio waves from ionospheric layers, high up in the sky, that makes radio reception possible over long distance. Radio waves beyond a certain high frequency in the short-wave radio band get so much refracted in the ionosphere that these bend enough to turn back towards earth to reach locations thousands of km away from the transmitting antenna, with enough strength to enable clear reception on radio receivers.

In conclusion, it may be noted that radio waves travel in straight paths in free space, take very small time to travel, get reflected from obstructions in their path and are refracted depending on the density of media through which they travel.

2.2 Generation of Radio Waves

Radio waves are generated as radiation from electric and magnetic fields present in space around current carrying elements called antenna when a high frequency alternating current flows through them. The source that feeds the antenna is the oscillatory (tank) circuit which forms part of an electronic oscillator. The manner in which varying electric and magnetic fields are generated at the frequency of interest is illustrated in Fig. 2.1.

At instant t_0 [Fig. 2.1 (a)], when capacitor 'C' is fully charged, the energy supplied to the tank circuit through the associated active device (tube/transistor) is stored in the form of electric field in the region between the capacitor plates. As such, current in the tank circuit at this instant is zero. Just then the capacitor starts discharging and the stored electric energy causes a current flow in the inductor 'L' as shown in Fig. 2.1 (b). The capacitor discharges to transfer stored energy in the form of rising magnetic field around the coil. When the capacitor is fully discharged at t_1 [see Fig. 2.1 (b)] the magnetic field around 'L' commences to collapse, thus causing current 'i' to continue flowing in the same direction. This charges the capacitor in the opposite direction and current returns to zero at t_2. This condition is shown in Fig. 2.1 (c), in which the tank circuit has been modified to show its connection to an antenna where the radiation actually takes place.

As shown in Fig. 2.1 (d), the current flow during the above explained sequence is sinusoidal in nature and the period t_0 to t_2 constitutes one half of the oscillatory cycle. The process then reverses and capacitor 'C' discharges causing current flow and magnetic field variations around 'L' in opposite direction. The next half-cycle (t_2 to t_4) is completed when the capacitor charges back with polarity shown in Fig. 2.1 (a).

The two half-cycles put together form one complete cycle of period $T = t_0$ to t_4 with current change from positive to negative direction as shown in Fig. 2.1 (d). Note the current variations in 'L' and voltage changes across 'C' are both sinusoidal in nature.

This oscillatory process continues and the frequency of oscillations $f = \frac{1}{T} = \frac{1}{2\pi\sqrt{LC}}$

ignoring small I^2R losses in the tank circuit.

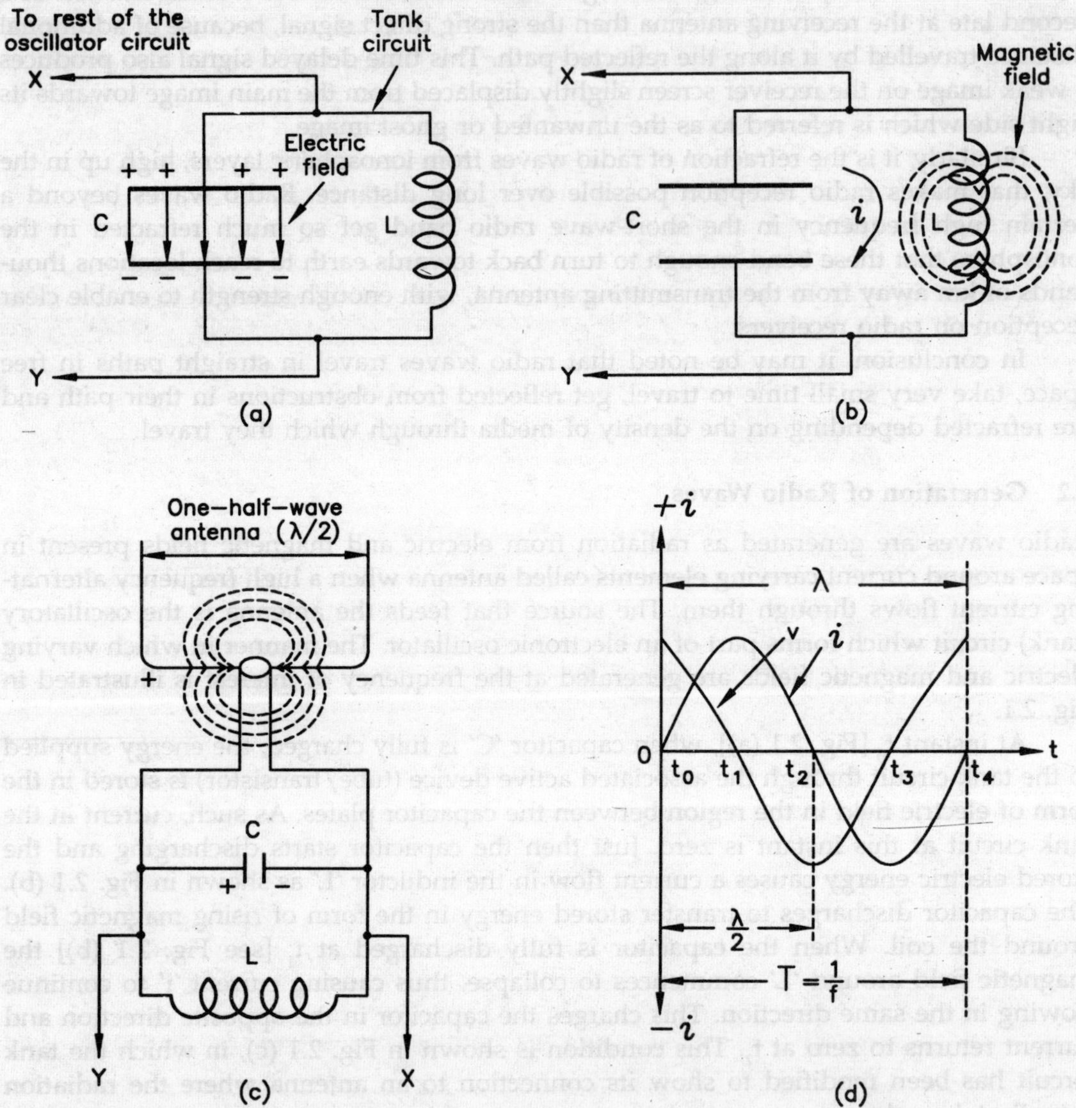

Fig. 2.1. Generation of electric and magnetic fields in a tank circuit.

2.3 Radiation Mechanism

A part of energy associated with varying electric and magnetic fields escapes through the transmitting antenna in the form of electro-magnetic (EM) waves and does not return to the tank circuit. Such a removal of energy stems from effects provided by periodic expansion and contraction of the two fields, the result of which is a succession of electric and magnetic radiation fields at the rate of frequency of oscillations. The energy thus lost in the form of series of waves is made up to the tank circuit from the dc supply fed to the oscillator configuration.

For smooth radiation of EM waves from the antenna, the rise and fall of current that occurs in its two elements due to field changes in the associated tank circuit must be in step or phase with current variations in the L-C circuit. For this to happen, each conductor element of the antenna must be a quarter wave *i.e.* $\frac{\lambda}{4}$ long from tip to tip and so it is chosen accordingly. Such a radiation structure forms the basis of all transmitting and receiving antennas [see Fig. 2.2 (a)] and is called a half-wave dipole or 'HERTZ' antenna after the name of its inverter. As shown in the figure, magnetic and electric fields that develop across the antenna are mutually perpendicular to each other.

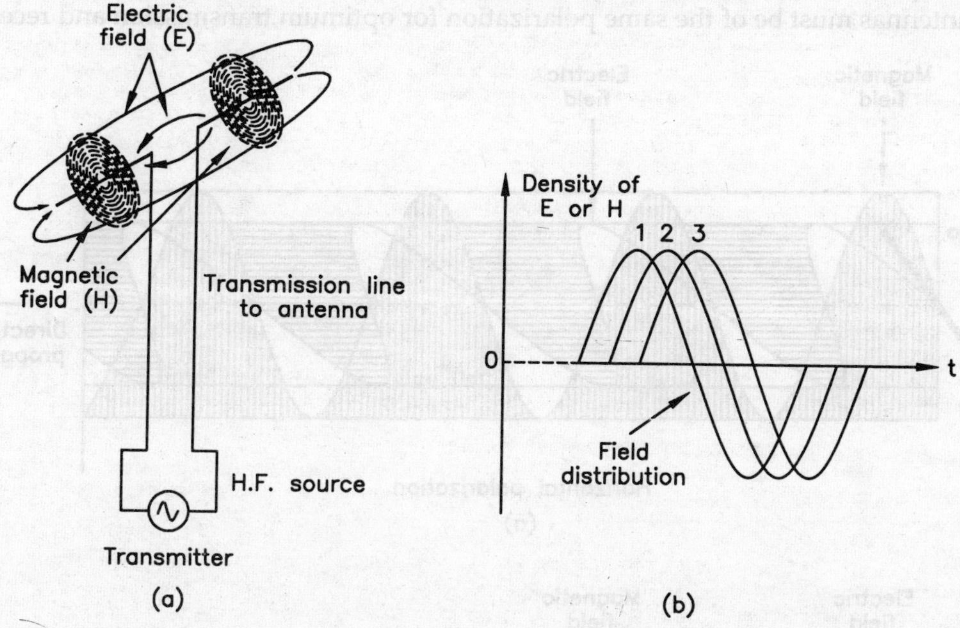

Fig. 2.2. (a) Magnetic and electric fields produced around an antenna.
(b) Nature of E and H field distributions at several successive instants of time.

Wave Motion: Each newly generated field extends into a region that is progressively more distant from the circuit and the field from which it is derived. The transfer process corresponds, therefore, to an outward movement of energy in the form of expanding radiation fields as a wave motion in much the same manner as a disturbance at a given point on the body of water is propagated over its surface as a water wave.

With a sinusoidal current flow in the radiation circuit, the magnitude of each radiating field at any instant will vary sinusoidally with distance along the direction of propagation. Correspondingly, at a given point in space, the field density will vary sinusoidally with time. Such a field density distributions of E and H waves for several successive instants of time is illustrated in Fig. 2.2 (b). Notice that such a motion has the characteristics of a travelling wave.

Polarization of E-M Waves: A close approximation to what an electro-magnetic field may look like is illustrated in Fig. 2.3. The two fields are perpendicular to each other and both are also perpendicular to the direction of propagation of the wave. An important consideration in transmission and reception of radio waves is the orientation of its magnetic and electric fields with respect to earth. The direction of electric field specifies polarization of an antenna. If the electric field is parallel to earth as in Fig. 2.3 (a), the electromagnetic wave is said to be horizontally polarized. However, if the electric field is perpendicular to earth as in Fig. 3.2 (b), then the wave is vertically polarized. As can be expected, the antennas that are horizontal to the earth produce horizontal polarization, and antennas that are vertical to the earth surface produce vertical polarization. The most important factor is that both the transmitting and receiving antennas must be of the same polarization for optimum transmission and reception.

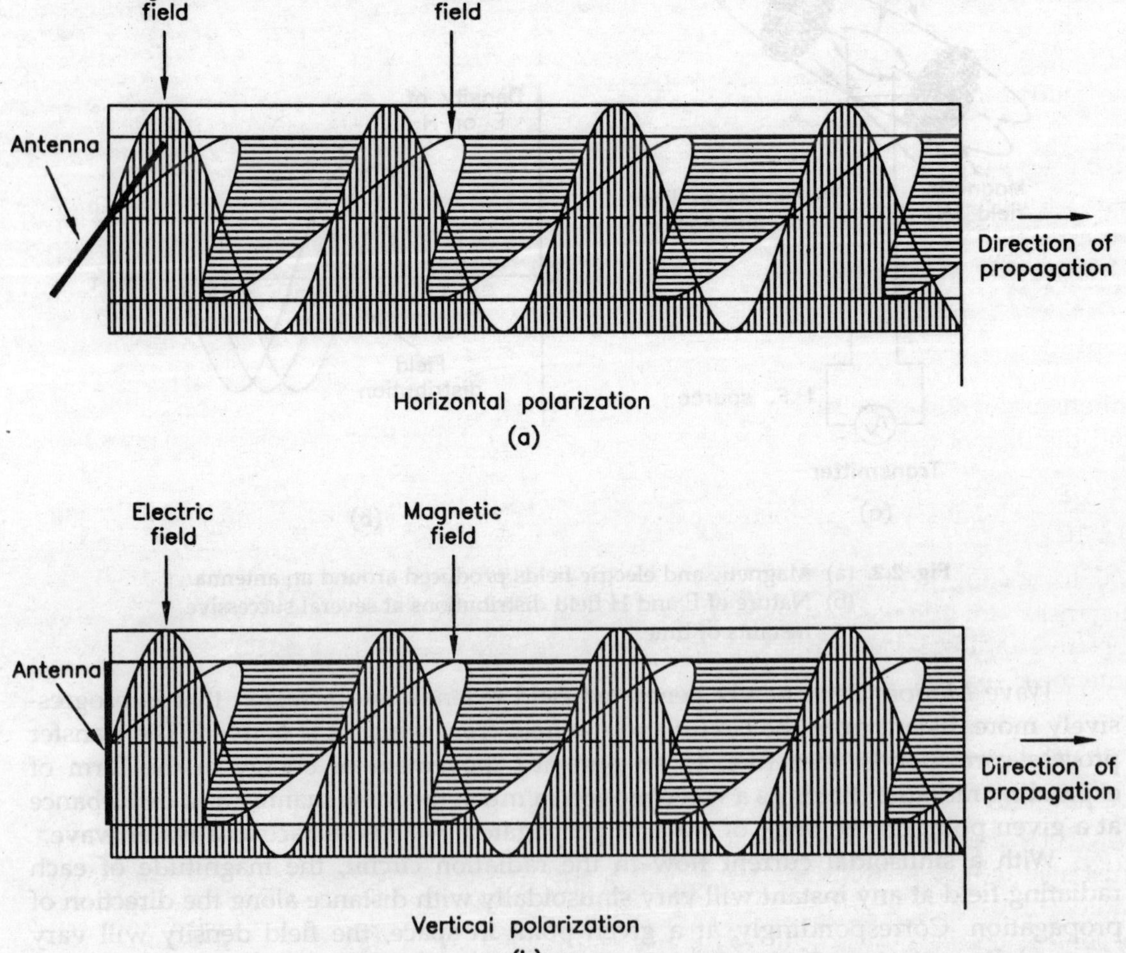

Fig. 2.3. The orientation of electric and magnetic fields determines signal polarization, (a) horizontal polarization, (b) vertical polarization.

Note that in both the cases shown in Fig. 2.3, the electric field E is in a plane which contains the direction of travel. Such a wave is called linearly polarized. Radio waves are most commonly generated with linear polarization. There are some cases, however, when the E field is circular in shape. Such EM waves are called circularly polarized.

Field Intensity: In the absence of any directional radiation characteristics, an oscillatory circuit located in free space will release radio waves radially in all directions and the consequent wave fronts extend over increasing surface area as the wave travels outwards. The density of each field is inversely proportional to its distance from the source and as such the field intensity gets attenuated (reduced) as the wave travels away from the transmitting source. Based on this, a quantitative measure of field intensity of a radio wave is given by the voltage magnitude it can induce in a one meter length of conductor located in a plane transverse to the magnetic field. The field intensity is, therefore, expressed in volts per meter or more commonly as microvolts per meter (,μV/m) because of its very small magnitude.

Spreading Loss: The amount of power transmitted by a travelling radio wave to a unit area of a plane situated at a given distance from a radiating source is dependent on the field intensity at that distance. Since the field intensity decreases inversely with increase in distance from the source, the power available decreases as the distance increases. This is called spreading loss.

Consider a point source of energy which radiates radio waves in all directions from it. Since the radiation pattern will be in the form of expanding spheres, energy passing, through a square meter of the sphere's surface at a radius distance R meters from the point source will be $\frac{1}{4\pi R^2}$ watts where $4\pi R^2$ is the total surface area of the sphere at this distance. Assuming that the radiated power at the point source is 2 watts and the distance R is 1 km, then the power passing a unit area at this distance will be $\frac{2}{4\pi(1000)^2}$ watts ≈ 0.16 μW and at a distance of 10 km, it will get reduced by a factor of 100 to become 0.0016 μW. Notice the great reduction in received power even at distances not too far away from the transmitter. This explains the need for very sensitive radio and television receivers and the use of high gain directional receiving antennas.

2.4 RF Band Allocations

As radio waves spread out, besides spreading losses, energy is also absorbed from the waves by the ground and ionized atmospheric layers. Even rain, moisture, snow and dust also reduce field strength. All such losses are also dependent on the frequency of radio waves. The resulting situation is quite complex and so a suitable classification becomes necessary to indicate frequency bands suitable for specific applications. This is summarized in the table drawn on next page.

Band	Frequency Range	Typical Application(s)
Very low Frequency (VLF)	3 to 30 KHz	All day long-distance communication.
Low Frequency (LF)	30 to 300 KHz	Marine communication and navigational aids.
Medium Frequency (MF)	300 to 3000 KHz	Radio broadcasting.
High Frequency (HF)	3 to 30 MHz	Short wave radio and long-distance communication.
Very High Frequency (VHF)	30 to 300 MHz	Television, FM and radar services.
Ultra High Frequency (UHF)	300 to 3000 MHz	TV and short distance communication.
Super High Frequency (SHF)	3000 to 30000 MHz - *i.e.* 3 to 30 GHz	TV and satellite communication.
Extra High Frequency (EHF)	30 to 300 GHz	Military applications.

Note: Frequencies higher than 1000 MHz (1 GHz) are referred to as microwaves.

Frequency Allocations for Special Applications: The frequencies from VHF band onwards are further classified as under, based on special applications in radar, television, satellite and military.

Frequency Range	Band	Frequency Range	Band
220 - 390 MHz	P	10.95- 12.75 GHz	Ku
350 - 530 MHz	J	19.7 - 21.2 GHz	Ka
1530 - 2700 MHz	L	27.5 - 31.0 GHz	K
2500 - 2700 MHz	S	36.0 - 46.0 GHz	Q
3200 - 4200 MHz	C	46.0 - 56.0 GHz	V
7250 - 8400 MHz	X	56.0 - 100.0 GHz	W

Our main interest is in the VHF and UHF bands allotted for CATV and TV transmission and SHF bands used for satellite communication.

2.5 Propagation of Radio Waves

Once a radio signal leaves the transmitting antenna it travels or propagates through space and ultimately reaches the receiving antenna. As already explained, energy level of the signal decreases rapidly with increase in distance due to spreading and other losses. Furthermore, the electromagnetic signal can take one or more of several different paths to the receiving antenna. The path that a radio wave takes depends on many factors including frequency of the signal, atmospheric conditions and time of the day. But mostly, depending on the frequency of signal, there are three distinct methods by which propagation takes place through space. These are:

(1) Ground or Surface wave Propagation
(2) Sky Wave Propagation
(3) Space Wave Propagation.

1. *Ground Wave Propagation:* Ground waves can exist when transmitting and receiving antennas are close to the surfaces of earth. These follow the curvature of earth

[see (Fig. 2.4)] and can therefore travel hundreds and sometimes thousands of km at frequencies in the 30 KHz to about 3 MHz range. The waves are vertically polarized because any horizontal component of electric field in contact with the earth is short-circuited by the earth. Amplitude modulation broadcast signals are propagated primarily by ground waves. At frequencies higher than about 2 MHz the earth begins to attenuate radio signals. Objects on the earth or terrain features then become the same order of magnitude in size as the wavelength of signal and, therefore, absorb and also otherwise affect the signal. Such an attenuation then becomes so high that no useful transmission is possible beyond 2 MHz by surface waves. The losses are, however, reasonably low below 2 MHz and hence, all medium-wave radio broadcasts are carried out by ground wave propagation.

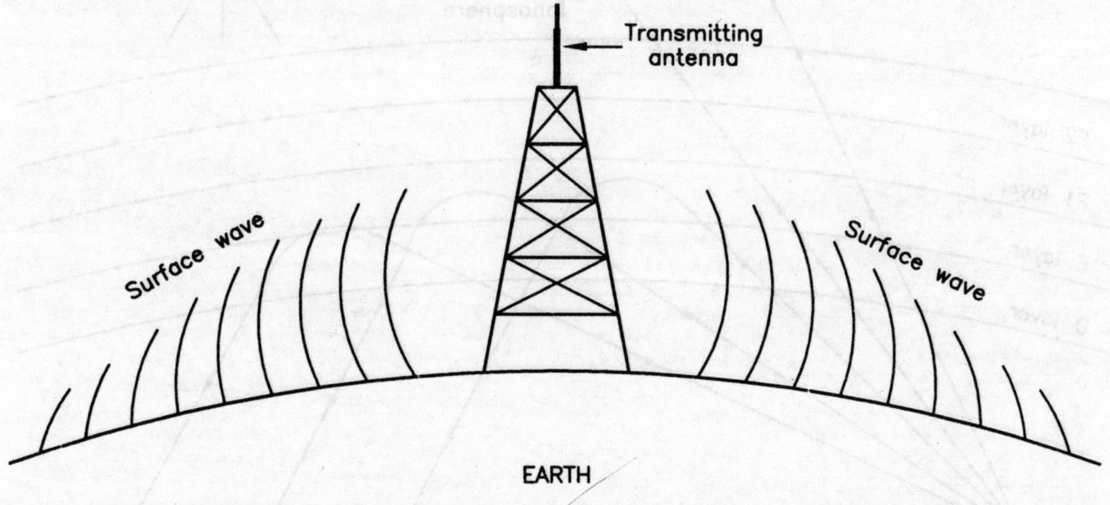

Fig. 2.4. Ground or surface waves radiated from an antenna.

Some waves that travel towards the sky are absorbed during daytime by lower layers of ionosphere and hence, do not contribute towards radio reception. At night when ultra-violet (U-V) radiation from the sun is not there, lower layers disappear while upper layers which do not absorb low and medium band frequencies reflect them back towards the earth. As such, considerable reflection occurs at night and radio waves reach back to earth at long distances. That is why, programmes from medium wave radio stations can be heard clearly at far-off distances at night but not during day-time.

2. *Sky Wave Propagation:* When radio waves are directed high up in the sky, they travel in straight lines until the ionosphere is reached. This region which begins about 120 km above the surface of earth contains large concentrations of charged gaseous ions, free electrons and neutral molecules due to ultra-violet radiations from the sun. Depending on the radiation effect, the ionosphere is generally considered to be divided into three basic layers designated as D layer, E layer [Fig. 2.5] and F layer. The D and E layers are only weakly ionized primarily because these are farthest from the sun, and exist only during daylight hours. As stated earlier, these layers absorb radio wave

signals in the medium wave frequency range and hence, do not contribute to any radio reception when sunlight is there.

Since ground wave propagation, as explained earlier, is not possible above about 2 MHz, most radio wave communication in the short wave bands is carried out by sky waves. When such waves are directed high up in the sky, they travel in straight lines until the ionosphere is reached where they are bent due to refraction in the upper layers as shown in Fig. 2.5.

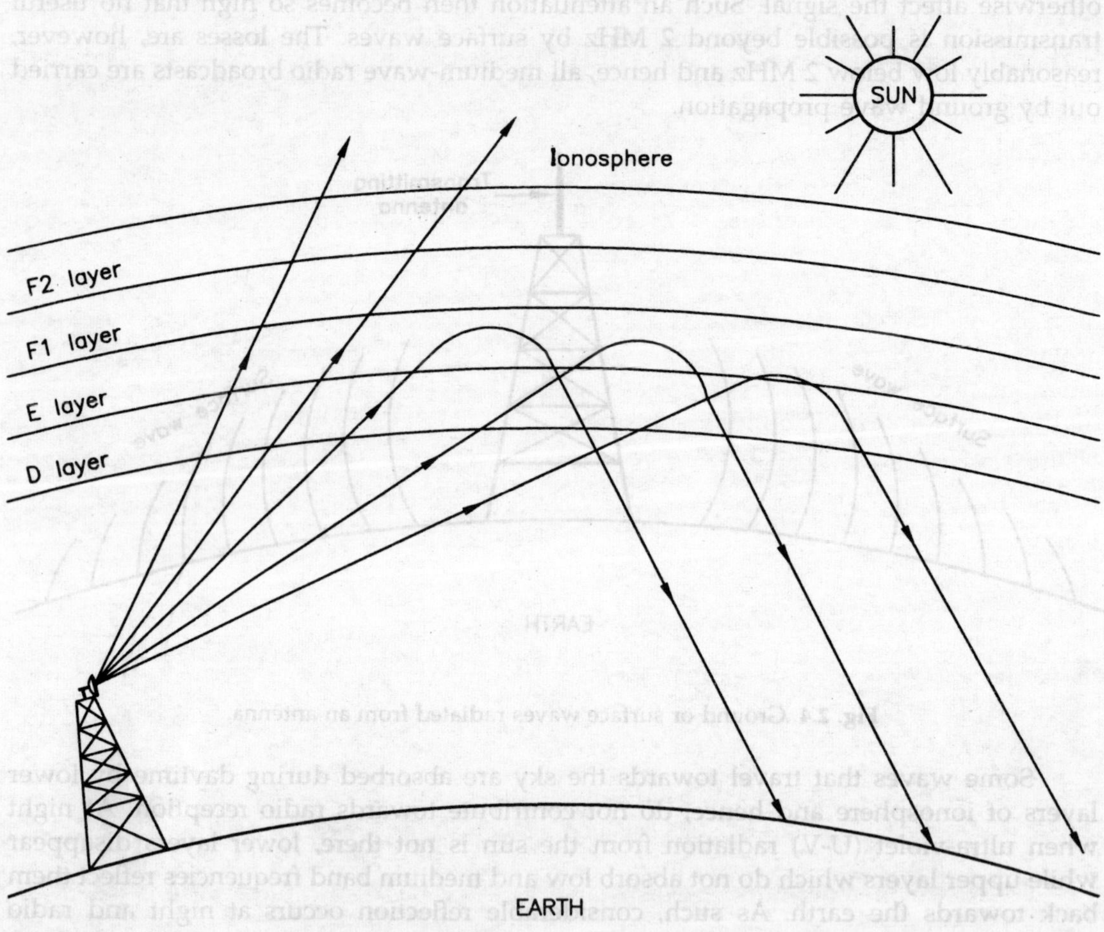

Fig. 2.5. Sky wave propagation.

The angle by which any wave deviates from its straight path depends on (i) frequency of radio wave, (ii) angle of incidence at which it enters the ionosphere, (iii) density of charged particles at that particular moment, and (iv) thickness of ionized layers. Fig. 2.5 shows the path of several waves entering the ionosphere at different incident angles. With increase in frequency, the allowable incident angle at the ionosphere becomes smaller until finally a frequency is reached when it becomes impossible to deflect by refraction the rays back to earth. For ordinary ionosphere conditions this frequency occurs around 35 to 40 MHz and hence sky waves cannot be used for radio communication between distant points on the earth beyond about 30 MHz. This is the reason for not allocating any frequency band beyond 11 meters (≈ 30 MHz) for radio communication.

3. *Space Wave Propagation:* Keeping in view the limitations of ground and sky waves, radio wave propagation above about 40 MHz, (which is the beginning of television broadcasting) is not possible by these modes of radiation. Thus, the only alternative for transmission in the VHF and UHF bands, despite large attenuation, is by waves which travel at a reasonable height above ground in straight lines from transmitting to receiving antennas. Such a mode of wave travel is known as direct or space wave propagation. Since direct waves are not refracted and also do not follow the curvature of earth, space wave propagation is also referred to as line-of-sight communication.

For not too long distances the surface of earth can be assumed to be flat and rays of propagation can reach the receiver from the transmitter. However, for long distances, earth's curvature limits the maximum distance between transmitting and receiving antennas. If the signal is to be received beyond the horizon, then the antenna must be high enough to intercept the straight line radio waves. In practice, in order to extend the distance of communication, both the transmitting and receiving antennas are located as high as possible as illustrated in Fig. 2.6. Note that the antenna tower heights ht and hr are shown exaggerated in comparison with distances 'd' and 'D'. Since receiving TV antennas cannot be located very high, television transmitting antennas are mounted on high towers or on tall buildings and on top of mountains to increase the range of reception.

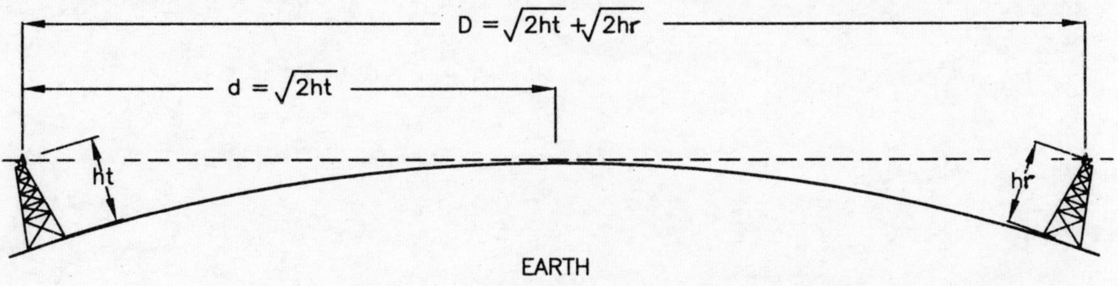

Fig. 2.6. Line-of-sight communication for direct or space waves.

To extend the communication distance at VHF, UHF and microwave frequencies, special techniques have been adopted. The most important of these is the use of repeater stations mounted on high towers a good example of which is the use of a series of microwave relay stations used by telephone companies for long distance communications. The ultimate high-up repeater or relay station, of course, is the communication satellite. It is this which has made TV transmission over long distances possible and has opened the way for meaningful cable television systems.

Frequency Band for Satellite Communication

For radio waves to travel straight from ground to satellites located high up in the sky, the frequencies must be chosen to be above 40 MHz so that these are not bent by refraction by ionospheric layers. Frequency band from 40 MHz to about 900 MHz is used mostly for terrestrial TV broadcasting and hence to avoid any mutual interference, frequencies chosen for satellite communication are above about 1 GHz (*i.e.* 1000 MHz). The spreading and other losses, because of absorption, moisture, rain, snow and dust increase with frequency and become very high beyond 10 GHz. Therefore, the frequency band from 1 GHz to 10 GHz is used for satellite communication and is often referred to as the 'Atmospheric Window'.

3

ANTENNAS AND TRANSMISSION LINES

Introduction

In order to provide good quality picture and sound output on television receivers of cable TV subscribers, it is essential that receiving antennas are carefully chosen and properly installed at CTV stations. It is often necessary to select high gain directional aerials. In locations far-away from desired TV stations, a booster RF amplifier is provided close to the antenna to ensure noise-free reception. For receiving signals from satellites, special antennas called dish antennas become necessary. While conventional receiving antennas for VHF channels are covered in this chapter, dish antennas are described in the next chapter along with microwave techniques.

A transmission line is a two-wire cable that connects a transmitter to its antenna or an antenna to the receiver. In CTV, cables are used to carry RF energy picked up by different antennas to corresponding signal processing equipments installed in control rooms. Cables are also used for distributing different channel signals to end-users who subscribe to the cable system. As necessary, special cables have been designed to carry high frequency signals with minimum energy loss. Such cables and their characteristics are described in later sections of this chapter.

3.1 Half-Wave Dipole

Any receiving antenna may be thought to be similar, though inverse in output, to that of a transmitting antenna. The travelling E-M waves induce a small voltage into the receiving antenna causing voltage and current variations as shown in Fig. 3.1 (a). The electric charge or voltage is maximum at the ends but the accompanying current is minimum there causing a phase difference of 90° with respect to each other. The polarity of voltage and current reverses every half cycle and this repeats at the frequency of E-M waves.

Resonant Length: It is desirable that an antenna behaves as a resonant circuit because at its resonant frequency, maximum signal voltage is induced in it by the travelling waves. Though not apparent, any conductor length has a very small distributed inductance and capacitance along its length which can cause resonance at a particular frequency. Thus, it is necessary to find out conductor length that will resonate at the desired frequency. For this, a relation between conductor length and frequency must be found out. In order to do so, consider the basic antenna structure shown in Fig. 3.1 (b).

Ignoring the small gap between the two conductor rods each of length L/2 meter, it can be visualised that resonance shall occur when electrons that cause current flow on induction of voltage in length L, travel from centre to each end of the rod and back in a time 't' seconds equal to that occupied by one-half of the induced voltage cycle. Thus

$t = \frac{1}{2f}$, where f is the frequency of induced voltage.

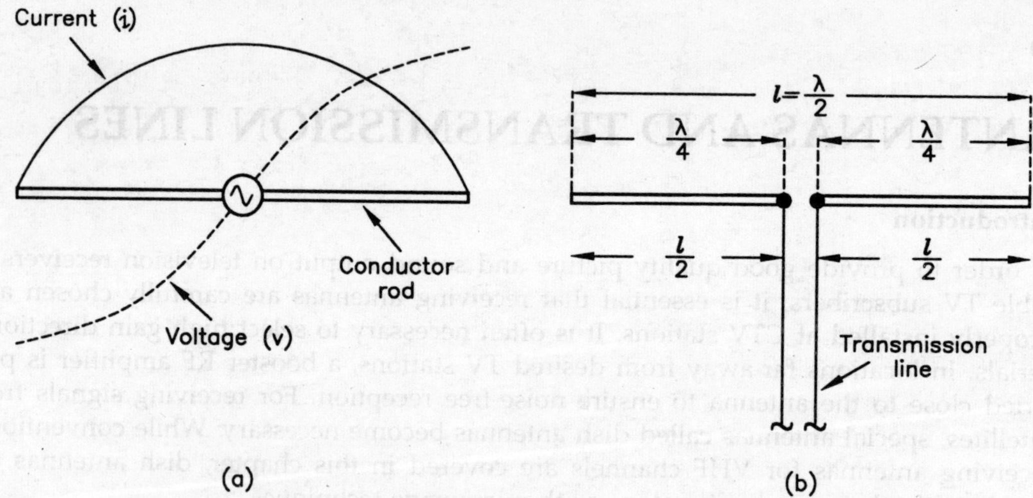

Fig. 3.1. (a) Voltage and current variations along the length of an antenna conductor rod. (b) Basic antenna structure and its resonant length.

The time 't' is also equal to L/c where 'c' is the velocity of electron travel in free space and is same as the speed of light waves. As such, if resonance criteria is to be met $L/c = \frac{1}{2f}$ is the desired condition. Since $f = c/\lambda$ where λ is the wavelength of the induced voltage, we can write :

$$L/c = \frac{\lambda}{2c} \quad \text{or} \quad L = \frac{\lambda}{2}$$

where both L and λ are expressed in meters. This then forms the basis of Half-wave dipole antenna also known as HERTZ's antenna.

The two conductors are often metallic pipes, preferably of aluminium, each of length $L/2 = \lambda/4$ and mounted horizontally on an insulator fixed on a supporting structure. The small gap between the two pipe lengths is about one-fourth of a centimeter and is meant to connect the lead-in cable or transmission line at the two open ends as shown in the figure.

Notice with reference to Fig. 3.1 (a) that the small gap between the two λ/4 length pipes does not disturb the induced voltage wave because at the gap point it passes through zero.

An important characteristic of this dipole antenna is that it receives signals with greatest intensity when oriented at right angles to the desired approaching signal wave. Radiations coming from other sides are poorly received *i.e.* these induce very little voltage in the antenna and hence, cause least interference.

3.2 Antenna Parameters

Practical television antennas have more than one element (pipes or rods) for obtaining more desired characteristics. However, before proceeding to discuss such arrays it is useful to be familiar with main parameters of any antenna.

(i) *Polarization:* As explained in sections 2-3, the direction of electric field of E-M waves determines polarization of any wave. For TV signal transmission, horizontal polarization is preferred and hence, all receiving antennas are mounted horizontally and oriented towards the wanted station's transmitting tower for a strong and noise-free signal pick-up. Besides, receiving antennas are mounted many wavelengths above ground level because space-wave propagation is used for television signal transmission.

(ii) *Antenna Impedance:* Antenna impedance is defined as the ratio of voltage and current induced in it. It is measured at the central point where transmission line is connected [see Fig. 3.1 (b)]. Note that impedance is the ratio of RF ac values of v and i and thus, cannot be measured with a dc meter. For a single half-wave dipole the impedance is ≈ 72 ohms.

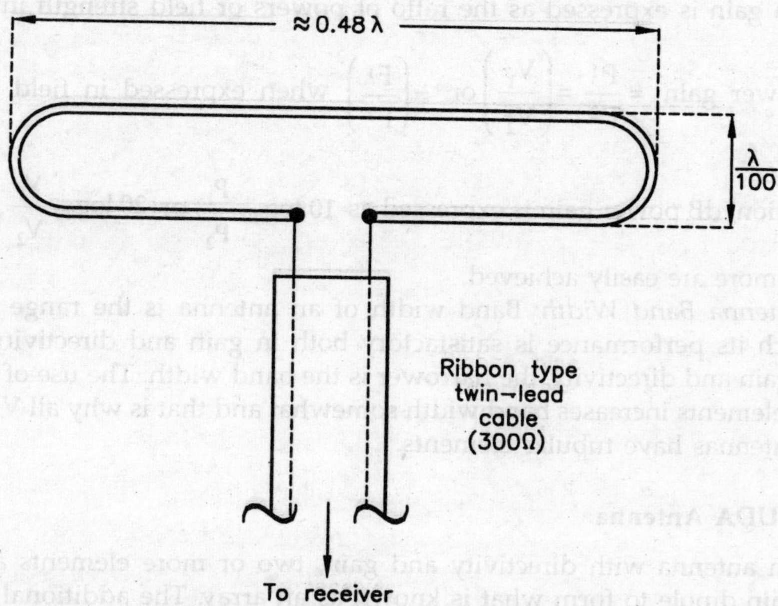

Fig. 3.2. Folded dipole construction.

A popular version of this antenna is the folded dipole shown in Fig. 3.2. This is obtained by joining two half-wave antennas at the ends with an open at the centre where transmission line is connected. The reason for its popularity is the fact that its impedance is 300 ohms and, therefore, makes an excellent match to the widely available 300 ohm, twin-lead transmission line [see Fig. 3.5 (c)]. Such an impedance match ensures maximum transfer of signal power to the receiver. The spacing between the two conductors is small ($\approx \lambda/100$) compared with the half-wave length. The folded dipole has the same directional characteristics and signal pick-up as that of a straight dipole and forms central element of the often used 'YAGI' antenna described in the next section.

(iii) *Antenna Directivity:* Directivity refers to the ability of an antenna to receive signals over a narrow horizontal directional range, thus eliminating interference from signals arriving from other directions. Receiving antennas can be made highly directive by adding more elements to the basic antenna. The measure of an antenna's directivity is its beam-width which refers to the angle of radiation pattern over which energy radiated by a transmitting antenna is received. The narrower the beam-width the better the selectivity.

(iv) *Antenna Gain:* When a highly directional transmitting antenna is used, all the transmitted power is focused in one direction. Because power gets concentrated into a narrow beam, the effect is as if antenna has amplified the transmitted signal. In the same way a receiving antenna that is constructed to be highly directive also exhibits gain by capturing more signal energy. Antenna gain is defined as the ratio of voltage induced per unit length in the directional antenna to that induced in an omnidirectional antenna at the same location with respect to the transmitting antenna.

An omnidirectional antenna receives or radiates energy equally in all directions. The antenna gain is expressed as the ratio of powers or field strength intensity. Since $P = \frac{V^2}{R}$, power gain $= \frac{P1}{P2} = \left(\frac{V_1^2}{V_2^2}\right)$ or $= \left(\frac{F1}{F2}\right)^2$ when expressed in field strengths. In decibel rotation, dB power gain is expressed as $10 \log_{10} \frac{P_1}{P_2}$ or $20 \log_{10} \frac{V_1}{V_2}$. Power gains of 10 dB or more are easily achieved.

(v) *Antenna Band Width:* Band width of an antenna is the range of frequency within which its performance is satisfactory both in gain and directivity. In general, higher the gain and directivity, the narrower is the band width. The use of metallic rods as antenna elements increases band width somewhat and that is why all VHF television receiving antennas have tubular elements.

3.3 YAGI-UDA Antenna

To create an antenna with directivity and gain, two or more elements are combined with the main dipole to form what is known as an array. The additional elements are called Reflector and Director. The antenna itself, usually a folded dipole is then referred to as the driven element and additional conductors *i.e.* reflectors and directors which are neither connected to the dipole nor to the transmission line are called parasitic elements. An antenna made up of a driven element, a reflector and one or more directors is known as YAGI-UDA or simply YAGI antenna, so named after its Japanese inventors. The simplest YAGI antenna with a reflector and one director is shown in Fig. 3.3 (a). Its directive characteristics [Fig. 3.3 (b)] are most suited for reception from television transmitters of moderate power.

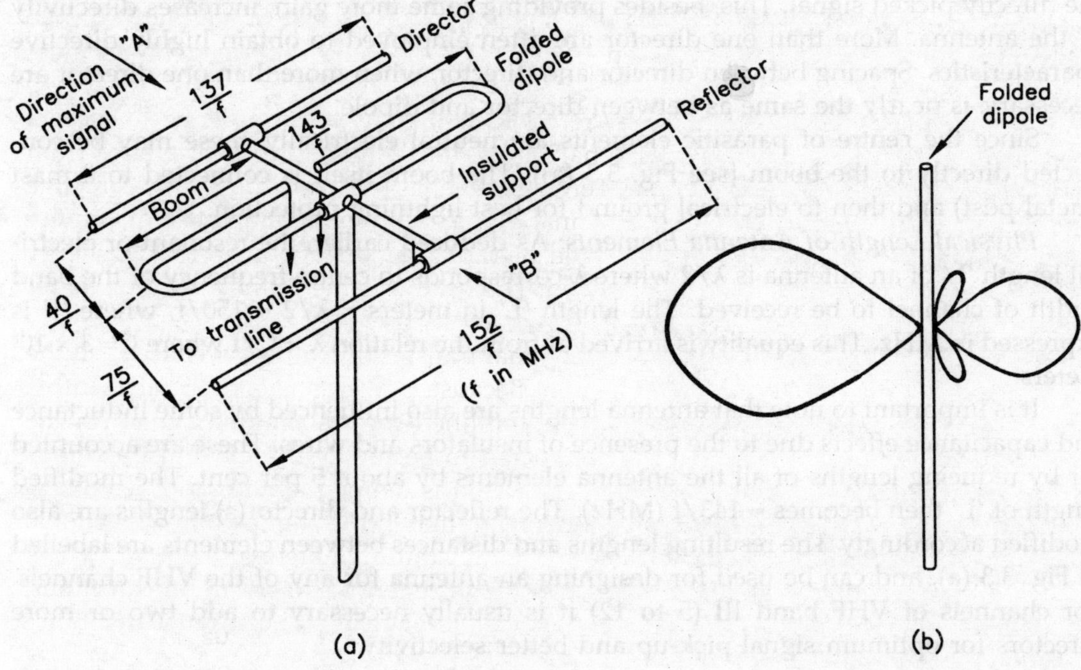

Fig. 3.3. YAGI antenna (a) dimensions, (b) directive response.

The reflector (Fig. 3.3 (a)) is typically 5 per cent longer than the driven element and is spaced from it by a distance ≈ 75/f (MHz) meters corresponding to centre frequency of the channel to be received. The signal arriving from direction 'A' will induce voltage in all the three elements of the array. Since the reflector has no electrical connection, the current that flows into it causes re-radiation of energy which in turn induces additional voltage in the folded dipole. With the chosen length and distance from the main dipole, the reflector's radiation will be mostly in phase with the original signal in the driven element, thus reinforcing it. This accounts for the gain of this antenna over the simple dipole.

If a signal arrives from the opposite direction indicated by arrow 'B' in the figure, it will also induce a signal voltage, both in the main dipole and reflector. However, the distances involved for more travel in terms of wavelength are such that signal induced in the driven element due to re-radiation from the reflector will be out of phase with the directly induced signal in it. Thus, the two out of phase signals will nearly cancel each other and there will be hardly any signal voltage across the folded dipole from the unwanted signal arriving from opposite direction. Thus, it is seen that the reflector not only increases output of the antenna but also makes it more directive.

The second parasitic element called 'director' is approximately 4 per cent shorter than the dipole and is mounted in front of it, at a distance equal to 40/f (MHz) meters where f is the centre frequency as defined earlier. It picks up part of the arriving signal from direction 'A' and re-radiates it towards the folded dipole. The spacing between the dipole and director is such that signal induced in the driven element is in phase with the directly picked signal. This, besides providing some more gain, increases directivity of the antenna. More than one director are often employed to obtain highly directive characteristics. Spacing between director and director, when more than one director are necessary, is nearly the same as between director and dipole.

Since the centre of parasitic elements are neutral electrically, these may be connected directly to the boom [see Fig. 3.3 (a)]. The boom itself is connected to a mast (metal post) and then to electrical ground for best lightning protection.

Physical Length of Antenna Elements: As deduced earlier, the resonant or electrical length 'L' of an antenna is $\lambda/2$ where λ corresponds to centre frequency of the band width of channel to be received. The length 'L' in meters = $\lambda/2 \approx 150/f$, where 'f' is expressed in MHz. This equality is arrived at from the relation $\lambda = C/f$ where $C = 3 \times 10^8$ meters.

It is important to note that antenna lengths are also influenced by some inductance and capacitance effects due to the presence of insulators and wires. These are accounted for by reducing lengths of all the antenna elements by about 5 per cent. The modified length of 'L' then becomes $\approx 143/f$ (MHz). The reflector and director(s) lengths are also modified accordingly. The resulting lengths and distances between elements are labelled in Fig. 3.3 (a), and can be used for designing an antenna for any of the VHF channels. For channels of VHF band III (5 to 12) it is usually necessary to add two or more directors for optimum signal pick-up and better selectivity.

3.4 Combination and Fringe Area Antennas

It is not necessary to erect a separate antenna for each channel if these are close to one another because the resonant circuit formed by the antenna is of low 'Q' (quality factor) and as such has a broad-based response. However, channel allocations are normally staggered to prevent adjacent channel interference.

For receiving channels which are not close to each other, a combination or 'In-Line' antenna is often used. It consists of a half-wave folded dipole with reflector for the lower VHF band channel and on the same boom and in line with it is mounted the short folded dipole meant for the upper VHF band channel. The distance between the two folded dipoles is approximately one-quarter wavelength of the high-band dipole centre frequency. This is the length of line connecting the short dipole to the long dipole where the transmission line to the receiver is connected. The directivity pattern of such a combination (in-line) antenna is relatively uniform on all the VHF channels. Its input impedance is about 150 ohms. An antenna of this type is shown in Fig. 3.4 (a) which is popular for receiving channel 4 (national) and channel 7 (metro) programmes.

Fringe Area Antennas: In fringe areas where the signal level is very low, high gain antenna arrays are needed. The gain of antenna increases with the number of elements employed. A Yagi antenna with a large number of directors is commonly used with success in fringe areas for stations in the VHF bands. As already mentioned, a large parasitic element resonant at a lower frequency than the driven element will act as a mild 'reflector' and a shorter parasitic element will act as a mild 'concentrator' of radiation. As a parasitic element is brought closer to the driven element, then regardless of its precise length, it will load the driven element more and, therefore, reduce its impedance. This is perhaps the main reason for the use of folded dipole as the driven element of such arrays. A gain of more than 10 dB with a forward to back ratio of about 15 dB is easily obtained with such an antenna. Such high gain combinations are sharply directional and so must be carefully aimed while mounting otherwise the captured signal will be much lower than it should be. However, while designing such a high gain directional antenna, its band width aspect is not to be compromised. In such antennas, the gain not only varies from channel to channel but also from one end of the channel to the other. Since the colour signal spectrum is near the end of channel width, any reduction of antenna gain would affect colour components in the reproduced picture.

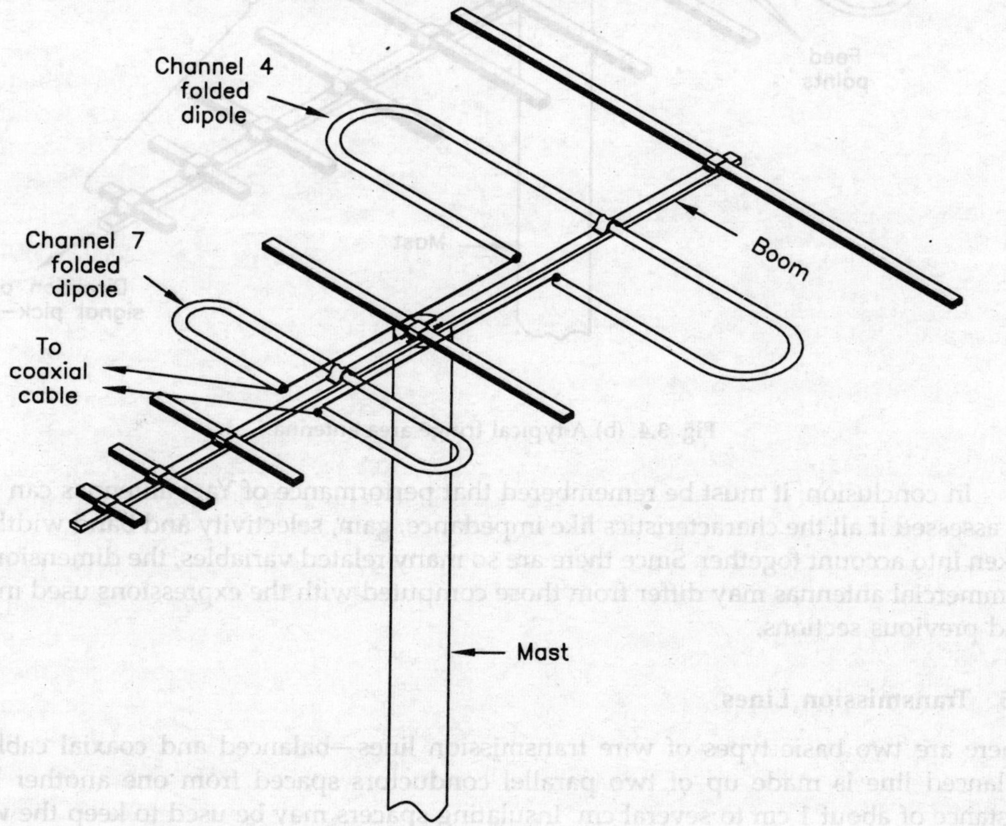

Fig. 3.4. (a) Combination antenna for channels 4 and 7.

A typical YAGI antenna for use in fringe areas is shown in Fig. 3.4(b). In such an antenna the reflector is usually 5 per cent longer than the dipole and may be spaced from it at 0.15 to 0.25 wavelength depending on various design requirements. Similarly the directors may be 4 percent shorter than the dipole element, but where broad band characteristics are needed, successive directors are usually made shorter, as shown in the figure, to be resonant for the higher frequency signals of the spectrum.

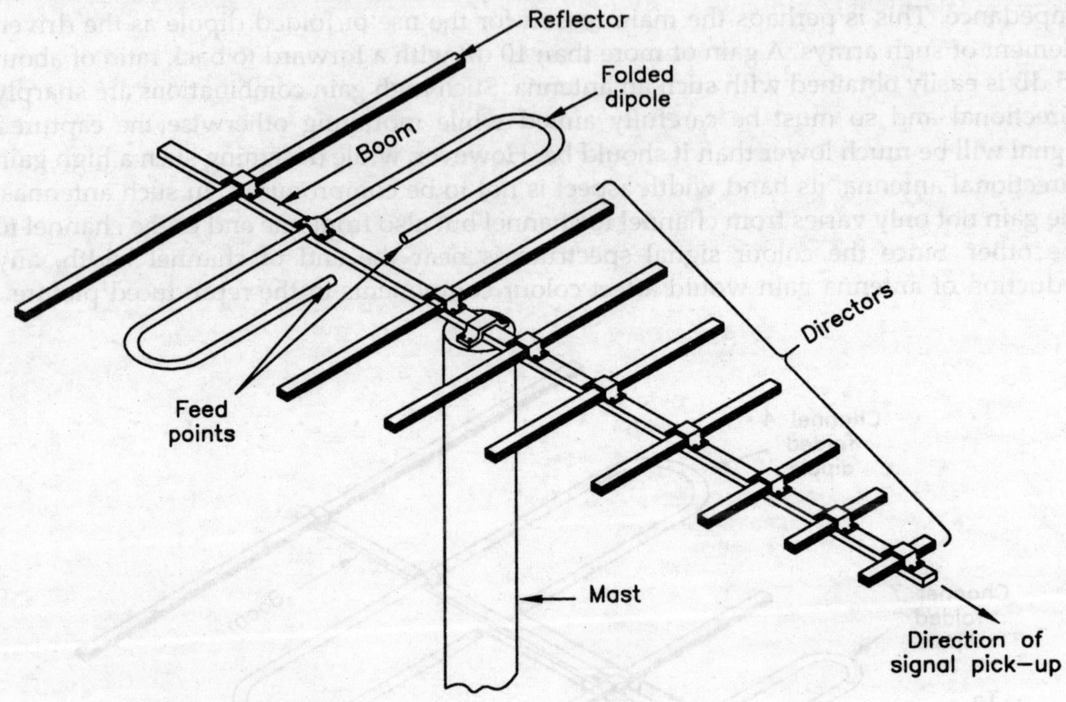

Fig. 3.4. (b) A typical fringe area antenna.

In conclusion, it must be remembered that performance of Yagi antennas can only be assessed if all the characteristics like impedance, gain, selectivity and band width are taken into account together. Since there are so many related variables, the dimensions of commercial antennas may differ from those computed with the expressions used in this and previous sections.

3.5 Transmission Lines

There are two basic types of wire transmission lines—balanced and coaxial cable. A balanced line is made up of two parallel conductors spaced from one another by a distance of about 1 cm to several cm. Insulating spacers may be used to keep the wires separated, or the spacing may be maintained by a continuous plastic insulator that is part of conductor insulation. Fig. 3.5(a), shows a two-wire balanced line. The term 'Balanced line' means that same current flows in each wire with respect to ground, though 180° out of phase with respect to each other. Neither wire is connected to ground. In an unbalanced line, one conductor is connected to ground.

The other type of transmission line is called coaxial cable [Fig. 3.5(b)]. It consists of a solid centre conductor surrounded by a plastic insulator such as Teflon. Over the insulator is a second conductor, a tubular braid or shield made up of fine wires. An outer plastic sheath protects and insulates the braid. Coaxiable cable is an unbalanced line since current in the centre conductor is referenced to the braid which is connected to ground. Coaxial cables are manufactured in a variety of sizes from about 0.5 cm to several centimeters in diameter and are widely used in VHF applications.

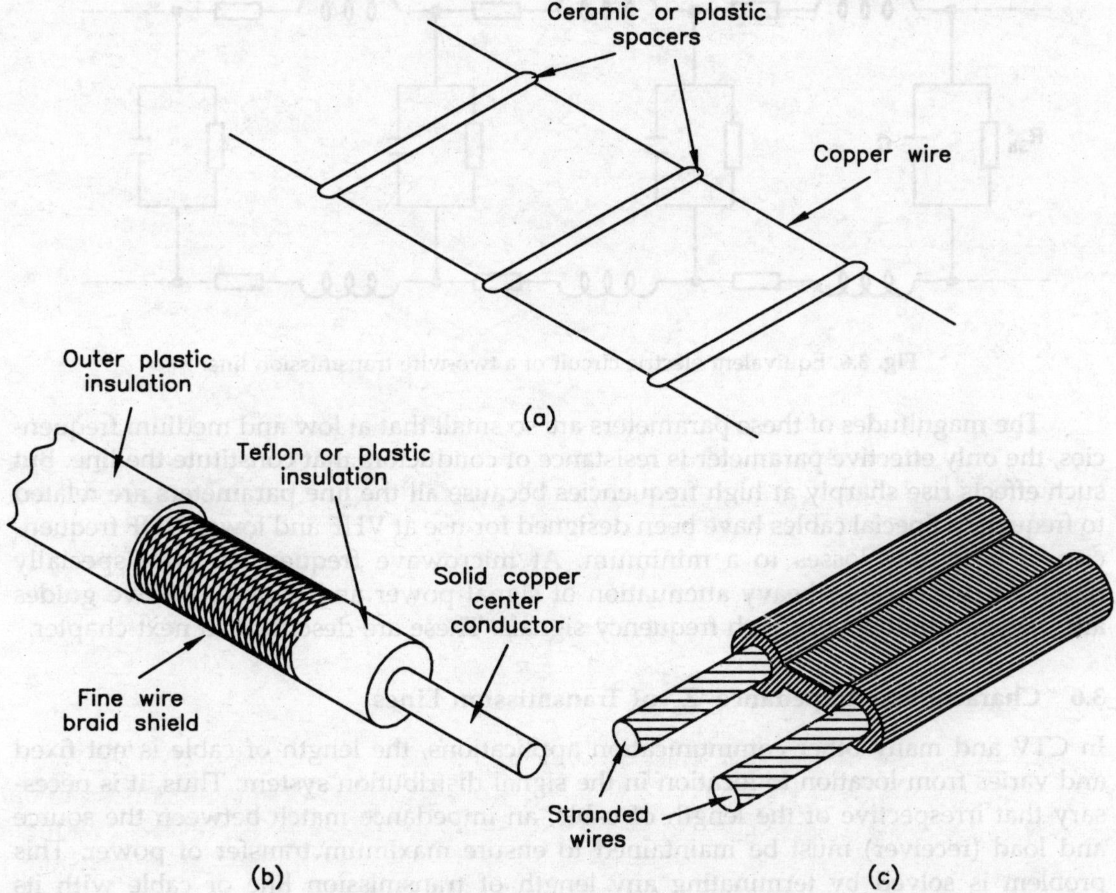

Fig. 3.5. Common types of transmission lines (a) two-wire balanced line, (b) Coaxial cable, (c) 300 ohms twin lead balanced line.

A widely used balanced transmission line is the twin-lead in which two stranded wires are separated by a flat continuous plastic insulator as shown in Fig. 3.5(c). Its characteristic impedance is 300 ohms and is commonly used as lead-in-cable from receiving antennas to TV receivers.

Impedance Parameters: All forms of transmission lines have certain impedance parameters or constants uniformly distributed along their length. Since all conductors offer a finite resistance to current flow, there is a series resistance along both the lines. Also, because the lines are linked by a magnetic field when current flows, there is also a series inductance along the line. Similarly, as the conductors are separated by a dielectric material, the lines become electrically charged when a potential is impressed across them which accounts for a shunt capacitance along the line. Finally, if insulation between the conductors is not perfect, there will also be resistance path and hence, a finite shunt resistance across the two conductors. Such a representation of transmission line is shown in Fig. 3.6.

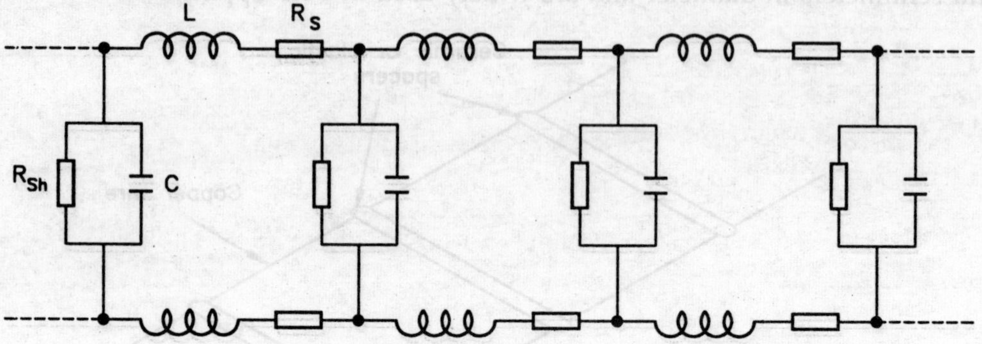

Fig. 3.6. Equivalent electric circuit of a two-wire transmission line.

The magnitudes of these parameters are so small that at low and medium frequencies, the only effective parameter is resistance of conductors that constitute the line. But such effects rise sharply at high frequencies because all the line parameters are related to frequency. Special cables have been designed for use at VHF and lower UHF frequencies to keep line losses to a minimum. At microwave frequencies even specially designed cables cause heavy attenuation of signal power and therefore, wave guides are used to conduct very high frequency signals. These are described in next chapter.

3.6 Characteristic Impedance 'Z_0' of Transmission Lines

In CTV and many other communication applications, the length of cable is not fixed and varies from location to location in the signal distribution system. Thus, it is necessary that irrespective of the length of cable, an impedance match between the source and load (receiver) must be maintained to ensure maximum transfer of power. This problem is solved by terminating any length of transmission line or cable with its characteristic impedance (Z_0), which in effect is the impedance the line would present to a voltage source connected at its input terminals as if length of the line is infinitely extended.

Determination of Z_0: As explained in previous section, line parameters are distributed over entire length of the line. For a line 'L' meters long, its parameters may be represented by a set of lumped constants, each for a tiny length ΔL of the line. Such a concept, that is convenient for analysis is illustrated in Fig. 3.7, where the series parameters *i.e.* resistance and inductance of both the lines have been shown on one side and the shunt resistance (R_{sh}), being too large as found in most cables, has been neglected.

If it is assumed that the line extends to infinity, the number of sections, each of length ΔL will also be infinite in number. To understand the behaviour of such a line, consider the circuit arrangement shown in Fig. 3.8 (a) where a signal source of internal resistance R_s is connected at its input terminals 1 and 2. On closing switch 'S', the inductor (L) sections will tend to oppose the flow of current in the line while the capacitors (C) will start charging. However, the charging takes a finite time as each capacitor charges the next with the voltage it attains from the previous capacitor. Since the line has been assumed to be infinitely long, the charging process will continue for ever and voltage that develops across the line will travel along as a wave motion with a velocity close to the speed of light in free space. However, a steady state is soon reached because of the balancing action of charging and partial discharging of capacitors. As such, the voltage and current readings will settle down at certain values of V and I. The ratio of V/I is then the characteristic impedance (Z_0) of the line measured at its input terminals 1 and 2.

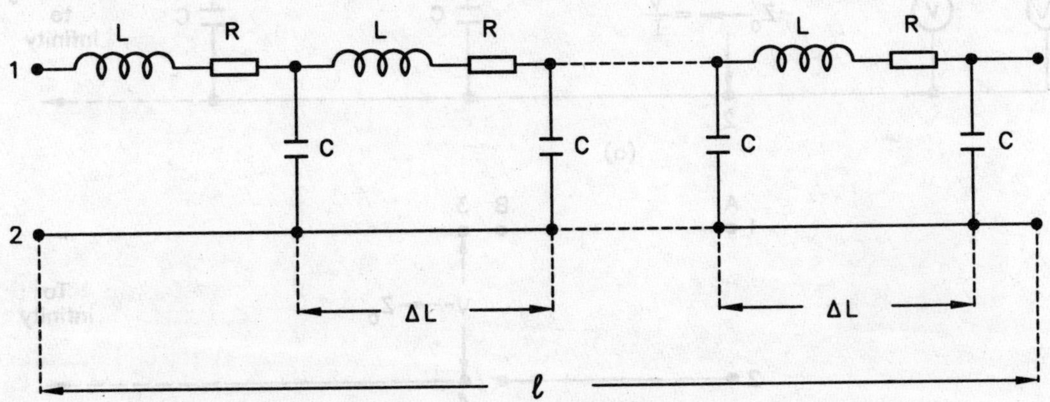

Fig. 3.7. Lumped constant model of a transmission line of length 'L' meters.

Now consider that a short section A-B of the line at its input side is cut away [Fig. 3.8 (b)] such that the line now starts at terminals 3 and 4. The impedance looking at terminals 3-4 will still be Z_0 since the removal of section A-B does not affect infinite nature of the line *i.e.* it still remains an infinite line.

Since the line has no means of knowing its length, if the short section removed earlier is now terminated by an impedance equal to Z_0 as shown in Fig. 3.8 (c), it will mean that the line is still infinitely long and impedance measured at input terminals 1-2 will again be Z_0. It therefore follows that a short line terminated in Z_0 behaves as if it is an infinitely long line. It therefore becomes evident that signal that flows along the line from the sending end is absorbed in the load connected at its termination and no part of it is reflected back thus ensuring maximum transfer of power from source to load. This important factor explains why distribution cables in CATV must always be terminated in their characteristic impedance which is generally 75 ohms.

Further on, (see Fig. 3.7) it can be shown that $Z_0 = \sqrt{\dfrac{R_s + j_{\omega L}}{R_{sh} + j_{\omega c}}}$ ohms. When R_{sh} being too large is neglected and R_s is also neglected in comparison with inductive reactance (ωL) at high frequencies $Z_0 \approx \dfrac{L}{C}$ ohms where L is in henrys and C in fareds per unit length. As an illustration, if 'L' of the line = 0.12 μH and C = 21 μμ F, then

$$Z_0 \approx \sqrt{\frac{0.12\times10^{-6}}{21\times10^{-12}}} \approx 76 \text{ ohms.}$$

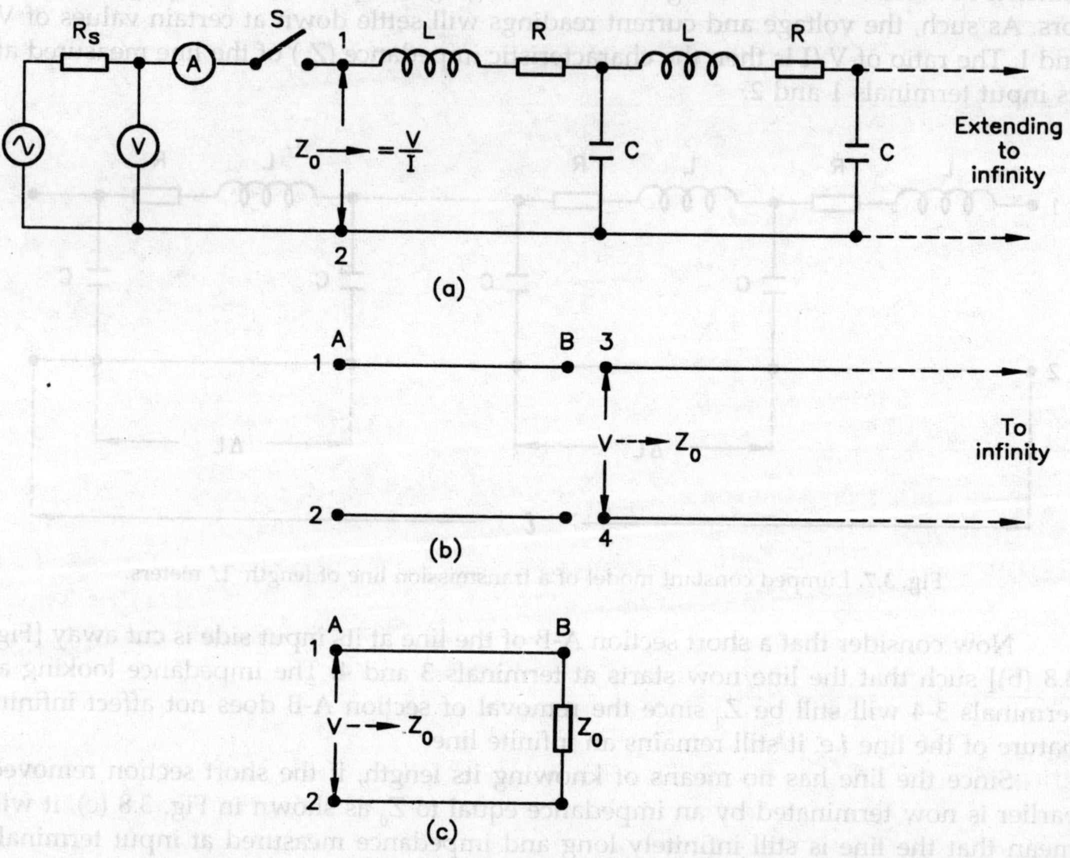

Fig. 3.8. Determination of characteristic impedance 'Z_0' of a finite line.

Since it is difficult to measure line parameters, Z_0 is generally determined by physical characteristics of the line such as conductor size, spacing and dielectric between them. It is found out that for air dielectric cables $Z_0 = 138 \log \frac{D}{d}$ where 'D' is the diameter of outer conductor and 'd' that of the inner conductor. For example if D = 0.25 and d = 0.04, both expressed in same units, $Z_0 = 138 \log \frac{0.25}{0.04} = 138 \log 6.25 = 138 \times 0.796$ = 110 ohms. Use of any plastic or foam material as insulator between the conductors reduces Z_0 to about 66% of the air dielectric value.

3.7 Coaxials for CATV

Coaxial cables are commonly available with characteristic impedances of 50 and 75 ohms. The chart drawn below gives a summary of characteristics of several popular types of transmission lines. Most cable manufacturers designate cables by an alphanumeric code beginning with letter RG. The primary specifications are Z_0 and attenuation.

Type	Z_0	Attenuation dB/ 30 metres length	Applications
RG 59	75 ohms	100 MHz → 3.4 dB 400 MHz → 7.1 dB	TV antennas, cable TV HF antennas.
RG 11	75 ohms	100 MHz → 2.5 dB 400 MHz → 5.5 dB	VHF antennas.
RG 214	50 ohms	100 MHz → 2.0 dB 400 MHz → 4.7 dB 900 MHz → 7.8 dB	Satellite TV antennas, UHF and microwave.
Twin-lead	300 ohms	100 MHz → 0.55 dB	TV antennas.

Specifications given next are for coaxial cables specially designed for distribution of CATV signals. These have polythene insulation. The code numbers are as used in Germany.

Type	Z_0 (Ohms.)	Attenuation (dB) per 100 metres (f) (dB)	DC resistance/KM
LCM-13 (TRUNK CABLE)	75	100 MHz — 5 200 MHz — 8 300 MHz — 10	25 ohms.
LCM-15 (TRUNK CABLE)	75	100 MHz — 4.0 200 MHz — 5.5 300 MHz — 6.9	13 ohms.
LCG-21 (BRANCH CABLE)	75	100 MHz — 9.0 200 MHz — 13.0 300 MHz — 16.0	50 ohms.

4

MICROWAVE TECHNIQUES AND ANTENNAS

Introduction

Microwaves are signals with a frequency greater than 1 GHz and the range generally extends to 30 GHz. The use of microwaves have found wide applications which include satellite communication relay links for very high frequency waves, radar, weather forecasting and high frequency heating. The greatest advantage of the use of microwaves stems from the fact that carrier frequencies chosen in this region enable wider band widths than is possible in lower frequency bands. For example, a standard AM broadcast station operating on a carrier of 1000 KHz has a maximum band width of 10 KHz (± 5 KHz) which is one per cent (10/1000 = 0.01 = 1 per cent) of the spectrum space at that frequency. But, in the microwave range at a carrier frequency of say 3 GHz and with the same one per cent total frequency deviation, an incredibly large band width of 30 MHz (± 15 MHz) becomes available. In fact, it represents low frequency, medium frequency and high frequency portions of the spectrum. Such a band width can easily accommodate many video and thousands of telephone and computer data channels by multiplexing techniques. Television stations located far away from each other use microwave links via satellites to transmit TV signals between each other, thus enabling a large number of television channels to subscribers of cable TV networks. There are even consumer satellite receivers capable of picking up TV transmissions directly from available satellites.

While the use of microwaves offers so many advantages, it also presents numerous problems. At these frequencies it is difficult to analyze electronic circuits in terms of current-voltage relations and therefore, it becomes necessary to use electric and magnetic field strengths to determine signal power levels at various stages of microwave circuitry. Another problem is that at microwave frequencies, conventional circuit components cannot be used since resistors, inductors and capacitors do not behave as such but look like L-C-R circuits because of tiny inductance of component leads, skin effect and stray capacitance effects with chassis (PCB) ground and other objects.

In addition, to create resonant circuits at microwave frequencies, the required values of inductance and capacitance become extremely small and as such, physical limits become a problem. For example, even a 1 cm length of component lead wire represents a significant amount of inductance at these frequencies. Further on, conventional semiconductor devices like diodes and transistors do not function properly at these high frequencies because of capacitances and inductances associated with them.

Another serious problem is the transit time in a transistor. Transit time refers to the extremely small time it takes for current carriers (holes or electrons) to move through the device. At not too high frequencies, transit time is negligible in comparison with the time-period of applied signal and hence is of no consequence. But at microwave frequencies, transit time becomes comparable with that of the applied signal to the transistor and this causes serious phase relation problems between different components of the complex input signal.

In addition to all this, there are problems of wave propagation at microwave frequencies as explained in the chapter devoted to RF waves. Besides, microwave signal strengths are very small and to receive these it becomes necessary to use specially designed parabolic dish antennas. Even specially designed HF cables cause heavy attenuation and microwave signals have to be carried through waveguides which are hollow metallic sections of suitable dimensions.

While solutions to these problems have been found out, but these are quite complex and thus working with microwave equipments requires special knowledge and skills considerably different from those needed for understanding conventional electronic circuits. This chapter, is therefore devoted to ingenious techniques evolved to circumvent special HF problems in order to make it easier to understand the working of microwave circuits.

4.1 Standing Waves on Transmission Lines

As explained in sections 3-6, a transmission line terminated by a resistive load equal to Z_0 of the line becomes effectively an infinitely long line and its length is not critical. Since no energy is reflected back, maximum power transfer takes place from source to load. Such a line is called non-resonant line. However, if the load impedance is different from the line Z_0, all the power is not absorbed by the load and the remaining gets reflected back towards the source. Much of this is dissipated in the transmission line and this results in loss of power. Such a line is called resonant line.

The forward and reflected signals on incorrectly terminated transmission lines produce Standing Waves as shown in Fig. 4.1. The magnitude of standing waves is determined by the ratio of maximum current to minimum current along the line or the ratio of maximum voltage to minimum voltage across the line. This ratio, called Standing Wave Ratio (SWR), is defined as:

$$\text{SWR} = \frac{\text{Imax}}{\text{Imin}} = \frac{\text{Vmax}}{\text{Vmin}}$$

On a properly terminated line, there are no standing waves, the voltage and current values are steady or say maximum and minimum values of I and V are the same. Therefore, SWR = 1, which is an ideal condition. In practice, there will be some mismatch and a SWR value below 2 is usually acceptable. The greater the mismatch, higher will be SWR and increased power loss along the transmission line.

Transmission Line as Resonant Circuit: The two worst cases of mismatch on a transmission line are (a) Short circuit, and (b) Open circuit at the end of line. These conditions have significance in a different way and are considered separately.

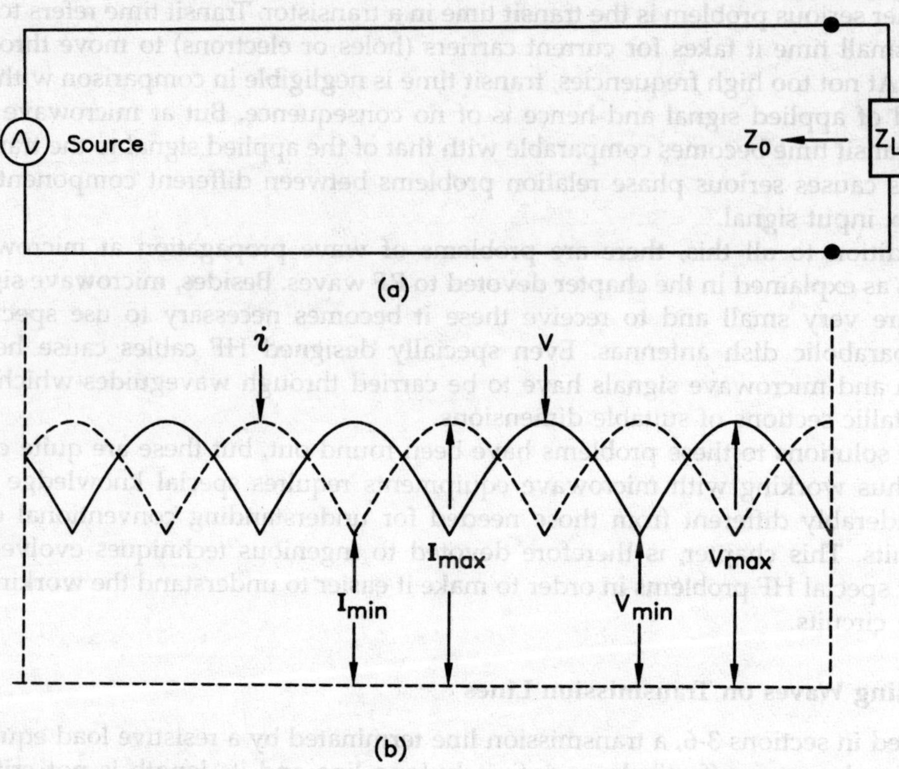

Fig. 4.1. Transmission line (a) with mismatched load ($Z_L \neq Z_0$) and the resulting, (b) standing waves.

(a) *Shorted Line:* Standing waves on a shorted transmission line are shown in Fig. 4.2. The voltage is zero when current is maximum. Since minimum voltage goes to zero, SWR is infinity and hence all the power is reflected back towards the source. Working back towards the source, the voltage and current variations distribute themselves according to the wavelength* (λ) of applied signal. The pattern repeats every one-half wavelength and thus current and voltage values at the source end depend both on signal wavelength and actual length of line.

When the line is only λ/4 long [Fig. 4.3 (a)], despite the fact that voltage is zero and current maximum at the shorted end, back at the source side the voltage is maximum and current is zero. Therefore, to the source, the line appears as an open circuit or

---

* Wavelength (λ) in metas $= \dfrac{3 \times 10^8}{\text{f (cycles)}}$. For example, for a frequency f = 1 GHz (1000 MHz),

$$\lambda = \frac{3 \times 10^8}{1 \times 10^9} = 0.3 \text{ meters} = 30 \text{ cm.}$$

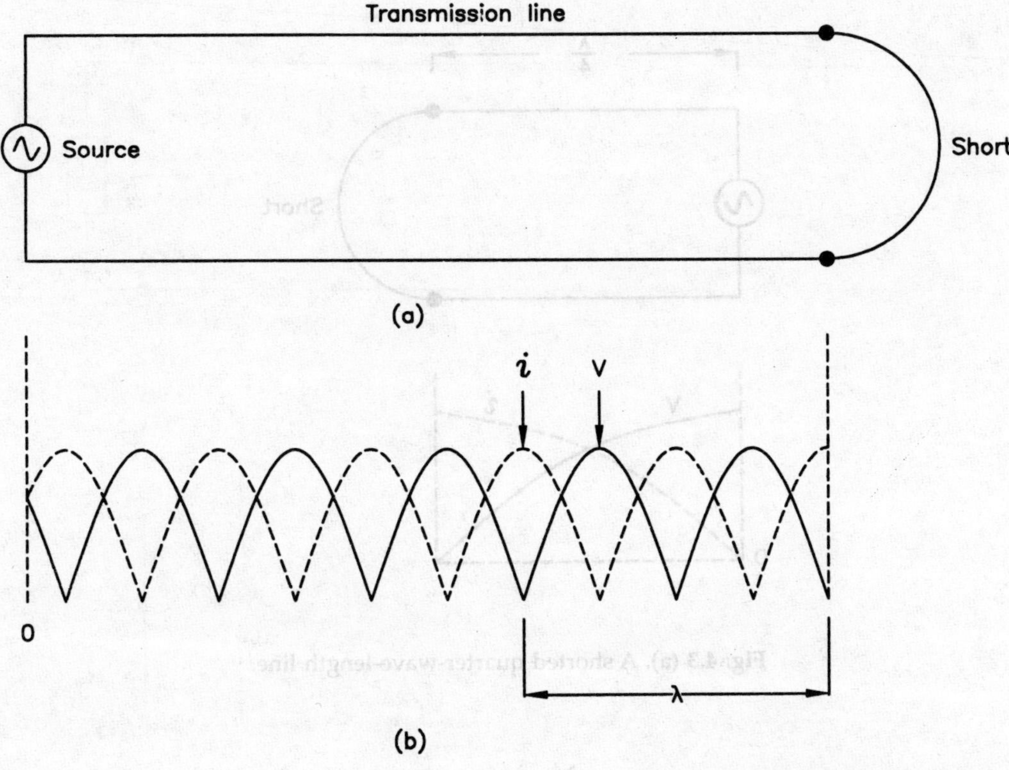

Fig. 4.2. Standing waves on a shorted transmission line.

at least a very high impedance. The key point here is that this condition exists at only one frequency, *i.e.* the frequency at which line length is exactly one-quarter of the wavelength. Because of this frequency sensitivity, $\lambda/4$ line length acts like an L-C tuned or resonance circuit, in this case a parallel resonant circuit because of its very high impedance at the source end.

If the line is made one-half wavelength ($\lambda/2$) the standing wave pattern will be as shown in Fig. 4.3 (b). The voltage source sees the same conditions as at the end of line *i.e.* zero voltage and maximum current. This represents a short, or a very low impedance. Note once again that such a condition occurs only if the line is exactly one-half wavelength long at the source voltage's frequency. In this case, the line looks like a series resonant circuit to the source.

If the line length is less than $\lambda/4$ at the operating frequency, the shorted line looks like an inductor to the source generator. Similarly, if the shorted line length is between one-quarter and one-half wavelength, the line looks like a capacitor to the source. All these conditions repeat with multiple quarter or half wavelengths of shorted line.

(b) *Open Circuited Line:* The standing waves on an open transmission line are shown in Fig. 4.4. The current is zero when voltage is maximum and as in the case of a shorted line, SWR is infinity and all the sent power is reflected back towards the source. The V and L patterns repeat as shown and are a function of source frequency.

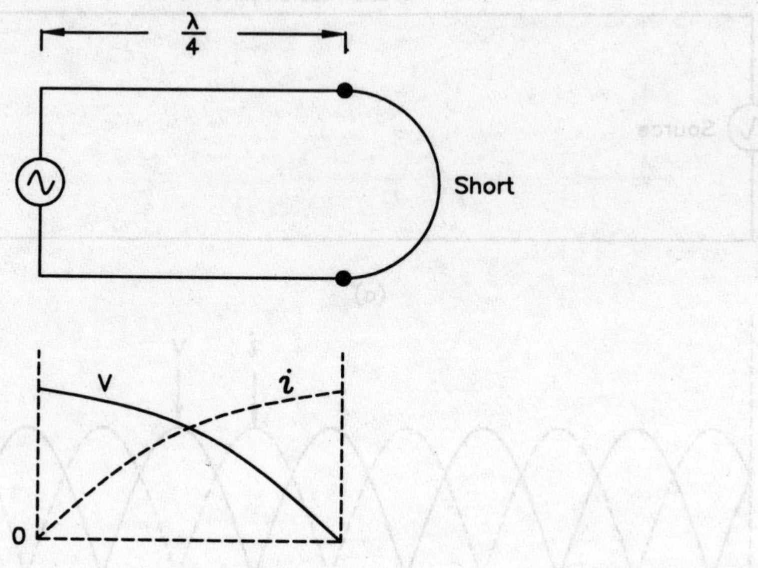

Fig. 4.3 (a). A shorted quarter-wave-length line.

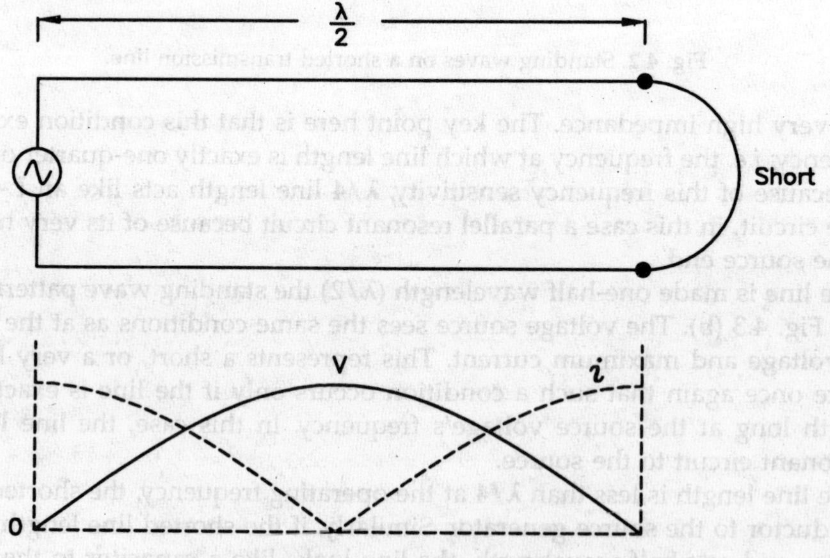

Fig. 4.3 (b). A shorted half-wave-length line.

When the line is made λ/4 and λ/2, the standing wave patterns will be as shown in Fig. 4.5. To the source, a λ/4 line looks like a series resonant circuit, whereas a λ/2 line looks like a parallel resonant circuit, just the opposite of conditions obtained on a

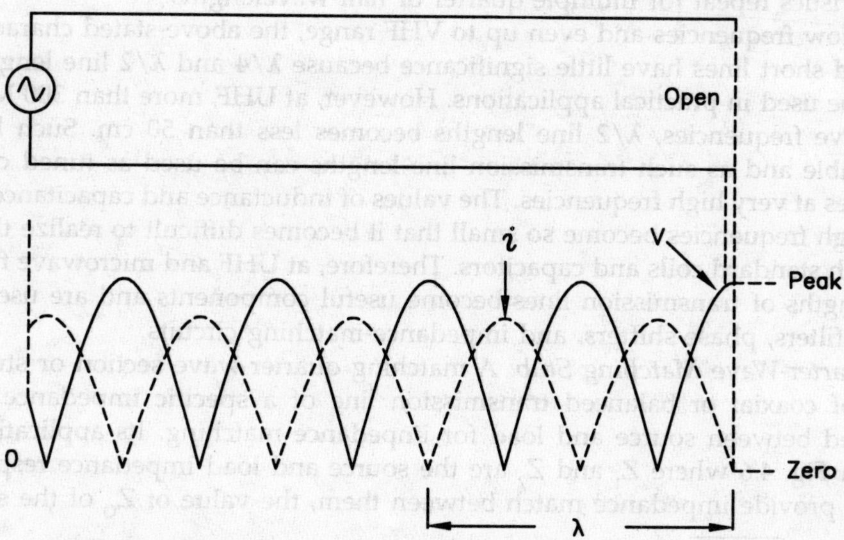

Fig. 4.4. Standing waves on an open-circuited transmission line.

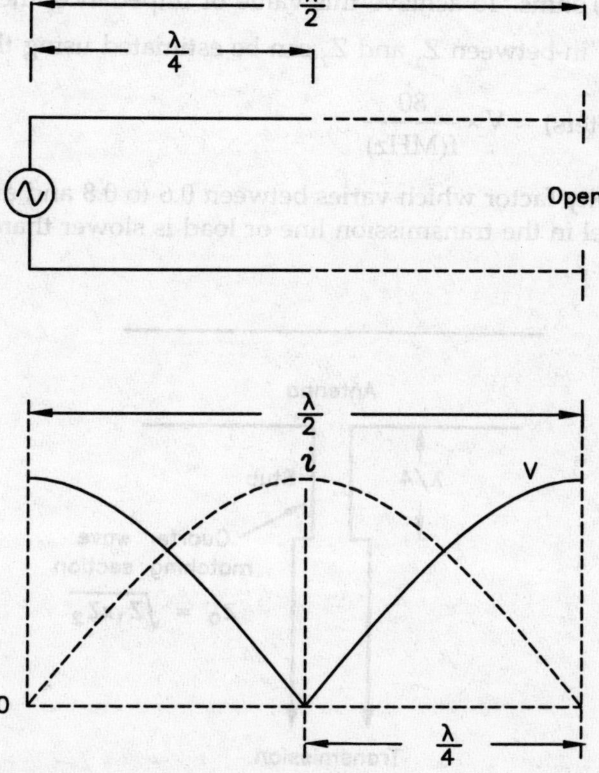

Fig. 4.5. One-quarter and one-half wavelength open lines also look like resonant circuits to a generator.

shorted line. If the line is less than $\lambda/4$, the source sees an inductance. Again all these characteristics repeat for multiple quarter or half wavelengths.

At low frequencies and even up to VHF range, the above-stated characteristics of open and short lines have little significance because $\lambda/4$ and $\lambda/2$ line lengths are too long to be used in practical applications. However, at UHF, more than 300 MHz and at microwave frequencies, $\lambda/2$ line lengths becomes less than 50 cm. Such lengths are manageable and as such transmission line lengths can be used as tuned circuits and reactances at very high frequencies. The values of inductance and capacitance needed at super high frequencies become so small that it becomes difficult to realize them physically with standard coils and capacitors. Therefore, at UHF and microwave frequencies, short lengths of transmission lines become useful components and are used in tuned circuits, filters, phase-shifters, and impedance-matching circuits.

Quarter-Wave Matching Stub: A matching quarter-wave section or stub is a $\lambda/4$ length of coaxial or balanced transmission line of a specific impedance Z_Q that is connected between source and load for impedance matching. Its application is illustrated in Fig. 4.6 where Z_1 and Z_2 are the source and load impedance respectively. In order to provide impedance match between them, the value of Z_Q of the stub section should be $\sqrt{Z_1 \times Z_2}$. If Z_1 is the antenna impedance equal to 75 ohms and Z_2 of the transmission line equal to 300 ohms as shown in the figure the required value of $Z_Q = \sqrt{75 \times 300} = 150$ ohms. To achieve this value of impedance, the length of $\lambda/4$ section to be connected in-between Z_1 and Z_2 can be estimated using the expression:

$$\lambda/4 \text{ (meters)} \approx V \times \frac{80}{f(\text{MHz})}$$

where V is the velocity factor which varies between 0.6 to 0.8 and arises because of the speed of signal travel in the transmission line or load is slower than the speed of radio signals in free space.

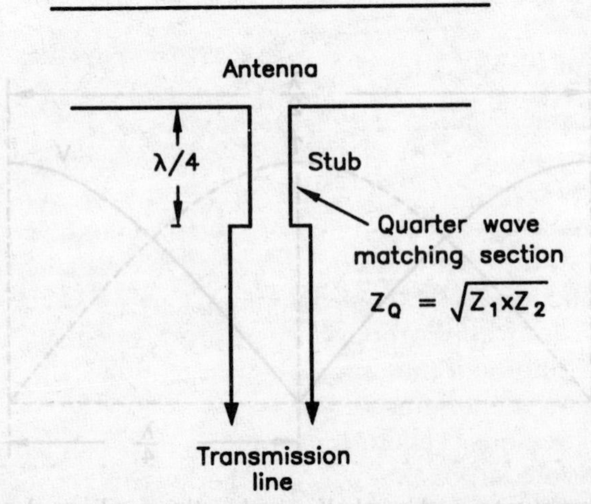

Fig. 4.6. Application of a stub for impedance matching.

Balun or Balancing Unit: Two quarter wave ($\lambda/4$) sections of stub lines can be combined to make a balancing and impedance transforming unit called BALUN. An illustration of its use for matching a 300 ohms twin-lead cable coming from antenna to the 75 ohm TV receiver input impedance is shown in Fig. 4.7 (a). Two quarter wave lines, each having an impedance equal to 150 ohms are connected in parallel at one end resulting in 75 ohms impedance across points C and D. Either C or D can be grounded to obtain an unbalanced impedance at the ungrounded point with respect to ground. At the other end the two $\lambda/4$ sections are connected in series to provide 300 ohm impedance between points A and B. The quarter-wavelength of the line isolates ground point from A or B, allowing a balanced impedance with respect to ground. Either side of the balun can be used as input with the other side as output.

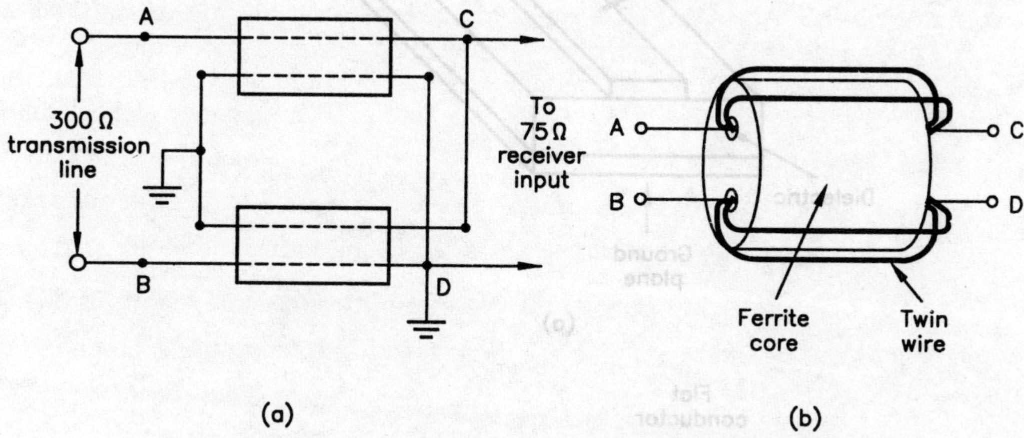

Fig. 4.7. Balun to match 300 Ω balanced to 75 Ω unbalanced impedance. (a) with $\lambda/4$ matching sections, (b) constructional details.

Most baluns are made of ferrite core, either a toroid or rod and windings of copper wire as shown in Fig. 4.7 (b). Baluns with impedance matching ratios of 4:1, 9:1 and 16:1 are also created for matching purposes in microwave equipments.

4.2 Special Lines and Wave Guides

At frequencies below microwaves, parallel wire transmission lines and coaxial cables are the primary means of carrying radio signals. As the frequency of operation gets higher, radiation losses from parallel wire lines become so high that very little energy reaches at the end of such transmission lines. Coaxial cables, though eliminate radiation losses, but have significant attenuation at microwave frequencies and in fact become virtually unusuable above 10 GHz. Special cables have however, been developed for use up to 6 GHz and find applications in consumer satellite TV and to interconnect pieces of microwave equipment that are close to each other. But for longer runs special methods of transmission must be used.

As explained in previous section, transmission lines can be used as tuned circuits, filters and even reactive components because the lengths of one-quarter and one-half wave lengths at microwave frequencies are of the order of a few centimeters making them of practical size. But instead of coaxial or balanced lines, special transmission lines are constructed by using printed circuit board (PCB) or IC manufacturing techniques.

Microstrips and Striplines: The PCB technique can be used to create special microwave transmission lines called Microstrip and Stripline. Microstrip is a flat conductor separated from a large conducting ground plane by an insulating dielectric as shown in Fig. 4.8(a). The length of microstrip is usually one-quarter or one-half wavelength. The ground plane is the circuit common.

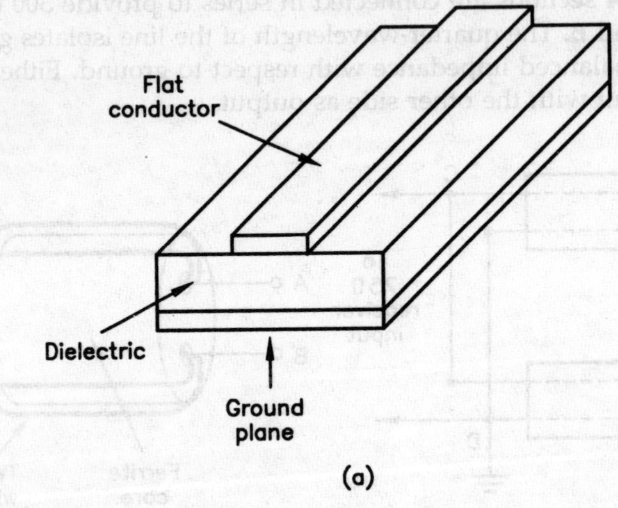

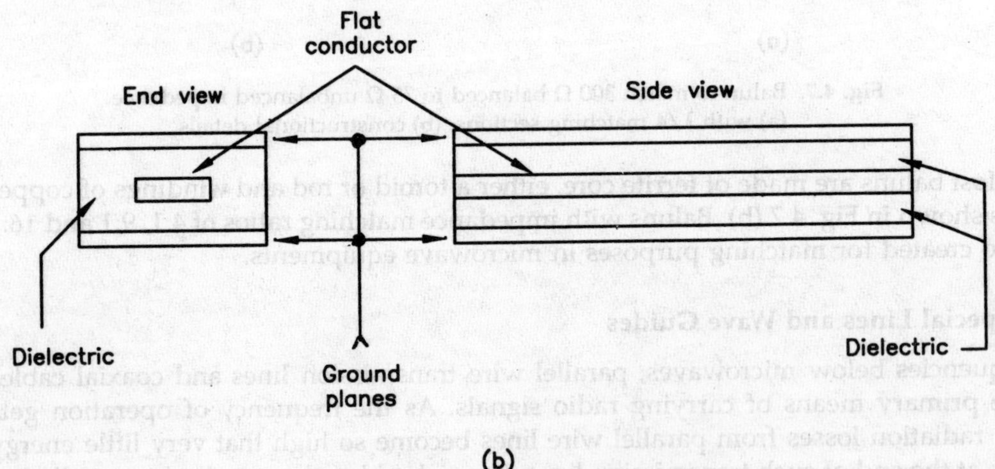

Fig. 4.8. (a) Microstrip. (b) Stripline

Stripline is a flat conductor sandwitched between two ground planes as shown in Fig. 4.8(b). It is more difficult to make but it does not radiate as a microstrip does. The length is same—$\lambda/4$ or $\lambda/2$. Both forms are widely used in microwave receiver front-ends and low noise microwave amplifiers. Diodes, transistors and other components are mounted right on the PCB and are connected directly to the formed microstrip or stripline. Even tinier microstrips and striplines can be evolved to form part of microwave integrated circuits.

Wave Guides: Microwave energy transmission is mostly done through waveguides which are hollow tubes of either rectangular or circular cross-section, the dimensions of which depend on the frequency of signal to be handled. These are made from copper, aluminium or brass and their insides are often coated with silver to reduce resistance losses. Such waveguides are used to transfer microwave energy from one piece of equipment to another and for carrying EHF signals from a transmitter to its antenna or from an antenna to a receiver.

The operating characteristics of a wave-guide are similar to those of transmission lines and have a wave impedance equal to the ratio of electric to magnetic field intensities that is analogous to the characteristic impedance of a coaxial cable or open wire line. Therefore, a waveguide must be terminated by an impedance equal to its wave impedance if standing waves along its axis are to be avoided.

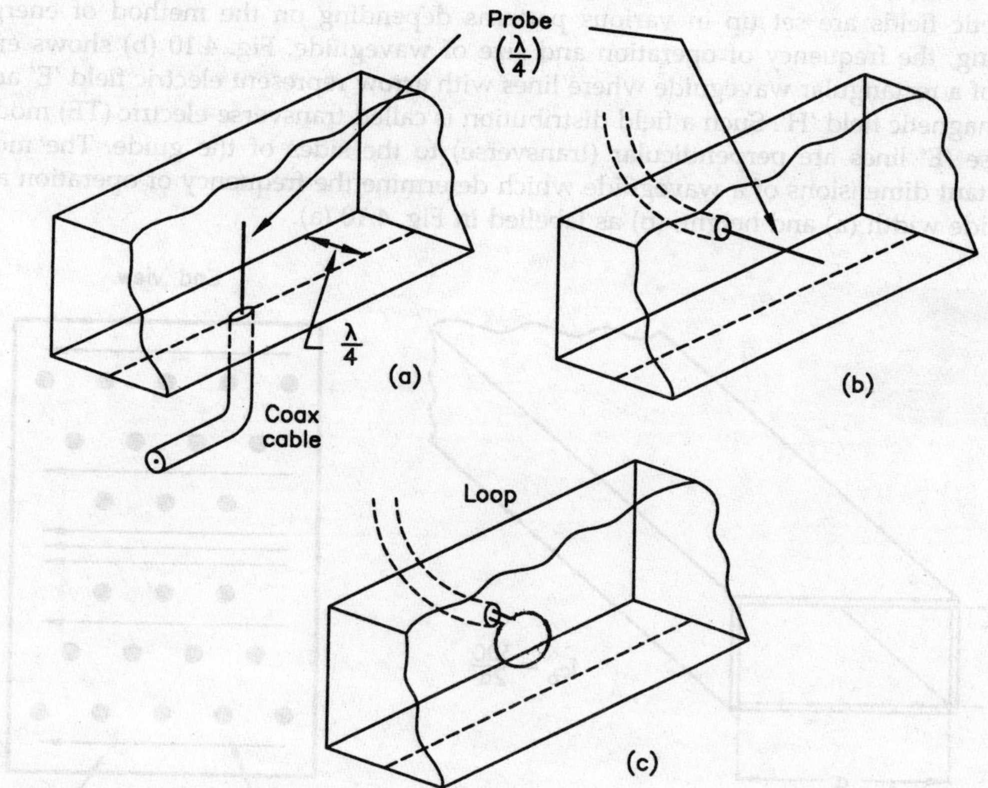

Fig. 4.9. Coupling of RF energy to a waveguide.

The microwave signal to be carried by a waveguide is injected into one of its ends with an antenna-like device called probe which creates an electromagnetic wave that propagates through its length. The electric and magnetic fields associated with the signal bounce off the inside walls back and forth as it progresses down the waveguide. It thus completely contains the signal so that no energy escapes by radiation. A common method of applying a microwave signal to the waveguide is shown in Fig. 4.9(a). The probe used to do so is a quarter-wavelength vertical antenna at the signal frequency and is inserted in the waveguide one-quarter wavelength from the end which is closed. The signal is usually coupled to the probe through a coaxial connector. An alternative probe position is shown in Fig. 4.9(b). The position of probe determines whether the signal is horizontally or vertically polarized. The electric and magnetic fields associated with the signal are at right-angles to one another and propagate through the waveguide axis that is at right angles to both electric and magnetic fields. As shown in Fig. 4.9(c) a loop can also be used to introduce electromagnetic radiation in the waveguide. Extracting a signal from a waveguide is also done with a probe or loop. When the waves strike the probe or loop, a signal will appear across the coaxial connector which is then fed to the connecting circuitry.

Dimensions of Waveguides: Most waveguides are of rectangular variety as shown in Fig. 4.10 (a). When a microwave signal is launched into the waveguide, electric and magnetic fields are set up in various patterns depending on the method of energy coupling, the frequency of operation and size of waveguide. Fig. 4.10 (b) shows end view of a rectangular waveguide where lines with arrow represent electric field 'E' and dots magnetic field 'H'. Such a field distribution is called transverse-electric (TE) mode, because 'E' lines are perpendicular (transverse) to the sides of the guide. The most important dimensions of a waveguide which determine the frequency of operation are its inside width (a) and height, (b) as labelled in Fig. 4.10 (a).

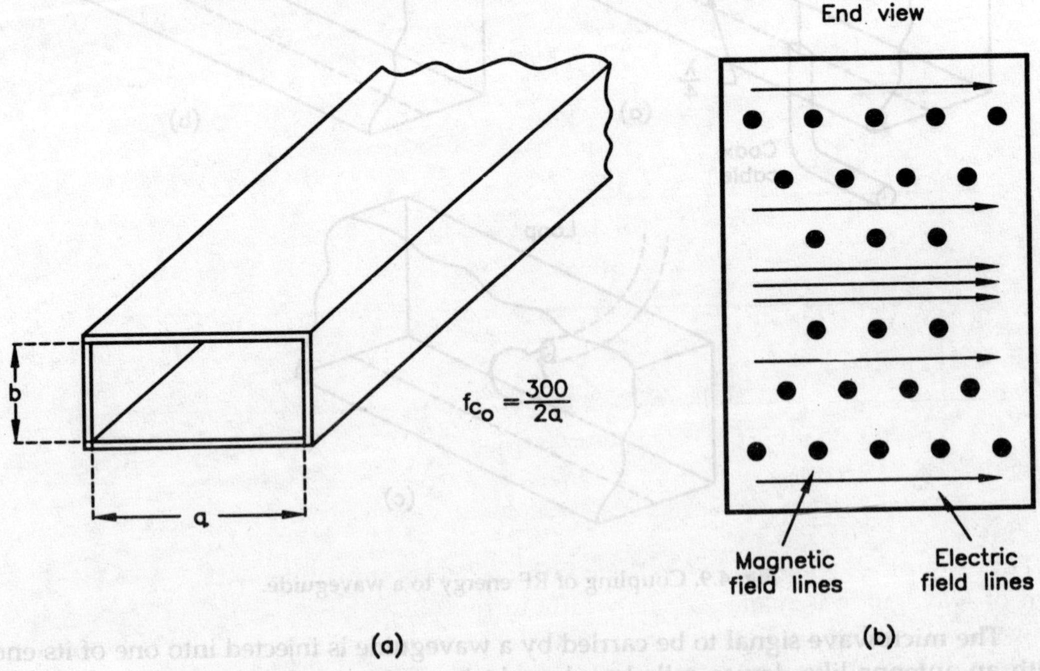

Fig. 4.10. (a) Dimensions of a rectangular waveguide. (b) Electric (E) and magnetic (H) fields in the waveguide.

Cut-off Frequency: When energy is fed into the waveguide, the electromagnetic fields bounce off its side walls as illustrated in Fig. 4.11. The angles of incidence and reflection depend on the operating frequency. At high frequencies the angles are large and therefore, path between opposite walls is relatively long as shown in Fig. 4.11 (a). As the operating frequency gets lower, the angle decreases and path between sides shortens [see Fig. 4.11 (b)]. At a particular frequency the signal will simply bounce back and forth between the side walls of the waveguide [Fig. 4.11 (c)] and thus will not have any forward motion. This frequency is called cut-off frequency (f_{co}) of the waveguide. Thus, it acts as a high-pass filter and no signal below f_{co} can pass through it, or say no energy is propagated at this and signals below this frequency. The cut-off frequency can be determined with the expression given below:

$$f_{sc}\text{ (MHz)} = \frac{300}{2a\text{ (metres)}}$$

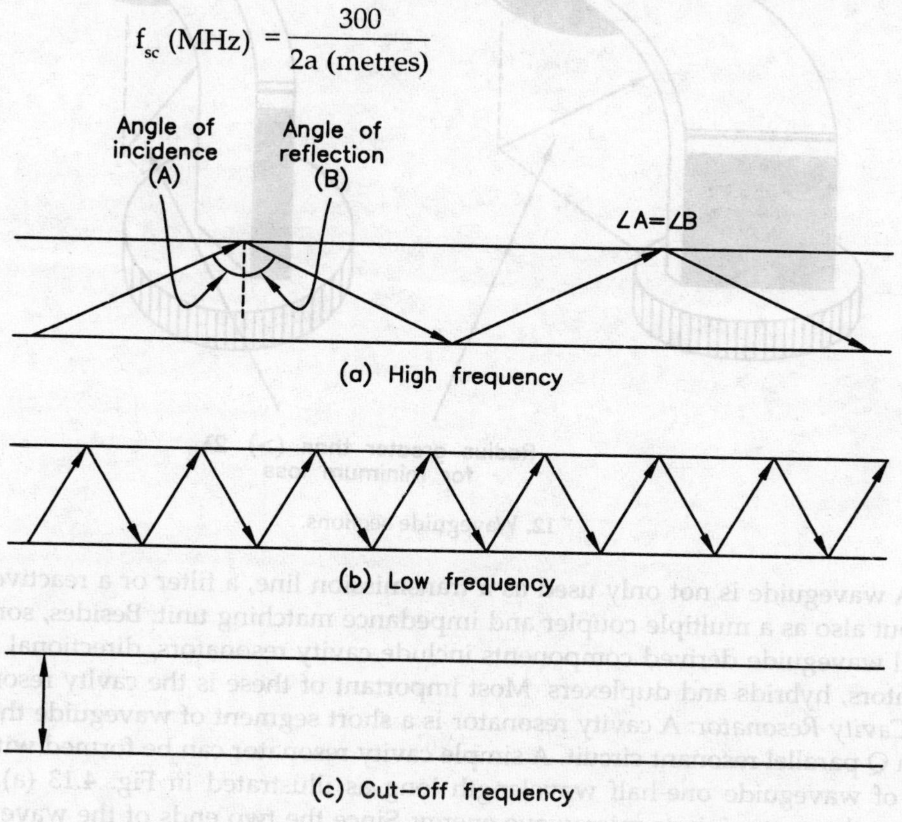

Fig. 4.11. Wavepaths in a waveguide at various frequencies.

For example, a waveguide having width *i.e.* 'a' = 2 cm will have f_{sc} = 7.5 GHz. The height of waveguide is made half its width and in the above example it will be 1 cm. Such waveguides find applications in microwave communication as H.P. filters to provide separate paths to up-link and down-link signals at satellite ground stations.

Waveguides are available in different lengths and shapes. As many sections as necessary can be interconnected. When a waveguide is used as a transmission line between a transmitter or receiver and associated antenna, relatively long straight sections are commonly used. However, in most installations it is necessary to make turns in-between for which special curved waveguide sections are available for making 90° bends. Fig. 4.12 shows two such sections. Curved waveguide sections are also available for specific applications. Most of these sections are accompanied by flanges as shown in the figure and are used to interconnect them with straight sections.

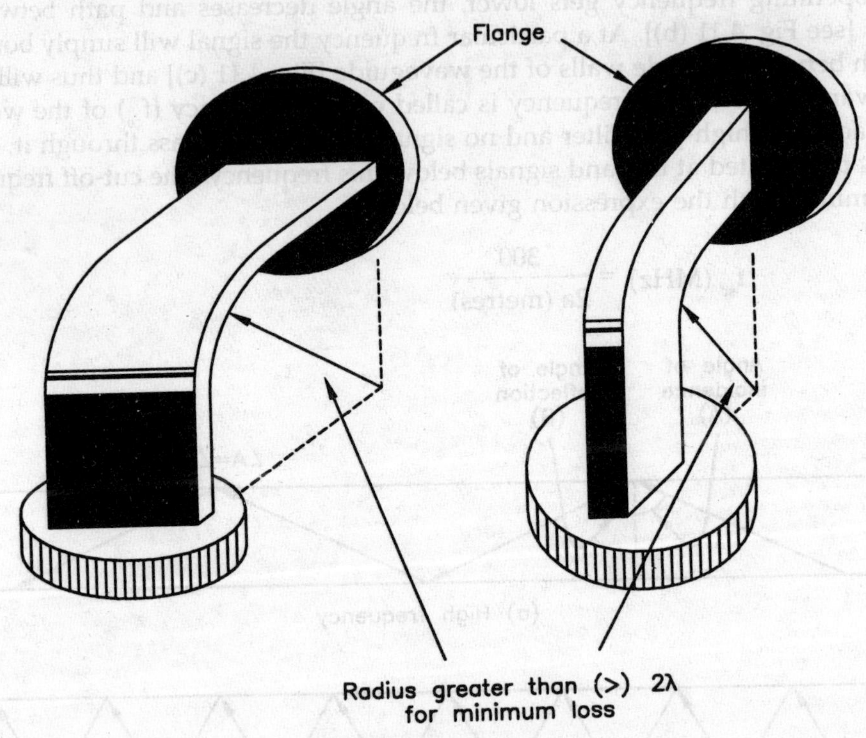

12. Waveguide sections.

A waveguide is not only used as a transmission line, a filter or a reactive component but also as a multiple coupler and impedance matching unit. Besides, some of the special waveguide derived components include cavity resonators, directional couplers, circulators, hybrids and duplexers. Most important of these is the cavity resonator.

Cavity Resonator: A cavity resonator is a short segment of waveguide that acts as a high Q parallel resonant circuit. A simple cavity resonator can be formed with a short piece of waveguide one-half wavelength long as illustrated in Fig. 4.13 (a). A small probe at the centre injects microwave energy. Since the two ends of the waveguide are closed (shorted), the signal bounces off closed ends and reflects back towards the probe. The result is that the signal bounces back and forth off the shorted ends. This effect is pronounced at a frequency where the length of waveguide is exactly one-half of its wavelength. The cavity is then said to resonate and acts like a parallel resonant circuit. A small burst of energy applied to the probe will cause the cavity to oscillate until losses cause it to die out. Cavities such as these have Q as high as 30,000 and because of this merit are used to create resonant circuits and filters at microwave frequencies.

A cavity can also be formed by using a shorted section of circular waveguide as shown in Fig. 4.13 (b). The diameter of the guide is chosen to be $\lambda/2$ at the operating

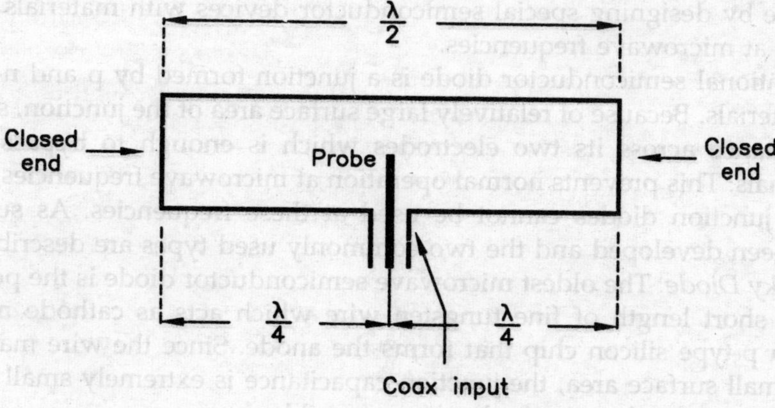

Fig. 4.13. (a) Side view of a λ/2 waveguide used as a cavity resonator.

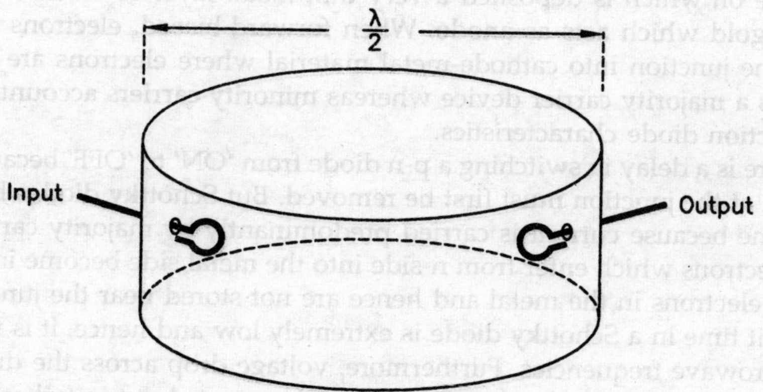

Fig. 4.13. (b) Circular resonant cavity with input-output loops.

frequency. Other cavity shapes are also possible where one wall of the cavity is made movable and an adjustment screw moves the end-wall in and out to adjust the resonant frequency. Further on, a coupling loop rather than a probe is more commonly used in cavity resonators. In some resonators two probes or two coupling loops are used, one for input the other for output. In this way, the cavity resonator can serve as a filter.

4.3 Microwave Devices

Most conventional diodes and transistors do not operate properly at microwave frequencies. The lengths of diode and transistor leads present significant amount of inductance at microwave frequencies and therefore, introduce losses. Their internal capacitances are also relatively high and hinder operation at extra high frequencies. Additionally, transit time of carriers in a diode or transistor is significantly high and prevents normal operation at microwave frequencies. As defined earlier, transit time refers to the small time it takes for electrons or holes to move from cathode to anode in a diode and from emitter to base and to collector in transistors. These problems have been overcome by designing special semiconductor devices with materials other than silicon for use at microwave frequencies.

A conventional semiconductor diode is a junction formed by p and n-type semiconductor materials. Because of relatively large surface area of the junction, such diodes exhibit capacitance across its two electrodes which is enough to bypass very high frequency signals. This prevents normal operation at microwave frequencies. Therefore, standard p-n junction diodes cannot be used at these frequencies. As such, special diodes have been developed and the two commonly used types are described:

1. *Schottky Diode:* The oldest microwave semiconductor diode is the point contact type where a short length of fine tungsten wire which acts as cathode make a fine contact with a p-type silicon chip that forms the anode. Since the wire makes contact over a very small surface area, the junction capacitance is extremely small and hence, operation at microwave frequencies becomes possible.

An improved version of point-contact diode is Schottky diode made with n-type silicon cathode on which is deposited a very thin metal layer of either aluminium or chromium or gold which acts as anode. When forward-biased, electrons from n-type silicon cross the junction into cathode-metal material where electrons are plentiful. In this sense, it is a majority carrier device whereas minority carriers account for conventional p-n junction diode characteristics.

Also, there is a delay in switching a p-n diode from 'ON' to 'OFF' because minority carriers stored at the junction must first be removed. But Schottky diodes have negligible storage time because current is carried predominantly by majority carriers. This is so, because electrons which enter from n-side into the metal side become indistinguishable from the electrons in the metal and hence are not stored near the junction. Therefore, the transit time in a Schottky diode is extremely low and hence, it is very suitable for use at microwave frequencies. Furthermore, voltage drop across the diode is much less than a p-n junction diode for the same forward current. A cut-in (where conduction starts) voltage of about 0.3V is necessary for such a metal-semiconductor diode against about 0.6V for a p-n junction barrier. This aspect of Schottky diode is also suitable for microwave signals whose amplitude is usually quite low. The formation of such a diode by integrated circuit technique (IC version) is shown in Fig. 4.14. The aluminium and the lightly doped n region form a rectifying contact (labelled 1) whereas the metal and heavily doped '+' region form an ohmic contact (2). Since the junction area is very small, high frequency shunting effects are also very low.

Because of very low threshold (cut-in) voltage, Schottky diodes are ideal for heterodyning (mixing), signal detection and other low level signal operations in microwave equipments. These are installed as part of a waveguide or cavity resonator tuned to the incoming signal frequency. A local oscillator (Lo) signal is also injected with a loop or probe whereupon the diode mixes the two signals to produce sum and difference frequency products. The difference frequency components are usually selected with another cavity resonator or separated with a lower frequency L-C tuned circuit. Such a mixer circuit usually forms the input section of the low noise block converter (LNBC) in the front-end section, installed for receiving down-link satellite signals and converting them to an intermediate frequency (IF) as early as possible for convenience of amplification and further processing with conventional active devices and circuitry.

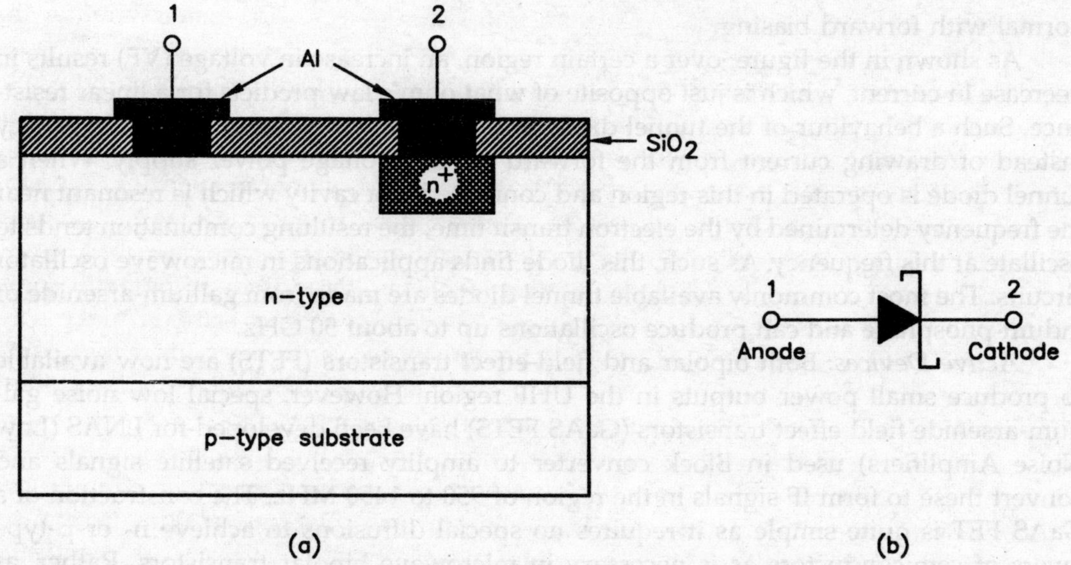

Fig. 4.14. Schottky diode (a) fabrication details. (b) symbol.

2. *Tunnel Diode:* It is a special type of diode that employs heavily doped materials and exhibits a negative resistance characteristics region [see Fig. 4.15 (a)] for an increasing forward-bias voltage. In standard p-n junction diodes, the impurity concentration is

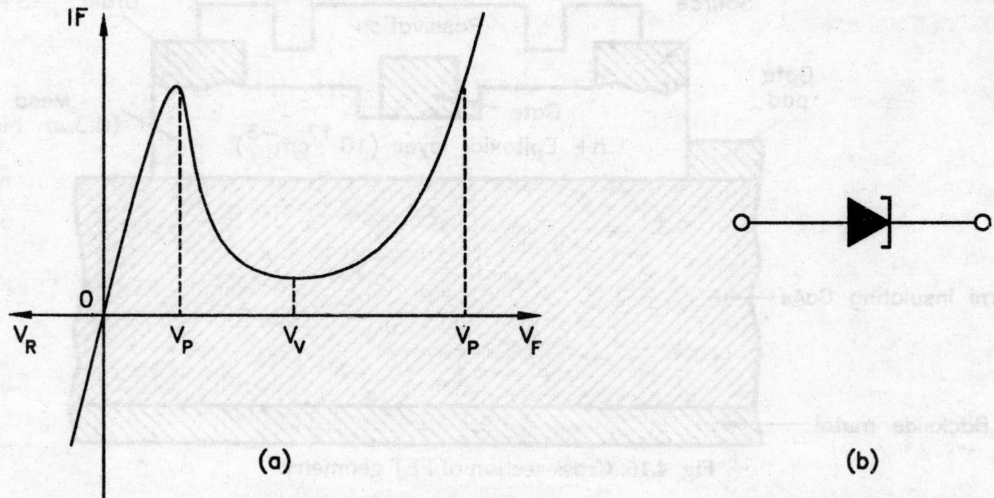

Fig. 4.15. Tunnel diode (a) V-I characteristics, (b) symbol.

about 1 part in 10^8 but if it is greatly increased, to say 1 part in 10^3, the diode characteristics are completely changed. Because of heavy doping, the thickness of junction barrier becomes extremely small-close to one-fifth of the wavelength of visible light. With such a thin barrier, there is a large probability that electrons will penetrate through the barrier because of their own excess energy. This is referred to as tunneling and accounts for decrease of forward current at low levels of forward bias instead of its increase as is normal with forward biasing.

As shown in the figure, over a certain region, an increase in voltage (VF) results in decrease in current, which is just opposite of what ohm's law predicts for a linear resistance. Such a behaviour of the tunnel diode in this region amounts to supplying energy instead of drawing current from the forward biasing voltage power supply. When a tunnel diode is operated in this region and connected to a cavity which is resonant near the frequency determined by the electron transit time, the resulting combination tends to oscillate at this frequency. As such, this diode finds applications in microwave oscillator circuits. The most commonly available tunnel diodes are made from gallium-arsenide or indum-phosphide and can produce oscillations up to about 50 GHz.

Active Devices: Both bipolar and field-effect transistors (FETS) are now available to produce small power outputs in the UHF region. However, special low noise gallium-arsenide field-effect transistors (GaAS FETS) have been developed for LNAS (Low Noise Amplifiers) used in Block converter to amplify received satellite signals and convert these to form IF signals in the region of 950 to 1450 MHz. The construction of a GaAS FET is quite simple as it requires no special diffusions to achieve n- or p-type layers of semiconductors as is necessary in microwave bipolar transistors. Rather, as shown in Fig. 4.16, it consists of a layer of semi-insulating GaAS on which a lightly doped 'n' layer of GaAS, called epitaxial layer is grown. Then in a vacuum chamber under proper temperature conditions, a layer of metal film, usually gold is evaporated on the epitaxial layer to form a Schottky-Barrier junction. Source, gate and drain

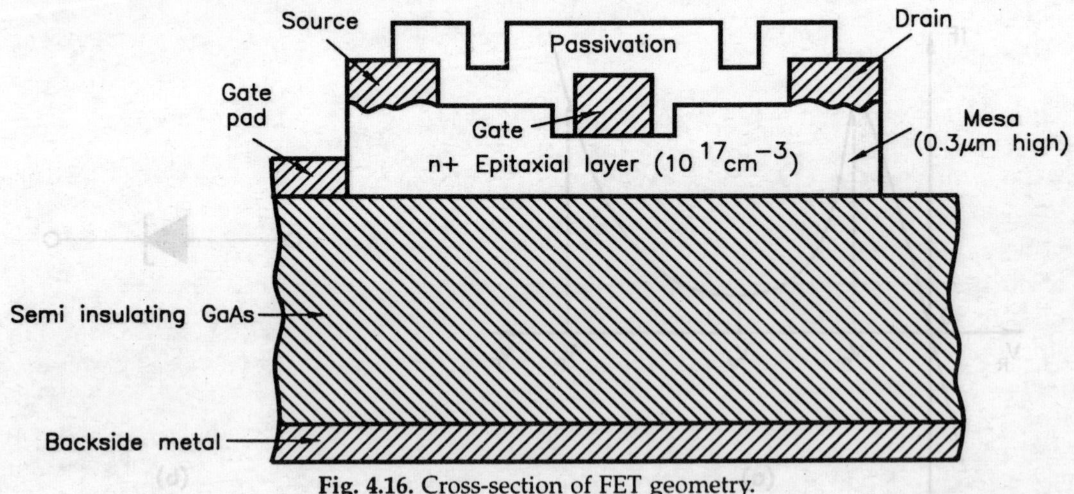

Fig. 4.16. Cross-section of FET geometry.

contacts are etched by photolitographic techniques. Finally, a passivation layer is added. To achieve lowest-noise FETS, the epitaxial layer under the gate must be etched into a recess, as shown in the figure. For a microwave FET, the gate 'finger' is extremely narrow—as narrow as 0.5 micron (10^{-6} m). This 0.5 micron width is equal to a wavelength of blue light.

Electrically, the FET resembles a vacuum triode with the gate corresponding to grid, the source to cathode and drain to anode. Since it corresponds to a triode, it has relatively high input impedance. It is this fact which makes the FET so useful at microwave frequencies. Bipolar transistors are difficult to use as low-noise amplifier above 4 GHz, because their matching impedance is low. GaAS FETS currently available are capable of noise figures as low as 0.6 dB at 4 GHz with corresponding gains of greater than 14 dB, although in production a spread of somewhat lower values of gain and noise figure must be accepted.

Microwave Tubes: While transistors can be used to generate power in the microwave region up to about 50 watts, which is sufficient for needs at the receiving end but for transmission and other high power applications vacuum tubes are still in use. In satellite earth stations, TV stations and in some navigational and military equipments very high power outputs are needed. For this, special vacuum tubes have been developed which include KLYSTRON, MAGNETRON and TRAVELLING WAVE tube. A Klystron uses cavity resonators to produce velocity modulation of the electron beam to cause signal amplification. Klystrons are available in a wide range of sizes. Small hand-held units produce only milliwatts of power amplification while, larger human-sized Klystrons produce many thousand watts power.

A magnetron is the combination of a simple diode vacuum tube with built-in cavity resonators and an extremely powerful permanent magnet. It can generate extremely high levels of microwave power.

A travelling wave tube (TWT) is the most versatile microwave RF power amplifier because of its extremely wide band width of operation. As such, these are commonly used as power amplifiers in satellite transponders.

4.4 Microwave Antennas

Theoretically, antennas designed for transmission and reception of VHF and UHF signals can also be used at microwave frequencies. However, it is not possible to do so in practice because these are not very directive and their physical sizes will be too small and thus, inefficient. For example, a half-wave dipole at 5 GHz will be only 3 cm long. Therefore, special antennas which are both efficient and highly directive are preferred because these do not waste energy and also provide gain to the relatively low power microwave signals.

Horn Antennas: The most widely used microwave antenna is HORN which is nothing more than a flared waveguide as shown in Fig. 4.17 (a). In this configuration, it exhibits both gain and directivity. The longer the horn, the greater the gain and directivity. The length of a typical horn is usually 2 to 15 wavelengths at the operating frequency. Different kinds of horn antennas can be created by flaring the end of waveguide in different ways. If circular waveguide is used, the flare produces a conical horn as shown in Fig. 4.17 (b).

Parabolic Reflector Antennas: Horn antennas are used by themselves in many microwave applications, but many times, higher gain and directivity is desirable. This can be easily obtained by using a horn in conjunction with a parabolic reflector. A parabolic reflector is a large dish-shaped structure made of screen mesh or metal. The energy radiated by the horn is pointed at the reflector which focuses the radiated energy into a narrow beam and reflects it towards its destination. A special characteristic of a parabola is that the distance (see Fig. 4.18) between the focal point to its surface and then to the vertical axis line X-Y is a constant value. For example, the sum of line FA to

AB and FC to CD are equal. This characteristic causes the parabolic-shaped surface to collimate *i.e.* concentrate electromagnetic waves into an extremely narrow beam. Placing a horn antenna at the focal point 'F' will cause it to radiate waves from the parabola in parallel lines. Similarly, if used as a receiver, the parabola will pick-up electromagnetic waves and reflect them to the antenna located at the focal point. Fig. 4.19 illustrates both sending and receiving conditions. The result in either case is an extremely high gain narrow beam width antenna.

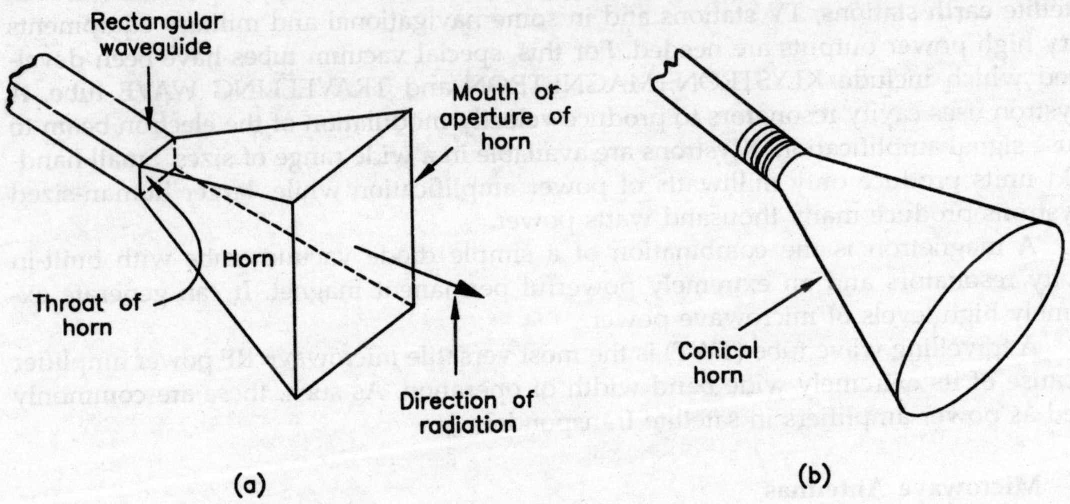

Fig. 4.17. (a) Basic horn antenna, (b) conical horn.

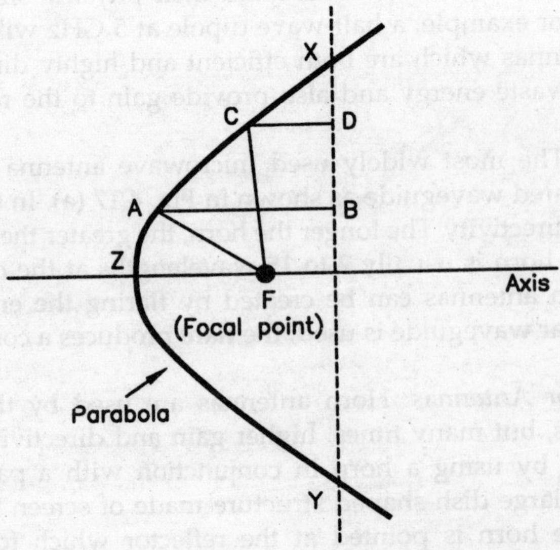

Fig. 4.18. Special characteristics of a parabolic reflector.

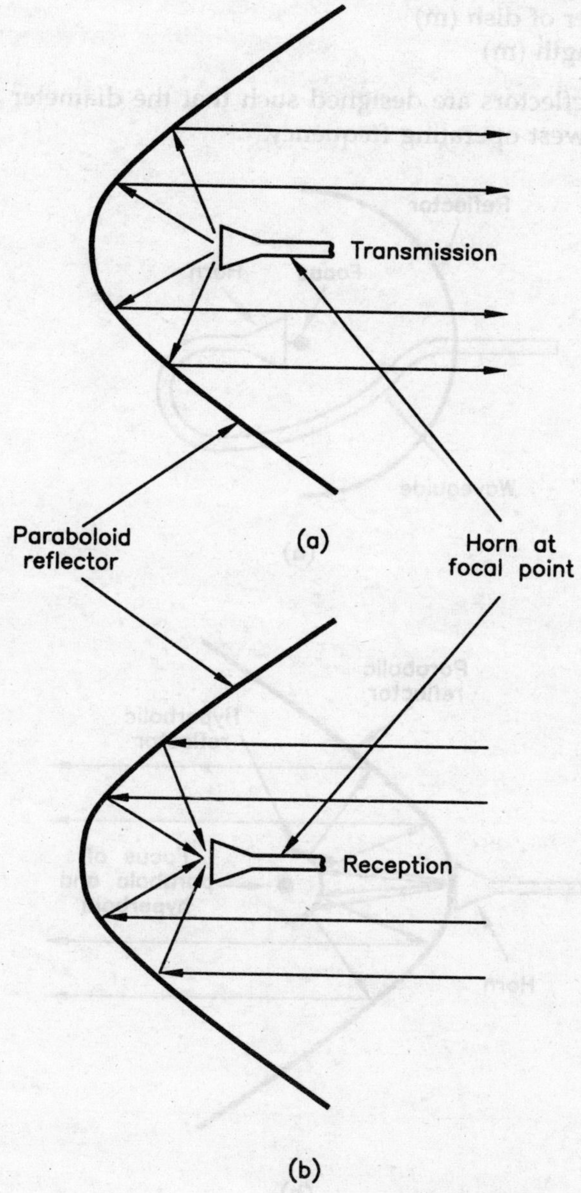

Fig. 4.19. (a) Sending, and (b) receiving with a parabolic reflector-antenna.

The gain of a parabolic antenna is directly proportional to the aperture of the parabola. The aperture is the area of the outer circle of the parabola. This area is:

A = 3.14R² *i.e.* (π R²) where R is the radius of outer circle. Similarly, the gain of a parabolic antenna is given by the simple expression:

$$G = 6\left(\frac{D}{\lambda}\right)^2$$

where G = gain expressed as a power ratio
D = diameter of dish (m)
λ = wavelength (m)

Most parabolic reflectors are designed such that the diameter is not less than 10 wavelengths at the lowest operating frequency.

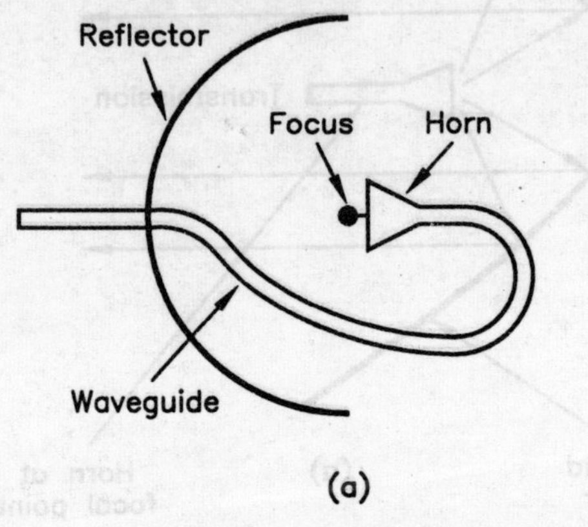

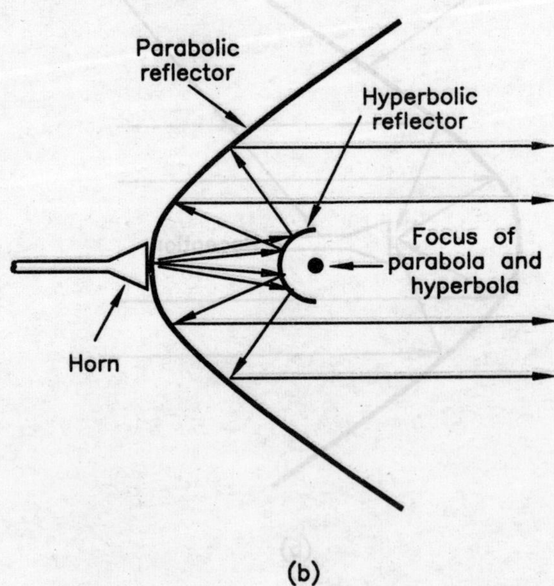

Fig. 4.20. Waveguide and horn feed systems.

Assuming that D = 5 meters, and λ = 0.03 meter (*i.e.* f = 10 GHz), the gain will be

$$G = 6\left(\frac{5}{0.03}\right)^2 = 166673$$

In decibels (dB) it will be:

$$10 \log_{10} 166673 = 10\,(5.22) = 52.2$$

Both transmitting and receiving antennas are usually of parabolic design and thus have extremely narrow beam widths. For this reason, the dish must be accurately pointed if correct contact is to be made. Naturally, the beam spreads out and grows in size with distance.

Signal Feed Methods: Keeping in mind that actual antenna is the horn and not the dish, there are many arrangements used in positioning the horn at the focal point. One of the most common method is shown in Fig. 4.20 (a). The waveguide feeds through the centre of parabolic dish and is curved around so that horn is positioned exactly at the focal point. In an another arrangement [Fig. 4.20 (b)], a second reflector is positioned at the focal point as shown in the figure.

Here the horn antenna is positioned at the centre of parabolic reflector. At the focal point is another small reflector with either a parabolic or hyperbolic shape. The electromagnetic radiation from the horn strikes the smaller reflector, which then reflects the energy towards the larger dish. The larger dish then radiates the signal in parallel beams. This arrangement is known as CASSEGRAIN feed system.

In large antenna systems, various mechanical arrangements are used to rotate or change its position physically for correct orientation towards the wanted satellite. Some sophisticated systems use multiple horns on a single reflector. Such multiple feeds enable several signals on different frequencies to be received with a single large reflecting structure.

Only Receive Dish Antennas: In dish antennas designed for receiving TV signals from satellites, a waveguide is not used. Instead a horn antenna is mounted at the point which collects signals from the parabolic reflector (dish) and feeds it to the LNBC as explained in Section 1.1. More details on such antennas are given in Chapter 14 which is devoted to reception of TV signals from satellites and their processing.

5

PERFORMANCE CHARACTERISTICS OF AMPLIFIERS

Introduction

Electrical signals generated by various transducers and sensors and those picked up by various antennas are very weak and need large amplification. Therefore, keeping in view circuit requirements amplifiers are designed to be voltage or power amplifiers. In addition, depending on frequency range of the input signal, amplifiers can be further classified into audio, video and RF amplifiers.

In any case, the basic function of an amplifier is to magnify the amplitude of input signal without altering any of its properties. However, in a practical amplifier, there are factors like gain, instability and different types of distortions which affect the fidelity of reproduction. For optimum results, change in gain and various distortions have to be kept within specified limits which themselves are dictated by specific applications. Thus, in order to be able to assess the suitability of a given amplifier to meet certain specifications, it is necessary to gain familiarity with performance measures of amplifiers. This chapter is written with this aim in view.

5.1 Amplifier Classifications

In present-day voltage and power amplifiers, bipolar junction transistors (BJTS) and field effect transistors (FETs) are used as active devices because of their superiority over vacuum tubes. One terminal of the amplifier is generally common to both input and output circuits and is grounded to chassis or earth lead on the printed circuit board (PCB). Any of the three electrodes or leads of the transistor can be made common to both input and output circuits thus giving rise to (i) common emitter (CE), (ii) common base (CB) and common collector (CC) configurations. In any of these configurations, the emitter-base junction is always forward biased and collector-base junction reverse biased.

The common emitter version is used in most voltage amplifier circuits because it provides high voltage gain and has fairly high (medium) input and output impedance levels. The common base amplifier performs better at high frequencies, provides high voltage gain but has very low input and very high output impedance levels. The common collector configuration, also called emitter follower has high input impedance but low output impedance. As such it is commonly used for matching high impedance signal sources to very low impedance loads like loudspeakers, and small electric motors. The price paid for this convenience is that voltage gain of an emitter follower is less than one though can be kept close to unity. The impedances referred to above are mostly resistive in nature at low and medium frequencies but become complex at very high frequencies.

Usually in any amplifier system, a number of stages are used to obtain more gain and power output. These can be cascaded (connected to each other) by direct coupling or resistance - capacitance (R-C) coupling or transformer coupling. Sometimes inductance - capacitance (L-C) coupling is also used. Depending on the ultimate purpose, amplifiers can also be classified as voltage (small signal) and power (large signal) amplifying types. Further, a power amplifier can be single ended using one active device or push-pull type using two active devices, depending on power output requirements. Power amplifiers are invariably preceded by one or more stages of voltage amplification.

Class of Operation: The amplifiers may also be classified according to where the quiescent (operating) point 'Q' is fixed and how much is the magnitude of input signal. The four classes of operation that are commonly used are described.

(i) *Class A:* [Fig. 5.1 (a)]. In class A operation the transistor stays in the active region of its transfer characteristics throughout the ac cycle. The Q point and magnitude of input signal are so chosen that collector current flows all the time *i.e.* for 360° as shown in the figure. More than 90 per cent of amplifiers used in electronic circuits operate under class A condition. If the characteristic curve around the point Q is quite linear and input signal amplitude is not very large, the output is nearly an exact amplified replica of the input signal.

The efficiency of class 'A' operated amplifiers is quite low, around 25 per cent, but this is of not much consequence since power drawn from the dc supply source is quite low.

(ii) *Class B:* [Fig. 5.1 (b)]. In class B operation, the transistor stays in the active region for half the ac cycle. The Q point is chosen at the cut-off point of characteristics and hence, collector current flows for 180° during every 360° signal cycle. The power drawn from the dc power supply is quite low since during quiescent condition periods, collector current stays at almost zero. Thus, the efficiency of class 'B' operated amplifiers is fairly high-around 35 per cent. In audio and other similar applications, where more power output is required, push-pull configuration is preferred. A push-pull amplifier is operated under class 'B' condition where each transistor conducts for 180° sequentially *i.e.* one after the other, but the two outputs get combined in the output circuit to provide amplified version of the input signal.

(iii) *Class AB:* [Fig. 5.1 (c)]. This operation is in between classes A and B conditions. The transistor stays in active region for more than half-the-cycle but less than 360°. For this, the Q point is chosen a little above the cut-off point to prevent operation in the region which is quite non-linear. As expected, the efficiency of class AB operated amplifiers is close to 30 per cent.

(iv) *Class C* [Fig. 5.1 (d)]. In a class C amplifier, the Q point is fixed beyond the extreme (zero current) end of transfer characteristics. As such, the transistor stays in the active region for less than half the ac cycle and hence, output current flows in pulses as shown in the figure. The efficiency of class C amplifiers is quite high—often more than 50 per cent. This class of operation finds application in tuned amplifiers and oscillator circuits where, though the collector current flows in short pulses (less than 180°), the output is sinusoidal (0° - 360°) because of resonant action of the parallel L-C combination load connected in the output circuit.

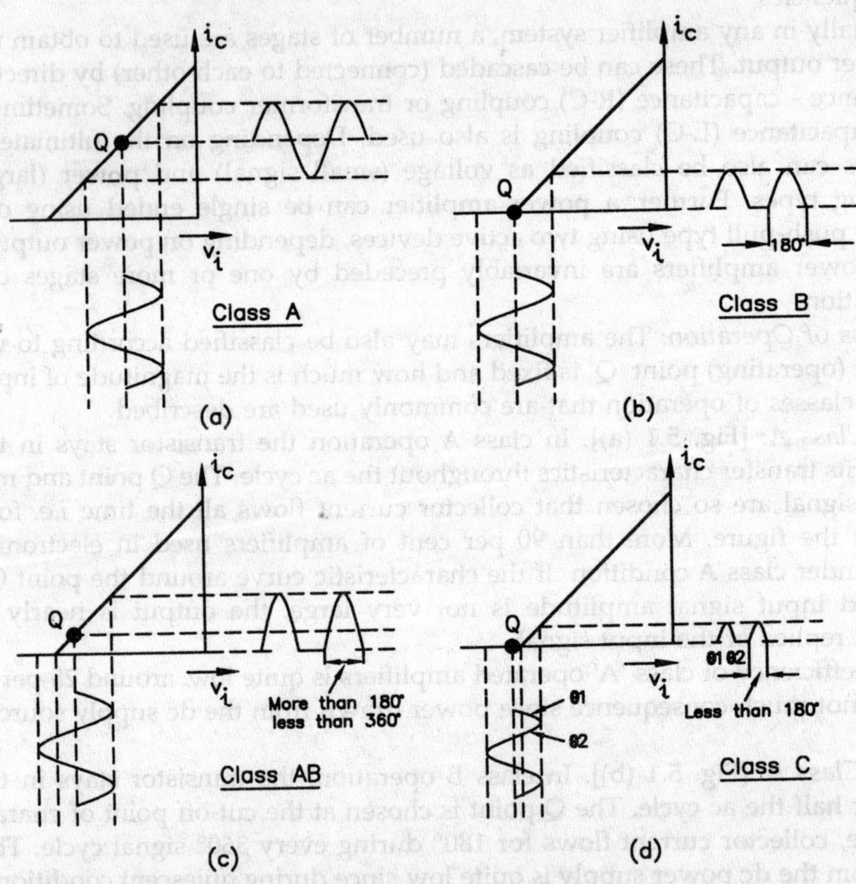

Fig. 5.1. Classification of transistor amplifiers based on biasing conditions. The transfer characteristics have been assumed to be linear.

Voltage and Power Gains: Any amplifier provides both voltage and power gain but voltage amplifiers are designed for large voltage gain and provide little power gain. Similarly, power amplifiers have high power gain and low voltage gain. The voltage gain of an amplifier is expressed as:

$$A_v = \frac{\text{output ac voltage}}{\text{input ac voltage}} = \frac{v_0}{v_i}$$

Two or more amplifier stages are cascaded to obtain more gain. For a three-stage amplifier the overall gain can then be expressed as $A_1 \times A_2 \times A_3$ where A_1, A_2 and A_3 are individual gains of the three stages.

The power gain of an amplifier is the ratio of output and input power levels. It is expressed as:

$$P_G = \frac{\text{output power}}{\text{input power}} = \frac{P_0}{P_{in}}$$

Both voltage and power gains have significance depending cn the type of amplifier and its application.

5.2 Power and Voltage Gains in Decibels

In many electric applications it is found convenient to express two powers on a logarithmic scale rather than on a linear scale. To begin with a logarithmic unit called BEL was proposed after Alexander Graham Bell, the inventor of telephone. The number of BELS by which a power P_2 exceeds P_1 is expressed as:

$$\text{Number of BELS} = \log_{10} \frac{P_2}{P_1}$$

Usually the suffix 10 is dropped and expression written as: * BELS = $\log \frac{P_2}{P_1}$

However, it was soon realised that for practical purposes, BEL was too large a unit and therefore another unit, one-tenth as large, came into use. The smaller unit called decibel is abbreviated as dB. Therefore, number of dB_s = 10 × number of BELS *i.e.* $= 10 \log \frac{P_2}{P_1}$.

In a passive network where output is less than input power due to losses, the dB value will be negative. In practice, when output value is less than input value, it is found convenient to use $\frac{P_1}{P_2}$ and express dB loss as $-10 \log \frac{P_1}{P_2}$

Reference Power: It is to be noted that dB denotes a power ratio and therefore a reference power level must be adopted for obtaining relative value of power gain or loss. The agreed reference level for most communication applications is 6 mW. However, in CATV systems it is convenient to use corresponding voltage ratios instead of power ratios to express gain or loss in decibels. The reference level chosen is 1 mV. If P_1 and V_1 are input power and voltage values and P_2 and V_2 corresponding output values of an

---

* In general $\log_{10} A = B$ *i.e.* log of A to the base 10 means $A = 10^B$. For example log 100 = 2 means $100 = 10^2$, which is true. It is also to be noted that with the suffix dropped log XY = log X + log Y.

Also $\log \frac{X}{Y} = \log X - \log Y$ and $\log X^n = n \log X$, where both X and Y are magnitudes of the same product. The use of log tables is necessary to calculate dB values.

amplifier, then $P_1 = \frac{V_1^2}{R_i}$ and $P_2 = \frac{V_1^2}{R_o}$ where R_i and R_o are the input and output resistance (impedance) values. The power gain in dB $= 10 \log \frac{P_2}{P_1} = 10 \log \frac{V_2^2/R_o}{V_1^2/R_i}$. If $R_i = R_o$ or difference between the two being small is ignored, power gain is $dB_s = 10 \log \left(\frac{V_2}{V_1}\right)^2 = 20 \log \frac{V_2}{V_1}$.

The reference used for CATV is 1 mV across 75 ohms. With this reference the units are indicated as dBmV. This reference is an arbitrary value but 1 mV happens to be just about the minimum signal voltage measured across 75 ohms that a TV receiver needs for a noise free picture. Across 300 ohms, the minimum voltage will accordingly be 2 mV.

The signal received from satellites are very weak—of the order of micro-volts. Therefore, instead of carrying a negative sign it is found convenient to use 1 μV as the reference voltage level in such situations and associated equipments.

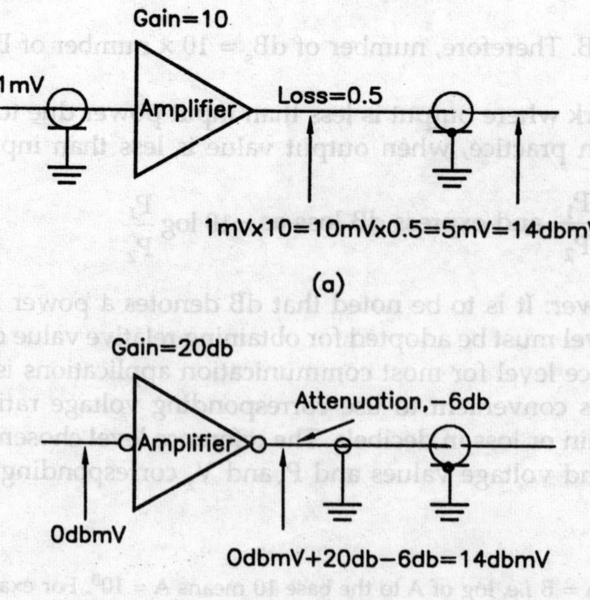

Fig. 5.2. Calculations of gain (a) in millivolts, and (b) in dBmV units.

The dBmV method is convenient for calculations because logarithmic units can be added or subtracted for voltage gains or losses. Consider the amplifier and cable run shown in Fig. 5.2 (a). The voltage gain of 10 multiplies the 1mV signal input to get 10mV output. The cable reduces the signal level by one-half. Thus the final output is $10 \times 0.5 = 5$ mV. In dB units it is:

$$20 \log 5 = 20 \times 0.7 = 14 \text{ dBmV}$$

In Fig. 5.2 (b) the same values are indicated in dBmV and can be just subtracted to obtain net dBmV value. As 0 dBmV is the input signal level of 1 mV and thus 20 dBmV is the amplifier gain ($20 \log 10 = 20 \times 1 = 20$) and –6 dBmV is the cable attenuation of 0.5 since $20 \log 2 = 20 \times 0.3 = 6$ dBmV.

The end result is $0 + 20 - 6 = 14$ dBmV. Thus, gains or losses expressed in dBs can be directly added or subtracted as necessary. Another advantage of expressing gains or losses in dB units is that very small and very large values of gain or loss can be expressed by much smaller figures. For example, a power gain of 456000 is only 65.9 dBs and a voltage gain of 0.000001 (actually it is a loss) is –120 dB ($-20 \log 10^6$). Finally, the use of decibel in acoustic measurements is very appropriate because human ear which is the final receiver of sounds has a logarithmic response to loudness variations.

Power and Voltage Gain Relations: While it is convenient to use voltage ratios for obtaining power gain in dBs, the significance of power readings remains equally important. As an illustration, a power gain of 40 dB means a voltage gain of 100 ($20 \log 100 = 40$) but the same dB value means a power gain of 10,000 ($10 \log 10{,}000 = 40$). Thus, the same numerical dB value does not mean equal voltage and power gains. Actually a voltage gain of X and a power gain of X^2 yield the same dB value. Similarly, a power loss of 20 dB means a fall of voltage gain by a factor of 10 but power reduction by a factor of 100.

5.3 Distortions in Amplifiers

In an amplifier, if the waveshape of output is not an exact replica of input waveshape, it is said to have been distorted. A number of factors may be responsible for causing distortion. It may be due to reactive components in the circuit or due to non-linear characteristics of the active device (transistor). There are three types of distortions. These may exist either separately or simultaneously in an amplifier.

1. *Frequency Distortion:* The signals handled by amplifiers are quite often non-sinusoidal *i.e.* these consist of several frequencies and their harmonies, each having different amplitude and even phase. Thus the sinusoidal constituents of the signals to be amplified extend over a wide range of frequencies. For example, audio amplifiers have to handle frequencies from about 20 Hz to nearly 20 KHz and video amplifiers from a few Hz to as high as 5 MHz. Hence, the study of gain response of an amplifier as a function of frequency is very important. Fig. 5.3 (a) shows such a frequency response plot for an R-C coupled common emitter amplifier. In the ideal case, the magnitude of voltage gain plotted against frequency should be flat. However, in practice, gain falls off at both high and low frequency ends due to varying reactance with frequency of the capacitors associated with amplifier circuit.

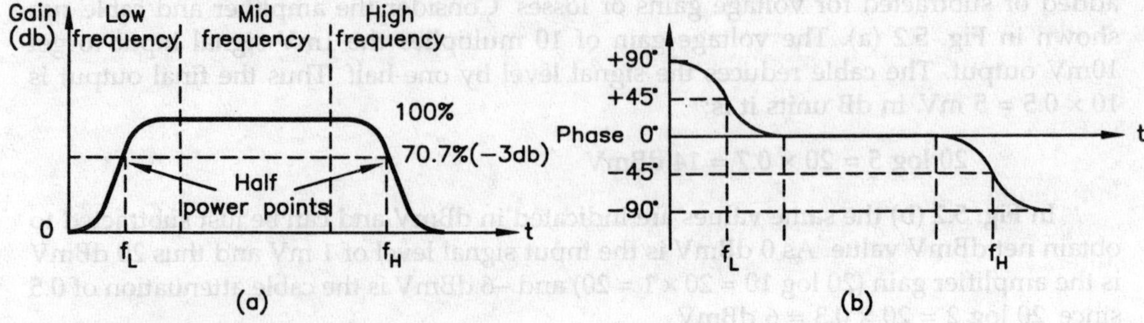

Fig. 5.3 (a). Frequency and (b) phase response of a practical R-C coupled amplifier.

In the R-C coupled amplifier, gain at high frequencies falls off because of the shunting effect of junction capacitances of the transistor and input capacitance of next stage or load connected across the amplifier. Similarly, gain falls off at low frequencies on account of high increasing series reactance of coupling (Cc) capacitor(s). At middle and high frequencies, the reactance of Cc is quite small and hence acts like a short without any voltage loss across it.

Bandwidth: As shown in Fig. 5.3 (a), the magnitude of voltage gain falls off to 70.7% of the midband value at $f = f_L$. This frequency is called lower cut-off or corner frequency and is chosen as the useful lower limit of the amplifier's frequency response. Similarly, the gain drops to 70.7% *i.e.* $\left(1/\sqrt{2}\right)$ of the midband value at $f = f_H$ called the upper cut-off or corner frequency.

Thus, due to combined effect of shunt and series capacitances associated with an amplifier, it behaves like an active band-pass circuit for the frequencies between f_L to f_H, the two corner frequencies. Therefore, the useful frequency span $(f_H - f_L)$ is called bandwidth of the amplifier. Notice that a fall of gain to 70.7% *i.e.* $\left(1/\sqrt{2}\right)$ of midband value means a power loss to 1/2 (50%) of its maximum value. Accordingly, the frequencies f_H and f_L are also called half-power or 3 dB down $(-10 \log 2 = -10 \times 0.3 = -3)$ frequencies.

In conclusion if a large bandwidth is desired, a transistor with very low junction capacitances is chosen and coupling capacitors as large as possible are used. For obtaining very large bandwidths, special compensating networks are employed.

2. Phase Distortion

Phase distortion is said to occur if the phase relationship between various frequency components of the composite signal waveform is not the same in the output as at input of amplifier. It means that time of transmission or delay introduced by the amplifier is not the same for different frequencies. The reactive components that cause frequency distortion are also responsible for causing phase or time-delay distortion. This is shown in Fig. 5.3 (b). The phase shift relative to midband frequencies becomes quite large at very high and very low frequencies. It can be shown that it is $\angle +45°$ at f_L and $\angle -45°$

at f_H relative to midband shift of 180° and then increases sharply beyond corner frequencies.

Phase distortion is not important in audio systems because our ears do not detect changes in relative phase shifts of various frequency components present in any sound signal.

However, phase shift control is important in video amplifiers, since any phase shift implies time shift which in turn causes position shift of certain elements in the reproduced visual image (picture). Thus phase-shift distorts the picture and its quality suffers. That is why video amplifiers are provided with compensating networks to obtain large bandwidth and minimum phase shift effects as shown in fig. 5.4. The inductors used in the load and output circuit compensate for the decrease in effective load at high frequencies due to shunting effect of capacitances and thus the gain remains steady upto about 5 MHz. The resulting increase in bandwidth reduces phase-shift effects at higher video frequencies. Similarly special decoupling network before the load resistance R_L boosts gain and reduces phase-shift distortion at very low frequencies.

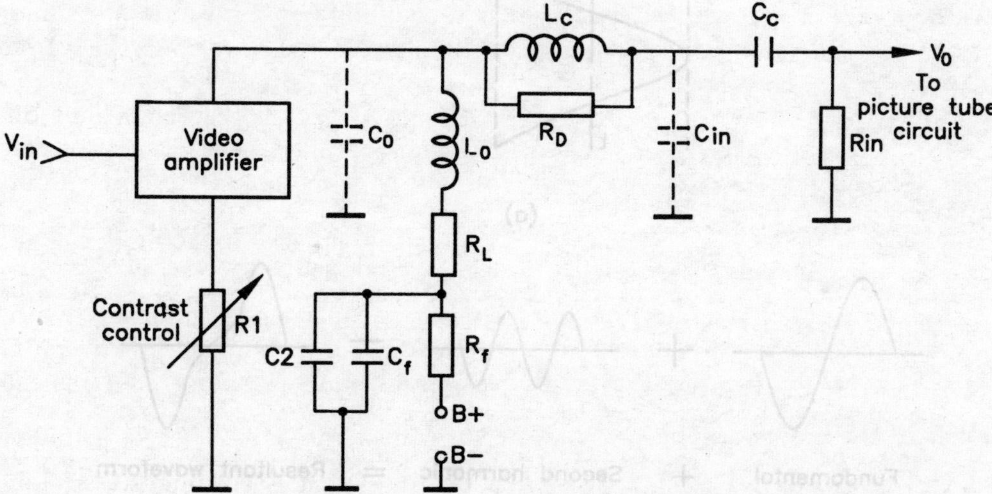

Fig. 5.4. Video amplifier employing both shunt and series peaking for high frequency compensation and a special decoupling circuit to boost low frequency response.

3. Harmonic or Non-Linear Distortion

Harmonic distortion occurs due to non-linearity of dynamic (transfer) characteristics of the amplifier. When output current or voltage amplitude is relatively small, as in small signal amplifiers, the operation is restricted to a small portion of the characteristic curve which can be taken as linear and hence, no noticeable distortion occurs. However, when signal amplitudes are large, as in power or large signal amplifiers, different parts of input signal are amplified differently due to operation along large part of the transfer curve where the curvature cannot be ignored. Such a condition (harmonic distortion) is illustrated in Fig. 5.5 (a). This type of distortion where a portion of output voltage or current gets either compressed or expanded is also called amplitude distortion because it is the large amplitude of output signal that results in its distortion.

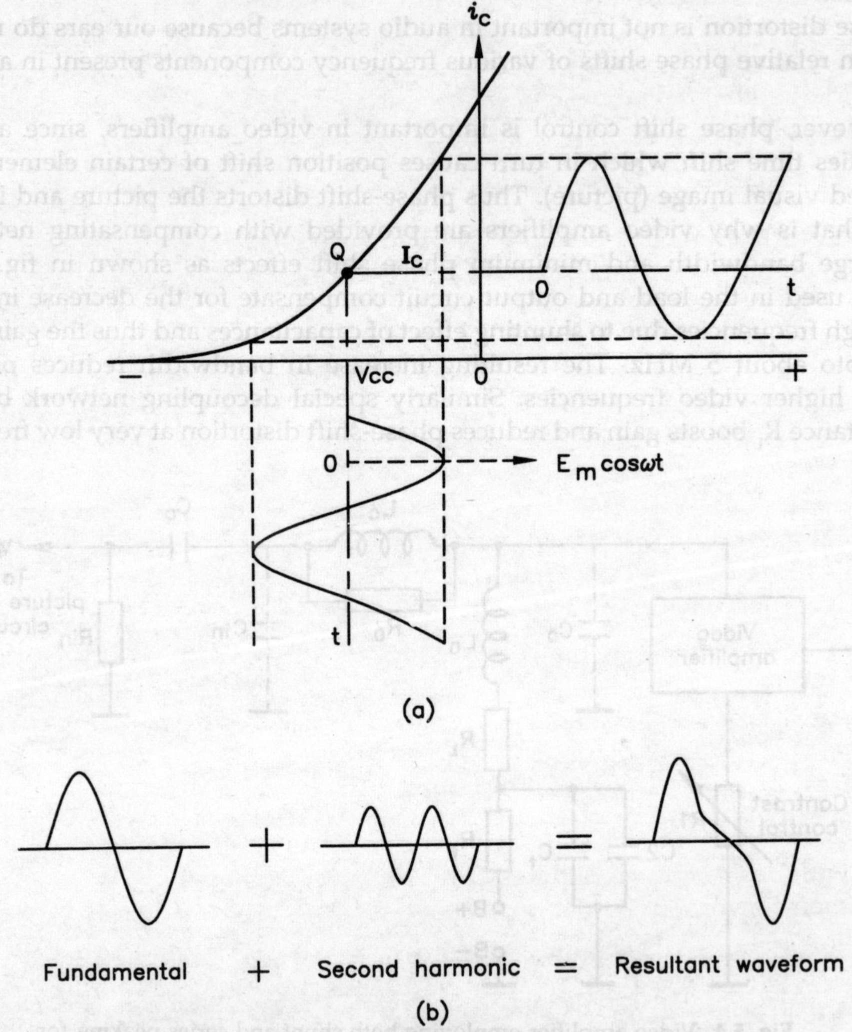

Fig. 5.5. (a) Distortion due to non-linearity of dynamic curve.
(b) Illustration of harmonic generation.

Generation of Harmonics: Amplitude distortion of the output signal amounts to presence of new frequency components in the output current or voltage not present in the input signal. This can be established by developing an equation for the dynamic curve of the amplifier. About the operating point Q [see Fig. 5.5 (a)], the instantaneous current can be expressed by a power series as:

$$ic = a_0 + a_1 e_i + a_2 e_i^2 + a_3 e_i^3 + a_4 e_i^4 .....$$

where e_i is the corresponding variational component of output voltage around the operating point. It can be seen from the expression that if only first two term on the right are used, the dynamic curve is approximated by a straight line; if the third term is added, the approximation is that of a straight line plus a parabola. Closer approximation to the actual dynamic curve is obtained by the addition of higher power terms as shown in the above expression. If the applied signal e_i is zero, then operation is at the Q point and i_c = Ic and a_o = Ic.

Considering that the input signal e_i is sinusoidal and equal to E_m cos ωt, the output current ic can be written as:

$$ic = I_c + a_1 E_m \cos \omega t + a_2 E_m \cos^2 \omega t + a_3 E_m \cos^3 \omega t + a_4 E_m \cos^4 \omega t + \ldots\ldots$$

By use of trigonometric identities for multiple angles and by combining the resultant coefficients into a new set, the above equation can be written as:

$$ic = Ic + A_0 + A_1 \cos \omega t + A_2 \cos 2\,\omega t + A_3 \cos 3\,\omega t + A_4 \cos_4 \omega t + \ldots.$$

Ignoring the relative phase relation of different output components, the result obtained above can be visualized as illustrated in Fig. 5.5 (b). The new frequencies that get generated are the harmonies of the fundamental *i.e.* of the input signal frequency. Higher the harmonic number the lesser is its amplitude.

The input signal which is usually complex may consist of many different frequency components and of different amplitudes. For example if the input consists of two frequency components, say at 500 Hz and 600 Hz and the amplifier gives rise to harmonic distortion, the output will contain:

500 Hz	(f_1 the fundamental)
600 Hz	(f_2 the fundamental)
1000 Hz	(second harmonic of f_1 = 2 × 500 Hz.)
1200 Hz	(second harmonic of f_2 = 2 × 600 Hz.)
1500 Hz	(third harmonic of f_1 = 3 × 500 Hz.)
1800 Hz	(third harmonic of f_2 = 2 × 600 Hz.)

and so on with decreasing amplitudes of higher order harmonics.

Total Harmonic Distortion. As deduced above—

$$ic = Ic + A_0 + A_1 \cos \omega t + A_2 \cos 2\,\omega t + A_3 \cos 3\,\omega t + \ldots\ldots$$

In this expression Ic and A_0 are the dc components, A_1 the peak value of the fundamental current component (may be called the first harmonic), A_2 the peak value of second harmonic component and so on. The harmonic distortion due to each component is then defined as:

$$\text{2nd harmonic component } D_2 = \frac{A_2}{A_1}$$

$$\text{3rd harmonic component } D_3 = \frac{A_3}{A_1}$$

and so on for higher harmonics.

If the load across the amplifier is R_L, the output power can be written as follows knowing that: Power in general = I^2R where I is the RMS value of the current and thus

$$= \left(\frac{I\text{ peak}}{\sqrt{2}}\right)^2 \times R_L = (I\text{ peak})^2 \frac{R_L}{2}$$

Thus with current A_1, A_2 etc. (peak) components in the above expression

$$P\text{ (output)} = (A_1^2 + A_2^2 + A_2^3 + \ldots\ldots)\frac{R_L}{2}$$

$$= \left(1 + \left(\frac{A_2}{A_1}\right)^2 + \left(\frac{A_3}{A_1}\right)^2 + \ldots\ldots\right)\frac{R_L \times A_1^2}{2}$$

$$= (1 + D_2^2 + D_3^2 + \ldots\ldots)P_1$$

since P_1 *i.e.* power due to fundamental component $= \frac{A_1^2 R_L}{2}$ and D_2 and D_3 are distortions as defined above.

Also the total distortion D is defined as:

$$D = \sqrt{D_2^2 + D_3^2 + \ldots\ldots}$$

Total power P can then be written as:

$$P = (1 + D^2)\,P_1$$

Thus if total distortion is 10% then the total power is equal to:

$$P = (1 + (0.1)^2) \times P_1 = 1.01\,P_1$$

This shows that 10% distortion represents a power of only 1 per cent of the fundamental. Thus only a small error is made in using only the fundamental term P_1 for calculating power output of the amplifier. In fact 10% distortion is accepted as tolerable in sound systems since 1% of the near total power at harmonic frequencies does not make any perceptible (noticeable) impact in loudness as perceived by the human ear.

Push-Pull Amplifier: A good part of distortion introduced by the nonlinearity of dynamic transfer characteristics can be eliminated by the circuit shown in Fig. 5.6, known as push-pull configuration.

In this circuit the excitation (signal) is introduced through a centre-tapped transformer T_1. Thus when the signal on transistor Q_1 is positive, the signal in Q_2 is negative by an equal amount. Any other circuit that provides two equal voltages which differ in phase by 180° may be used in place of input transformer T1. This in fact is the normal practice in practical amplifiers to eliminate the use of transformers.

Consider an input signal (base current) of the form $ib_1 = Ib_m \cos \omega t$ applied to Q_1. The output current of this transistor can be expressed by a power series as done for a single-ended amplifier in the previous section. Thus:

$$i_1 = Ic + B_0 + B_1 \cos \omega t + B_2 \cos 2\,\omega t + B_3 \cos 3\,\omega t + \ldots\ldots\ldots\ldots$$

The corresponding input signal to Q_2 is $ib_2 = -ib_1 = Ib_m \cos(\omega t + \pi)$ ($+\pi$ *i.e.* 180° accounts for inversion as compared to $1b_1$).

Thus output current of transistor Q_2 can be obtained by replacing ωt by $(\omega t + \pi)$ in the expression for ib_1. That is:

$i_2(\omega t) = i_1(\omega t + \pi)$ whence:

$$i_2 = Ic + B_0 + B_1 \cos(\omega t + \pi) + B_2 \cos 2(\omega t + \pi) + B_3 \cos 3(\omega t + \pi) + \ldots\ldots$$

which is:

$$i_2 = Ic + B_0 - B_1 \cos \omega t + B_2 \cos 2\omega t - B_3 \cos 3\omega t + \ldots\ldots$$

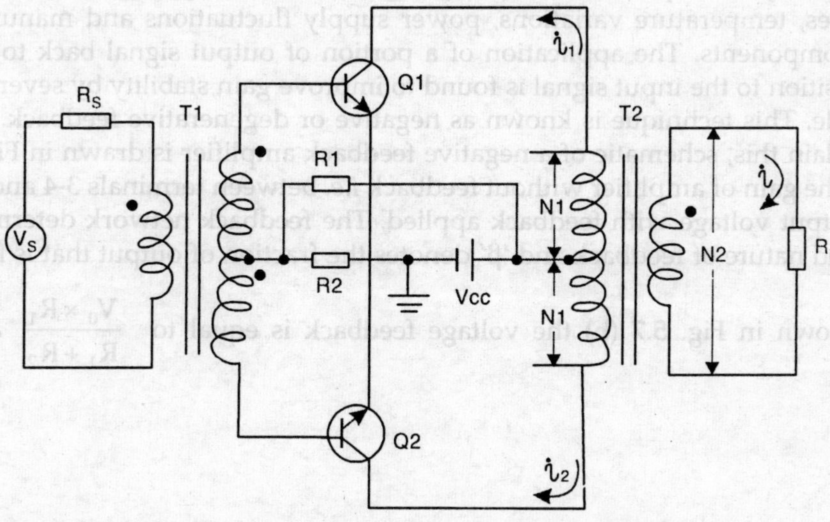

Fig. 5.6. Basic circuit of a two-transistor push-pull amplifier.

As illustrated in Fig. 5.6, the currents i_1 and i_2 flow in opposite direction through the primary winding of output transformer T_2. The total output current is then proportional to the difference between the collector currents in the two transistors. That is:

$$i = k(i_1 - i_2) = 2k(B_1 \cos \omega t + B_3 \cos 3\omega t + \ldots\ldots)$$

This expression shows that a push-pull circuit will balance out all even harmonics in the output and will leave odd harmonics as the principal source of distortion. This is with the assumption that the two transistors Q_1 and Q_2 are chosen to have identical characteristics. Notice that the 2nd harmonic which is the main source of distortion gets eliminated in such an amplifier configuration.

Because no even harmonics are present in the output of a push-pull amplifier, such a circuit will give more output per active device for a given amount of distortion. For the same reason, a push-pull arrangement may be used to obtain less distortion for a given power output per transistor.

In practical circuits, it is possible to eliminate both input and output transformers of the basic push-pull arrangement. One such circuit called complementary symmetry uses two opposite types of transistors *i.e.* if Q_1 is chosen to be PNP type then Q_2 must be NPN type. For large power requirements class B operation is used for which both the transistors can be of same type and hence, only one dc supply source is needed. Such amplifier configurations are commonly used in trunk and line amplifiers of CATV systems. In order to further improve the performance of amplifiers where even small percentage of distortion is not desirable, feedback arrangement is used to achieve this. This is explained in the following section.

5.4 Gain Stability

The gain stability of amplifiers is not very high because of ageing or replacement of active devices, temperature variations, power supply fluctuations and manufacturing spread of components. The application of a portion of output signal back to input in phase opposition to the input signal is found to improve gain stability by several orders of magnitude. This technique is known as negative or degenerative feedback.

To explain this, schematic of a negative feedback amplifier is drawn in Fig. 5.7 (a) where A is the gain of amplifier without feedback *i.e.* between terminals 3-4 and 5-6 and V_0 is the output voltage with feedback applied. The feedback network determines the quantum and nature of feedback and 'β' denotes the fraction of output that is feedback.

Thus, as shown in Fig. 5.7 (b) the voltage feedback is equal to $\dfrac{V_0 \times R_1}{R_1 + R_2}$ and thus

$$\beta = \frac{R_1}{R_1 + R_2}.$$

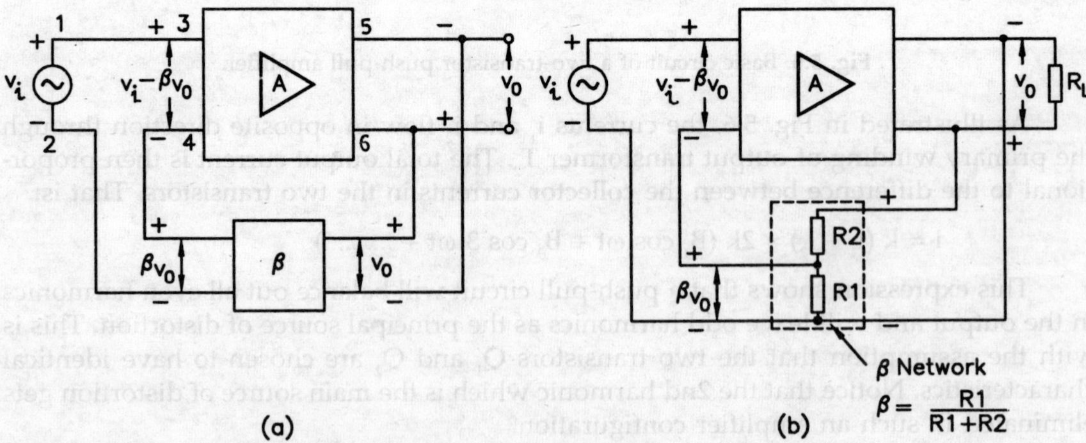

Fig. 5.7. (a) Application of negative feedback to an amplifier, (b) network for voltage feedback

Since the voltage that is fed back opposes the applied signal v_i, the voltage at terminals 3-4 will be $v_i - \beta v_0$. The gain between terminals 3-4 and 5-6 of the amplifier is still 'A' and therefore output voltage

$$v_0 = A\,(v_i - \beta\, v_0)$$
$$= Av_i - A\beta\, v_0$$
$$\therefore\ v_0 + A\beta\, v_0 = Av_i \text{ or } v_0\,(1 + A\beta) = Av_i$$

$$\text{and } v_0 = \frac{Av_i}{1+A\beta}$$

Thus, overall gain A′ *i.e.* gain with feedback between terminals 1-2 and 5-6 is equal to:

$$\frac{v_0}{v_i} = \frac{A}{1+A\beta}$$

and this is less than the gain without feedback. However, this reduction of gain improves stability of the amplifier considerably. This can be illustrated by an example. Consider that gain of an amplifier without feedback is 500 and 1/50 of the output voltage is feedback in phase opposition to the input side. Since $\beta = 1/50$, gain with feedback $A' = \frac{A}{1+A\beta} = \frac{500}{1+500\times 1/50} = \frac{500}{11} = 45.45$. This when expressed in dB equals 33.15.

Now suppose that for any reason the inherent gain A of the amplifier drops to 250, then the new value of A′ will become $\frac{250}{1+250/50} = 41.6$ and in $dB_s = 32.3$. This clearly shows gain stability with negative feedback but the price paid for it is the reduction in overall gain of the amplifier. This, however, can be made up to the required level by increasing number of stages in the amplifier set-up. Furthermore, if $A\beta$ is made quite large *i.e.* $A\beta >> 1$, then $A' = \frac{A}{1+A\beta} \cong \frac{1}{\beta}$.

This indicates that if inherent gain of an amplifier is large and β is so chosen that $A\beta >> 1$, the amplifier gain can be made independent of all variations associated with the amplifier.

Negative feedback also reduces distortion and noise generated within or picked up by the amplifier. If the β network is chosen to be independent of frequency as in Fig. 5.7 (b), it can be shown that the reduced total harmonic distortion $D' = \frac{D}{1+A\beta}$. For example, if A = 70, β = 1/10 and total distortion is 10% in a power amplifier, then on application of negative feedback $D' = \frac{10}{1+70/10} = \frac{10}{8} = 1.25\%$ only.

Similarly, if β is made to be a function of frequency *i.e.* β network contains inductance and or capacitance and their values are so chosen that effective value of β reduces with rise in frequency, the bandwidth of amplifier will increase since A′ will increase with increase in frequency. It can also be shown that both input and output impedances of any amplifier can be modified *i.e.* increased or decreased by choosing the voltage to be feedback to be either proportional to output voltage or current and returning it to the input side either in parallel with the external input voltage or in series with it. As an illustration of this, the feedback arrangement shown in Fig. 5.7 (b) is a voltage-series negative feedback.

6

RESONANT LOAD RF AMPLIFIERS

Introduction

Video amplifiers cannot be used at very high frequencies because as signal frequency goes up, their gain falls down rapidly. Moreover, there is no provision in video amplifiers to select a band of frequencies for amplification and reject other components that may be present in the input signal. However, such a provision is very important for amplifiers to be used in RF and IF sections of high frequency communication systems. Tuned RF amplifiers have been designed to meet this requirement that provide large gain to a select band of high frequency signals.

This class of amplifiers is widely used as RF and IF amplifiers in radio and television receivers. In CATV, these serve the same purpose in satellite receivers, channel modulators and signal distribution network. The technique used to select and amplify desired band of radio frequency signals is explained in this chapter.

6.1 Resonance

When an ac voltage is impressed across an inductive-capacitive (L-C) circuit, connected in series or parallel, resonance occurs at a particular frequency say, f, where inductive reactance $X_L = 2\pi f_L$ of the circuit becomes equal to its capacitive reactance $XC = 1/2\pi f_c$. The two reactive effects then cancel each other and the circuit draws just enough energy from the supply source to overcome its internal resistive losses which are quite small. Such a condition is the fundamental meaning of resonance.

Resonance Frequency: As explained above, resonance occurs when X_L equals X_C at a particular frequency called the resonant frequency (f_0).

Equating $X_L = X_C$ *i.e.* $2\pi f_0 L = \dfrac{1}{2\pi f_0 c}$

we get $f_0^2 = \dfrac{1}{4\pi^2 LC}$ or $f_0 = \dfrac{1}{2\pi\sqrt{LC}}$

6.2 L-C Resonant Circuits

An inductor and capacitor may be connected in series or parallel to form a resonant circuit.

(1) *Series Resonant Circuit:* A series resonant circuit is shown in Fig. 6.1 (a). The resistance R shown in series with L is not a physical component but represents effective resistance of the coil. It is usually quite small and may be neglected here while understanding the phenomena of resonance. The same current I flows through L and C and voltage drop V_L across the inductor leads I by 90° (quarter cycle). This is shown in the phasor diagram drawn in Fig. 6.1 (b). At resonance when $X_L = X_c$, V_L and V_c also become equal to each other in magnitude but being 180° out of phase cancel each other. As such, the current I will tend to become very large at f_0 but is limited to a safe value by the internal resistance of the coil and also by the load in practical circuits. In conclusion when series resonance occurs:

(i) $X_L = X_C$, (ii) $f_0 = \dfrac{1}{2\pi\sqrt{Lc}}$ (iii) Z_0 (impedance at resonance) = R (minimum) and (iv) $I = \dfrac{V_s}{R}$ (maximum).

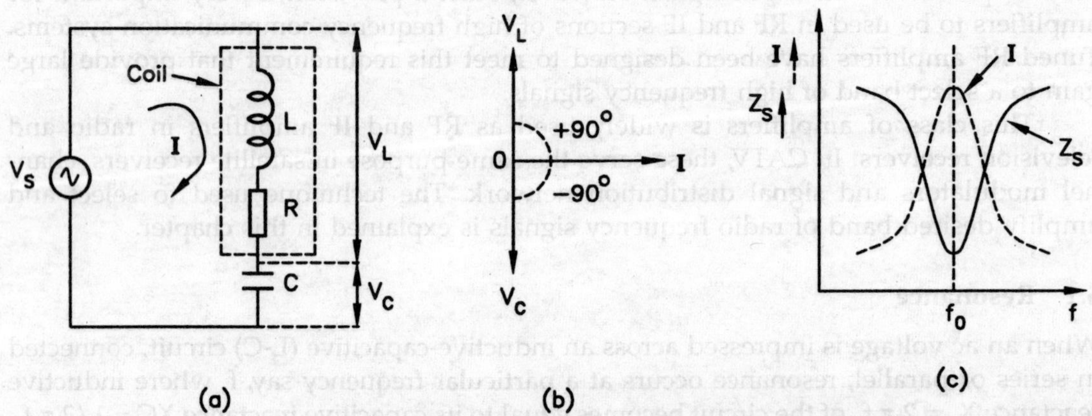

Fig. 6.1. Series resonant circuit, (a) circuit diagram, (b) phasor diagram where R is neglected, and (c) variation of I and Z_S (impedance) with frequency.

Variation of Current and Impedance with Frequency: The magnitude of impedance of the circuit at any frequency is given by $Z_s = \sqrt{R^2 + (XL - X_C)^2}$. At resonance $X_L = X_C$ and therefore, Z_0 = R only. Below resonance, the impedance is mainly determined by X_C and increases with decrease in frequency. Similarly at frequencies above resonance, z_S is mostly due to X_L and increases with increase in frequency. Such a behaviour is shown in Fig. 6.1 (c) from which the selective property of the circuit is evident.

(2) *Parallel Resonant Circuit:* A parallel resonant circuit is shown in Fig. 6.2 (a), where R in series with L is the coil resistance which being quite small may be neglected in this analysis. The phasor diagram of the circuit is drawn in Fig. 6.2 (b) to show that I_C leads the applied voltage V_s by 90° while I_L lags it by 90°. Thus, the two current components are 180° out of phase with respect to each other. At resonant frequency

f_0, X_L becomes equal to X_C and hence I_L and I_C also become equal in magnitude and thus cancel each other. The current I from source V_S will then tend to drop to zero but in practice a small current flows due to the presence of R and load across the circuit if present. In conclusion, the main features of a parallel resonant circuit are:

(i) $X_L = X_C$, (ii) $I_L = I_C$ (in magnitude), (iii) $f_0 = \dfrac{1}{2\pi\sqrt{Lc}}$, (iv) Z_P is maximum and resistive, and (v) $I = \dfrac{V_S}{Z_P}$ (minimum and in phase with V_S).

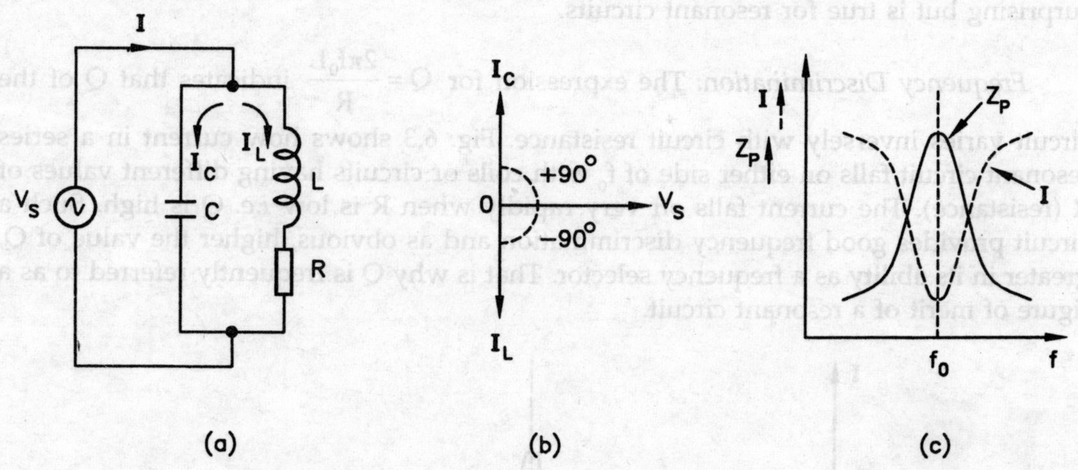

Fig. 6.2. (b) Parallel resonance circuit: (a) circuit diagram, (b) phasor diagram when R is neglected, and (c) variation of I and Z_P with frequency.

Variation of Current and Impedence with Frequency: The variation of current I and impedance Z_P with frequency are shown in Fig. 6.2 (c). Note that Z_P is maximum at resonance but becomes less and less as frequency moves away from f_0 on both sides. Similarly, I is minimum at resonance but increases at off-resonant frequencies.

6.3 Bandwidth and 'Q' of Resonant Circuits

The bandwidth of a tuned (resonant) circuit is associated with the quality factor Q of that circuit. It is defined as the ratio of inductive reactance to the resistance of the circuit coil or circuit. Since at resonance, $X_L = X_C$, either can be used to calculate the magnitude of Q. Thus, we can write $Q = \dfrac{XL}{R} = \dfrac{2\pi f_0 L}{R}$. To understand its significance consider the series resonant circuit of Fig. 6.1 (a), where $V_L = I \times X_L$ but at resonance $I = \dfrac{V_S}{R}$

$$\therefore V_L = \frac{V_S}{R} \times X_L = \frac{V_S}{R} \times 2\pi f_0 L$$

$$= V_S \times \frac{2\pi f_0 L}{R} = V_S \times Q \text{ since as defined earlier } Q = \frac{2\pi f_0 L}{R}$$

It can also be shown that $V_C = V_S \times Q$. In effect it means that at resonance the voltage that develops across L-C circuit is Q times the applied voltage V_S. Most coils/circuits have Q of the order of 20 to 50 and thus when the applied voltage V_S is only one volt, $V_L = V_C$ will be of the order of 20 to 50 volts at resonance. This appears quite surprising but is true for resonant circuits.

Frequency Discrimination: The expression for $Q = \frac{2\pi f_0 L}{R}$ indicates that Q of the circuit varies inversely with circuit resistance. Fig. 6.3 shows how current in a series resonant circuit falls on either side of f_0 with coils or circuits having different values of R (resistance). The current falls off very rapidly when R is low *i.e.* Q is high. Such a circuit provides good frequency discrimination and as obvious, higher the value of Q, greater in its ability as a frequency selector. That is why Q is frequently referred to as a figure of merit of a resonant circuit.

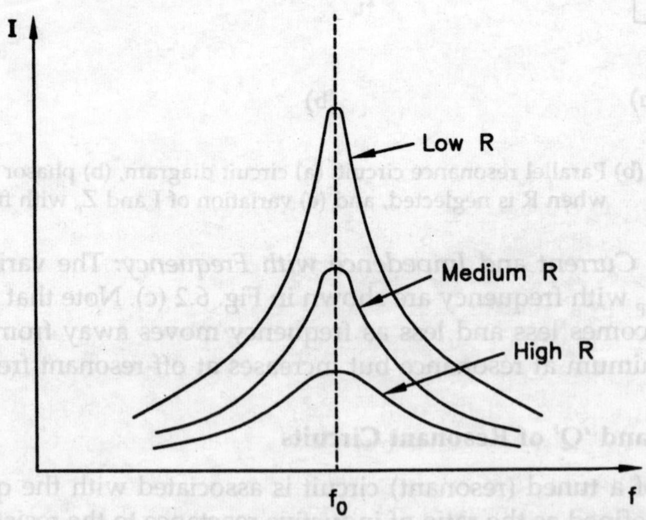

Fig. 6.3. Effect of resistance on current in a series L-C circuit.

Bandwidth: The property of both series and parallel resonant circuits to discriminate between the frequency at resonance and those not at resonance is expressed in terms of their bandwidth. In a small region around the resonant frequency where the response is nearly flat, is called pass-band or bandwidth of the circuit. In a series resonant circuit, the response can be seen in terms of line current I, and for a parallel resonant circuit in terms of its impedance Z_P as illustrated in Fig. 6.4. The effective limits of pass-band or bandwidth are taken in the same way as for defining bandwidth of audio and video amplifiers. The frequencies f_1 and f_2 are the 3 dB down points where the current or impedance value falls to 0.707 of the maximum value. As shown in Fig. 6.3, the bandwidth can be varied by choosing coils having different values of quality factor Q.

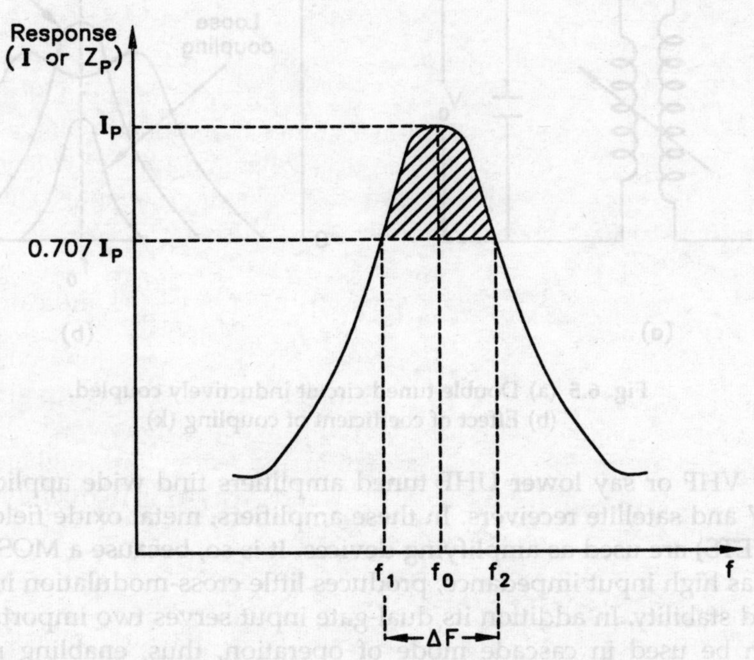

Fig. 6.4. Bandwidth of a single tuned resonant circuit.

Coefficient of Coupling: Double tuned circuit [Fig. 6.5 (a)] that are inductively coupled are often used in RF and IF amplifiers. For loose coupling, as shown in Fig. 6.5(b), the response curve has a single peak at f_0 like a single tuned circuit but with low amplitude. As the coupling is increased *i.e.* coils are brought closer, the output amplitude increases and at a middle value of coupling called critical coupling, the secondary has its maximum amplitude and greatest bandwidth without double peaks. When coupling is further increased to what is known as over-coupling, two peaks result in the secondary output to broaden the output response. Thus, varying the coupling beyond critical coupling results in greater bandwidth of coupled circuits that are tuned to the same resonant frequency. Also for the over-coupled case, drop in response is sharper on either edge of the pass-band, thus enabling better selectivity.

6.4 Band Pass RF Amplifiers

Band-pass or tuned RF amplifiers can be broadly classified into VHF and UHF amplifiers.

(1) *VHF Amplifiers:* Lower VHF tuned voltage amplifiers use transistors as amplifying devices with bifilar coupling coils to obtain desired bandwidth and selectivity. Such amplifiers were commonly used in the IF section of earlier TV receivers. However, modern television receivers employ an integrated IC chip in the IF section where band selection and selectivity is done with a 'SAW' filter that precedes the integrated circuit.

Such amplifiers are also used in IF sections of satellite receivers.

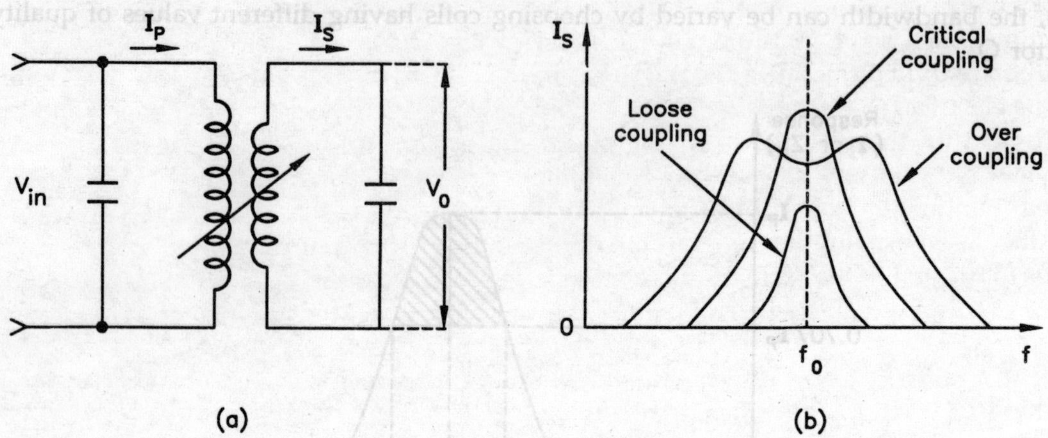

Fig. 6.5 (a) Double tuned circuit inductively coupled.
(b) Effect of coefficient of coupling (k)

Higher VHF or say lower UHF tuned amplifiers find wide applications in UHF tuners of TV and satellite receivers. In these amplifiers, metal oxide field-effect transistors (MOS FETS) are used as amplifying devices. It is so, because a MOS-FET generates less noise, has high input impedance, produces little cross-modulation interference and enables good stability. In addition its dual-gate input serves two important purposes—firstly it can be used in cascade mode of operation, thus, enabling more gain and secondly it provides separate means of applying RF signal and AGC voltage to the amplifier. Such circuits are explained in books that describe TV receivers.

In cable TV, several variants of VHF tuned amplifiers are used as trunk and distribution amplifiers. These amplifiers are basically push-pull band-pass RF amplifiers that provide enough voltage and power gain. The additional features that are incorporated in these amplifiers are automatic gain control (AGC) to keep the output signal level constant and automatic slope control (ASC) which compensates the loss of high frequency components of the signal during transmission on lengthy coaxial cables. In addition, special ferrite cores are used for winding coupling coils to obtain necessary bandwidth of the cable signal that extends from 40 MHz to 450 MHz. More details of trunk amplifiers are given in Chapter 15.

(2) *UHF Amplifiers:* UHF tuned amplifiers used in CATV systems may be designated as higher UHF and Super UHF amplifiers. Higher UHF amplifiers also called super IF tuned or bullet amplifiers are used to amplify IF signal of 950 to 1450 MHz which is obtained as output from low noise block converters (LNBCs) provided at the dish antennas for down conversion of satellite signals. At these frequencies, physical components can still be used. Transistors that operate satisfactorily at such high frequencies and generate very little noise are now available and are used in such amplifiers. Tuning techniques at these frequencies are different from those used in VHF amplifiers and in many designs micro-strips are used as tuning elements.

Super UHF Amplifiers: These are specially designed low-noise super high frequency tuned RF amplifiers and are used to amplify C-band down-link satellite signals from 3700 to 4200 MHz (*i.e.* 3.7 to 4.2 GHz). As explained in section 4.3, special Ga-As FETS have been developed for use as active devices in such amplifiers. The equivalent circuit of such an FET is shown in Fig. 6.6. However, its use at microwave frequencies is extremely difficult because the package in which the device is placed has a great effect on its electrical characteristics. The tiny wire used between the gate bonding pad and the gate lead on the package has a significant inductance at 4 GHz. Consequently, such devices are best characterised by S-parameters. S-parameters are two-port parameters like h-parameters in terms of transmission and reflection coefficients. These parameters are then used in computer programmes for designing super UHF amplifiers. Transmission lines fabricated in the form of micro-strips and micro-lines as explained in chapter 4, are used as tuned circuit elements, filter units and even as reactive components in associated circuitry of such amplifiers. The amplifier load circuit contains a cavity resonator that acts as a high 'Q' parallel resonant circuit and its dimensions decide its pass-band. For making connections to the cavity resonator, two probes or loops are used, one at the input side and the other for taking out the amplified signal.

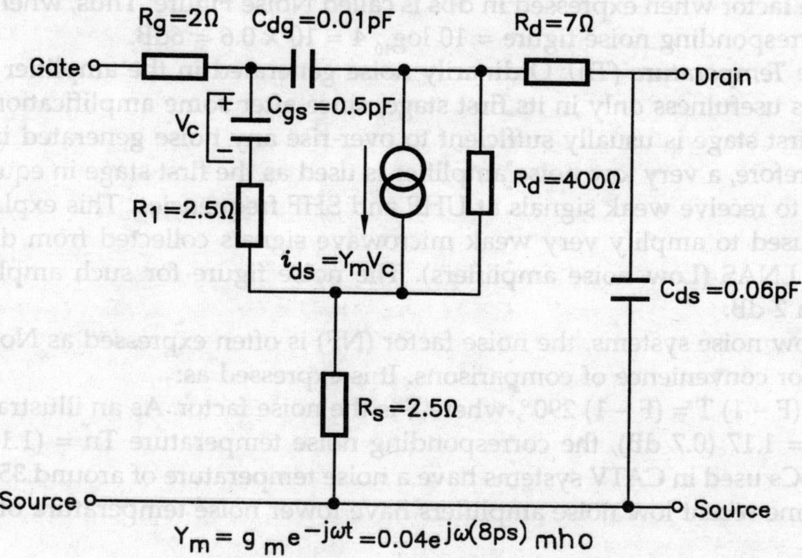

Fig. 6.6. Ga-AS MOSFET equivalent circuit.

Besides amplification, the main criterion of such amplifiers is high signal to noise ratio because microwave signals to be amplified are extremely small in amplitude. As such these amplifiers are called Low-Noise Amplifiers (LNAS).

6.5 Noise Factor in UHF and Super HF Amplifiers

UHF and SHF amplifiers are limited in their useful sensitivity by erratic currents (noise) generated in various circuits of the amplifier or by noise introduced in its input with the signal. The noise generated within the amplifier is caused by random motion of electric charges in circuit elements and in tubes and transistors. Such thermal noise is present at all times unless the temperature is absolute zero (0° kelvin *i.e.* 0° K or – 273° C) and it increases with increase in temperature of the environment. It is usually this internal noise which limits receiver sensitivity in UHF and SHF amplifiers, because thermal noise causes currents that mostly lie in the high frequency spectrum. Thus, in the design of amplifiers meant to amplify very weak signals collected from UHF antennas or parabolic dishes, the presence of noise becomes a major constraint in obtaining higher gains. To account for the level of thermal noise present in a device or amplifier, a factor called Noise Factor has been evolved. Noise Factor (NF) = $\frac{P_{in}}{KTB}$ where: P_{in} is the power made input to the amplifier under test from an external noise generator to obtain an output power that is double of which develops as output without any input from the noise generator.

K is called Boltzman's constant and is 1.38 × 10 × 10^{-23} watt-sec/degree K.

T at the normal room temperature = 273°C + 17°C = 290° and B is the bandwidth in Hz of the amplifier under test.

Noise factor when expressed in dBs is called Noise Figure. Thus, when noise factor = 4 the corresponding noise figure = 10 $\log_{10}$ 4 = 10 × 0.6 = 6dB.

Noise Temperature (Tn): Ordinarily noise generated in the amplifier is a limiting factor to its usefulness only in its first stage, since after some amplification, the output from the first stage is usually sufficient to over-rise any noise generated in the second stage. Therefore, a very low noise amplifier is used as the first stage in equipments that are meant to receive weak signals at UHF and SHF frequencies. This explains why the amplifier used to amplify very weak microwave signals collected from dish antennas are called LNAS (Low noise amplifiers). The noise figure for such amplifiers can be lower than 2 dB.

For low noise systems, the noise factor (NF) is often expressed as Noise Temperature (Tn) for convenience of comparisons. It is expressed as:

Tn = (F – 1) T = (F – 1) 290°, where F is the noise factor. As an illustration, if noise factor (F) = 1.17 (0.7 dB), the corresponding noise temperature Tn = (1.17 – 1) 290° ≈ 50°K. LNBCs used in CATV systems have a noise temperature of around 35 to 40 degree Kelvin. Some recent low noise amplifiers have lower noise temperature of the order of 30°K.

7

SINUSOIDAL WAVE GENERATION AND FREQUENCY CONTROL

Introduction

Any circuit that generates an alternating voltage is called a wave generator. There are two types of wave generators—sinusoidal and nonsinusoidal. The circuit configurations that produce sinusoidal outputs are called oscillators and the ones that develop non-sinusoidal waveshapes like square wave, or pulses are usually referred to as multivibrators and pulse generators. For any type of wave generator, the required energy is supplied from an external dc supply source.

Essentially an oscillator is an overdriven amplifier that generates ac voltage by positive feedback from its output to input side. The frequency of oscillations is determined by the resonant circuit connected in the feedback loop. An important use of high and super high frequency oscillators is in communication systems. For example, in radio and television broadcasting, high voltage fixed frequency oscillators become carrier waves to audio and video signals which are modulated on them before radiation by transmitting antennas. Low voltage H.F. oscillators are also used in TV and radio receivers for translating signals picked up by antennas to an intermediate frequency (IF) for convenience of processing and later in demodulators for recovering baseband video and audio signals. In cable television, oscillators form part of Front-End and Head-End units and are also used in channel modulators.

In order to prevent frequency drift in oscillators, special circuits called automatic frequency control (AFC) are used to keep the frequency at a fixed value. Electronic tuning with veractor diodes is used to change the frequency of an oscillator to another value. When a large number of accurate frequency outputs are required as in television and satellite receiver tuners, a technique called phase-locked loop (PLL) is used to obtain several exact frequencies from a single crystal controlled oscillator instead of providing for separate oscillators. In CATV where a large number of channels are to be processed and frequent changing of channels is common, PLL circuits are combined with micro-computer control to form frequency synthesizers which enable frequency selection by remote control units. The use of frequency synthesizers is common in conventional colour receivers and in satellite receivers at ground stations. This chapter is written to convey basics of sinusoidal oscillators, methods of frequency tuning control and generation of a large number of accurate frequency outputs.

7.1 Positive Feedback Amplifier as an Oscillator

A resonant L-C circuit [Fig. 7.1 (a)] is the controlling component of most H.F. oscillators. The manner in which it resonates to generate sine-wave output is explained in section 2.2 of chapter 2 that is devoted to generation of RF waves. In order to prevent damping of oscillations due to circuit losses, an amplifier is made part of the set-up which feeds and controls dc energy to the tank circuit to ensure sustained oscillations. The positive feedback from output to input is through the β network which contains L-C circuit, which in turn controls frequency. The frequency of oscillation $f_0 = \frac{1}{2\pi\sqrt{LC}}$.

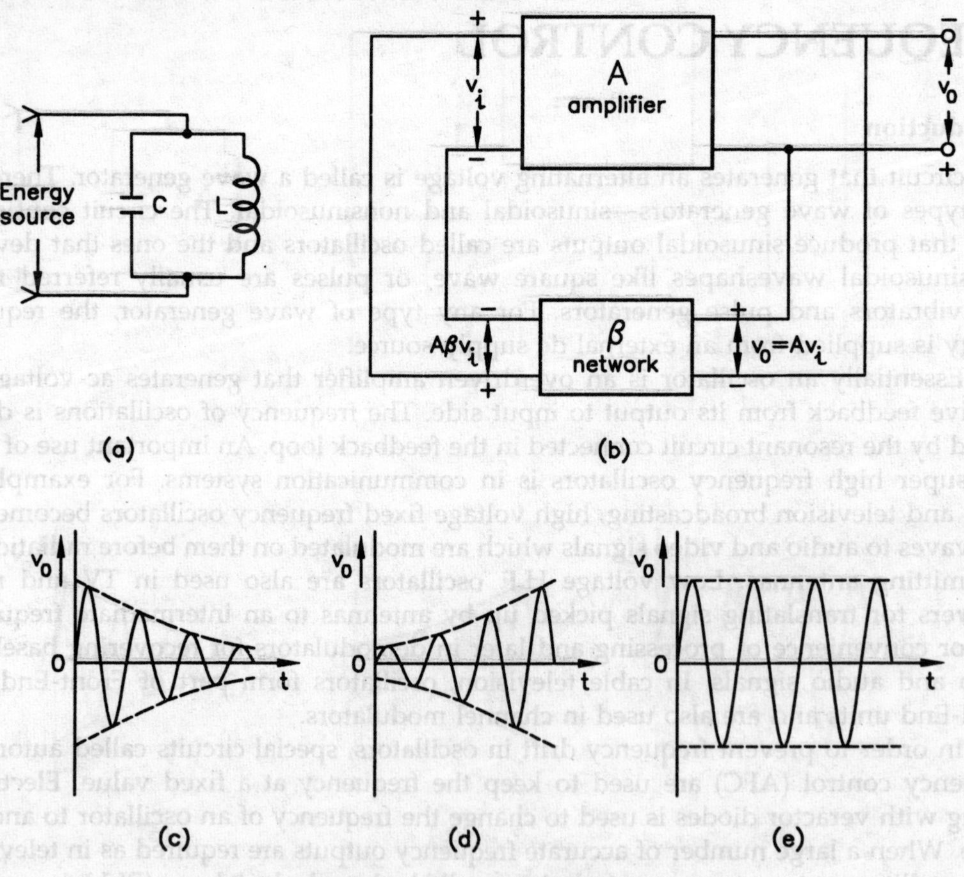

Fig. 7.1. Block schematic of an oscillator.
(a) parallel LC resonant circuit,
(b) circuit with positive feedback,
(c) output when $A\beta < 1$,
(d) output when $A\beta > 1$, and
(e) constant amplitude output when $A\beta = 1$

Criteria for Continued Oscillations: To understand the condition that must be satisfied for continuous oscillations, consider block schematic of the oscillator circuit shown in Fig. 7.1 (b). As in the case of negative feedback in amplifiers, assume to begin with that for a moment v_i is present at the input and v_0 is the output with feedback. The fraction of output voltage *i.e.* $\beta v_0 = A\beta v_i$ that is fed back to the input side is inverted and so input to the amplifier equals $v_i + \beta v_0$. With this it is easy to deduce the gain with feedback *i.e.* $A' = \dfrac{A}{1-A\beta}$. Three possible conditions arise from this. These are:

(i) $A\beta < 1$. This means that $A\beta v_i$ becomes less than assumed v_i and output signal will die as shown in Fig. 7.1 (c).

(ii) $A\beta > 1$. In this case, $A\beta v_i$ becomes greater than v_i and output voltage builds up as shown in Fig. 7.1 (d). This, however, is prevented from exceeding a certain limit due to fall in gain 'A' of the amplifier as input voltage rises to become large. The output falls and rises periodically with the result that output voltage is not a constant amplitude sine wave. This naturally is not desirable.

(iii) $A\beta = 1$. With this condition, $1 - A\beta$ becomes zero and A' tends towards infinity. While this cannot happen in practice, but the condition indicates that there can be output without any external input source. In other words, the amplifier becomes as oscillator where β network decides the frequency of oscillations (see Fig. 7.1 (e)).

Starting Voltage in Oscillators: In the absence of any external starting voltage v_i, the noise voltage generated by random motion of electrons in the active device (tube or transistor) of the oscillator serves this purpose. The tiny noise voltage is complex in nature and consists of a large number of fundamentals and their harmonic frequencies. On amplification, the noise voltage becomes input to the tuned circuit in the β network. It selects only the frequency for which it is tuned *i.e.* $f_0 = \dfrac{1}{2\pi\sqrt{LC}}$ and attenuates all other components present in the noise voltage. The selected frequency signal soon builds up by loop gain action to attain sufficient amplitude necessary for starting oscillations thus making the external input voltage source unnecessary.

7.2 L-C Oscillators

L-C oscillators or resonant circuit oscillators are widely used for generating high frequencies. There are many types of L-C oscillators like tuned collector, tuned base, Hartley, Colpitts but all essentially consist of an amplifier, on L-C resonant circuit and a positive feedback arrangement. Each type has its own merits and is chosen on suitability or specific applications. In order to gain familiarity with this class of oscillators, the circuit of a tuned collector oscillator is shown in Fig. 7.2. It is called tuned collector oscillator because the tuned circuit is in the collector lead of the amplifying transistor 'Q'.

Fig. 7.2 (a) is drawn from ac point only to explain positive feedback arrangement. The inductance L of the primary of transformer and capacitor C form the tuned or tank circuit which decides frequency of oscillations. The lower winding on the secondary side and primary constitute β network. The turns ratio between the two is so chosen that the required condition $A\beta = 1$ is satisfied where A is the gain of the transistor amplifier. Since a phase difference of 180° is provided by the amplifier, another phase shift of 180° is obtained by choosing necessary direction of the feedback winding. This then satisfies the second requirement for sustained oscillations *i.e.* total phase shift equal to 0° = 180° + 180° = 360°. This makes the feedback signal in-phase with v_i as indicated by dots (•) on the two windings.

A practical tuned collector oscillator with biasing network is shown in Fig. 7.2 (b). Resistors R_1, R_2 and R_E provide necessary bias to the transistor and capacitors C_2 and C_E act as shorts to the high frequency ac component of circuit currents thus making the circuit essentially the same as shown in Fig. 7.2 (a). The oscillator output is taken from a second winding on the feedback transformer as shown in the figure.

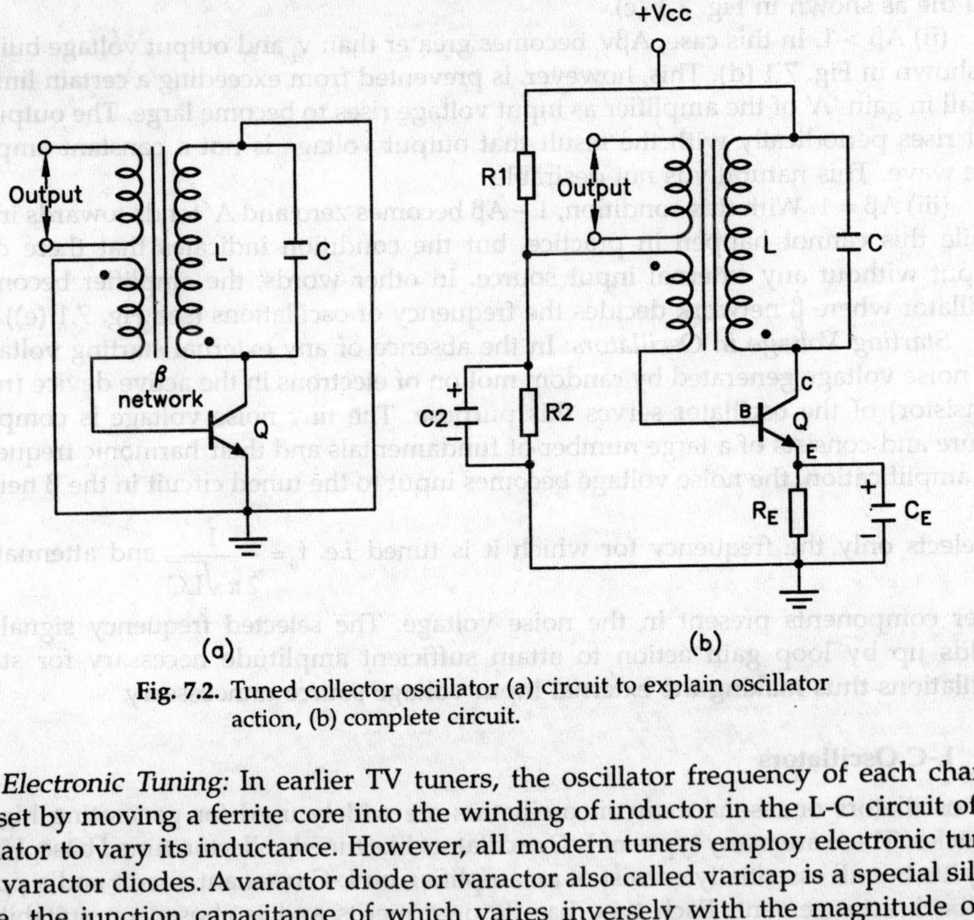

Fig. 7.2. Tuned collector oscillator (a) circuit to explain oscillator action, (b) complete circuit.

Electronic Tuning: In earlier TV tuners, the oscillator frequency of each channel was set by moving a ferrite core into the winding of inductor in the L-C circuit of the oscillator to vary its inductance. However, all modern tuners employ electronic tuning with varactor diodes. A varactor diode or varactor also called varicap is a special silicon diode, the junction capacitance of which varies inversely with the magnitude of dc reverse bias applied across it and has negligible leakage current. It is connected across the tank circuit capacitor and reverse bias applied across it from a stable dc source through a potentiometer. Necessary circuit details are given in Fig. 11.15 of the chapter titled "Terrestrial Television".

Automatic Frequency Tuning: Any drift in the set oscillator frequency is prevented by a technique called automatic frequency tuning, where IF output from the tuner is sampled and proportionate dc voltage developed. The magnitude of this voltage is zero when there is no drift in frequency and varies to become positive or negative depending on whether the frequency of oscillations increases or decreases. The error voltage thus developed is applied across a varactor diode connected in parallel with C of the tank circuit as shown in Fig. 11.17 of Chapter 11. The corresponding variation in junction capacitance returns the oscillator frequency to its correct value.

7.3 Crystal Oscillators

A crystal oscillator is basically a tuned oscillator. It uses a crystal (usually quartz) to control the frequency of oscillations. When an ac voltage is applied across it, mechanical vibrations are set up which have a natural resonant frequency depending on its size, thickness, angle of cut and mounting. Its resonant behaviour can be represented by a series R, L, C circuit. For use in electronic circuits the crystal is mounted between two metal plates [Fig. 7.3 (a)], which then act as a capacitor across it. The resultant equivalent circuit is shown in Fig. 7.3 (b). Conversely, if the crystal is made to vibrate mechanically it generates ac voltage across it, the frequency of which depends on the component values of equivalent circuit shown in the figure. The crystal has a series resonant frequency because of L and C in series in one branch. It will also have a parallel resonant frequency with C_m in parallel with the R, L-C circuit. Since $C_m >> C$, the parallel resonant frequency is somewhat higher than the series resonant frequency. In crystal oscillators, it is the higher resonant frequency that is availed off. The outstanding feature of crystals as compared with discrete L-C tank circuits is their incredibly high 'Q'. Values of 'Q' higher than 10^4 can be achieved by using crystals whereas L-C tank circuits seldom have 'Q' more than 100. The extremely high 'Q' of the crystal makes the oscillator frequency very stable.

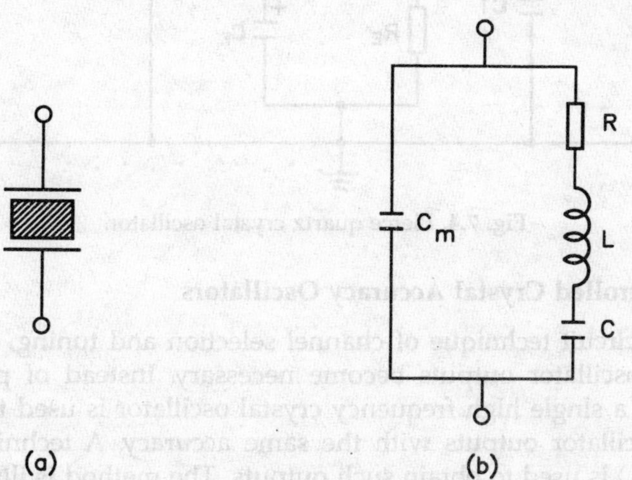

Fig. 7.3. Quartz crystal (a) mounting between plates, (b) its equivalent electrical circuits.

Oscillator Circuit: The circuit of a crystal controlled oscillator known as Pierce Oscillator is shown in Fig. 7.4. Since the crystal is connected to the collector of the transistor at one end, the application of ac voltage makes it to vibrate mechanically. The vibrations in turn, produce electrical signal at a frequency determined by the parallel resonant circuit formed by the crystal and its electrodes. Because of very high 'Q' the controlling frequency is that of the crystal which by feedback at the base produces sustained oscillaíions at the output. The RF choke prevents the flow of high frequency ac voltage that develops at the collector of transistor from getting shorted to ground via dc supply source. The feedback ratio is determined by relative values of C_1 and C_2. Suitable orientation of plane faces with respect to crystalline structure of the crystal makes the resonant frequency independent of temperature over a reasonable wide frequency range. Amplifier stages are used to isolate the oscillator from variations in load. Carefully designed crystal oscillators provide extremely stable and accurate frequencies with an accuracy of 1 part in 100 million.

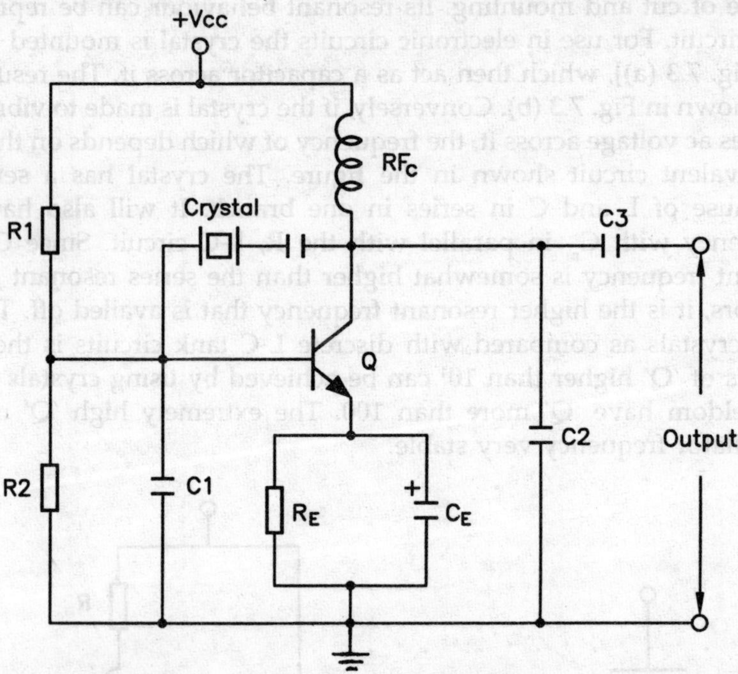

Fig. 7.4. Pierce quartz crystal oscillator.

7.4 Voltage Controlled Crystal Accuracy Oscillators

In the integrated circuit technique of channel selection and tuning, a large number of crystal accuracy oscillator outputs become necessary. Instead of providing separate crystal oscillators, a single high frequency crystal oscillator is used to generate several low frequency oscillator outputs with the same accuracy. A technique called Phase-Locked-Loop (PLL) is used to obtain such outputs. The method is illustrated in Fig. 7.5 by showing how a 50 KHz output is obtained from a crystal oscillator operating at 50 MHz. The output of 50 MHz crystal stabilized oscillator feeds into a divide by 1000 digital frequency divider. The output is a 50 KHz crystal stable signal and becomes reference input to the phase comparator. If the two signals are synchronized, no correction is required. However, if the Voltage Controlled Oscillator (VCO) drifts off-frequency, the phase comparator develops a correction voltage proportional to phase difference of its input signals. The output from the comparator is filtered and dc control voltage thus obtained applied to the 50 KHz VCO. The control voltage is such that it forces the oscillator to lock with the crystal stable frequency of 50 KHz. The PLL control thus enables crystal stable accuracy of a voltage controlled oscillator. It may be mentioned that as explained in section 7.2, the control voltage to the VCO is applied to the varactor diode connected in parallel with its tank circuit. By changing the divide ratio different VCO outputs can be obtained as is the normal practice in digital frequency synthesizer controlled channel selection circuits.

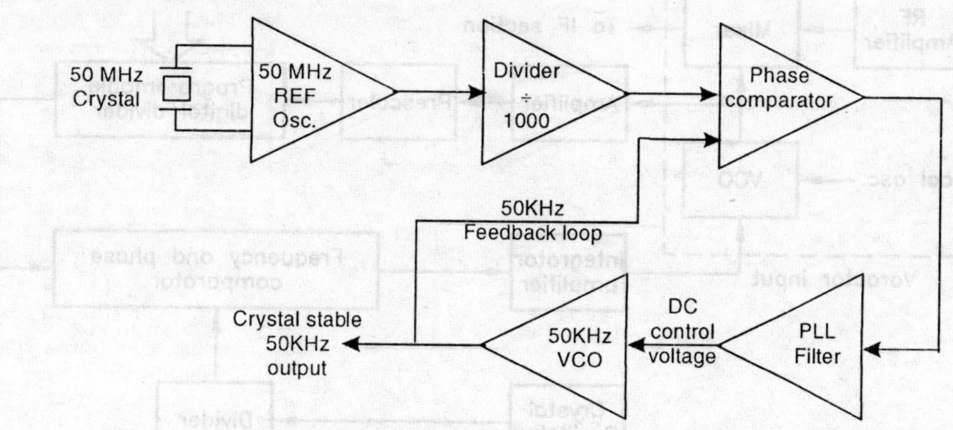

Fig. 7.5. Phase-locked loop (PLL) control of a 50 KHz voltage controlled oscillator (VCO).

7.5 Frequency Synthesizer

Developments in high speed IC counters have made the frequency synthesizer approach practicable for TV receiver tuning and channel selection in satellite receivers and table top converters. A frequency synthesizer is a closed loop system, self-compensating for varactor tolerances and component drifts. It provides easy interface with digital displays and simplifies remote control design. It is highly reliable and has no alignment problems.

The block diagram of Fig. 7.6 illustrates frequency synthesizer system for electronic channel selection. The dotted chain block represents a standard VHF or UHF varactor tuner. The output of local voltage controlled oscillator (VCO) is amplified to an acceptable level for driving a digital emitter coupled logic (ECL) prescaler (divider). The purpose of prescaler is to divide down the local oscillator high frequencies to a range that can be processed and counted by standard TTL logic. The output of prescaler is fed to the programmable divider which is the heart of the system. Its output is passed to a frequency and phase comparator where it is compared with a crystal controlled reference frequency. Then the comparator output is fed to an integrator amplifier, which in turn provides voltage control to the varactor tuner input.

The divide ratio of the programmable counter determines the desired channel. At the comparator there are two significant signals—the counted output and the reference signal. When these two signals differ in frequency or phase, a dc error voltage is applied to the tuner VCO with a suitable polarity in order to reduce the frequency and phase error to zero. At this time, the VCO is at the selected channel frequency. A touch tuning or remote control type entry system is used for channel selection and a seven segment display is employed for channel indication.

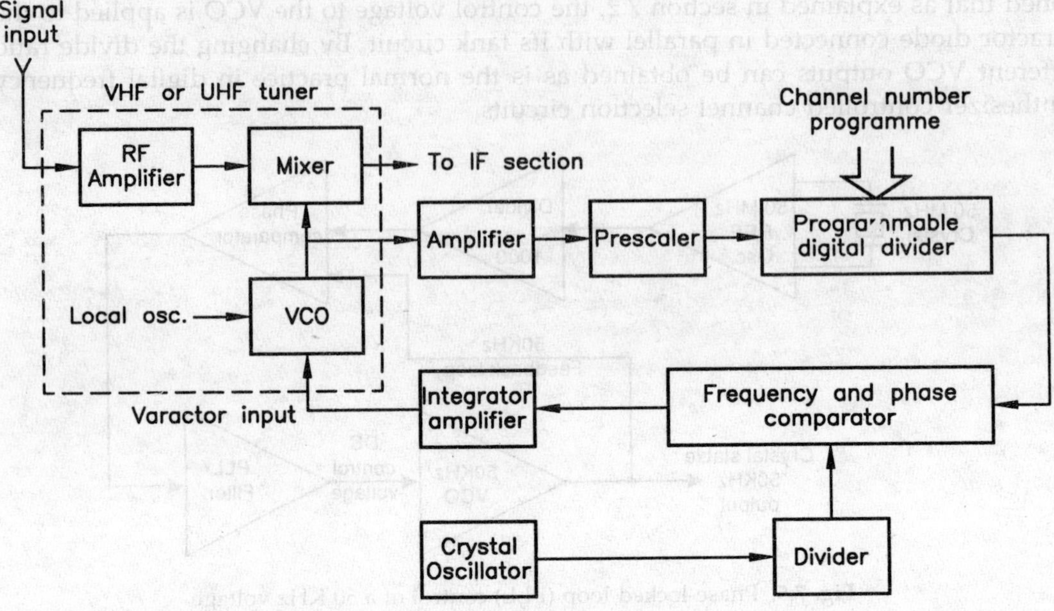

Fig. 7.6. Block diagram of a frequency synthesizer receiver tuner.

7.6 Frequency Multiplier

In radio and TV broadcasting, high voltage fixed frequency oscillators are built around high voltage vacuum tubes. It is often found convenient to use frequency multipliers to obtain high frequency outputs from relatively low frequency oscillators. For this, a class C amplifier is used because its output has a large harmonic content. The oscillator output is made input to a class C operated amplifier, the load circuit of which is tuned to be resonant at the desired multiple of the input frequency.

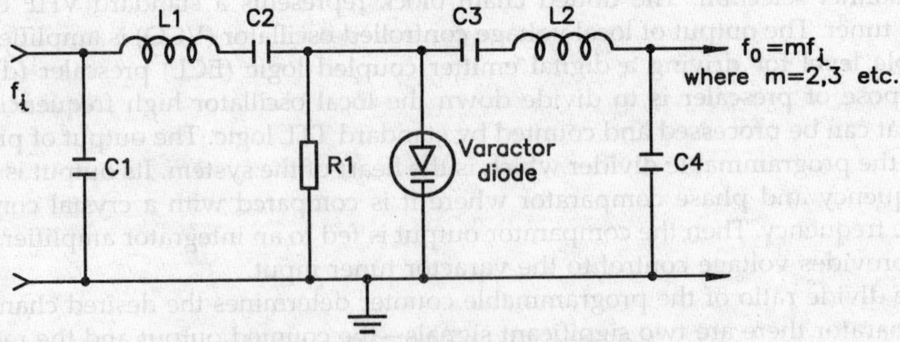

Fig. 7.7. A varactor frequency multiplier.

In satellite receivers the incoming signals have downlink frequencies in the GHz range. A double conversion is often used to translate them to an IF frequency of 70 or 140 MHz. The first mixer (converter) usually employs a local oscillator having frequency of 1.5 GHz. It is not possible to have a crystal controlled oscillator to generate such a high frequency. Therefore, the oscillator is set to generate 60 MHz output and this is fed to a multiharmonic generator or frequency multiplier to obtain 60 × 25 = 1.5 GHz. For this one or more stages of varactor frequency multiplier is used. Fig. 7.7 shows the circuit of such a multiplier. When an input RF signal is applied across the varactor diode, it alternately conducts and cuts off. The result is a non-linear or distorted output containing many harmonic components. By using a tuned circuit in the output, the desired harmonic frequency is selected and others are rejected. Capacitors C_1 and C_2 are used for impedance matching.

Since the lower harmonics produce the greatest amount of energy, varactor multipliers are only used for doubling or tripling operations. Several such multipliers can be used in tandom to obtain desired output frequency. In practice, the tuned circuits are not actually made up of individual inductors or capacitors. Instead there are microstrip, stripline or cavity resonators that form tuned circuits.

8

WAVEFORM MODULATION AND DEMODULATION

Introduction

It is not possible to transmit video and audio signals directly because both picture and sound signals from different television stations are concentrated within the same range of frequencies. If radiated directly, signals from different stations will get hopelessly mixed up and it would be impossible to separate one from the other at the receiving end. Thus, in order to separate intelligence from different stations, video and audio signals are superimposed on different RF carrier waves. Each station is, therefore, assigned separate carrier frequencies and at the receiving end, the tuner of TV receiver selects desired stations.

The process by which any intelligence is superimposed on the carrier wave is called modulation. To affect modulation, either amplitude or frequency of the carrier wave can be varied in accordance with the characteristics of modulating signals. When amplitude of the carrier wave is varied in accordance with amplitude changes of the modulating signal, the modulation is called amplitude modulation (AM). If, however, frequency of the carrier wave is varied in accordance with variations in amplitude of the modulating signal, the modulation is called frequency modulation (FM).

At the receiver, the selected channel signal is suitably processed and then detected or demodulated to recover modulating signal. Separate methods are used for demodulating AM and FM signals. An AM detector is essentially a rectifier-cum-high frequency filter but FM detection is carried out in two steps. The constant amplitude FM signal is first converted into a proportionate voltage that is both frequency and amplitude modulated. It is then applied to a detector which detects amplitude changes but ignores frequency variations. The amplitude changes with time is then the required modulating signal. Both amplitude and frequency modulation and demodulation techniques are briefly described in this chapter.

8.1 Amplitude Modulation

Amplitude modulation is not a process where two or more different frequency signals are mixed but it involves frequency translation of the low frequency modulating signal to the region of high frequency carrier wave. To affect this, the active device (tube or transistor) employed in the modulator is operated in the non-linear region of its transfer characteristics. As explained in section 5.3, operation in non-linear region generates harmonics of input signal frequencies. However, when input consists of a large amplitude high frequency signal called carrier and one or more low amplitude, low frequency components called modulating signal, the modulator output predominantly consists of carrier frequency and sum and difference frequencies of the carrier and modulating signal frequencies. This amounts to translating the modulating signal to the region of carrier frequency. The objective of superimposing the modulating signal on the carrier is thus achieved.

Amplitude Modulated Signal: An amplitude modulated signal is shown in Fig. 8.1 (a) where $e_c = E_c \cos 2\pi f_c t$ *i.e.* $E_c \cos \omega_c t$ is the carrier wave and $e_m = E_m \cos 2\pi f_m t$ *i.e.* $E_m \cos \omega_m t$ is the modulating signal. Note that modulating signal, video or audio is quite complex in nature but a single modulating frequency f_m has been chosen for convenience of analysis.

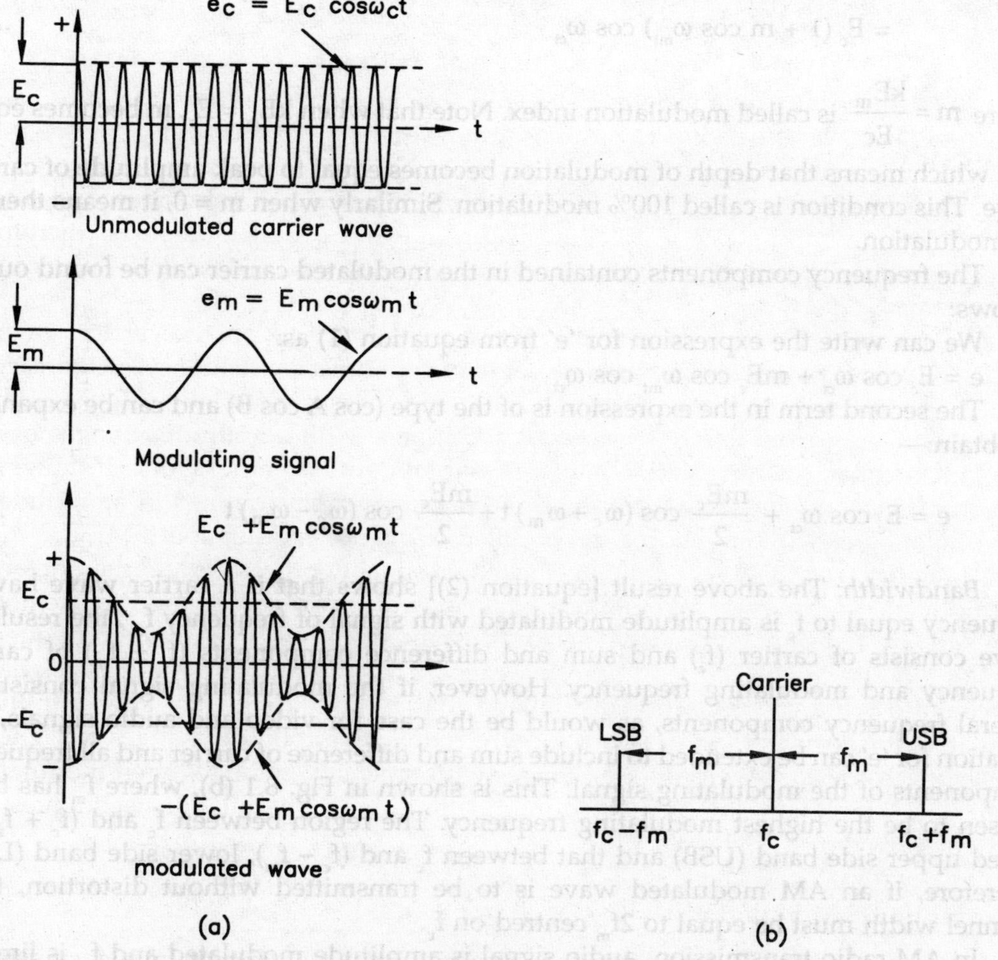

Fig. 8.1. (a) Modulation of RF carrier with a single low frequency sinusoidal signal, (b) Frequency spectrum of AM wave.

Since in amplitude modulation, the amplitude of carrier should vary in accordance with instantaneous value of modulating voltage, the modulated carrier can be expressed as:

$$e = A \cos \omega_{ct}$$

Where $A = (E_c + KE_m \cos \omega_{mt})$

and k is a constant of modulation.

On substituting the value of A we get:

$$e = (E_c + kE_m \cos \omega_{mt}) \cos \omega_{ct}$$

$$= E_c = (1 + \frac{kE_m}{Ec} \cos \omega_{mt}) \cos \omega_{ct}$$

$$= E_c (1 + m \cos \omega_{mt}) \cos \omega_{ct} \quad ...(1)$$

where $m = \frac{kE_m}{Ec}$ is called modulation index. Note that when $kE_m = E_c$, m becomes equal to 1, which means that depth of modulation becomes equal to peak amplitude of carrier wave. This condition is called 100% modulation. Similarly when $m = 0$, it means there is no modulation.

The frequency components contained in the modulated carrier can be found out as follows:

We can write the expression for 'e' from equation (1) as:

$e = E_c \cos \omega_{ct} + mE_c \cos \omega_{mt} \cos \omega_{ct}$

The second term in the expression is of the type (cos A cos B) and can be expanded to obtain:—

$$e = E_c \cos \omega_{ct} + \frac{mE_c}{2} \cos (\omega_c + \omega_m)\, t + \frac{mE_c}{2} \cos (\omega_c - \omega_m)\, t \quad ...(2)$$

Bandwidth: The above result [equation (2)] shows that if a carrier wave having frequency equal to f_c is amplitude modulated with signal of frequency f_m, the resultant wave consists of carrier (f_c) and sum and difference components ($f_c \pm f_m$) of carrier frequency and modulating frequency. However, if the modulating signal consists of several frequency components, as would be the case for video and audio signals, the equation for 'e' can be extended to include sum and difference of carrier and all frequency components of the modulating signal. This is shown in Fig. 8.1 (b), where f_m has been chosen to be the highest modulating frequency. The region between f_c and ($f_c + f_m$) is called upper side band (USB) and that between f_c and ($f_c - f_m$), lower side band (LSB). Therefore, if an AM modulated wave is to be transmitted without distortion, total channel width must be equal to $2f_m$ centred on f_c.

In AM radio transmission, audio signal is amplitude modulated and f_m is limited to 5 KHz. Therefore, bandwidth allotted to all radio stations is 10 KHz (± 5 KHz) centred around the carrier frequency (f_c) assigned to that radio station. Similarly, in TV transmission, where the highest modulating video frequency is 5 MHz, for AM transmission a bandwidth equal to 10 MHz (± 5 MHz) around the carrier would be necessary. However, in practice lesser bandwidth is allowed because information contained in both upper and lower sidebands is the same. This aspect is fully explained in chapter 11 that is devoted to terrestrial television.

Amplitude Modulation Circuit: A typical circuit arrangement used for amplitude modulation at transmitting stations is shown in Fig. 8.2. For obtaining higher circuit efficiency, the modulating amplifier is operated under class 'C' condition and the modulator as class B amplifier. The ac component of amplified modulating signal appears in series with dc supply to the amplifier used as modulator. Therefore, the plate or collector current pulses of class C amplifier will vary in accordance with instantaneous amplitude of the net power supply source to it. Thus, the output current will have amplitude variations as present in the modulating signal. The tuned circuit enables selection of carrier and its sidebands and all other higher frequency components get rejected. The output waveshape will be like that shown in Fig. 8.1 (a) if the modulating signal is a single frequency, sinusoidal voltage.

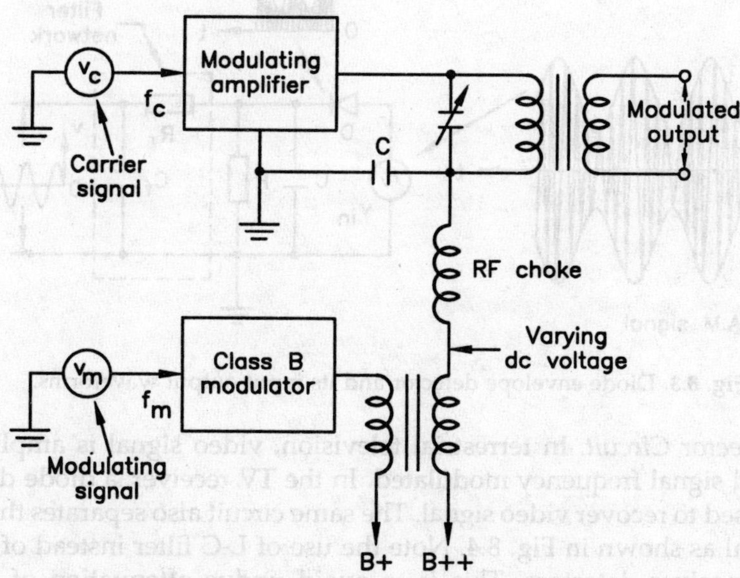

Fig. 8.2. Block diagram of an amplitude modulator.

8.2 Demodulation of AM Waves

Demodulators or detectors for amplitude modulation can be (i) square law type, or (ii) envelope type using a diode. The square law detector will produce sum and difference products of various components of the input AM signal. Thus, the output will contain signals having frequencies $f_c + (f_c + f_m)$ and $f_c - (f_c - f_m) = 2f_c + 2\ f_m$, besides other high frequency outputs. The carrier (f_c) is filtered out to obtain the modulating signal (f_m).

In radio and television receivers, a diode detector is preferred because it is more efficient. The basic circuit of a diode or envelope detector is shown in Fig. 8.3. During each cycle of input carrier the diode conducts for a short duration and the capacitor gets charged to the peak amplitude of carrier. Between peaks, the capacitor discharges through the resistor. If time-constant RC is much greater than the period of carrier, the discharge of capacitor will be very small. As such, the capacitor almost follows the modulation envelope which is the modulating signal. By adding a 'R_f cum C_f' filter circuit to the detector, the carrier can be grounded (filtered out) to obtain modulating signal as the output. By equating the maximum rate of change of envelope to the rate of discharge of capacitor, it can be shown that time-constant should satisfy the inequality:

$$RC << \frac{1}{m\omega_m}$$

for satisfactory detection. In the expression, m is the modulation index and ω_m *i.e.* f_m the maximum frequency of modulation.

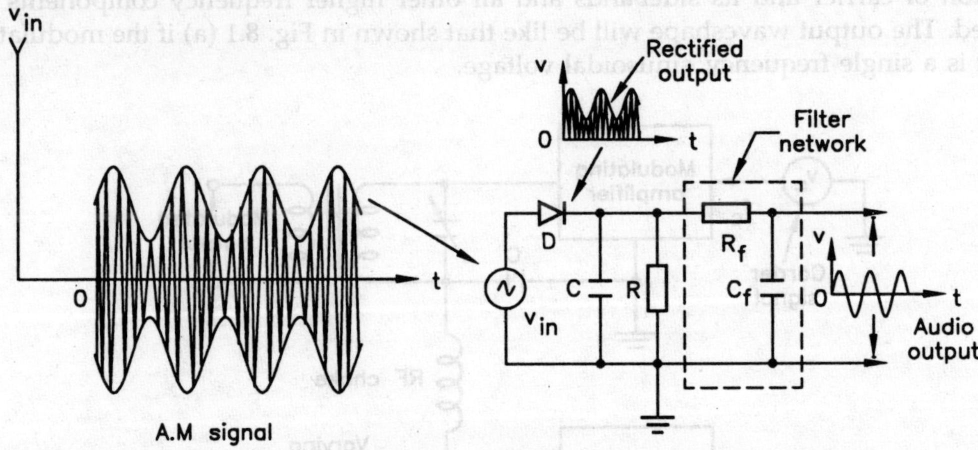

Fig. 8.3. Diode envelope detector and its input output waveforms.

Video Detector Circuit: In terrestrial television, video signal is amplitude modulated and sound signal frequency modulated. In the TV receiver, a diode detector-cum-filter circuit is used to recover video signal. The same circuit also separates the intercarrier FM sound signal as shown in Fig. 8.4. Note the use of L-C filter instead of R-C filter as used in radio receiver detectors. This is to avoid undue attenuation of video signal while filtering out carrier components. The circuit bandwidth must extend from 0 Hz to 6 MHz to include sound inter-carrier frequency of 5.5 MHz and its FM sidebands besides the video signal. For this, the detector load includes an inductor to obtain low and nearly constant attenuation over the entire frequency range up to 6 MHz.

8.3 Frequency Modulation

In frequency modulation, the amplitude of carrier remains constant but its frequency is varied to be proportional to the instantaneous value of modulating voltage. The rate at which frequency deviation takes place is equal to the modulating signal frequency. Assuming the modulating signal to be sinusoidal, the FM wave would look as shown in Fig. 8.5.

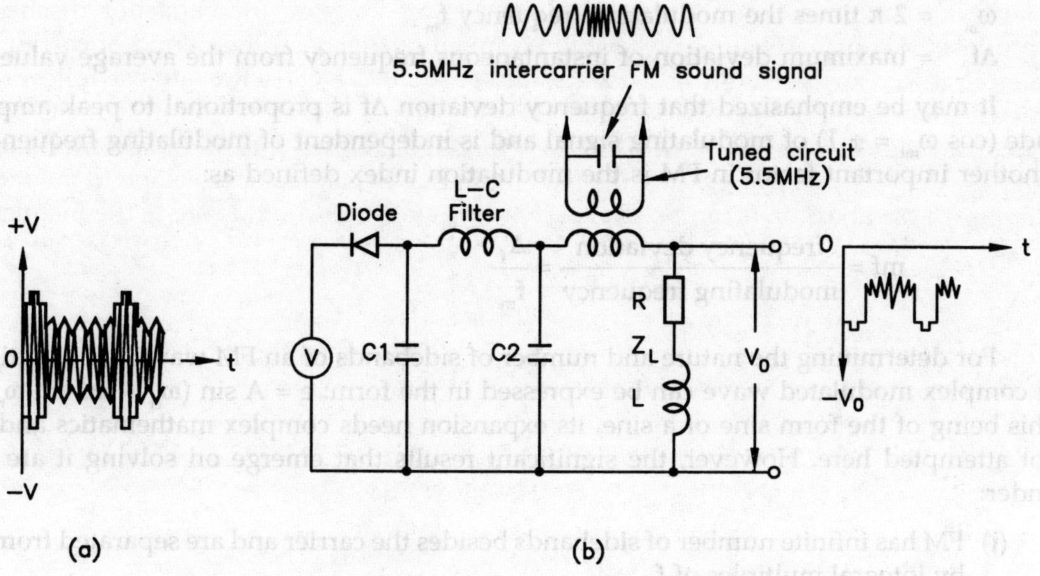

Fig. 8.4. Video detection and sound signal separation circuit.

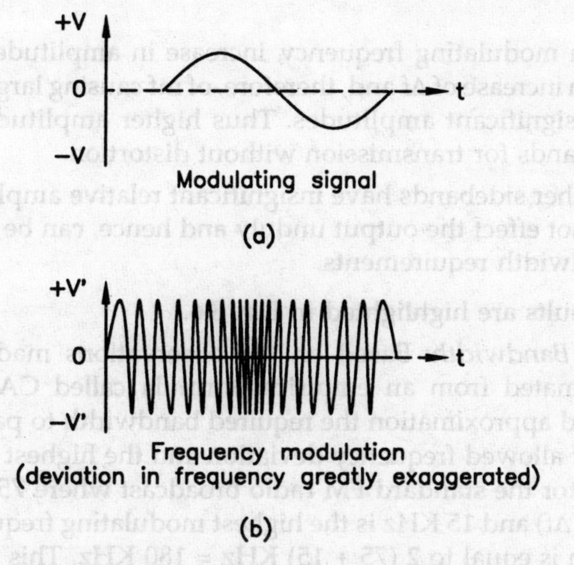

Fig. 8.5. Frequency modulation (a) modulating signal,
(b) frequency modulated carrier.

FM Side-Bands: A frequency modulated carrier wave with sinusoidal modulation can be expressed as:

$$\omega_i = \omega_c + 2\,\pi\,\Delta f \cos \omega_{mt}$$

where ω_i = instantaneous angular velocity

ω_c = angular velocity of carrier wave *i.e.* (average angular velocity)
ω_m = 2π times the modulating frequency f_m
Δf = maximum deviation of instantaneous frequency from the average value.

It may be emphasized that frequency deviation Δf is proportional to peak amplitude ($\cos \omega_{mt} = \pm 1$) of modulating signal and is independent of modulating frequency. Another important factor in FM is the modulation index defined as:

$$mf = \frac{\text{frequency deviation}}{\text{modulating frequency}} = \frac{\Delta_f}{f_m}$$

For determining the nature and number of sidebands of an FM wave, the equation of complex modulated wave can be expressed in the form: $e = A \sin (\omega_{ct} + mf \sin \omega_{mt})$. This being of the form sine of a sine, its expansion needs complex mathematics and is not attempted here. However, the significant results that emerge on solving it are as under:

(i) FM has infinite number of sidebands besides the carrier and are separated from it by integral multiples of f_m.

(ii) The sidebands on either side of the carrier at equal distance from f_c have equal amplitudes so that sideband distribution is symmetrical about the carrier frequency.

(iii) For a given modulating frequency, increase in amplitude of modulating signal results in an increase of Δf and, therefore, of mf causing large number of sidebands to acquire significant amplitudes. Thus higher amplitude signals would need more sidebands for transmission without distortion.

(iv) A lot of higher sidebands have insignificant relative amplitudes and their exclusion does not effect the output unduly and hence, can be neglected while calculating bandwidth requirements.

The above results are highlighted in Fig. 8.6.

FM Channel Bandwidth: Based on the observations made above, FM channel width can be estimated from an empirical formula called CARSON'S RULE which states that to a good approximation the required bandwidth to pass a FM wave is equal to twice the sum of allowed frequency deviation and the highest modulating frequency. As an illustration, for the standard FM radio broadcast where 75 KHz is the maximum allowed deviation (Δf) and 15 KHz is the highest modulating frequency (f_m), the required channel bandwidth is equal to 2 (75 + 15) KHz = 180 KHz. This nearly checks with the 200 KHz bandwidth allowed in FM radio broadcast systems.

Similarly in the 625 CCIR TV system adopted by India where maximum allowed Δf = 50 KHz and f_m = 15 KHz, the necessary bandwidth equals 2 (50 + 15) KHz = 130 KHz. In practice a deviation = ± 75 KHz *i.e.* channel width = 150 KHz is provided–for at FM transmitters.

Pre-Emphasis and De-Emphasis: The greatest advantage of FM is its ability to eliminate noise interference. This becomes possible because variations in amplitude of the modulating signal cause frequency deviations and not change in amplitude of the carrier wave. Any noise interference results in amplitude variations of the carrier wave and these can be easily removed by passing it through an amplitude limiter amplifier before detection.

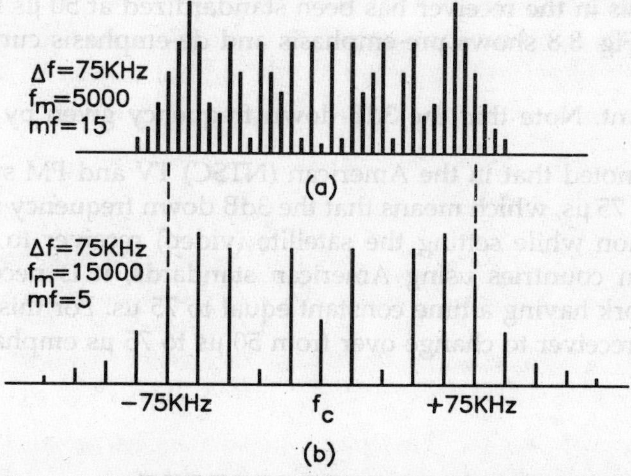

Fig. 8.6. Sideband spectrum of frequency modulated waves with same Δf but different f_m and mf.

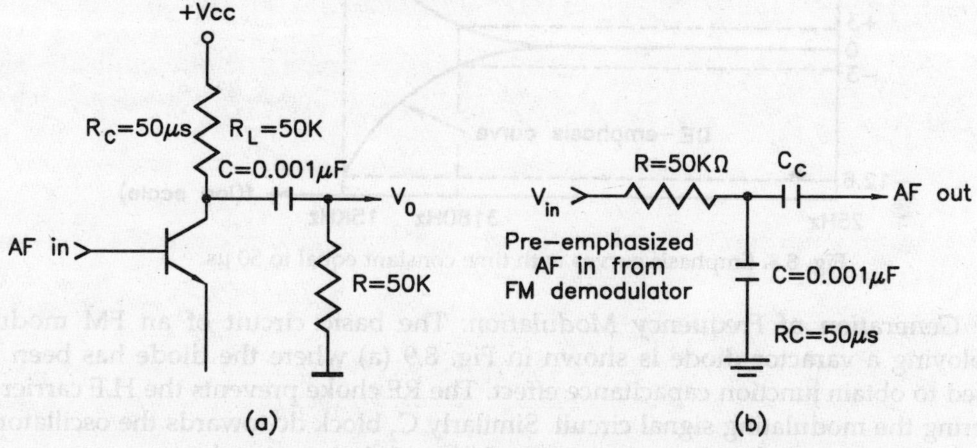

Fig. 8.7. Pre-emphasis and de-emphasis of audio signal components (a) pre-emphasis circuit, (b) de-emphasis circuit.

It is also possible to reduce noise by increasing frequency deviations which may be made as large as required because this does not cause any increase in transmitter power. Higher audio frequencies are mostly harmonies of low and mid-range spectrum and thus have low amplitudes. As such, these cause small deviations of the carrier on modulation. External noise power is also low in amplitude and causes frequency deviations in the same range as by higher audio frequencies. This makes higher frequency components of the audio signal most susceptible to noise effects. To prevent this, higher audio frequencies are artificially boosted in amplitude before modulation. Similarly, at the receiving end, these are reduced to their original level after detection. The two processes are known as pre-emphasis and de-emphasis respectively. Examples of networks used for each function are shown in Fig. 8.7. As obvious from these configurations, the pre-emphasis and de-emphasis circuits are high-pass and low-pass filters respectively. The time-constant RC of filters for pre-emphasis at the transmitter and later de-emphasis in the receiver has been standardized at 50 μs for CCIR TV and FM radio systems. Fig. 8.8 shows pre-emphasis and de-emphasis curves corresponding to this time constant. Note that the 3dB down frequency given by $\frac{1}{2\pi RC}$ is 3180 Hz.

It is to be noted that in the American (NTSC) TV and FM systems, the emphasis time-constant is 75 μs, which means that the 3dB down frequency is 2120 Hz. Therefore, in cable television while setting the satellite (video) receiver to tune-in TV channels originating from countries using American standards, it is necessary to choose de-emphasis network having a time constant equal to 75 μs. For this a switch is provided on the satellite receiver to change over from 50 μs to 75 μs emphasis network.

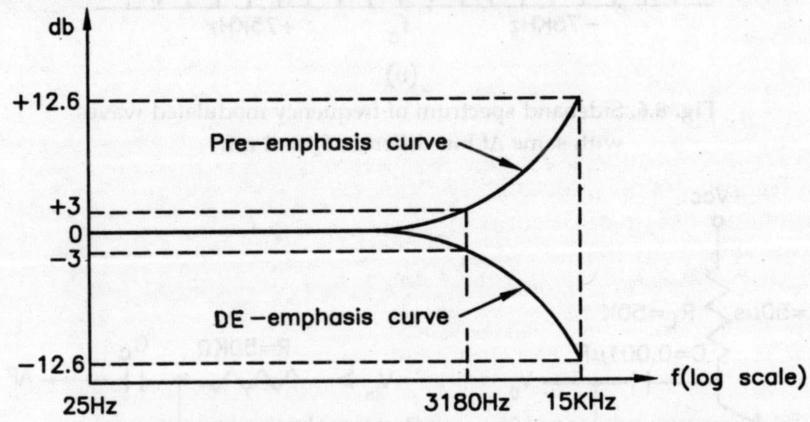

Fig. 8.8. Emphasis curves with time constant equal to 50 μs.

Generation of Frequency Modulation: The basic circuit of an FM modulator employing a varactor diode is shown in Fig. 8.9 (a) where the diode has been back-biased to obtain junction capacitance effect. The RF choke prevents the H.F carrier from entering the modulating signal circuit. Similarly C_2 block dc towards the oscillator tank circuit. The instantaneous variations in the modulating signal cause corresponding changes in junction capacitance of the diode which in turn varies the oscillator frequency to cause frequency modulation.

An active device can also be used in place of varactor diode. Fig. 8.8 (b) shows necessary connections around a field effect transistor (FET). It can be shown that if current i_g is negligible compared to 'i' and impedance $X_c >> R$, preferably by more than 5 : 1, the impedance 'Z' as seen at the terminals marked A – A', is almost entirely reactive. It is equal to a capacitance C_{eq} = gmRC. When used in the modulator circuit, variations in the modulating voltage applied across A–A cause variations in C_{eq} because gm, the mutual conductance of the device varies with the magnitude of applied voltage. An FM oscillator using such a reactance device is called 'Reactance modulator'.

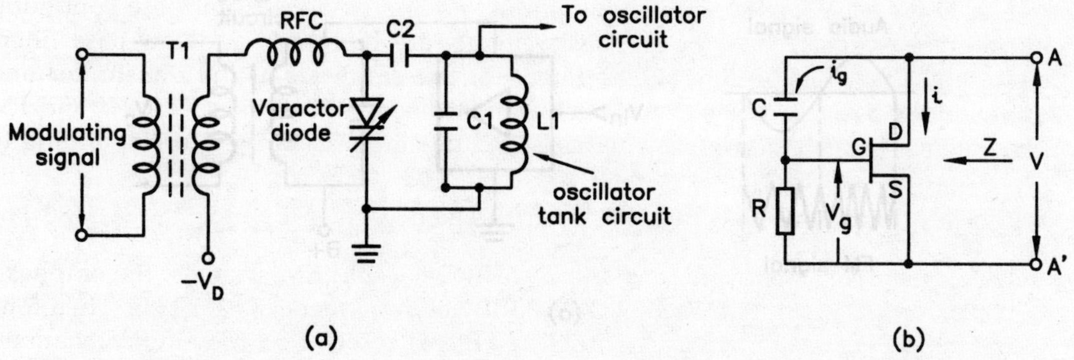

Fig. 8.9. Frequency modulation (a) modulator circuit using a varactor diode, (b) FET circuit that can replace the varactor diode.

8.4 FM Detection

The frequency modulated IF signal has some amplitude variations due to unequal amplification at RF and IF stages of the receiver. These are removed by passing the FM signal through a limiter amplifier which is designed to have a narrow dynamic range necessary for clipping action. Basically, FM detection is carried out in two steps. The frequency modulated IF signal of constant amplitude is first converted into a proportionate voltage that is both frequency and amplitude modulated. The latter voltage is then applied to a detector arrangement which detects amplitude changes but ignores frequency variations.

Amplitude variations in an FM signal can be caused by using one of the sloping sides of a tuned circuit. For this, the FM signal is fed to a tuned amplifier [see Fig. 8.10 (a)] whose resonant frequency is on one side of the centre frequency of FM signal. As shown in Fig. 8.10 (b), the circuit is detuned to bring the centre frequency (f_c) to point 'A' on the selectivity curve.

As the input signal frequency shifts, the frequency deviations of the carrier signal are converted into proportionate amplitude variations as illustrated in the figure. These AM variations result from the unequal IF gain above and below the carrier frequency. Thus, the IF output varies in amplitude at the modulating signal rate, in addition to its continuously changing frequency. The resulting AM signal can be coupled to a diode detector-cum-filter circuit to remove the carrier and retain the modulating signal as its output.

Basic Detector Circuit: The basic discriminator or detector circuit that operates as explained above is shown in Fig. 8.11. It uses three tuned circuits. The circuit L_1 C_1 is tuned to the centre frequency, L_2 C_2 to above centre frequency and L_3 C_3 to below centre frequency by an equal amount. At centre *i.e.* IF carrier frequency, equal voltages are developed across the tuned circuits and D_1 and D_2 conduct equally. The voltages across R_1 and R_2 are equal and opposite in polarity and thus, the circuit output is zero. If the input frequency increases above centre, as would happen on positive deviation, then L_2 - C_2 circuit develops a higher voltage and D_1 conducts more than D_2 and unequal

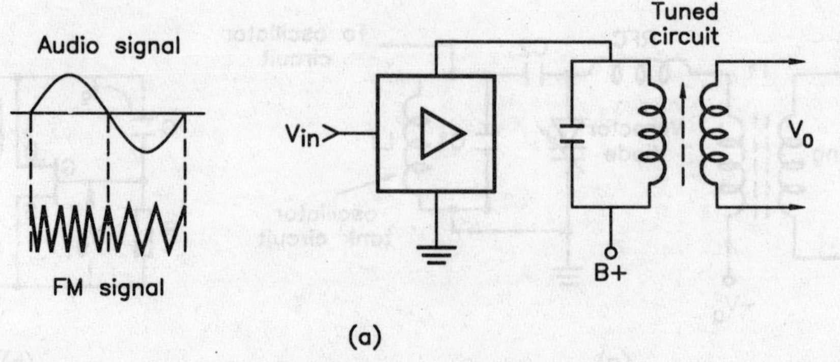

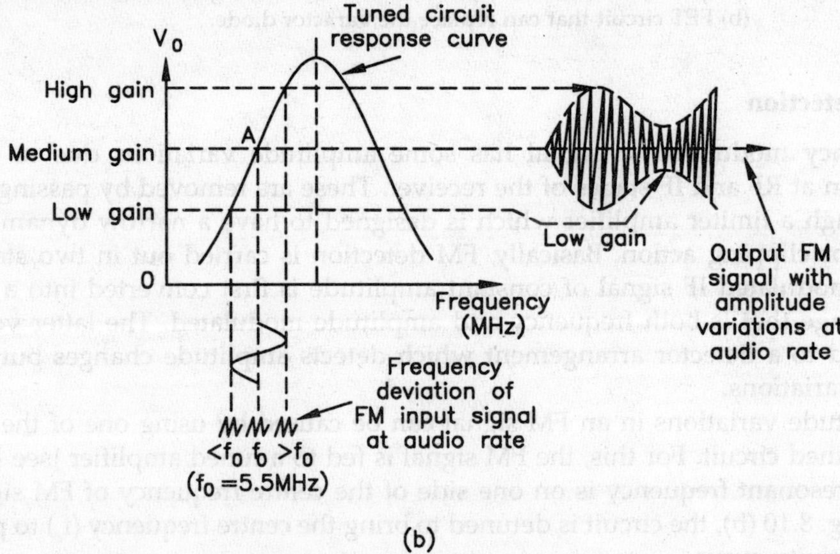

Fig. 8.10. (a) Tuned amplifier,
(b) Slope detection of an FM signal.

Note: The modulating signal in the illustration is a sinusoidal L.F. sound signal modulated on an inter-carrier frequency of 5.5 MHz as in TV receivers.

voltages develop across R_1 and R_2. Similarly, when the centre frequency deviates to become less, the unequal voltages developed across R_1 and R_2 will be of opposite polarities. The net voltage after filtering action with C_4 and C_5 will follow frequency deviations only and hence, will be the modulating voltage.

Other Detector Circuits: One of the earliest successful detector was the Foster-Seely discriminator and operates as explained above. It was soon followed by a balanced ratio detector which does not need a limiter. Later, a differential peak FM detector came into use in integrated circuit form where a new technique called differential peak detection is used to develop audio voltage from the frequency modulated signal.

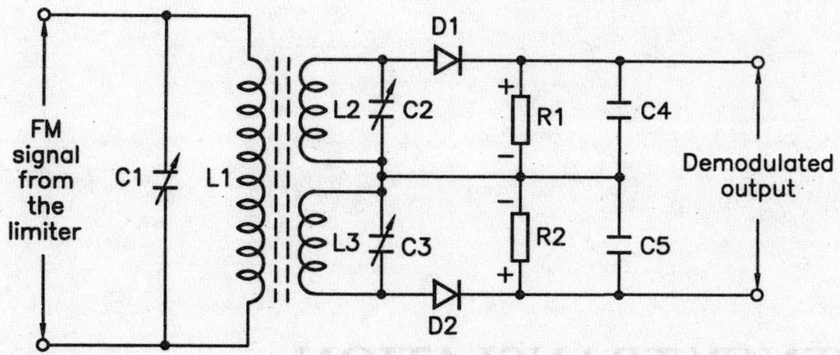

Fig. 8.11. A basic FM detector (discriminator) circuit using semi-conductor diodes.

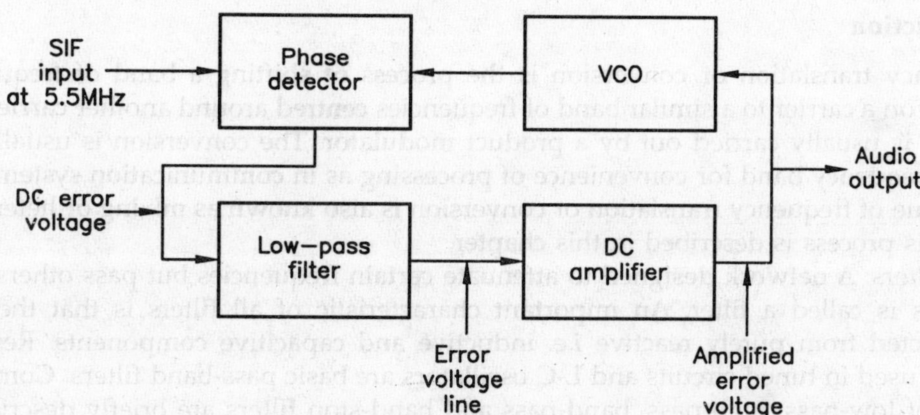

Fig. 8.12. Schematic block diagram of a phase-locked loop FM detector.

Phase-Locked Loop FM Detector: The latest in FM detection is the phase-locked loop detector that requires no alignment. Its block schematic is shown in Fig. 8.12. In this circuit, a dc error voltage is developed by a phase detector that compares the phase of input FM signal with that of a locally generated oscillator of the same frequency. The dc control voltage obtained at the output of low-pass filter is directly related to the degree of phase difference between the two signals. This control voltage which is returned to the voltage controlled oscillator (VCO) after amplification forces the oscillator to change its frequency thereby reducing the phase error. When the free running frequency of the VCO becomes close to the input signal frequency, the system locks at this frequency and the VCO tracks the input FM signal. Since the dc control voltage will vary in step with frequency deviations of the FM signal, it represents audio modulation. Thus, output from the dc amplifier is the demodulated audio signal. This detector is available in IC form. Another version of it called 'Digital FM detector' is also used in many receivers.

9

FREQUENCY TRANSLATION AND FILTERS

Introduction

Frequency translation or conversion is the process of shifting a band of frequencies centred on a carrier to a similar band of frequencies centred around another carrier. This process is usually carried out by a product modulator. The conversion is usually to a lower frequency band for convenience of processing as in communication systems. The technique of frequency translation or conversion is also known as mixing or heterodyning. This process is described in this chapter.

Filters: A network designed to attenuate certain frequencies but pass others without loss is called a filter. An important characteristic of all filters is that these are constructed from purely reactive *i.e.* inductive and capacitive components. Resonant circuits used in tuned circuits and L-C oscillators are basic pass-band filters. Composite forms of low-pass, high-pass, band-pass and band-stop filters are briefly described in later sections of this chapter.

9.1 Frequency Translation

In a number of important radio frequency applications, signals derived from two separate sources are impressed across a non-linear circuit element in order to move 'down' or 'up', the frequency band of one of the signals to another frequency spectrum without altering relative amplitude and phase relationships. Usually, the desired output of such conversions is at difference frequency of the two input frequencies as in tuners of satellite, TV and radio receivers.

As explained in previous chapter, if two signals are fed in series to the input of a square law detector (non-linear circuit) the output contains, among other products, the sum and difference components. The wanted difference components are filtered out by a tuned circuit. Such a mixer circuit employing a transistor is shown in Fig. 9.1. The constant amplitude signal of frequency f_0 fed at the emitter is $e_1 = E_0 \cos \omega_0 t$. The other signal e_2 made input at the base of transistor is an AM RF signal. As shown in section 8.1, such a signal can be expressed as:

$$e_2 = E_c \cos \omega_c t + \frac{mE_c}{2} \cos (\omega_c + \omega_m)\, t + \frac{mE_c}{2} \cos (\omega_c - \omega_m)\, t$$

where f_c is the carrier frequency and f_m the modulating frequency. The amplitude of e_1 is kept quite large as compared to that of e_2 and f_0 is usually higher than f_c. The two input signals, though isolated from each other, cause collector current variations collectively. The transistor is so biased by resistors R_1, R_2 and R_E that the operating point lies in non-linear region of its transfer characteristics. Thus, on application of these two signals, inter-modulation occurs and sum and difference products of the two signals get generated. The output at the collector will have the following important frequency components:

f_0	$f_0 - f_c$
f_c	$f_0 - f_c - f_m$
$f_0 + f_c$	$f_0 - f_c + f_m$
$f_0 + f_c - f_m$	
$f_0 + f_c + f_m$	

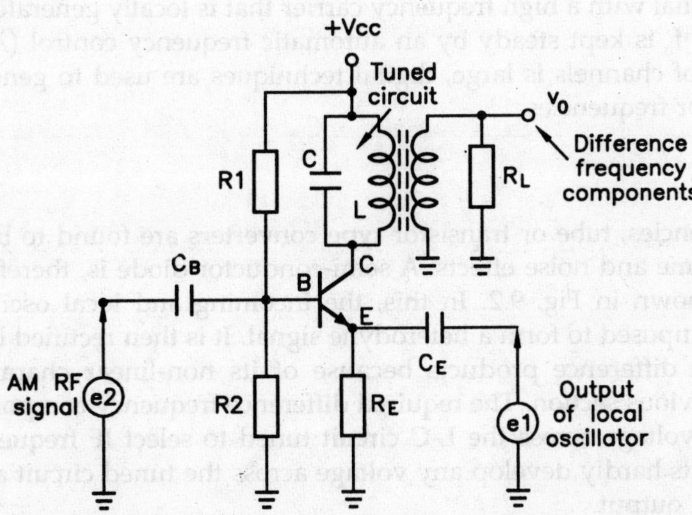

Fig. 9.1. Basic frequency converter circuit employing a transistor.

For down frequency translation as is the case in most applications, the frequencies of interest are those listed in the right hand column. These can be filtered out by a tuned circuit as shown in the figure.

IF Frequency: If frequency f_0 is so chosen that $f_0 - f_c = f_x$, where f_x is the desired new carrier, then it follows that the three terms listed in right side column represent original carrier and sidebands translated to the new carrier frequency f_x, where:

(i) $f_0 - f_c = f_x$, (ii) $f_0 - f_c - f_m = f_x - f_m$, (iii) $f_0 - f_c + f_m = f_x + f_m$, constitute the new carrier groups where $f_x - f_m$ and $f_x + f_m$ are lower and upper side-bands respectively.

In communication receivers of any type, the incoming high frequency AM or FM signal obtained from the output of tuned RF amplifier is heterodyned with the output of a local oscillator to produce a difference frequency called intermediate frequency (IF) and this then becomes the new carrier earlier called f_x. By suitably tracking the local oscillator frequency as is done in AM or FM receivers, the difference between carrier frequency of the station to which the receiver is tuned and local oscillator frequency remains constant and equal to IF. When the local oscillator signal is produced by the mixer itself, the circuit is known as frequency converter, and the receiver a superheterodyne receiver.

Merits of Conversion to IF: By converting the incoming station signals around a common much lower IF frequency makes the receiver more sensitive and more stable. This is so, because more stages for amplification can be built at this low frequency to obtain much higher gain and selectivity.

The IF frequency used in most AM radio receivers is 455 KHz and in FM radio receivers it is 10.7 MHz. In satellite receivers either 70 MHz or 140 MHz is used as the IF frequency. In television receivers using CCIR 625-B standards, IF for picture signal is 38.9 MHz and for sound signal it is 33.4 MHz. All IF frequencies are provided by mixing the incoming RF signal with a high frequency carrier that is locally generated. The local oscillator frequency f_0 is kept steady by an automatic frequency control (AFC) circuit. When the number of channels is large, digital techniques are used to generate crystal stable local oscillator frequencies.

9.2 Diode Mixer

At very high frequencies, tube or transistor type converters are found to be inefficient because of transit-time and noise effects. A semi-conductor diode is, therefore, used in mixer circuits as shown in Fig. 9.2. In this, the incoming and local oscillator signal voltages are super-imposed to form a heterodyne signal. It is then rectified by the diode to obtain sum and difference products because of its non-linear characteristics as explained in the previous section. The required difference frequency component current develops sufficient voltage across the L-C circuit tuned to select IF frequency signals. All other components hardly develop any voltage across the tuned circuit and are thus eliminated from the output.

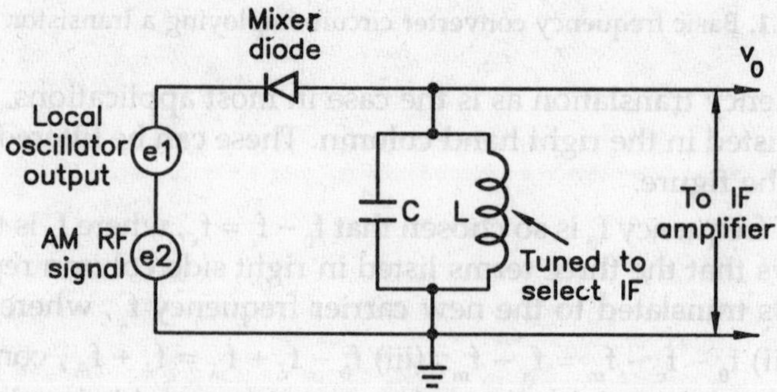

Fig. 9.2. Basic circuit of a UHF mixer employing a diode.

9.3 Attenuators and Pads

In transmission equipments it is frequently necessary to attenuate (reduce) current, voltage and power levels at certain stages. Attenuators are networks designed to meet these requirements. Since all frequency components in the signal must be attenuated to the same degree, the networks should consist of purely resistive components. If there is provision in this network to vary attenuation it is called an attenuator and if not it is called a Pad. In television, pads can be used at the input to protect receivers against overload and consequent damage. For example, if a TV receiver is in use close to the local TV transmitting tower, the signal strength can be reduced to prevent overloading by inserting a pad in series with the transmission line coming from the antenna. Similarly, in cable distribution networks, subscriber locations close to line or trunk amplifiers may need a reduction in signal strength.

Two different pad configurations are given in Fig. 9.3 together with charts for determining component values for specific attenuations.

Attenuation	R_1	R_2	Attenuation	R_1	R_2
6 dB	47 Ω	390 Ω	6 dB	22 Ω	100 Ω
10 dB	82 Ω	220 Ω	10 dB	33 Ω	51 Ω
20 dB	120 Ω	68 Ω	20 dB	56 Ω	15 Ω

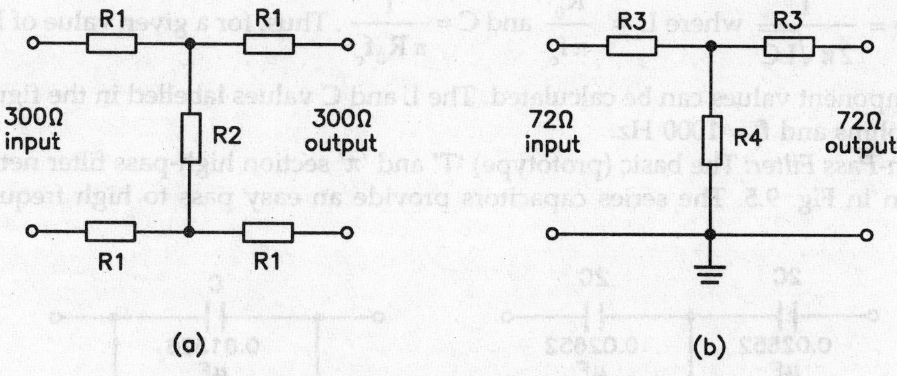

Fig. 9.3. Resistance pads: (a) balanced 'H' pad for 300 Ω twin lead, (b) 'T' pad for 72 Ω coaxial line.

9.4 Low-Pass and High-Pass Filters

The simplest type of filter has only one pass-band, one attenuation band and a single cut-off frequency. If it passes all frequencies up to the cut-off frequency and attenuates all frequencies above it, it is called 'low-pass' filter. If on the other hand, it attenuates all frequencies below the cut-off frequency and passes all frequencies above it, it is called a 'high-pass filter'. An ideal filter would have zero attenuation in the pass-band and infinite attenuation in the attenuation band. However, this cannot be achieved in practice because of losses in inductors and capacitors and due to impedance mismatch if present on either side of the filter network.

Low-Pass Filter: The prototype (basic) 'T' and 'π' section low-pass filter networks are shown in Fig. 9.4. The inductors prevent passage of high frequencies and capacitors allow these to pass through.

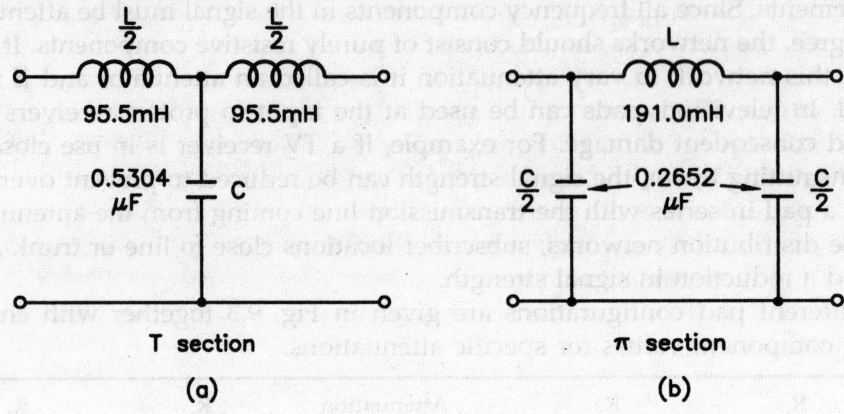

Fig. 9.4. Prototype low-pass filter section.

It can be shown that R_0, the characteristic impedance = $\sqrt{\frac{L}{C}}$ and f_c, the cut-off frequency = $\frac{1}{2\pi\sqrt{LC}}$ where $L = \frac{R_0}{\pi f_c}$ and $C = \frac{1}{\pi R_0 f_c}$. Thus, for a given value of R_0 and f_c, the component values can be calculated. The L and C values labelled in the figure for R_0 = 600 ohms and f_c = 1000 Hz.

High-Pass Filter: The basic (prototype) 'T' and 'π' section high-pass filter networks are shown in Fig. 9.5. The series capacitors provide an easy pass to high frequencies

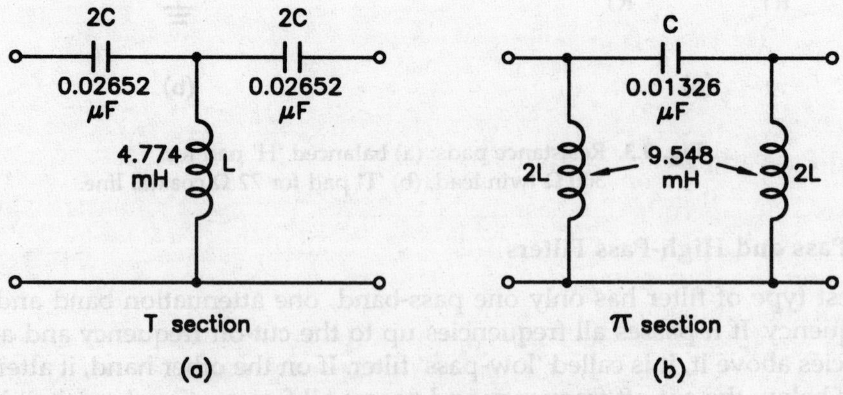

Fig. 9.5. Prototype high-pass filter sections.

while inductors block their passage to ground. The values of R_0, f_c, L and C can be calculated from the following expressions:

$$R_0 = \sqrt{\frac{L}{C}}, \qquad f_c = \frac{1}{4\pi\sqrt{LC}}$$

$$L = \frac{R_0}{4\pi f_c} \text{ and } C = \frac{1}{4\pi R_0 f_c}$$

The L and C values labelled in the figure are for R_0 = 600 ohms and f_c = 10 KHz.

m-Derived Filter Sections: There are two disadvantages of prototype filter sections described above. In the first place, the attenuation does not rise very rapidly after the cut-off frequency and hence, the pass-band is not well defined. Similarly, the impedance of such filter sections does not remain constant over the pass-band. To overcome these difficulties as far as possible, m-derived filter sections are used. The value of 'm' lies between 0 and 1. Fig. 9.6 shows such filter sections in 'T' configurations both for low-pass and high-pass filters.

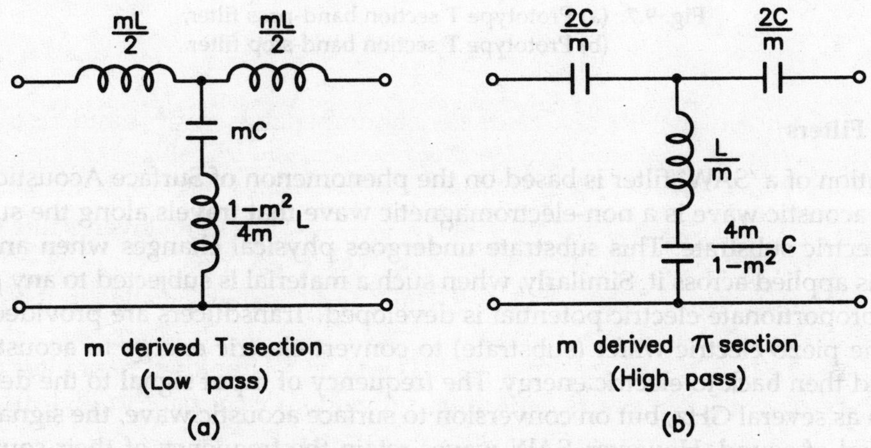

Fig. 9.6. m-derived filters (a) low-pass in T section configuration, (b) high-pass in T section configuration.

9.5 Band-Pass and Band-Stop Filters

A band-pass filter permits the passage of a desired band of frequencies while a band-stop filter prevents the passage of unwanted band of frequencies. Prototype 'T' sections of these filters are shown in Fig. 9.7. In band-pass filters the values of L_1 and C_1 are so chosen that their impedance to the frequencies to be passed is quite low. Similarly L_2 and C_2 are so chosen that their impedance to the frequencies below the pass-band is very high. At frequencies below the pass-band, C_1 offers high impedance while L_2 provides an easy shunt path. At frequencies above the pass-band, L_1 offers high impedance and C_2 becomes an easy shunt path to ground these frequency components. Analogous arguments enable understanding the functioning of a band-stop filter.

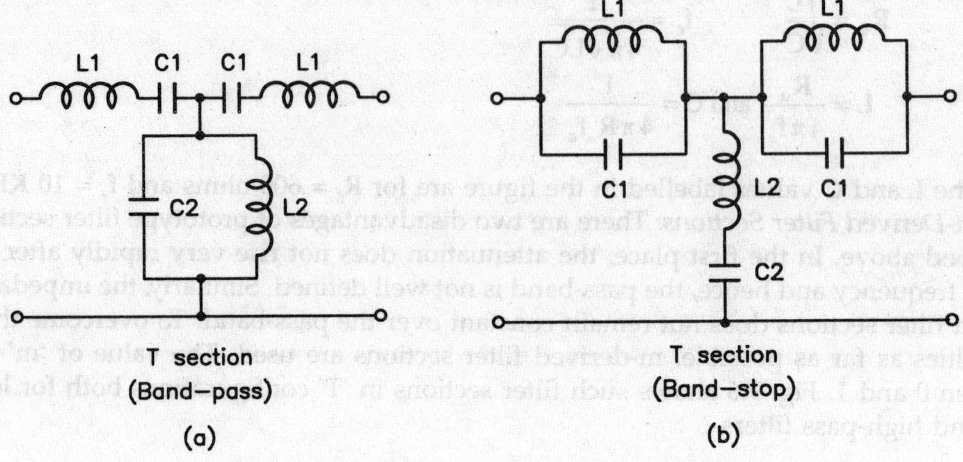

Fig. 9.7. (a) Prototype T section band-pass filter, (b) Prototype T section band-stop filter.

9.6 Saw Filters

The operation of a 'SAW' filter is based on the phenomenon of Surface Acoustic Waves. A surface acoustic wave is a non-electromagnetic wave that travels along the surface of a piezo-electric substrate. This substrate undergoes physical changes when an electric potential is applied across it. Similarly, when such a material is subjected to any physical stress, a 'proportionate electric potential is developed. Transducers are provided at two ends of the piezo-electric wafer (substrate) to convert electric energy to acoustic wave energy and then back to electric energy. The frequency of input signal to the device can be as high as several GHz, but on conversion to surface acoustic wave, the signal travels at the speed of sound. However, SAW waves retain the frequency of their source with possibility of affecting different attenuation to different bands of frequencies.

Wave Shaping: The bandwidth and frequency characteristics of a SAW filter depend on the geometric structure of the input transducer array because this determines the extent of attenuation to desired band of frequencies. Thus, the filter can be designed to have any group delay characteristics by changing the distance between electrodes of transducer array. A typical SAW filter that provides complete desired bandpass characteristics is shown in Fig. 9.8.

A typical application of SAW filter is in the IF section of satellite and TV receivers where effectively several filtering and band shaping requirements are met simultaneously. As an illustration of this, consider Fig. 9.9 which shows input circuit of the IF section of an earlier TV receiver.

The first section is a series trap that blocks the lower channel interfering frequency. Similarly the third section filters out upper channel beat frequency. Section two of the filter unit is designed to attenuate sound IF frequency to about 20 dB below the picture level and also to provide vestigial sideband correction. Coil L_4 is for frequency compensation against low input impedance of the common emitter first stage IF amplifier. In present-day receivers all this is achieved by using a SAW filter instead of different filter sections which need exact tuning and suitable mounting.

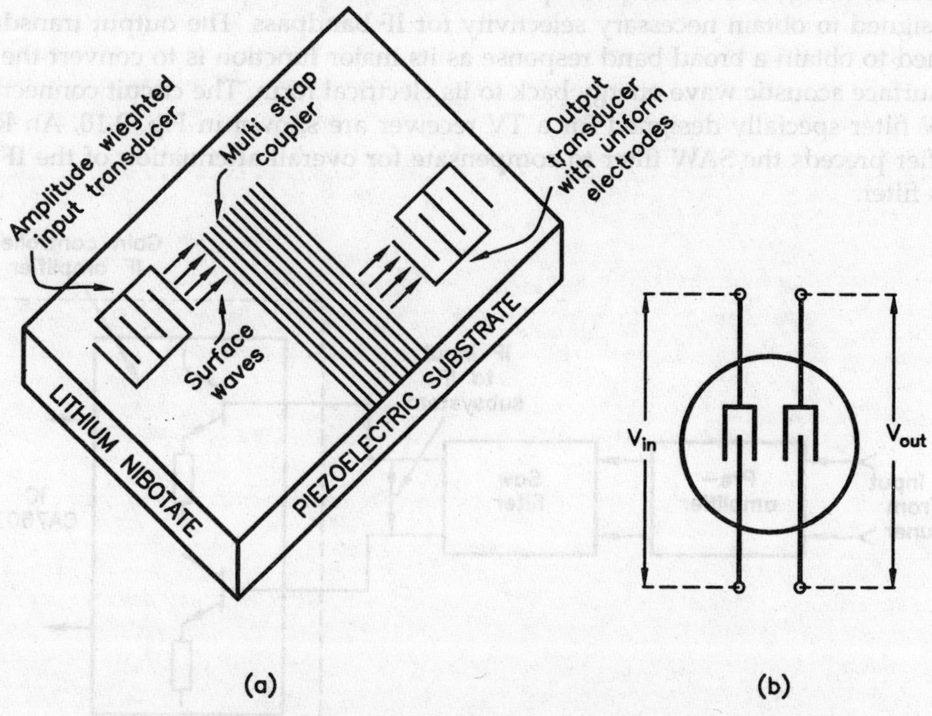

Fig. 9.8. A typical SAW filter having desired band-pass characteristics: (a) simplified constructional details, (b) symbol.

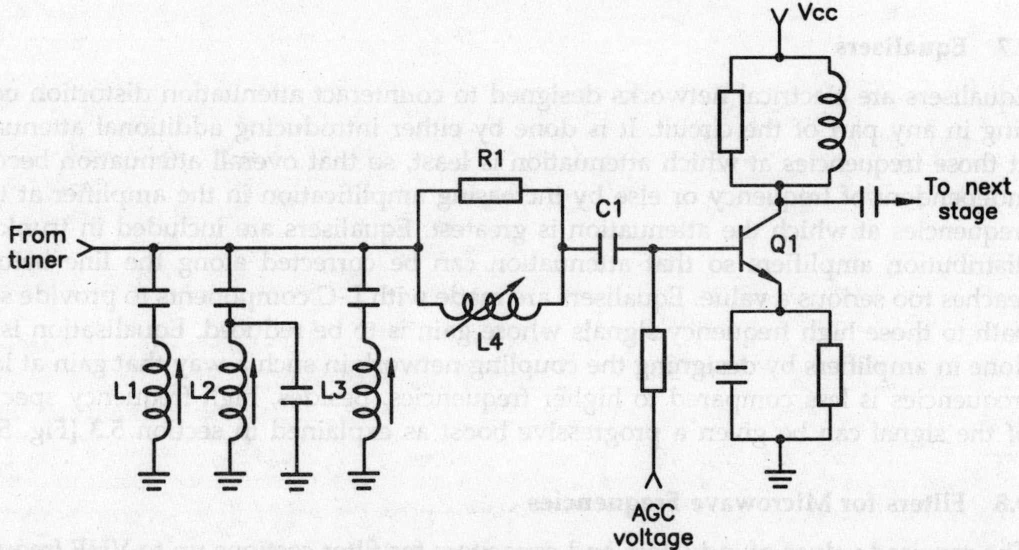

Fig. 9.9. Input circuit of a IF sub-system using discrete components and devices.

It consists of input and output transducer arrays [see Fig. 9.8] separated by a multistrap coupler which prevents distortion of surface wave transmission caused by reflection of energy from bottom to top of the substrate. The input transducer is weighted *i.e.* designed to obtain necessary selectivity for IF bandpass. The output transducer is designed to obtain a broad band response as its major function is to convert the signal from surface acoustic wave energy back to its electrical form. The circuit connections of a SAW filter specially designed for a TV receiver are shown in Fig. 9.10. An RF Pre-amplifier preceds the SAW filter to compensate for overall attenuation of the IF signal in this filter.

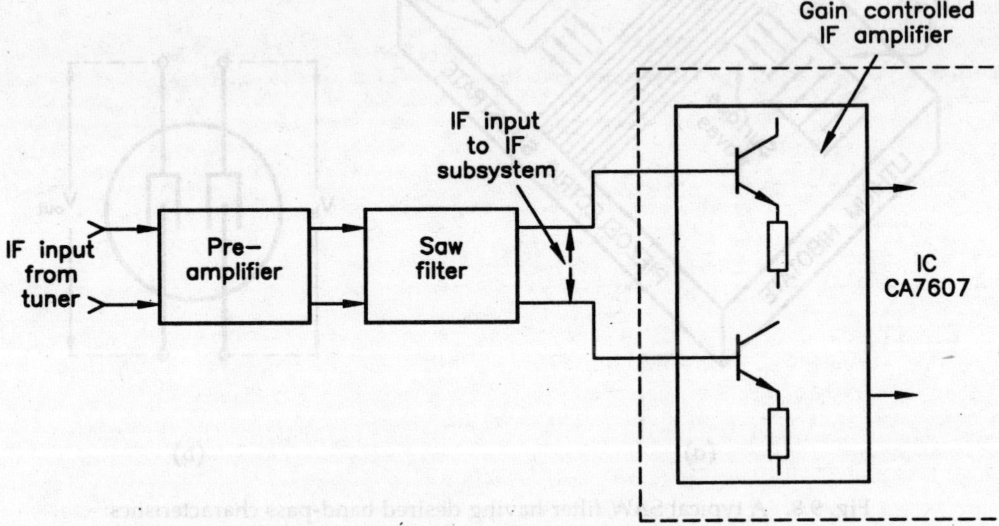

Fig. 9.10. SAW filter connections to a vision IF IC.CA7607.

9.7 Equalisers

Equalisers are electrical networks designed to counteract attenuation distortion occurring in any part of the circuit. It is done by either introducing additional attenuation at those frequencies at which attenuation is least, so that overall attenuation becomes independent of frequency or else by increasing amplification in the amplifier at those frequencies at which the attenuation is greatest. Equalisers are included in trunk and distribution amplifiers so that attenuation can be corrected along the line before it reaches too serious a value. Equalisers are made with L-C components to provide shunt path to those high frequency signals whose gain is to be reduced. Equalisation is also done in amplifiers by designing the coupling network in such a way that gain at lower frequencies is less compared to higher frequencies. Besides, high frequency spectrum of the signal can be given a progressive boost as explained in section 5.3 [Fig. 5.4].

9.8 Filters for Microwave Frequencies

The required values of inductors and capacitors for filter sections up to VHF frequency range can be obtained as discrete components. However, at UHF frequencies a single turn of a coil is enough and often stray, wiring and component capacitances provide necessary capacitance and no physical capacitor is necessary. But to create resonance at microwave frequencies the required values of inductance and capacitance become extremely small and as such physical limits become a problem. For example, even a very small length of component lead wire represents enough inductance at these frequencies. Therefore, in equipments that carry microwave signals, special transmission lines called micro-strips and strip-lines are used as tuned circuits, filters and even reactive components. Similarly, wave guides which are hollow tubes and are used for transmission of microwave signals are also used as high pass filters. Also cavity resonators which are made out of short segments of waveguides act as high 'Q' parallel resonant circuits as explained in chapter 4, devoted to microwave techniques.

10

VOLTAGE REGULATION AND CABLE POWERING

Introduction

The dc voltage requirements of various CATV equipments range from 5 to 25 volts. All low voltage dc supplies are obtained from ac mains through a step-down transformer and rectifier-filter circuits. The use of a voltage regulator after the filter circuit is necessary because transistors and ICs used in all equipments are very sensitive to over-voltage. Similarly, ac power supply from mains is often not steady and it becomes essential to use an automatic voltage regulator at the input point of ac mains to prevent any damage to various equipments. Furthermore, on account of frequent failures and load shedding of electric supply, it is necessary to equip the CATV station with an alternative source of ac supply. This need is met by installing inverter units run by lead-acid batteries or diesel run generator sets.

Cable powering is another important aspect of cable television plant. All the trunk, distribution and line extender amplifiers require ac supply for their operation to amplify CATV signals. For this, ac mains voltage is stepped down to 50 to 60V and then distributed to network amplifiers through the same cables that carry TV signals. A different type of ac regulation is used to keep this ac voltage close to the set value. In smaller distribution networks, power supply units are installed along the trunk routes. All aspects of CATV power supply are examined in this chapter.

10.1 Unregulated Power Supply

The use of bridge rectifier arrangement is now a standard practice for obtaining dc voltages from ac mains. A transformer on the input side steps down the 220V mains voltage as necessary and also provides isolation from supply line thus preventing shock hazards. Large electrolytic capacitors are used to filter out 100 Hz ripple. Small value high voltage capacitors are connected across the diodes to bypass high frequency transient pulses which may otherwise damage rectifier diodes.

A typical power supply suitable for most CATV equipments is shown in Fig. 10.1. The input circuit contains a 'hash' filter to prevent entry of undesired signals originating from motors, neon lights or any other sparking contacts. The inductors L_1 and L_2, block any line transient and other disturbing frequencies while capacitor C_1 and C_2 act as short to such line disturbances. The filter unit also helps in blocking unwanted signals from local oscillators if operative in close by circuit units. Besides, a quick burn-out fuse is also included on the input side to prevent damage to power supply due to partial or total short caused by failure of any component or device in the equipment circuitry.

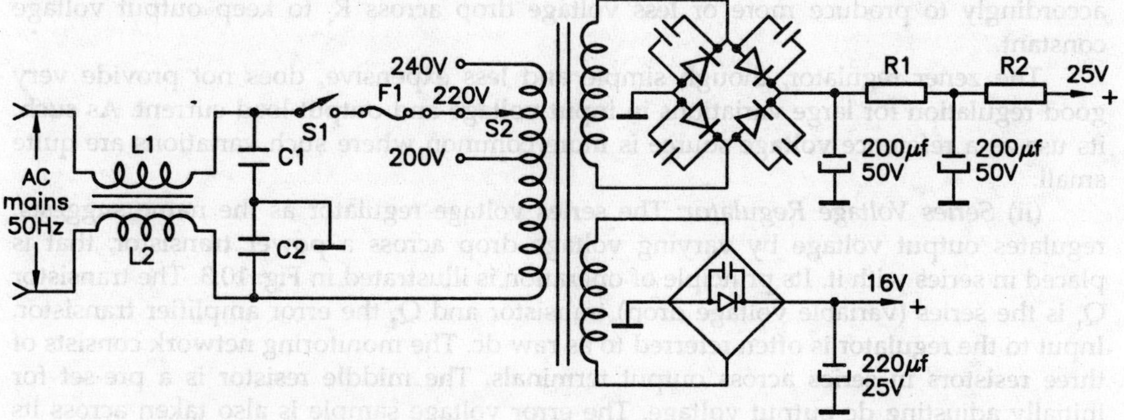

Fig. 10.1. Basic power supply circuit for obtaining low voltage dc supplies.

The circuit of Fig. 10.1 shows how two different low voltage dc supplies can be obtained. The number of turns on secondary winding determine dc output voltages. Another winding can be added to obtain a different dc voltage if necessary. In the figure, for the 16V dc supply the bridge rectifier unit has been shown by its accepted symbol often used in circuit diagrams.

10.2 Voltage Regulators

Voltage regulators are used for keeping rectified-filtered dc output voltage constant despite reasonable variations in load current and/or ac mains voltage fluctuations.

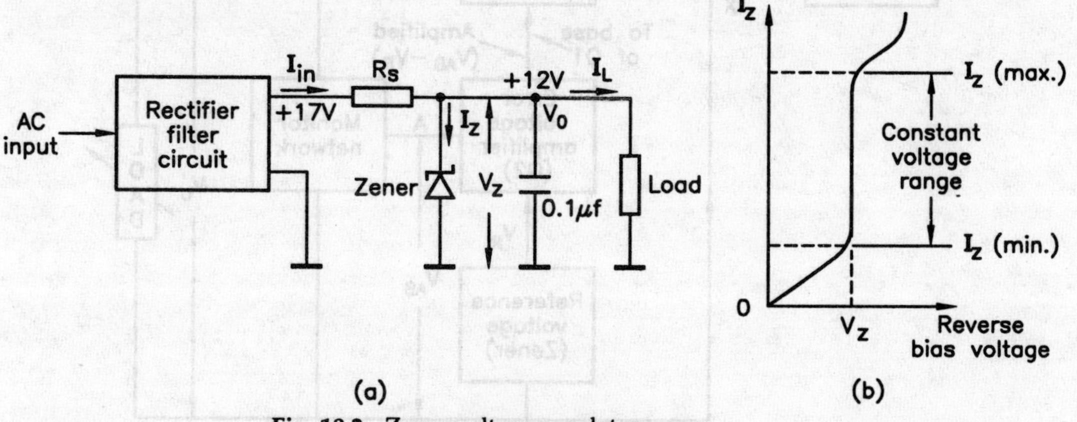

Fig. 10.2. Zener voltage regulator.
(a) circuit, (b) I-V characteristics of zener diode.

(i) *Zener Diode Regulator:* A zener diode when operated under reverse bias maintains a fixed voltage equal to its breakdown voltage across its terminals. The use of zener diode as a voltage regulator, is shown in Fig. 10.2 (a) and its characteristics in the operative region in Fig. 10.2 (b). The input voltage is kept higher than the required dc output voltage to enable the zener to operate in its constant voltage region. Any change in dc input voltage or variations in I_L, the load current, cause the zener current I_z to vary accordingly to produce more or less voltage drop across R_s to keep output voltage constant.

The zener regulator, though simple and less expensive, does not provide very good regulation for large variations in input voltage and output load current. As such, its use as a reference voltage source is more common where such variations are quite small.

(ii) *Series Voltage Regulator:* The series voltage regulator as the name suggests, regulates output voltage by varying voltage drop across a power transistor, that is placed in series with it. Its principle of operation is illustrated in Fig. 10.3. The transistor Q_1 is the series (variable voltage drop) transistor and Q_2 the error amplifier transistor. Input to the regulator is often referred to as raw dc. The monitoring network consists of three resistors in series across output terminals. The middle resistor is a pre-set for initially adjusting dc output voltage. The error voltage sample is also taken across its middle terminal and ground and this forms one of the inputs to error amplifier Q_2. The other input to error amplifier is the zener regulator output V_R which acts as reference voltage. The net input to Q_2 is thus the difference between the two inputs, where V_R is fixed. Any change in output voltage changes $V_{(A-B)}$ which in turn varies base-to-emitter voltage (V_{BE}) of error amplifier transistor Q_2. Even a small change in V_{BE} of Q_2 results in a large change in its collector current and this amounts to amplifying the error signal. Since collector current of Q_2 is the base current of regulator transistor Q_1, any change in $V_{(A-B)}$ changes its collector current accordingly. This in turn varies voltage between its

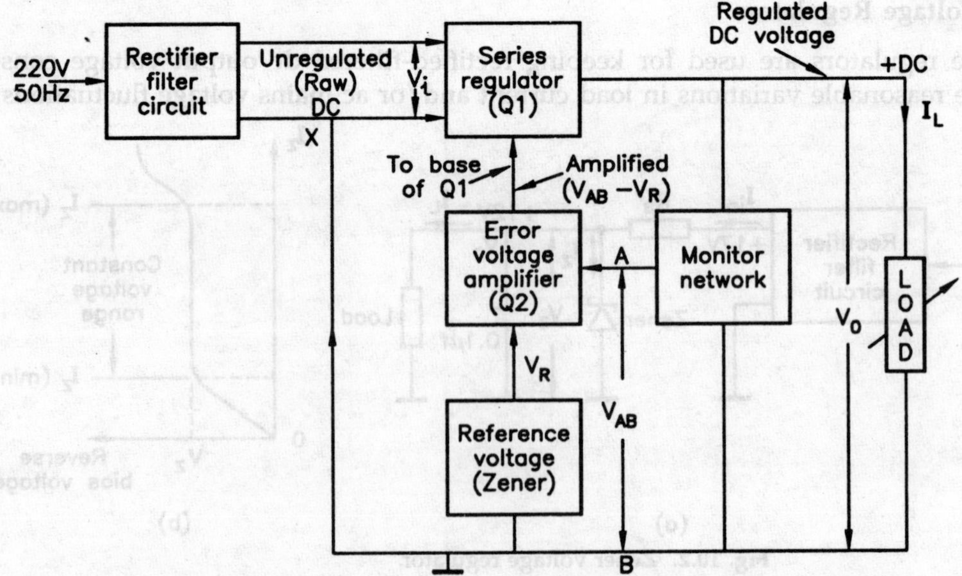

Fig. 10.3. Block schematic of a series voltage regulator.

collector and emitter and hence, drops more or less voltage to keep the output voltage constant. Thus the regulator transistor acts as a dynamic variable resistance, the magnitude of which changes in response to error signal to keep output voltage constant despite fluctuations in mains voltage or changes in load current. In some power supply designs, a protection circuit is also included as part of the regulator to almost shut the regulator when a serious fault amounting to a near short occurs on the load side.

10.3 Switched Mode Power Supply (SMPS)

The switched mode power supply is essentially a dc to dc converter where energy transfer is controlled to obtain output voltage regulation. The block diagram of such a system suitable for TV and satellite receivers is shown in Fig. 10.4. The single ended switching converter chops unregulated dc (V_i) at a very fast rate in the form of current pulses that flow through primary winding of the SMPS transformer T_1. The transfer of energy is by electromagnetic induction from primary to secondary side and thus there is complete isolation between regulated and unregulated supplies. The error voltage induced in the sensing winding is used to control ON-OFF periods of the converter to regulate transfer of energy from V_i side to load circuits and this results in regulation of output voltages. The voltage overload circuit is designed to block the converter in case output voltage tends to overshoot for any reason. Unlike the pass (series) transistor in the series regulator, the device (transistor) in the converter is switched 'ON' for periods depending on load side demands and this makes the SMPS highly efficient. In addition,

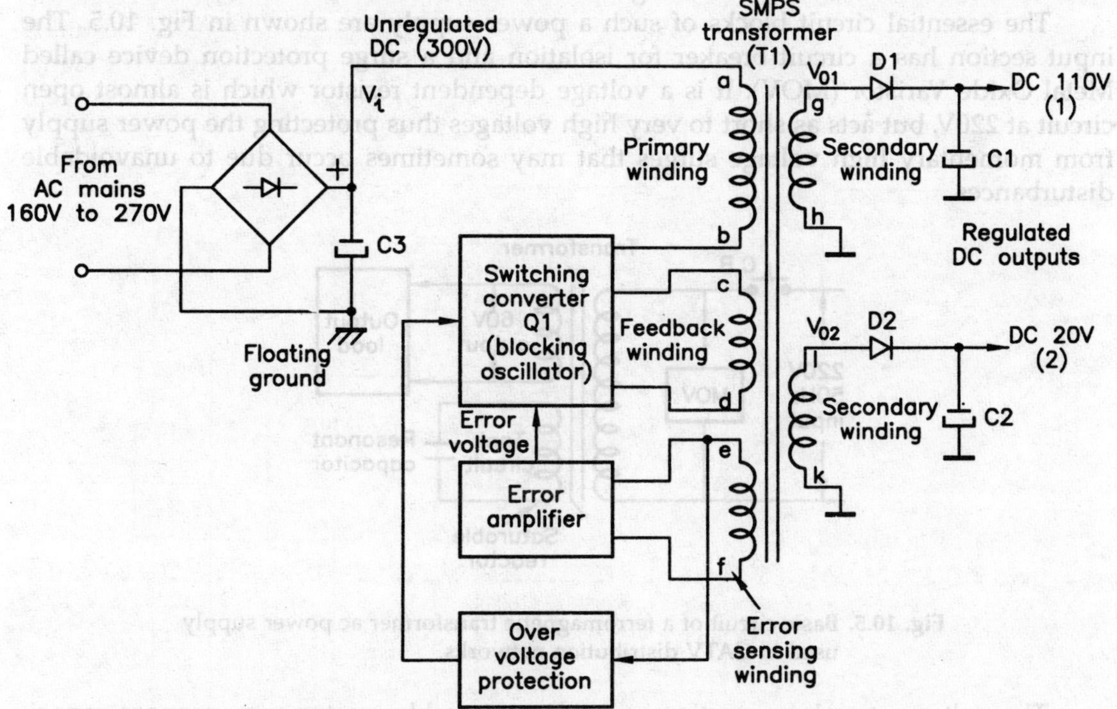

Fig. 10.4. Block diagram of an SMPS power supply.

the SMPS transformer is of much smaller size because the frequency of current pulses is chosen to be quite high.

Merits of SMPS: The main merits of switched mode power supplies include (i) small physical size because of high chopping rate (16 KHz to 50 KHz), (ii) greater efficiency, (iii) isolation from mains supply, and (iv) different stabilized dc output voltages. The output voltage labelled in the figure are suitable for TV receivers. But for satellite receivers, where picture tube and associated drive circuitry is not there, instead of 110V, a low dc output can be obtained by reducing the number of turns on corresponding secondary winding.

The rapid switching of chopping current generates strong interfering signals and therefore, rigorous shielding of SMPS is absolutely necessary. Similarly, to prevent entry of interfering HF pulses to tuner circuits, an elaborate mains filter section becomes necessary.

10.4 Cable Powering

Cable powering means supply of ac input to trunk, and distribution amplifiers of the cable distribution network. A CATV system serving nearly 1,00,000 customers will have about 1500 amplifiers installed at different locations. This will require nearly 500 ac power outlets (supplies) for feeding these amplifiers. For such large CATV installations, specially designed ferromagnetic power supplies are provided which step-down ac mains voltage to 60 volts and regulate it before feeding to various amplifiers through the same cable that carries channel signals.

The essential circuit blocks of such a power supply are shown in Fig. 10.5. The input section has a circuit breaker for isolation and a surge protection device called Metal Oxide Varistor (MOV). It is a voltage dependent resistor which is almost open circuit at 220V, but acts as short to very high voltages thus protecting the power supply from momentary high voltage surges that may sometimes occur due to unavoidable disturbances.

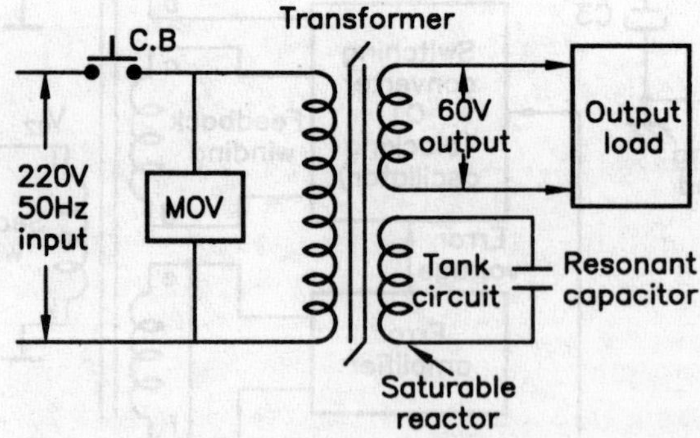

Fig. 10.5. Basic circuit of a ferromagnetic transformer ac power supply used in CATV distribution networks.

The voltage step-down section contains a saturable reactor-cum-resonant capacitor circuit along the secondary winding. On application of 220V ac to the primary winding of ferromagnetic transformer, resonance sets up in the tank circuit consisting of saturable reactor and capacitor. This saturates the core which in turn steps down the voltage to its design value of 60 volts. The resonant circuit acts as a flywheel and thus resists any change in input voltage. Similarly, the shunts in the transformer core laminations regulate load on the secondary side by limiting magnetic flux path in the air gaps between the two sets of laminations. If load on the secondary winding increases, the resonance of magnetic circulating paths decreases and allows it to receive more power from the primary winding *i.e.* from 220V ac mains. Specifications of such power supplies include (i) max input current = 4 amps, (ii) output voltage 60V ± 5% ac, (iii) output current 14 amps, and output power equal to 840 VA.

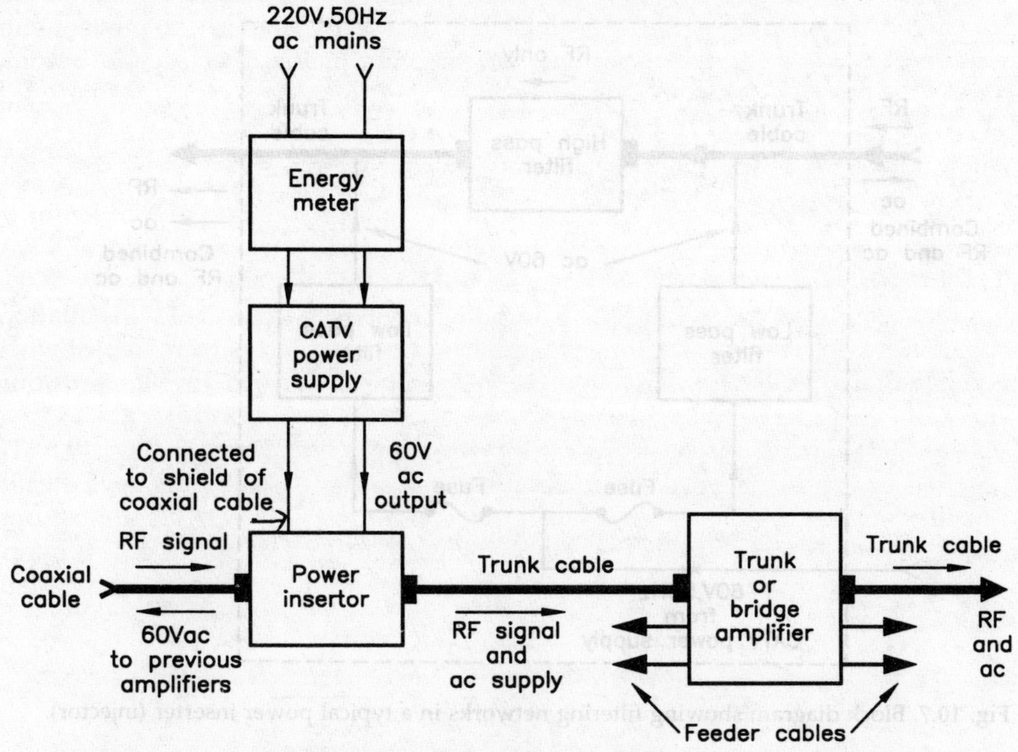

Fig. 10.6. Powering of trunk and bridge amplifiers through a power inserter.

Power Insertion Methods: The two methods of inserting or injecting ac power into the CATV trunk cables for onward transmission to trunk and line extender amplifiers are: (i) through a power inserter, and (ii) direct into the ac input point of the amplifier.

Amplifier Powering with Power Inserter: The circuit arrangement for feeding ac power to amplifiers through an Inserter or Injector is shown in Fig. 10.6. The power supply is connected to the supply lines through an energy meter for recording power consumption as required by the supply undertaking. The 60V ac output is connected to the trunk cable in the inserter. The neutral wire is earthed to the shield of cable. As shown in the figure, ac supply travels in both forward and backward directions on the same cable that carries RF signals. This enables feeding to several amplifiers that are located before and after the injector. The I^2R drop on the cable reduces the 60V voltage after 10 to 15 km to a value that is below the minimum voltage necessary for the ac to dc power supply in amplifiers. Another ac power supply is then provided to serve another length of about 15 km, or so.

The power inserter contains three filtering networks that permit the combining of RF signals and 60V ac while isolating the two RF/ac ports. The filter configuration is shown in Fig. 10.7. The low-pass filter network passes 60V ac and prevents RF signals from being transmitted through the power supply. The high-pass filter network passes RF signals between the two RF/ac ports but prevents the 60V ac from passing between the same two ports.

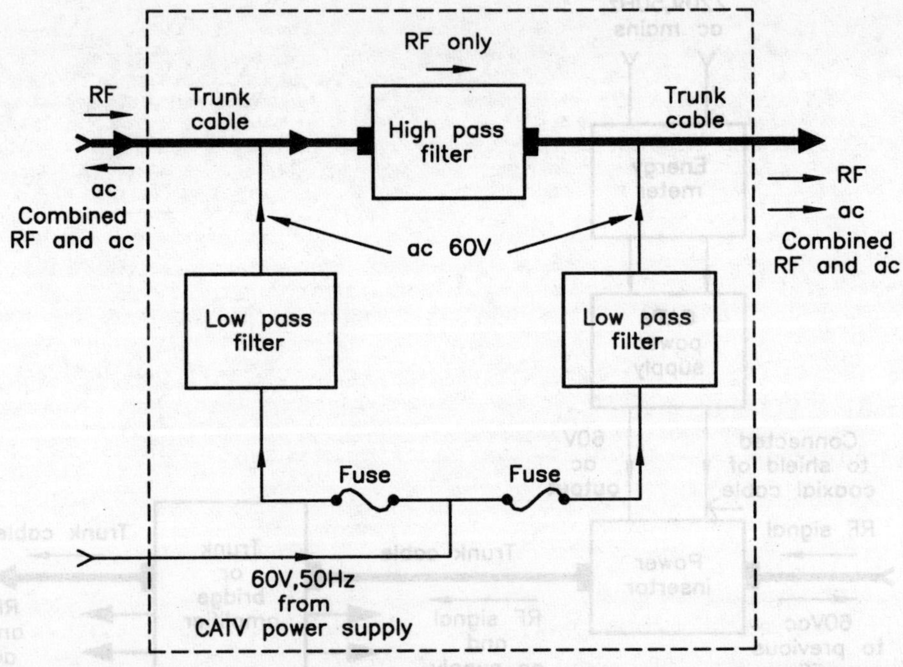

Fig. 10.7. Block diagram showing filtering networks in a typical power inserter (injector).

Direct Trunk Amplifier Powering: Direct trunk amplifier powering allows amplifiers to receive 60V ac supply directly from the ferromagnetic power supply's coaxial output cable without the use of a power inserter. The necessary connections are shown in Fig. 10.8. The RF signal in the cable leaving the combining network at the CATV station or main distribution amplifier is strong enough and does not need any amplification for several km of cable length. When the signal becomes weak and needs boosting, a trunk amplifier is provided and it is here where the ac power supply output is also injected. The cable length after the amplifier then carries both RF signal and ac voltage for feeding successive amplifiers on the line. This range will end when ac voltage falls to about 30V due to I^2R losses. Another ac power supply would then be injected as shown in Fig. 10.8. Note that the last trunk amplifier output will carry only RF signal over the length till another amplifier becomes necessary.

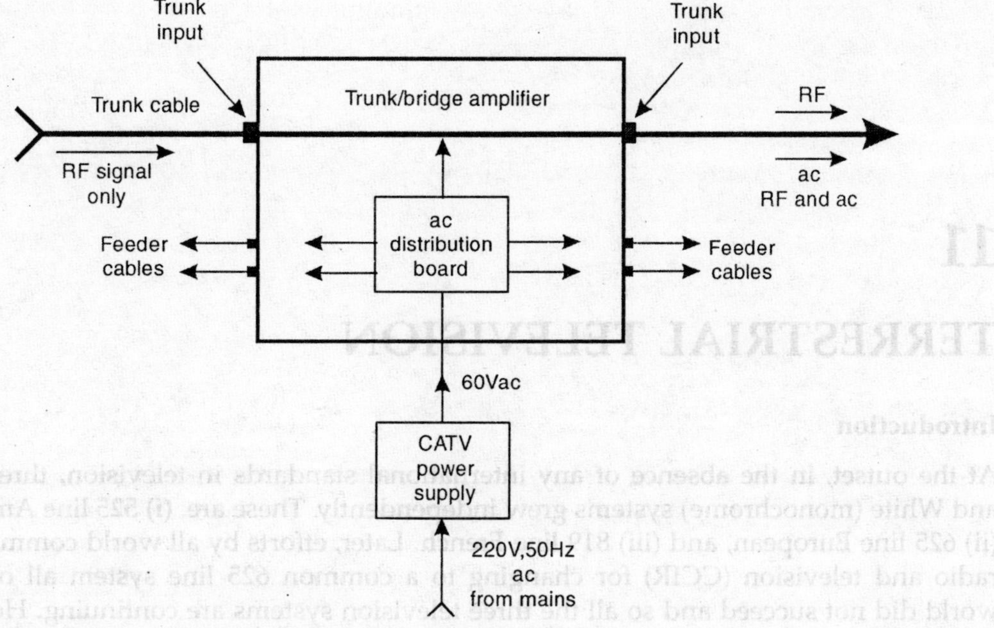

Fig. 10.8. Direct trunk amplifier powering.

10.5 Alternative AC Powering

In smaller CATV systems where cable length does not exceed 10 to 15 km, elaborate ac powering schemes explained in previous section are not considered necessary. Instead, trunk and line amplifiers have their own power supply units of the type shown in Fig. 10.1, to which a dc voltage regulator is added. The ac mains to it is taken from a suitable outlet close to the location of amplifier. After a few km of cable length when RF signal needs amplification, another similar amplifier is provided. This turns out to be economical for medium and smaller CATV networks. In some small CATV systems ac supply is carried to the amplifiers from the control room through a separate insulated wire. The cable sheath serves as the second wire.

In some CATV systems, chargeable storage batteries are provided as a pack along the amplifier to serve as dc supply source. These are charged through battery charges from ac mains supply. This ensures uninterrupted TV channel signal supply to customers even during periods of mains supply failure.

11

TERRESTRIAL TELEVISION

Introduction

At the outset, in the absence of any international standards in television, three Black and White (monochrome) systems grew independently. These are: (i) 525 line American, (ii) 625 line European, and (iii) 819 line French. Later, efforts by all world committee on radio and television (CCIR) for changing to a common 625 line system all over the world did not succeed and so all the three television systems are continuing. However, in due course of time, the French changed over to 625 lines but retained original modulation techniques. This system with minor modifications is now used in all Russian and many East European countries.

After intensive research for many years each B&W system developed into its own colour TV adaptation. The three colour TV system that thus emerged are: (i) NTSC, (ii) PAL, and (iii) SECAM compatible to the American, European and French monochrome systems respectively. This naturally prevents direct-exchange of programmes between countries using different television standard. However, equipments are now available for converting recorded material from one system to another.

India adopted PAL-B system because it is compatible to the 625 line CCIR B&W system chosen by it earlier. In these systems, both monochrome and colour signals are amplitude modulated (AM) and sound signal frequency modulation (FM) before transmission. While the carrier frequencies of picture and sound signals are different but these are chosen to be close enough to contain their modulation products within the allotted channel bandwidth. The combined modulation products of monochrome, colour and sound signals are transmitted through a common antenna. As the word 'terrestrial' suggests, radiation of modulated RF TV signals is done by space wave propagation from transmitting antenna to distant places where these are picked up by receiving antenna.

As in radio-communication, each TV station is allotted a different 7 MHz channel width to enable selection of desired station at the receiving end. The television receiver has tuned circuits in its input section called 'tuner'. It selects and amplifies selected channel signals out of the many picked up by the receiving antenna. The RF band thus selected is converted (frequency translated) to a common IF band for convenience of obtaining large amplification and application of automatic gain control (AGC).

In B&W receivers, the amplified IF signal is detected to obtain video (picture) and audio (sound) signals. The video signal after large amplification drives the single gun picture tube to reconstruct a black and white picture on its screen. Similarly, the audio signal is amplified and fed to the loudspeaker to produce sound output associated with the television scene.

In colour receivers, the detected signal contains luminance (brightness) signal, chrominance (colour) signal and intercarrier sound signal. The colour component of detector output is passed on to a 'decoder' to obtain three colour-difference signals. These are separately matrixed (combined) with the luminance signal to obtain three primary colour (red, green and blue) voltages. These on amplification drive respective guns of the tri-colour picture tube. The varying electron currents of these three guns combine to create a colour picture on the screen. The sound signal is processed in the same way as in B&W receivers.

PART—A
TELEVISION TRANSMISSION

11.1 Television Pictures and Cameras

The visual information of a B&W picture may be thought of as an assemblage of a very large number of bright and dark elements lying very close to each other. These picture elements are called 'pixels' which when viewed together convey information of the picture. Since there is motion in every scene, brightness from the pixels continuously changes with time. Because it is not possible to pick up and transmit brightness information from all the picture elements simultaneously, the scene is instead scanned element by element and line by line at a fast rate and this serves the purpose.

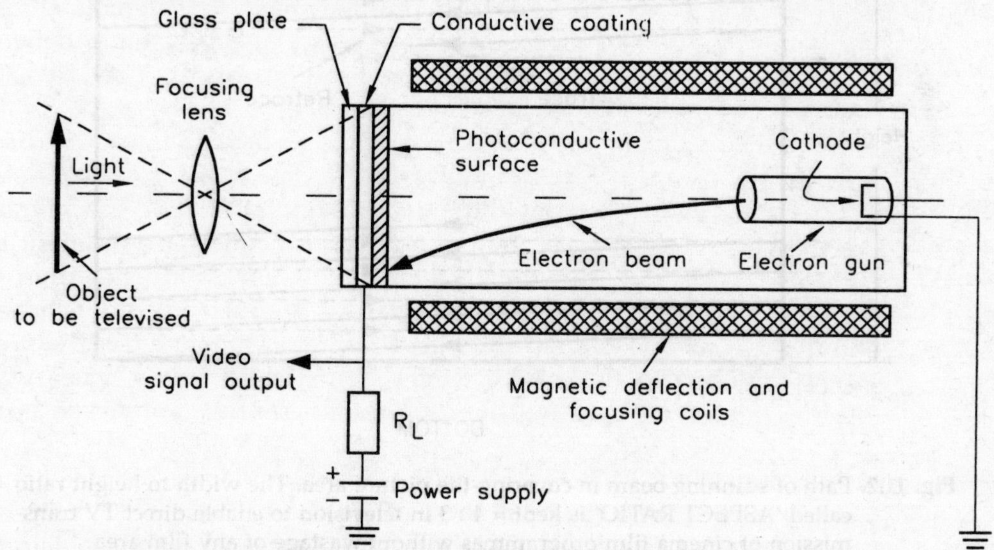

Fig. 11.1. Simplified cross-sectional view of a B&W TV camera employing a vidicon camera tube.

B&W TV Camera: A TV camera is used to convert optical information of the picture into corresponding proportionate electrical signal, the amplitude of which varies in accordance with variations in brightness. Fig. 11.1 shows very elementary details of a B&W camera employing a cathode-ray type of tube called 'vidicon'. An optical image of the scene to be transmitted is focussed by a lens assembly on the rectangular glass face-plate of the camera tube. The inner side of the glass face-plate has a transparent conductive coating on which is laid a very thin layer of photo-conductive material. The photo-layer has very high resistance when no light falls on it, but decreases as the intensity of light falling on it increases. Thus, depending on light intensity variations in the focussed optical image, the conductivity of each element of photolayer changes accordingly. An electron beam formed by the gun assembly of the picture tube (vidicon) is used to pick-up picture information now available on the target plate in terms of varying resistance at each point.

The camera has a built-in scanning circuitry based on magnetic deflection, which enables its beam to travel both in horizontal and vertical directions simultaneously, thus picking up information from the entire target plate as illustrated in Fig. 11.2. As shown, scanning is done in the same way as one reads a written page to cover all the words in one line and all the lines on the page.

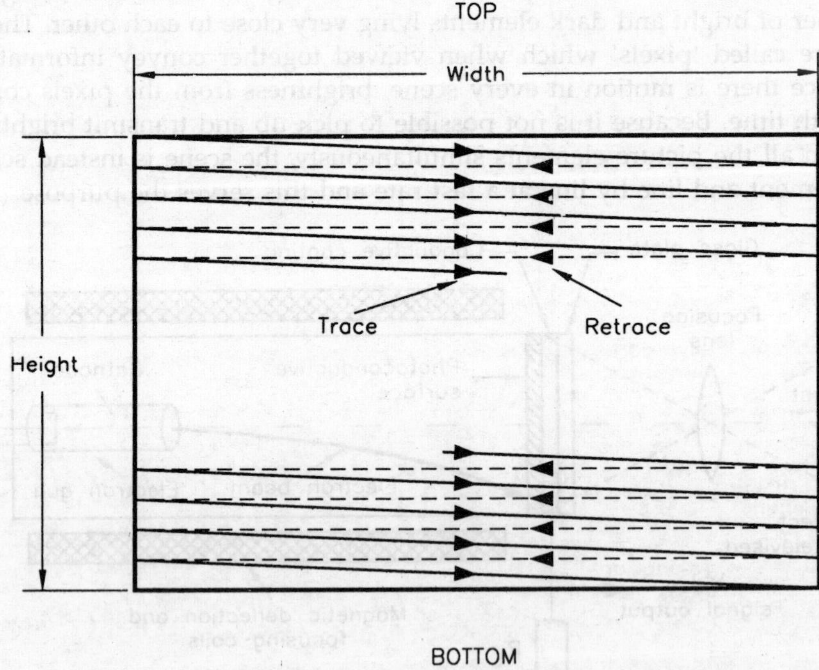

Fig. 11.2. Path of scanning beam in covering the picture area. The width to height ratio called 'ASPECT RATIO' is kept = 4 : 3 in television to enable direct TV transmission of cinema film programmes without wastage of any film area.

As the beam moves from element to element it encounters a different resistance across the face plate, depending on the resistance of photo-conductive coating. The result is a flow of current which varies in magnitude as the elements are scanned. The current passes through [see Fig. 11.1] a load resistance R_L connected to the conductive coating on one side and a dc supply source on the other. Depending on the magnitude of current, a varying voltage develops across R_L and this corresponds to optical information of the picture. The electrical information thus obtained from the TV camera is generally referred as 'Video' signal (video is Latin for see).

The scanning is done at such a fast rate that any portion of the scene content does not have time to move perceptibly in the time required for one complete scan of the image. Therefore, the resultant electrical signal collected across R_L contains true information existing in the picture during the time of scan. Since scanning process is repeated a large number of times per second, the eye due to its storage capability perceives it as a simultaneous pick-up and reception of picture details. Thus, an illusion of continuity is created as if the eye is directly viewing the scene being televised.

Colour Pictures: A colour picture is nothing but a black and white picture to which colours have been added. It is possible to create any colour including white by additive mixing of red, green and blue colour lights in suitable proportions. For example, yellow colour can be obtained by mixing red and green colour lights in intensity ratio of 30:59. Similarly light reflected from any colour picture element can be synthesised (broken-up) into red, green and blue colour light constituents. This forms the basis of colour television where Red (R), Green (G) and Blue (B) colours are called primary colours and all other colours are obtained by mixing them in definite proportions.

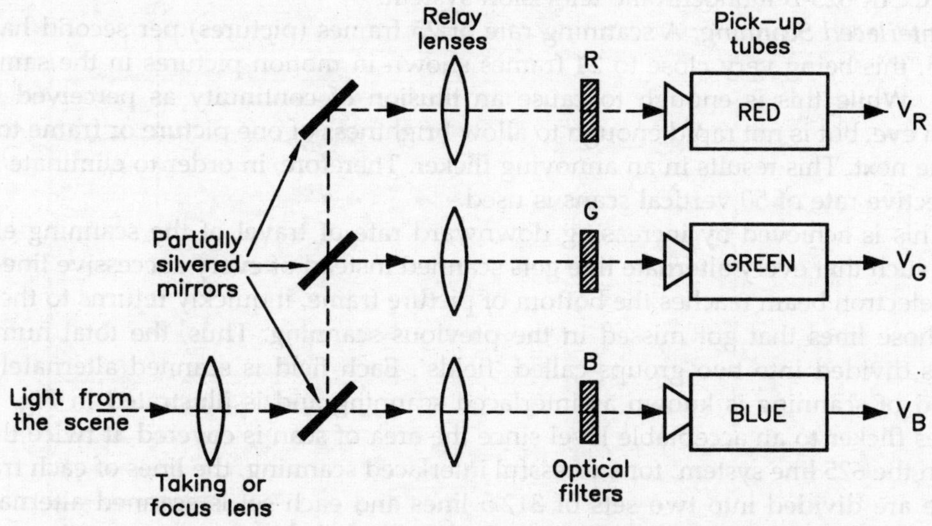

Fig. 11.3. Simplified block diagram of a colour camera set-up. Deflection coils around the pick-up (vidicon) tubes are not shown.

Colour Camera: A colour camera, the elements of which are shown in Fig. 11.3 is used to develop signal voltages proportional to the intensity of each primary colour light. It contains three camera tubes (vidicons) where each tube receives light of only one primary colour. Light from the scene falls on the focus lens and through that on special mirrors. Colour filters that receive reflected light via relay lenses split it into R, G, and B colour lights. Thus, each vidicon receives a single colour light and develops a voltage proportional to the intensity of one of the three primary colours. If any primary colour is not present in any part of the picture, corresponding vidicon does not develop any output when that picture area is scanned. The electron beams of all the three camera tubes are kept in step (synchronism) by deflecting them horizontally and vertically from common driving sources.

Luminance (Y) Signal: Any colour light has a certain intensity of brightness. Therefore, light reflected from any colour element of a picture also carries information about its brightness called luminance. A signal voltage 'Y' proportional to luminance at various parts of the picture is obtained by adding definite proportions of V_R, V_G and V_B in the ratio of 30:59:11 to yield $V_Y = 0.30\ V_R + 0.59\ V_G + 0.11\ V_B$. The luminance voltage V_Y thus obtained is the same as would be developed by a B&W camera when made to scan the same colour scene.

11.2 Interlaced Scanning and Picture Resolution

It is easy to imagine that greater the number of lines into which picture is divided in the vertical plane for scanning, better will be the quality of reproduced picture. However, keeping in view limitations of human eye to perceive any improvement beyond about 600 lines of scanning in the vertical plane, the total number of lines inclusive of those lost during vertical retrace (beam returning from bottom to top) were chosen to be 625 in the CCIR 625-B monochrome television system.

Interlaced Scanning: A scanning rate of 25 frames (pictures) per second has been chosen, this being very close to 24 frames shown in motion pictures in the same time period. While this is enough to cause an illusion of continuity as perceived by the human eye, but is not rapid enough to allow brightness of one picture or frame to blend into the next. This results in an annoying flicker. Therefore, in order to eliminate flicker, an effective rate of 50 vertical scans is used.

This is achieved by increasing downward rate of travel of the scanning electron beam, such that every alternate line gets scanned instead of every successive line. Then, when electron beam reaches the bottom of picture frame, it quickly returns to the top to scan those lines that got missed in the previous scanning. Thus, the total number of lines is divided into two groups called 'fields'. Each field is scanned alternately. This method of scanning is known as interlaced scanning and is illustrated in Fig. 11.4. It reduces flicker to an acceptable level since the area of scan is covered at twice the rate.

In the 625 line system, for successful interlaced scanning, the lines of each frame or picture are divided into two sets of 312.5 lines and each set is scanned alternately to cover the entire picture area. To achieve this, the horizontal sweep oscillator, which feeds the horizontal deflection coils, is made to work at a frequency of 15625 Hz per second ($312.5 \times 50 = 15625$) to scan the same number of lines per frame ($15625/25 = 625$ lines) and the vertical sweep oscillator that feeds vertical deflection coils of the camera tube is run at a frequency of 50 Hz instead of 25 Hz. Note that since the beam is now deflected from top to bottom in half the time and horizontal oscillator is still operating at 15625 Hz, only half the total lines *i.e.* 312.5 ($625/2 = 312.5$) get scanned during each vertical sweep. Since the first field ends in a half-line and second field commences at middle of the line on top of the camera target plate or screen [see Fig. 11.4] the beam is able to scan the remaining 312.5 alternate lines during its downward journey.

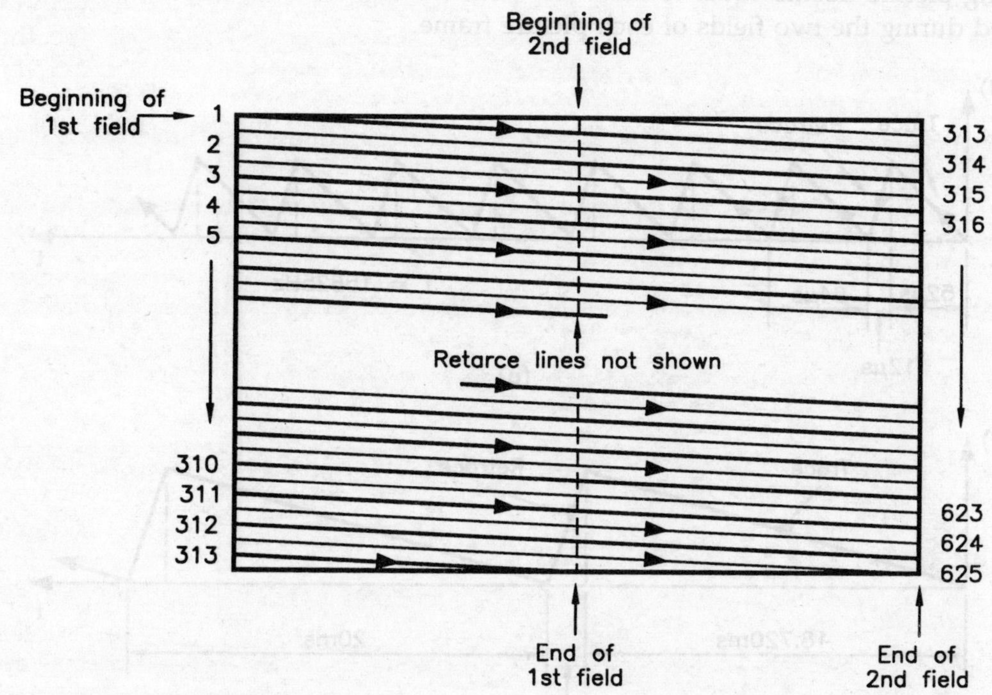

Fig. 11.4. Principle of interlaced scanning. Note that the vertical retrace time has been assumed to be zero. The retrace lines are also omitted for clarity of interlaced lines.

Scanning Periods: The waveshapes of both horizontal and vertical sweep* currents are shown in Fig. 11.5. The retrace times involved (both horizontal and vertical) are due to physical limitations of practical scanning systems (oscillators) and are not used for transmitting or receiving any video signal. The nominal duration of horizontal line is 64 µs ($10^6/15625 = 64$ micro-seconds), out of which the active line period is 52 µs and the remaining 12 µs, the line blanking period. The beam returns during the short blanking period to extreme left side of the picture frame to start tracing the next line.

Similarly, with field oscillator frequency set at 50 Hz, the nominal duration of vertical trace [see Fig. 11.5 (b)] is 20 ms (1/50 = 20 milli-second). Out of this period of 20 ms, 18.720 milli-seconds are spent in bringing the camera tube electron beam from top to bottom and the remaining 1.280 ms is taken by the beam to return back to top to commence next vertical trace cycle. Since both horizontal and vertical sweep oscillators operate continuously, 20 horizontal lines $\left(\frac{1280\ \mu S}{64\ \mu S} = 20 \text{ lines}\right)$ get traced during each vertical retrace interval. Thus, 40 scanning lines are lost per frame as blanked lines during retrace intervals of the two fields. This leaves the actual number of lines for

* Magnetic deflection is proportional to current flowing into the deflection coils.

scanning picture details equal to 625 – 40 = 585 instead of 625 lines that are actually scanned during the two fields of each picture frame.

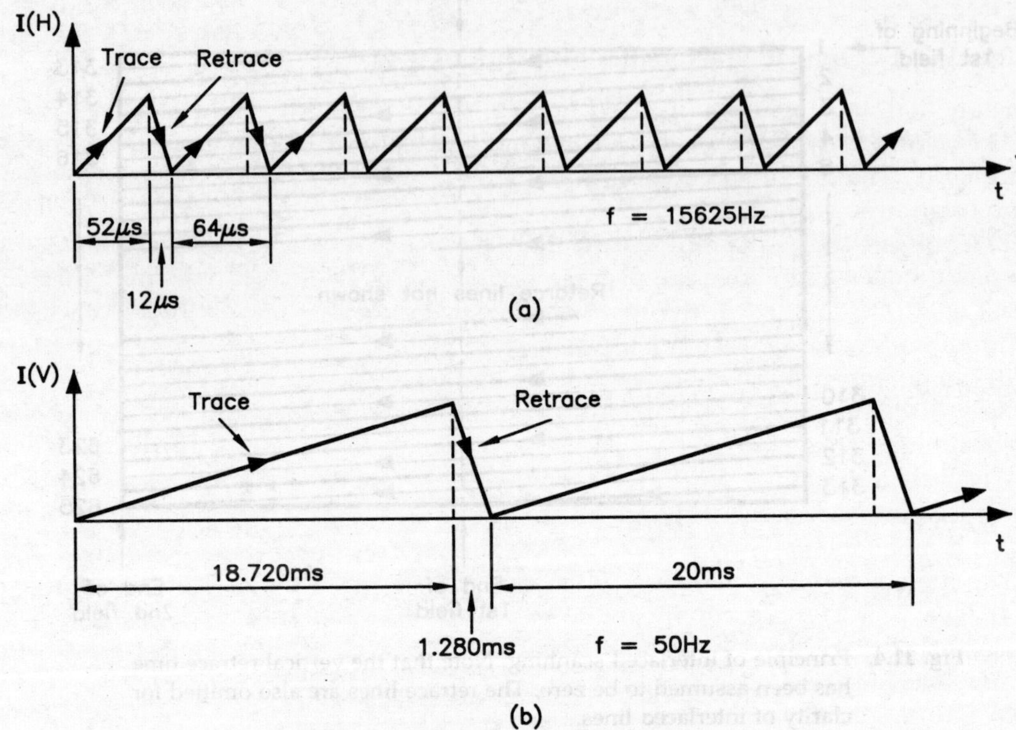

Fig. 11.5. (a) Horizontal deflection current,
(b) Vertical deflection current.

Picture Resolution: Vertical resolution is a function of scanning lines into which picture is divided in the vertical plane. It is found that 70 per cent of the total lines get separately scanned and the remaining 30 per cent get merged with adjoining lines due to beam spot falling on two consecutive (one after the other) lines. Thus, vertical resolution comes to $585 \times 0.7 \approx 400$ lines.

Similarly, capability of the system to resolve maximum number of picture elements along scanning lines determines horizontal resolution. It is realistic to aim at equal vertical and horizontal resolution. For this, the beam must be able to resolve adjoining pixels (picture elements) distinctively. The maximum demand would come when alternate pixels are black and white as squares in a chess board. While scanning such pixels the time duration 't_h' of each cycle of voltage that develops comes to nearly 0.2 μs for each set of black and white picture elements. This time is so small because the time taken to scan the entire line is just 52 μs. The frequency f_h of the generated signal is

then $\frac{1}{t_h}$ *i.e.* $\frac{10^6}{0.2}$ Hz or 5 MHz.

However, the other extreme would be when many consecutive lines are either white or black which would mean the frequency of generated voltage cycle will be just a few Hz. To be more specific for lines which are alternately white and black, the frequency of signal will be 25 Hz. Thus a bandwidth from almost 0 Hz (dc) to 5 MHz is provided for the video signal in the 625 line television system.

11.3 Tonal and Colour Characteristics of Pictures

The factors that need attention in B&W and colour pictures are:

(1) *Brightness:* It is the overall or average intensity of illumination in the reproduced picture. The brightness control in the receiver can be varied to obtain optimum average brightness of the scene.

(2) *Contrast:* This is the difference in light intensity between black and white parts of the picture over and above the average brightness level. Contrast can be varied by 'contrast control' provided in the receiver. Too much contrast makes the picture hard while too little gives impression of a washed-out picture.

(3) *Luminance:* This is the amount of light intensity as perceived by the eye regardless of colour in a colour picture. Like different shades from black to white, through different shades of grey in B&W pictures, colour pictures also have different shades of brightness in the sense that though equally illuminated appear more or less bright as determined by the response of human eye to various colours. As an illustration of this phenomenon, on a B&W, television receiver screen, dark red colour will appear to the eye as black, yellow as white and light blue colour as a grey shade.

(4) *Hue:* This is the predominant colour of the received light. Thus the colour of any object is distinguished by its hue, also called tint. For example, green leaves have green hue and red tomatoes have red hue.

(5) *Saturation:* This determines the pureness of colour light. A fully saturated colour has no white mixed with it. Level of saturation indicates amounts of other colours present in the received light or how much white light is present along with the colour light.

The hue and saturation of a colour when considered together is called chrominance or chroma of the colour light. It may be noted that chroma of a colour does not contain any information, about brightness of that colour.

11.4 Composite B&W and Colour Video Signals

In B&W television, composite video signal consists of camera signal corresponding to light variations in the picture, blanking pulses to make retrace invisible and synchronizing (sync) pulses to keep scanning at the receiver in step with scanning in the camera at the transmitting end. The location of blanking and sync pulses along with video signal is shown in Fig. 11.6 and nature of horizontal and vertical sync pulses in Fig. 11.7. A horizontal sync pulse [see Fig. 11.6] is needed at the end of each active line period whereas a vertical sync pulse is required after each field of scanning.

As seen in Fig. 11.7, the nature of sync pulses shows that horizontal sync pulses are narrow while vertical sync pulses are quite broad. This enables their easy separation at the receiver. A high-pass filter is used to separate horizontal sync pulses and a low-pass filter to separate vertical sync pulses. Notice, that the broad vertical sync pulse that is inserted after each field of scanning is notched (split) into five pulses to ensure its distortion free transmission.

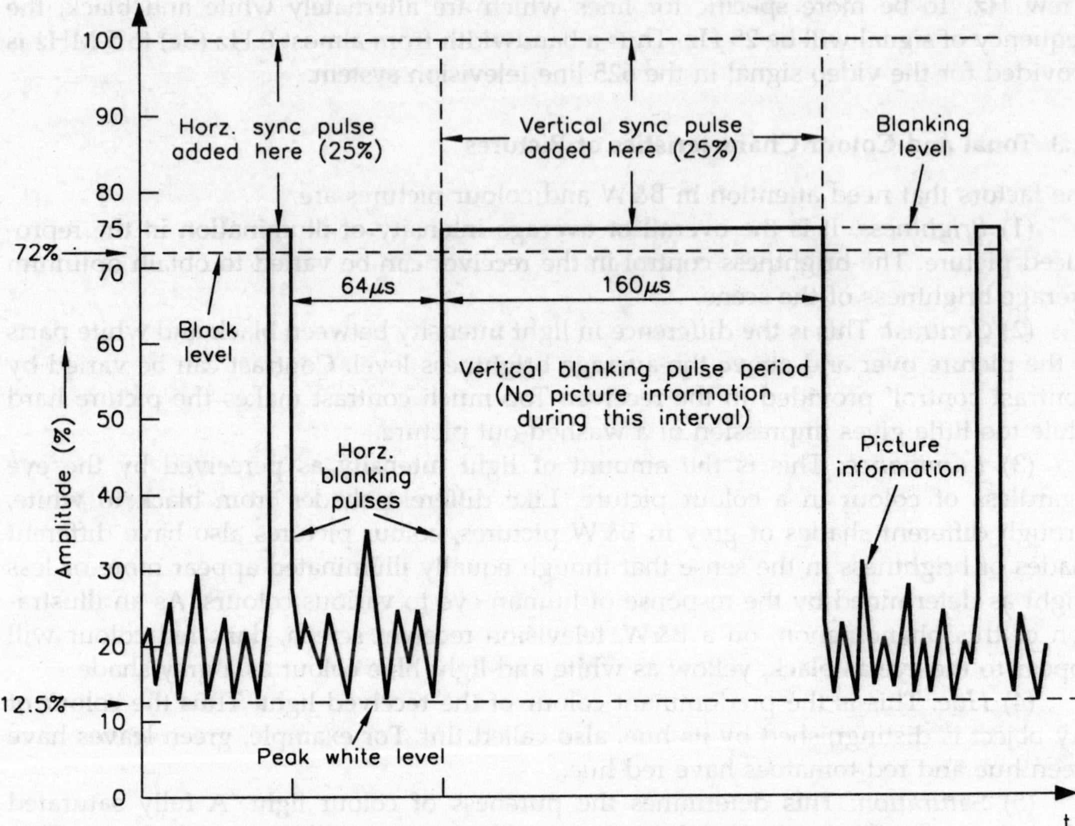

Fig. 11.6. Horizontal and vertical sync pulses are added above the blanking level of picture signal and occupy 25% of the total composite video signal amplitude.

Colour Signal Generation and Encoding: As explained in section 11.1, all colours can be produced by additive mixing of R, G and B colour lights in suitable proportions. Also, by additive mixing of primary colours (red, green and blue) the following complementary colours can be produced:

Red + Green = Yellow

Red + Blue = Magenta (purplish red shade)

Blue + Green = Cyan (greenish blue shade)

Luminance (Y) Signal: As shown in Fig. 11.8, a colour camera develops three voltages, each proportional to red, green and blue colour contents of the picture being scanned. From these, the luminance (brightness) signal 'Y' is developed as under:

$$V_Y = 0.30\ V_R + 0.59\ V_G + 0.11\ V_B$$

Dropping V as is usually done:

$$Y = 0.30\ R + 0.59 + 0.11\ B$$

Note that the percent proportion of each colour voltage is related to brightness sensitivity of the eye to those colours and their additive effect on the eye is that of brightness as obtained in a B&W camera. Also notice, that when R = G = B = say 1 volt, Y is also equal to one volt *i.e.* Y = 0.30 Y + 0.59 Y + 0.11 Y = 1Y.

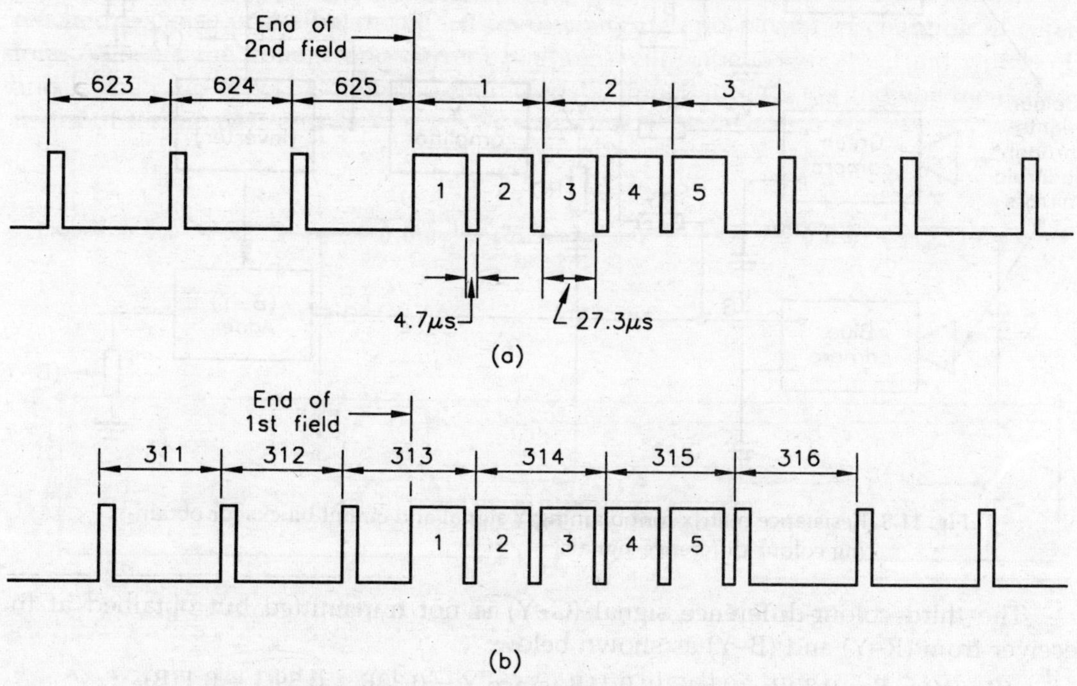

Fig. 11.7. Nature of horizontal and vertical sync pulses. Note the location of vertical sync pulses after 1st and 2nd field of scanning.

Y Matrix: The matrix (resistive network) used to develop 'Y' is shown in Fig. 11.8. The values of R_1, R_2 and R_3 are so chosen that for bright white light, the three voltages add up to yield Y equal to one or multiple of one unit as shown there. Similarly when R = G = B = 0V, which will mean darkness (black) Y = 0V. Naturally for intermediate values of brightness, from bright white to black, Y varies from its maximum value to zero.

Compatibility: The colour TV system is designed to be compatible with its monochrome system. This implies that a colour TV signal can produce a black and white picture on a B&W receiver and signals of B&W systems can produce a black and white picture on a colour receiver screen. In practice this means the use of same scanning rates (frequencies) and other standards.

Compatibility however, puts a constraint on transmission of colour signals. The three camera colour voltages developed by the camera cannot be transmitted separately because of channel width limitations. This problem is solved by generating two colour-difference signals (R–Y) and (B–Y) as shown in Fig. 11.8. These are later frequency translated and combined with Y signal for transmission.

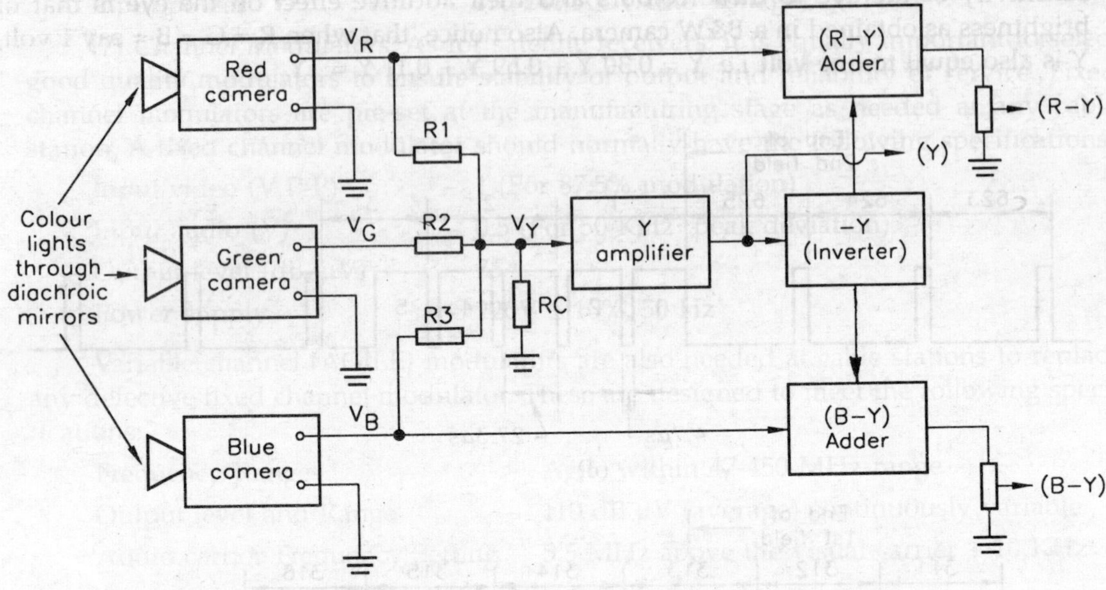

Fig. 11.8. Resistance matrix for obtaining Y signal and circuit blocks for obtaining colour-difference signals.

The third colour-difference signal (G–Y) is not transmitted but obtained at the receiver from (R–Y) and (B–Y) as shown below:

$$(R - Y) = R - 0.30R - 0.59G - 0.11B \text{ (since } Y = 0.30R + 0.59G + 0.11B)$$

$$= 0.70R - 0.59G - 0.11B$$

Similarly $(B\text{–}Y) = -0.30R - 0.59G + 0.89B$

Since $Y = 0.30R + 0.59G + 0.11B$ and also $Y = 0.30Y + 0.59Y + 0.11B$,

we can write $0.30\,(R\text{–}Y) + 0.59\,(G\text{–}Y) + 0.11\,(B\text{–}Y) = 0$

$$\text{or } (G\text{–}Y) = \frac{-0.30}{0.59}(R-Y) - \frac{0.11}{0.59}(B-Y)$$

$$= -0.51\,(R - Y) - 0.186\,(B - Y).$$

Thus, the third colour-difference signal can be obtained by setting a matrix that satisfies the above expression. On obtaining the three colour-difference signals, these are suitably combined with Y to obtain R, G and B signal voltages for feeding the three guns of colour picture tube to reconstruct the colour picture.

Encoding: The process of combining the two colour-difference signals chosen for transmission by first translating them to a higher frequency domain is known as encoding. The word 'encoder' has been taken from computers where it means translating or preparing a programme.

The necessary encoding circuit is shown in Fig. 11.9. The carrier frequency is called sub-carrier because the final carrier frequency is the picture carrier allotted to the station. As shown in the figure, the sub-carrier oscillator output (f_{sc} 0°) is given a phase shift of 90° to create another sub-carrier. Thus, the two carrier frequencies have the same numerical value but with a phase difference of 90°.

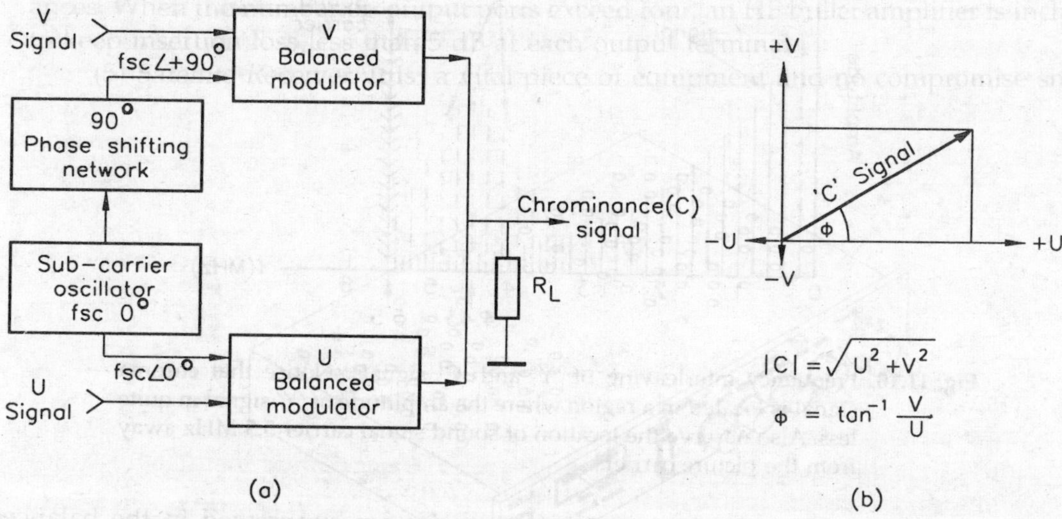

Fig. 11.9. Encoding of colour-difference signals (a) block diagram of quadrature modulator, (b) vectorial addition of modulation products to obtain 'C' signal.

Weighting Factors: To prevent over modulation and hence, distortion on certain hues (colours), the colour-difference signals are reduced in magnitude before applying them to respective modulators. The scaled down colour-difference signals are termed U and V in the PAL colour system where:

$$U = 0.493\ (B–Y) \quad \text{and} \quad V = 0.877\ (R–Y).$$

Since the sub-carriers are 90° out of phase, the modulation products are also at 90° with respect to each other. Such a modulator is called quadrature modulator because of the 90° phase difference between the two outputs. The two outputs are combined to obtain chrominance or chroma (C) signal as shown in Fig. 11.9 (b). In order to prevent any interference due to large amplitude of the two carriers, these are suppressed by operating the modulators as balanced modulators. The outputs then contain only the two sideband components. It may be noted that the magnitudes of U and V are restored back to their normal values *i.e.* (B–Y) and (R–Y) after detection in the colour receiver.

Frequency Interleaving: The chroma signal (C) is carried along with corresponding Y signal as shown in Fig. 11.10. The colour sub-carrier frequency in the encoder is chosen to be exact at 4.3361875 MHz to ensure that when 'C' signal is combined with Y signal, its sidebands fit within the gaps of the low amplitude upper sideband products of the Y signal as illustrated in the figure. Such an arrangement called frequency interleaving prevents any interference between luminance and colour signals. The sound signal carrier is located 5.5 MHz away from the picture carrier where there are practically no Y and C signal components.

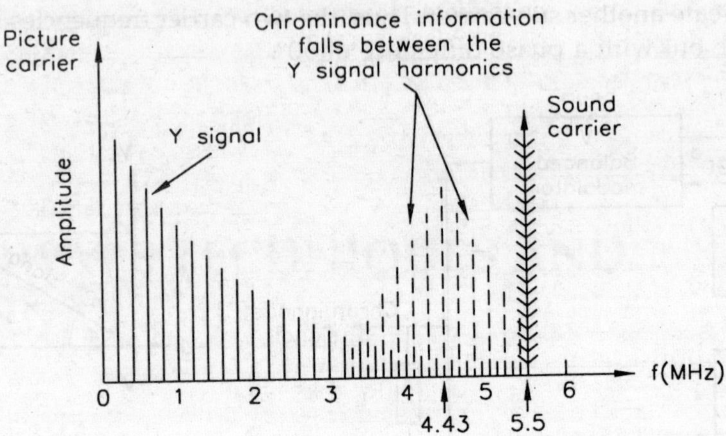

Fig. 11.10. Frequency interleaving of 'Y' and 'C' signals. Notice that chroma signal is located in a region where the amplitude of 'Y' signal in quite less. Also observe the location of sound signal carrier 5.5 MHz away from the picture carrier.

Colour Burst: Since the two carrier frequencies are suppressed in the balanced modulators of the encoder, it is necessary to regenerate them at the receiver for demodulation (detection) of the chroma signal. For this 8 to 10 cycles of the colour subcarrier oscillator output are transmitted along with sync pulses. These are placed at the end of each horizontal sync pulse at the blanking level. Since the colour sync pulses are transmitted in short bursts, it is called colour burst.

PAL Encoder: An oversimplified block diagram of the PAL encoder is shown in Fig. 11.11 (a). The 'C' signal can suffer a phase shift during transmission causing phase location errors of U and V signals which in turn result in reproduction of slightly different colours at the receiving end. The PAL system has a built-in protection against such errors. The quadrature modulation circuitry is so designed that polarity of 'V' modulation products gets reversed (*i.e.* phase shifted by 180°) on alternate line of scanning. This amounts to inverting the error on alternate lines. Since scanning is done at a very fast rate, the eye on seeing the picture, perceives an average value of two consecutive lines thus cancelling the effect of any phase error.

The shifting of 'V' phasor on alternate lines is a unique feature of PAL and infact the system derives it name (Phase Alteration by Line *i.e.* PAL) from this mode of transmission. As shown in Fig. 11.11 (a), a 7.8 kHz $\left(\text{actually } \dfrac{15625}{2}\right)$ square wave switching signal is used to shift the phase of colour sub-carrier to the V demodulator on alternate lines. A phase shift of +90° on one line and –90° on the next line amounts to 180° relative phase shift on alternate lines as shown in Fig. 11.11 (b). The delay line in the Y signal channel is to maintain time coincidence of Y and C signals because on account of large bandwidth of luminance signal as compared to 'C' signal, it arrives a bit too soon as compared to chroma signal.

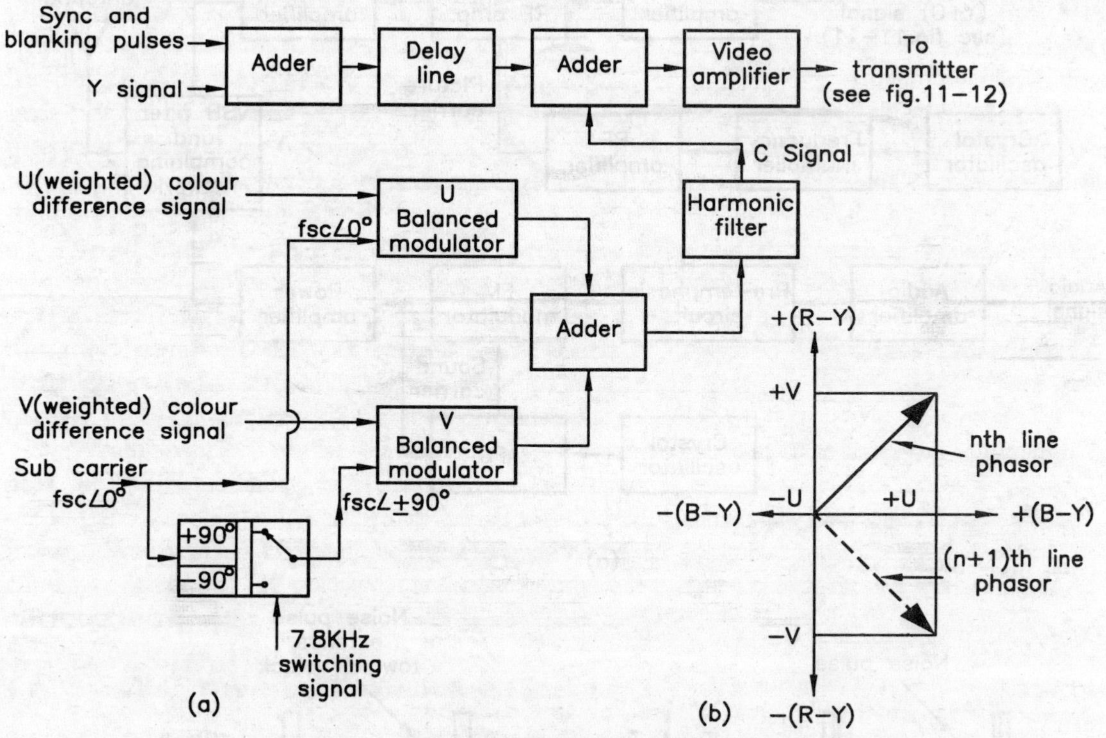

Fig. 11.11. Simplified diagram of PAL encoder (a) block diagram, (b) associated phasor diagram.

The outputs of U and V balanced modulators are added and passed through a harmonic filter to suppress high frequency sideband components and thus limit the colour signal bandwidth to about ± 1.2 MHz. The 'C' signal thus obtained is combined with the composite Y signal to form composite colour video signal. This, after necessary amplification, is passed on to the transmitter.

11.5 Colour TV Transmitter

A functional block diagram of a colour TV transmitter is shown in Fig. 11.12 (a). The composite colour video signal is amplified and modulated with picture carrier frequency of the channel allotted to the station. A crystal controlled oscillator together with a frequency multiplier and RF amplifier provide the carrier signal.

Negative modulation [see Fig. 11.12 (b)] is preferred where polarity of the modulating signal is so chosen that sync tips lie at 100% level of the carrier amplitude and increasing brightness reduces amplitude of the modulating envelope. Any noise pulses extend in black level direction of the modulating signal and thus their effect on picture details is less pronounced. Furthermore, in negative modulation, the peak sync pulse level, remains fixed at 100% of modulated signal amplitude and thus serves as a fixed reference level for automatic gain control (AGC) circuit in television receivers.

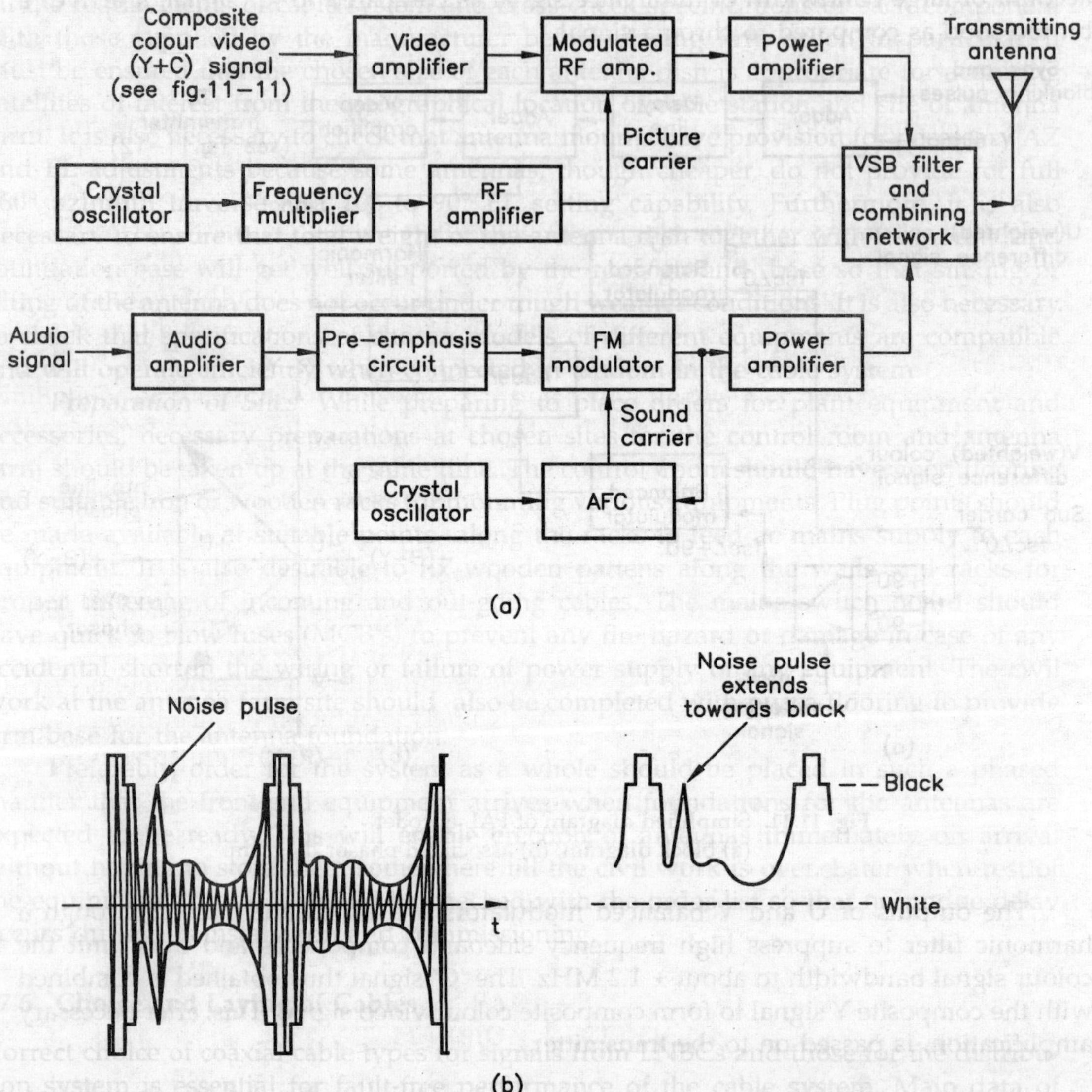

Fig. 11.12. Colour TV transmitter (a) functional block diagram, (b) nature of negatively modulated video signal.

Sound Signal Transmission: Sound signal is frequency modulated with station channel sound carrier frequency which is located at 5.5 MHz away from the channel picture carrier. As explained in section 8.3, for the 625 lines TV system, a bandwidth of 130 KHz is necessary for accommodating the frequency modulated sound signal. However, in practice a bandwidth equal to 150 KHz (+ 75 KHz) is allowed around the sound carrier frequency. This range is very much within the channel's upper edge and reasonably away from any significant video sideband components [see Fig. 11.13 (a)].

As shown in Fig. 11.12 (a), the FM sound signal on amplification is fed to the combining and vestigial sideband filter network which also receives amplitude modulated (Y+C) signal. The combined output is finally fed to the transmitting antenna network through a coaxial cable.

Channel Bandwidth: In the 625 line TV system, where video signal bandwidth is allowed to extend from d.c. (0 Hz) to 5 MHz, a double sideband AM transmission would occupy a bandwidth of 10 MHz (± 5MHz). Such a large bandwidth is too large and to reduce it, advantage is taken of the fact that the two sidebands (upper and lower) are images of each other and contain same information. However, instead of fully eliminating the lower sideband which can result in phase shift caused by filter design difficulties, a part of the lower sideband up to 0.75 MHz is also allowed to be transmitted. This is called vestigial sideband transmission and so a VSB filter in the combining network [see Fig. 11.12 (a)] cuts off the remaining lower sideband components.

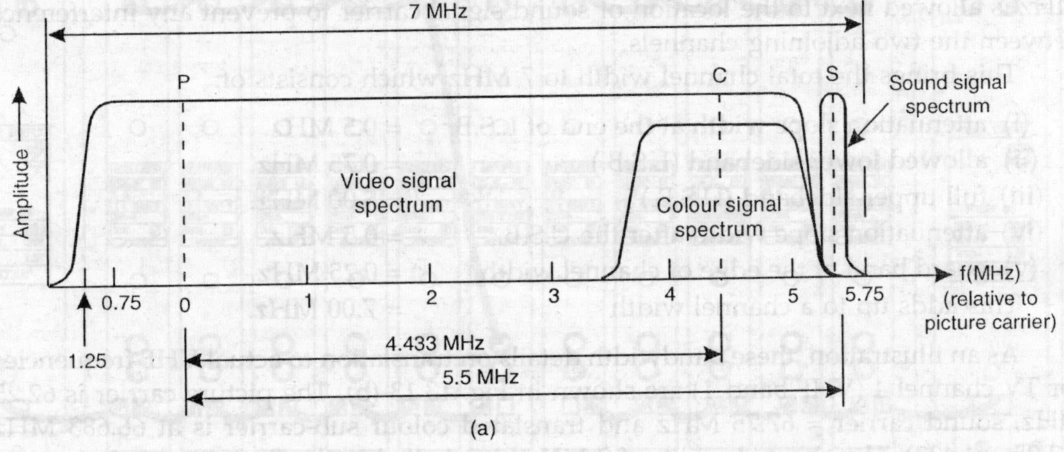

(a)

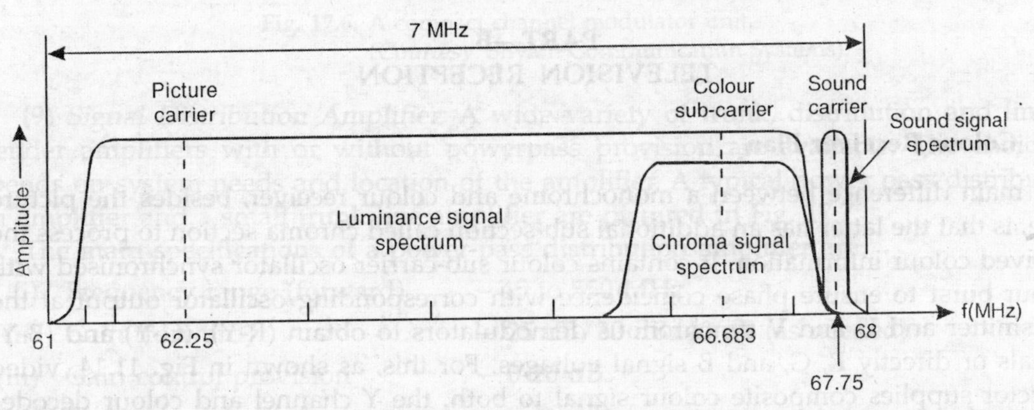

(b)

Fig. 11.13. (a) PAL colour TV channel bandwidth details.
(b) Frequency spectrum of VHF channel 4 (61 to 68 MHz).

Thus, the video signal is now allowed a bandwidth equal to 5 + 0.75 = 5.75 MHz, where 0.75 MHz is the width of the allowed lower sideband. In addition, attenuation slopes of 0.5 MHz are allowed on either side of the total video signal width because it is difficult to design filters that can abruptly terminate the bandwidth of the signal at the edges of chosen sideband widths. The resulting channel width details are shown in Fig. 11.13 (a). Notice that chroma signal (± 1.2 MHz) is around the colour sub-carrier frequency (≈ 4.43 MHz) is accommodated within the upper side band (U.S.B.) of Y signal and hence no additional channel width is necessary for the chroma signal. This not only saves channel width but also meets the compatibility requirements of keeping the same channel width for B&W and colour television system.

For colour signals, a bandwidth beyond about 1.2 MHz is not necessary because of limitations of human eye to perceive very fine colour details that need higher frequency colour signal components. Finally, at the edge of U.S.B., a guard band of width = 0.25 MHz is allowed next to the location of sound signal carrier to prevent any interference between the two adjoining channels.

This brings the total channel width to 7 MHz which consists of:

(i)	attenuation slope width at the end of L.S.B.	= 0.5 MHz.
(ii)	allowed lower sideband (L.S.B.)	= 0.75 MHz.
(iii)	full upper sideband (U.S.B.)	= 5.00 MHz.
(iv)	attenuation slope width after the U.S.B.	= 0.5 MHz.
(v)	guard band at the edge of channel width	= 0.25 MHz.
	This adds up to a channel width	= 7.00 MHz.

As an illustration, these bandwidth details on translation to actual VHF frequencies for TV channel 4 (VHF band 1) are shown in Fig. 13.13 (b). The picture carrier is 62.25 MHz, sound carrier = 67.75 MHz and translated colour sub-carrier is at 66.683 MHz (62.25 + 4.433). The channel width of 7 MHz, as labelled in the figure, extends from 61 to 68 MHz.

PART—B
TELEVISION RECEPTION

11.6 Colour Receiver Plan

The main difference between a monochrome and colour receiver, besides the picture tube, is that the latter has an additional sub-section called chroma section to process the received colour information. It contains colour sub-carrier oscillator synchronised with colour burst to ensure phase coincidence with corresponding oscillator output at the transmitter and U and V synchronous demodulators to obtain (R-Y), (G-Y) and (B-Y) signals or directly R, G, and B signal voltages. For this, as shown in Fig. 11.14, video detector supplies composite colour signal to both, the Y channel and colour decoder through a buffer amplifier. The luminance signal at the output of Y channel and U and V signals at the output of decoder combine in the matrixing circuit to provide R, G, and B colour video signals. The use of delay line in the Y channel enables exact time correspondence of colour and brightness informations of picture elements as explained in section 11.4 under PAL encoder. The automatic frequency tuning (AFC) circuit shown

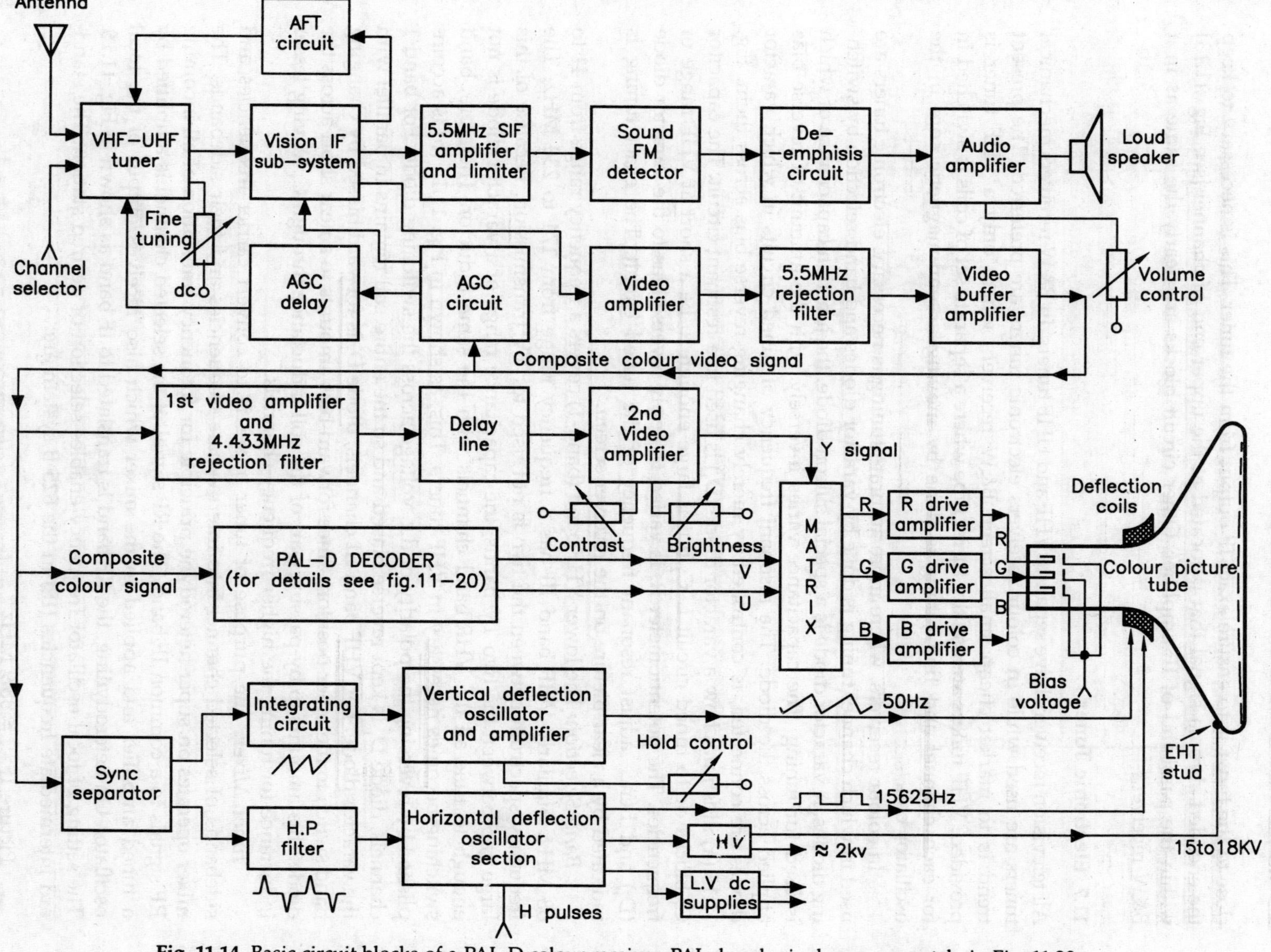

Fig. 11.14. Basic circuit blocks of a PAL-D colour receiver. PAL decoder is shown separately in Fig. 11.20.

close to the tuner is to ensure exact IF output from the tuner. This is necessary to keep the encoded chroma signal that is located at the upper end of channel [see Fig. 11.13] within the passband of IF amplifier. Other circuit blocks are nearly the same as in a B&W receiver.

11.7 Electronic Tuner

All television receivers have separate VHF and UHF tuners. In B&W receivers mechanical tuners are used while in colour receivers electronic tuners are preferred. The present trend is to market cheap small screen B&W receivers where only a VHF tuner is provided. VHF tuners are usually turret type where a separate set of coils is switched in for each channel and fine tuning is done by moving a semi-magnetic core in the oscillator coil.

In colour receivers, where more accurate tuning is necessary, electronic tuners are used in which channel tuning is done by varactor diodes and band selection by switching diodes. A varactor diode is a special silicon diode, the junction capacitance of which is used for tuning. The capacitance varies inversely with the amount of reverse bias applied across the diode. The resonant frequency of tuned circuits in which varactor diodes are connected, is controlled by merely changing reverse bias across them. Fig. 11.15 (a) illustrates how a varactor diode (D_1) is used in a resonant circuit. The capacitor C_1 blocks dc to tuned circuit L–C_2 but is large enough to be a short at VHF range of frequencies. The potentiometer R_3 is varied to apply reverse bias to the varactor diode (D_1) and hence adjust resonant frequency of the tuned circuit. The exact tuning is indicated by a clear picture on the receiver screen.

Band Selection: The lower VHF band (band-1) covers a frequency range from 41 to 68 MHz. Similarly VHF band III has a frequency range from 174 to 230 MHz. The frequency spectrum between the two is not used for TV transmission. Because of this large gap between the two TV bands, the capacitance range of a varactor diode is not enough to tune all the VHF band channels with the same inductor. Therefore, band switching becomes necessary in VHF tuners. This, as shown in Fig. 11.15 (b) is accomplished by reversing the polarity of d.c. voltage across the switching diode. For band 1 channels, diode D_1 is kept reverse biased and so the entire coil remains in parallel with the varactor diode. For VHF band III channels, diode D_1 is forward biased by changing band switch to the other position. Since forward bias amounts to a near short across the diode, this amounts to by-passing part of the coil inductance through C_1 leaving lesser inductance for tuning the higher frequency channels.

Tuner Mixer: The purpose of tuner mixer is to convert carrier frequencies and sidebands of selected channel into intermediate frequencies and their sidebands. The mixer operates on super-heterodyne principle for frequency translation of all incoming RF signals to a common IF band. The RF signal of the selected channel is amplified by a tuned amplifier and applied to the mixer which also receives output of the local oscillator. On heterodyning, the RF band is translated to IF band as shown in Fig. 11.16. The setting of local oscillator frequency enables selection of desired station. The standard intermediate frequencies (IF) in the 625-B system are:

Picture IF = 38.9 MHz
Sound IF = 33.4 MHz

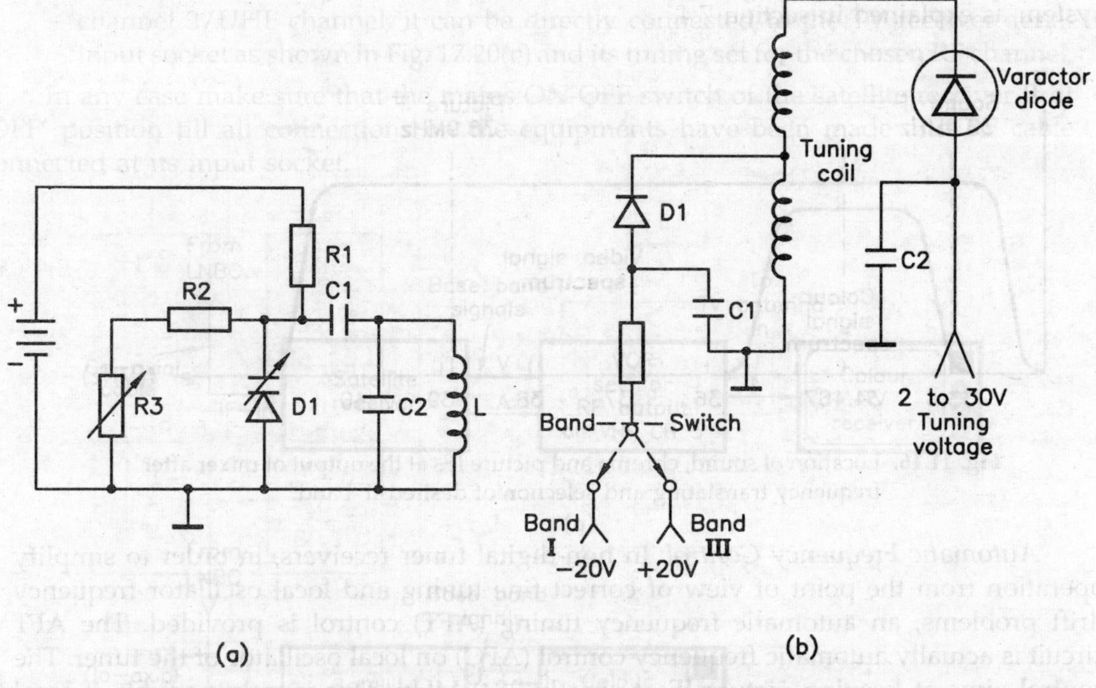

Fig. 11.15 (a) Basic circuit for varactor diode tuning of channels, (b) Simplified circuit showing switching diode operation for Band selection.

For channel 4 (61 to 68 MHz) where picture carrier is 62.25 MHz and sound carrier 67.75 MHz, the local oscillator frequency is set at 101.15 MHz. The correct setting is indicated by a clear picture on the screen. Once tuned, the channel can be selected just by pushing corresponding selector button (switch).

As explained in section 9.1, on heterodyning, both sum and difference (side band) frequencies are generated, but output circuit of mixer is, however, tuned to deliver difference *i.e.* IF frequencies and their sidebands. Thus, when the receiver is set for channel 4, we get

Picture IF = 101.15 – 62.25 = 38.9 MHz

Sound IF = 101.15 – 67.75 = 33.4 MHz

and chroma subcarrier IF = 101.15 – 66.683 = 34.467 MHz. These locations are shown in Fig. 11.16. Notice that frequency changing process has reversed relative positions of modulated picture, chroma and sound signals. This is so, because the oscillator frequency is chosen above the signal frequencies.

Digital Tuning: The transition from discrete to integrated circuits and availability of micro-computers has totally transformed the method of channel selection in modern television receivers. In new methods of channel selection, a frequency synthesizer is employed. It generates a number of crystal stable frequencies using only one crystal. These stable reference sources are used to lock tuner oscillator frequency for each channel at its correct value by a comparison method called phase-locked loop (PLL) system as explained in section 7.4.

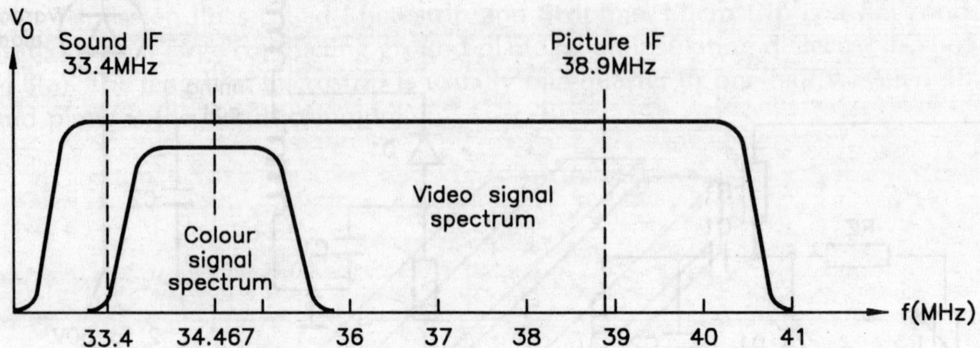

Fig. 11.16. Location of sound, chroma and picture IFs at the output of mixer after frequency translating and selection of desired IF band.

Automatic Frequency Control: In non-digital tuner receivers, in order to simplify operation from the point of view of correct fine tuning and local oscillator frequency drift problems, an automatic frequency tuning (AFT) control is provided. The AFT circuit is actually automatic frequency control (AFC) on local oscillator of the tuner. The control aims at keeping picture IF at exactly 38.9 MHz. This is only possible if local oscillator frequency for the selected channel is maintained at its correct value. To achieve this, the picture IF as obtained from the last stage of IF amplifier section is measured by a discriminator circuit whose output is a dc correction voltage that indicates deviation, if any, from its exact value of 38.9 MHz. Since this frequency depends on the oscillator frequency, the AFT output voltage indicates error of the local oscillator frequency.

A block diagram of the AFT circuit is shown in Fig. 11.17. The dc control voltage is zero at balance *i.e.* when IF frequency is exactly 38.9 MHz. For frequency deviations on either side of this, the dc control voltage is either positive or negative depending on whether the frequency is above or below the correct value. The dc control voltage thus developed is applied to a varactor diode that forms part of the local oscillator tank circuit [see Fig. 11.15]. The junction capacitance of varactor diode varies with the applied dc control voltage and thus changes resonant frequency of the local oscillator tuned circuit to shift the frequency to its correct value.

11.8 IF Subsystem

The IF subsystem also called vision IF section [see Fig. 11.14] consists of band shaping filter circuits, IF amplifiers, AGC circuit, AFC circuit, intercarrier sound IF detector, video detector and buffer video amplifier. In modern receivers all the above circuit blocks except filter network form part of an integrated circuit.

The band of frequencies to be handled by the IF section of both monochrome and colour receivers lies between 33.15 MHz to 40.15 MHz. With an input of about 0.5 mV from the mixer to deliver nearly 4V into the video detector needs an overall gain of about 8000. This is provided by atleast three cascaded stages of IF amplifiers.

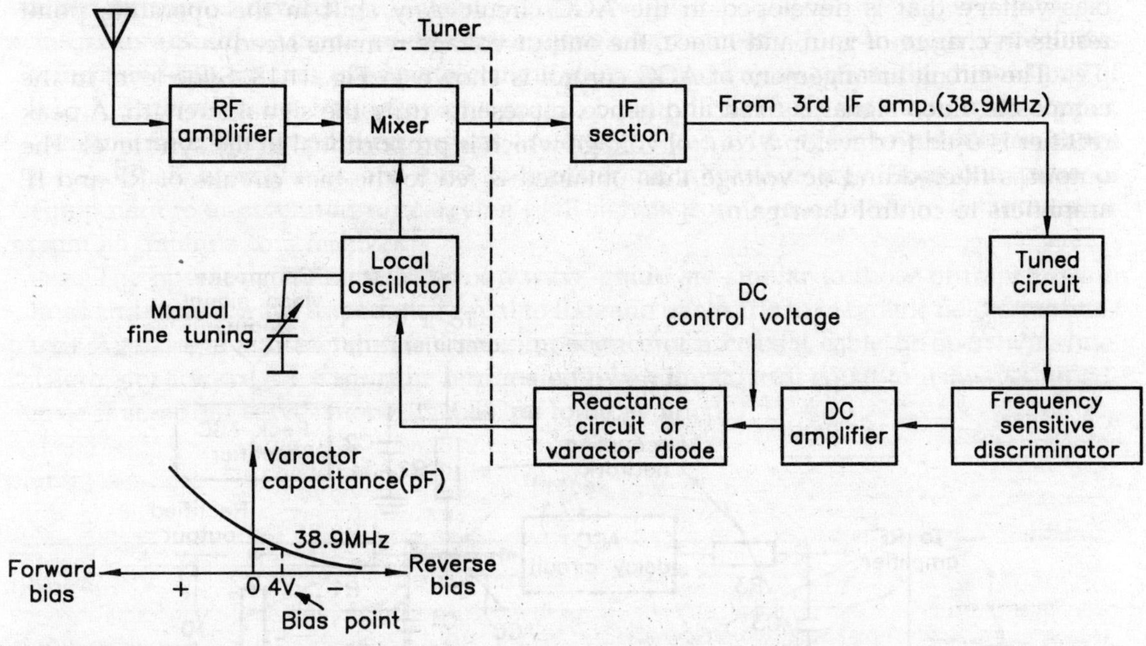

Fig. 11.17. Block diagram of an AFT circuit.

While amplifying the IF band, it is necessary (i) to provide bandwidth of nearly 7 MHz, (ii) vestigial sideband correction necessary for uniform gain over the entire range of video frequencies, and (iii) to keep gains of chroma and sound IF signals at levels below the picture carrier amplitude for their easy separation and detection. All this is done in modern receivers by a single section called 'SAW filter. The operation of a SAW filter is based on the phenomenon of surface acoustic waves. A surface acoustic wave is a non-electromagnetic wave that travels along the surface of a piezo-electric substrate which undergoes physical changes when an electric potential is applied across it. Transducers are provided at two ends of the peizo electric wafer (substrate) to convert electrical energy to accoustic wave energy, and then back to electric energy. The bandwidth and frequency characteristics of a SAW filter depend on the geometric structure of input transducer array because this determines the extent of attenuation to desired frequencies. Special SAW filters have been developed for use in television receivers thus making the use of L–C trap filter units unnecessary as was the practice in earlier receivers.

The last IF amplifier feeds into detector, AGC and AFT circuits as shown in Fig. 11.14. The video detector employs a diode as rectifier and a filter network to ground all unwanted high frequency components. The signal that becomes available across the detector has all the components that constitute the composite colour signal. Thus we get back luminance and chrominance signals along with sync pulses and colour burst.

Automatic Gain Control (AGC): AGC circuit controls gain of RF and IF stages to enable almost constant signal voltage at the output of video detector, despite changes in the signal picked up by the antenna. The change in gain is achieved by shifting the operating point of transistors used in IF amplifiers. The operating point is changed by a bias voltage that is developed in the AGC circuit. Any shift in the operating point results in change of gain and hence, the output voltage remains steady.

The circuit arrangement of AGC control is shown in Fig. 11.18. Sync level in the composite video signal is fixed and hence, represents truly the signal strength. A peak rectifier is used to develop a control voltage which is proportional to the sync level. The output is filtered and dc voltage thus obtained is fed to the bias circuits of RF and IF amplifiers to control their gain.

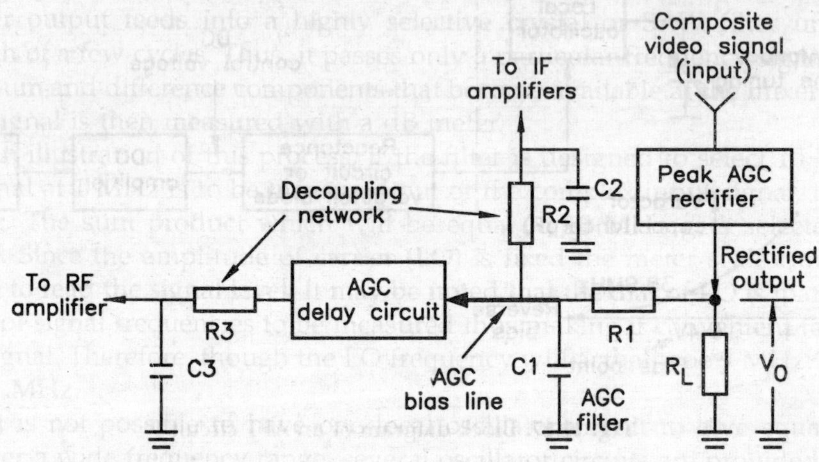

Fig. 11.18. Block diagram of AGC system.

Since AGC voltage is proportional to signal strength, even weak RF signals will also develop some control voltage. This when applied to RF amplifier will tend to reduce its gain, though the stage should have maximum gain for weak signals to maintain high signal to noise ratio. Therefore, the RF amplifier is not fed any AGC voltage till the signal attains a certain predetermined level. This is achieved by providing a voltage delay circuit in the AGC line as shown in the figure. Thus, a delayed AGC action occurs at the RF amplifier.

11.9 Receiver Sound System

The sound signal with SIF at 33.4 MHz is amplified along with picture signal though to a lesser extent and then detected as 5.5 MHz beat output between the picture carrier and FM sound signal around 33.4 MHz (38.9 – 33.4 = 5.5 MHz). All this happens within the IF subsystem IC as made obvious in Fig. 11.14. It may be noted that the 5.5 MHz intercarrier sound signal retains frequency modulation as was present in its IF spectrum.

The second frequency transformation of FM sound signal amounts to another heterodyning but with the advantage that this output does not drift from its centre frequency of 5.5 MHz, even when local oscillator frequency in the tuner tends to become more or less. The reason for this is that when local oscillator frequency drifts, both picture and sound IF frequencies vary by the same value thus maintaining their difference equal to 5.5 MHz.

In colour receivers, a separate detector is used to prevent any objectionable beat output between the 5.5 MHz sound carrier and 4.433 MHz chroma subcarrier. The intercarrier sound signal obtained from the IF IC is given one stage of amplification which also acts as a limiter to remove any amplitude variations in the signal. It is then detected by a differential phase detection technique to obtain audio signal. The characteristics of a typical FM detector are shown in Fig. 11.19. The audio output is proportional to deviations from the carrier frequency. The frequency of audio signal depends on the rate of frequency deviations. At the output of FM detector [see Fig. 11.14] a de-emphasis circuit is provided that has the same time-constant (50 μs) as that of the pre-emphasis circuit employed at the sound transmitter. This restores the amplitude of higher audio frequencies to their correct level. The audio signal receives atleast one stage of amplification before it is passed on to the audio (power) amplifier. The volume control and if necessary tone control are provided in the audio amplifier section. In present-day receivers, all the stages after detection are contained in a single IC usually called sound strip.

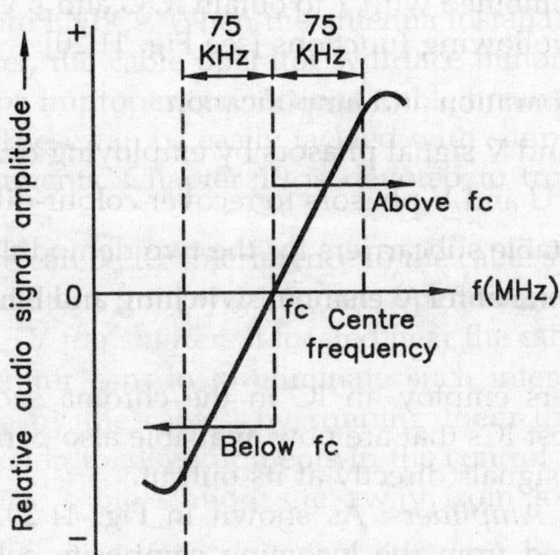

Fig. 11.19. Response curve of an FM detector.

11.10 Y-Signal Channel

The foundation on which the full colour picture gets built is the luminance or black and white image. In fact, the colours get added on the B&W image depending on the magnitude of R, G and B video signals. During monochrome reception or when colour control is turned to zero, only 'Y' signal gets applied to the three guns of colour picture tube. The reproduced picture is then a neutral *i.e.* B&W image.

The luminance and chroma signals take separate paths after the buffer amplifier [see Fig. 11.14] and later rejoin in the matrixing section. The 1st video amplifier includes a filter section that bypasses the 4.433 MHz chroma signal thus allowing only Y signal to get amplified. Since Y signal path is wideband that extends up to atleast 4 MHz and that of the chroma signal is narrow band restricted to nearly 1.5 MHz, a relative time delay occurs while the two signals pass through their respective paths in the receiver.

The colour signal is delayed, or say Y signal arrives a bit too soon with respect to R, G and B signals. Therefore, Y signal needs to be delayed so that brightness and colour informations are reproduced at the same time for any element of the picture. Based on the bandwidth of two channels, necessary time delay is approximately 0.06 μs. This is provided by the delay line that is inserted in-between 1st and 2nd video amplifiers. The delay line is constructed on the principle of distributed L-C elements and this results in the desired lag of the passing band of signals through it.

Contrast and Brightness Controls: Contrast of the picture can be varied by changing amplitude of Y signal. This is done by providing a manual control of gain of one of the video amplifiers in the Y channel. Similarly, average brightness of the scene can be changed by increasing or decreasing steady dc component of the Y signal. Such a provision is usually made in the output circuit of 2nd video amplifier.

11.11 Chroma Decoder

The main function of chroma or colour decoder is to recover U and V colour-difference signals which are later combined with Y to obtain R, G and B video signals. For this, the decoder has to perform following functions [see Fig. 11.20]:

(i) chroma signal separation and amplification.

(ii) separation of U and V signal phasors by employing delay line technique.

(iii) demodulation of U and V phasors to recover colour-difference signals.

(iv) generation of suitable subcarriers for the two demodulators.

(v) to develop ident signal for V channel switching and bias voltage for colour killer circuit.

All modern receivers employ an IC in the chroma section to carry out above functions. In fact, the latest ICs that are now available also perform matrixing functions and deliver R, G, and B signals directly at its output.

Chroma Band-Pass Amplifiers: As shown in Fig. 11.20, the chroma signal and colour burst are separated from the incoming composite colour video signal by the chroma signal separation circuit. It essentially consists of a bandpass circuit whose centre frequency is chosen to be equal to that of the chroma subcarrier itself *i.e.* 4.433 MHz and passband equal to approximately 3 MHz. In modern receivers, specially designed filter sections called 'comb filters' are used for separation of luminance and chroma signals. On separation, chroma and burst signals are amplified by the 1st chroma amplifier, the output of which goes to both the 2nd chroma amplifier and burst separation and amplifier circuit. The second chroma amplifier incorporates colour saturation control circuit the purpose of which is to form a variable attenuator to alter the magnitude of chroma signal and thus saturation of colours in the reproduced picture.

Separation of U and V Colour Phasors: As shown in the figure, the output of 2nd chroma amplifier stage is fed to the delay line and also simultaneously to addition and subtraction circuits. The technique of separating U and V signals is based on the assumption that picture information carried on two successive lines changes so little that it can be regarded as identical for both. This method is known as PAL-D (delay) and is used in all PAL receivers. Assuming that nth line phasor is (U + jV) *i.e.* V-signal is at 90°

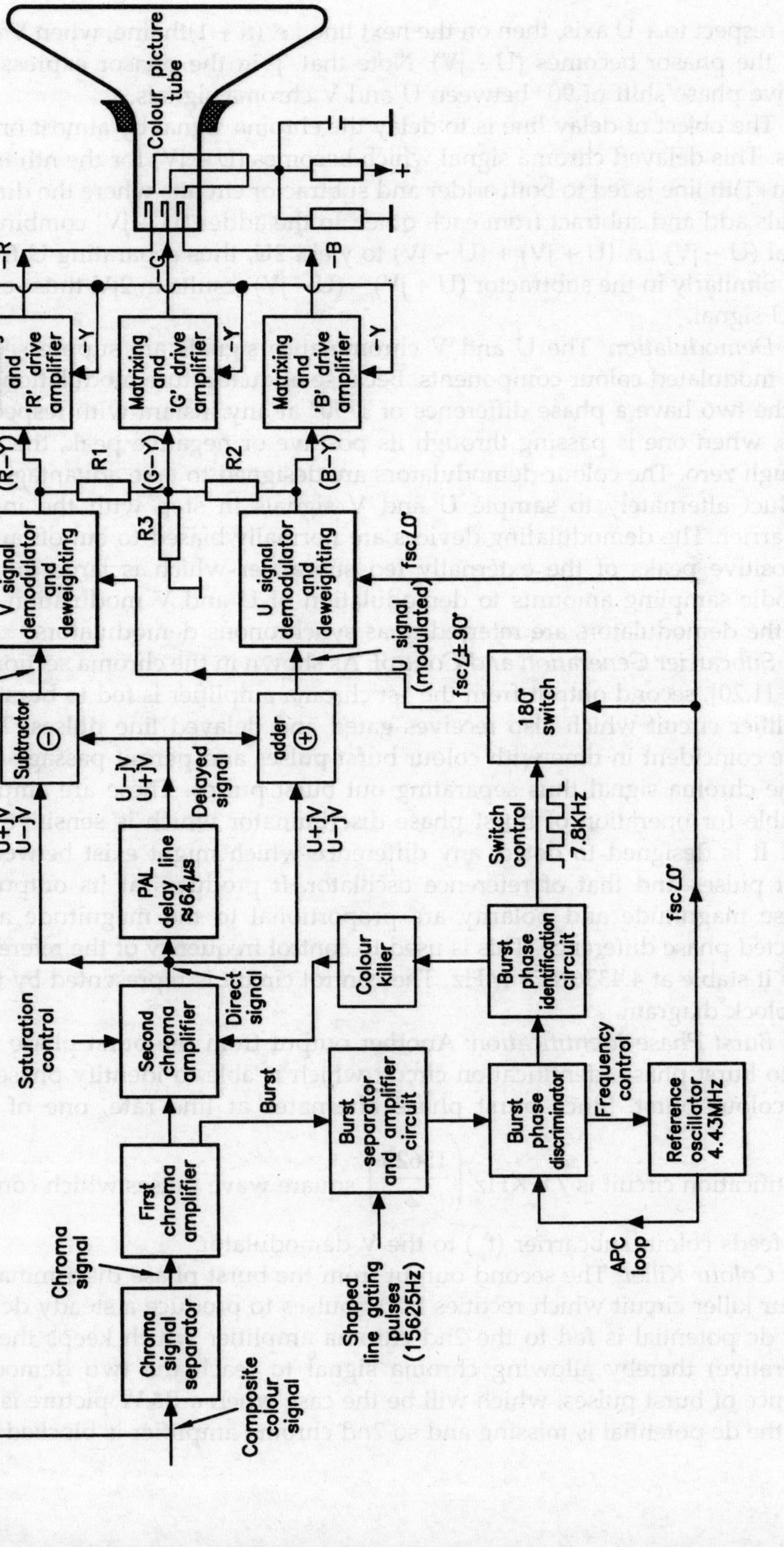

Fig. 11.20. Block diagram of a PAL-D decoder and matrixing circuit.

with respect to + U axis, then on the next line *i.e.* (n + 1)th line, when V gets switched to -90°, the phasor becomes (U - jV). Note that 'j' in the phasor expressions indicates a relative phase shift of 90° between U and V chroma signals.

The object of delay line is to delay the chroma signal by almost one line period of 64 µs. This delayed chroma signal which becomes (U – jV) for the nth line and (U + jV) for (n+1)th line is fed to both adder and subtractor circuits where the direct and delayed signals add and subtract from each other. In the adder (U + jV) combines with delayed signal (U – jV) *i.e.* (U + jV) + (U – jV) to yield 2U, thus separating U from V.

Similarly in the subtractor (U + jV) – (U – jV) results in 2jV, thus separating V from the U signal.

Demodulation: The U and V chrominance signals are suppressed carrier amplitude modulated colour components. Because of quadrature modulation at the transmitter, the two have a phase difference of ± 90° at any instant with respect to each other. Thus, when one is passing through its positive or negative peak, the other is passing through zero. The colour demodulators are designed to take advantage of this fact and conduct alternately to sample U and V signals in step with the injected reference subcarrier. The demodulating devices are normally biased to cut-off and conduct only at positive peaks of the externally fed subcarrier which is large in amplitude. This periodic sampling amounts to demodulation of U and V modulation product signals and the demodulators are referred to as synchronous demodulators.

Subcarrier Generation and Control: As shown in the chroma section block diagram [Fig. 11.20], second output from the 1st chroma amplifier is fed to burst separation and amplifier circuit which also receives gated and delayed line pulses. These pulses are made coincident in time with colour burst pulses and permit passage only to this part of the chroma signal thus separating out burst pulses. These are amplified to a level suitable for operation of burst phase discriminator which is sensitive to burst pulses only. It is designed to detect any difference which might exist between the phase of burst pulses and that of reference oscillator. It produces at its output, a dc voltage, whose magnitude and polarity are proportional to the magnitude and direction of detected phase difference. This is used to control frequency of the reference oscillator to keep it stable at 4.43361875 MHz. The control circuit is represented by the APC loop in the block diagram.

Burst Phase Identification: Another output from the burst phase discriminator is fed to burst phase identification circuit which is able to identify phase relationship of the colour burst. Since burst phase alternates at line rate, one of the outputs of identification circuit is 7.8 KHz $\left(\frac{15625}{2}\right)$ square wave pulses which control 180° switch that feeds colour subcarrier (f_{sc}) to the V demodulator.

Colour Killer: The second output from the burst phase discriminator is fed to the colour killer circuit which rectifies burst pulses to produce a steady dc output voltage. This dc potential is fed to the 2nd chroma amplifier which keeps the amplifier open (operative) thereby allowing chroma signal to reach the two demodulators. In the absence of burst pulses, which will be the case when a B&W picture is being transmitted, the dc potential is missing and so 2nd chroma amplifier is blocked thus preventing appearance of any colour noise on the picture tube screen when a monochrome picture is being received.

Matrixing for Drive Circuits: Deweighting circuits are provided in V and U demodulator circuits to obtain back (R–Y) and (B–Y) signals. As shown in section 11.4 under sub-heading 'compatibility', (G–Y) = –0.51 (R–Y) –0.186 (B–Y) and thus it can be obtained from (R–Y) and (B–Y) by a resistance matrix [see Fig. 11.20] by making $\frac{R_3}{R_1 + R_3} = 0.51$ and $\frac{R_3}{R_2 + R_3} = 0.186$. Thus 51% of (R–Y) combines with 18.6% of (B–Y) to develop (G–Y) across R_3.

R, G, B Matrix: In this matrix arrangement colour difference signals are combined with Y in the drive amplifier in such a way that the outputs are R, G and B video signals. These are negative going and of sufficient amplitude to drive cathodes of colour picture tube directly.

11.12 Raster Circuits

The deflection or raster forming circuits of a colour receiver are same as in a B&W receiver. With the more common use of integrated circuits to generate drives for deflection coils, the sync separator is a part of an IC where composite video signal is sliced to separate sync information. The pulse train thus obtained is suitably shaped before feeding to the vertical deflection section and to automatic frequency control (AFC) circuit in the line deflection section.

Vertical Deflection: The vertical deflection oscillator is designed to deliver output at 50 Hz that is synchronized with the incoming vertical sync pulses. The 50 Hz output is waveshaped in accordance with the impedance of vertical deflection coils to ensure linear vertical deflection of the scanning beam. All this is carried out in an IC, the output of which goes to the vertical drive amplifier. The amplifier also feeds vertical blanking pulses to the drive circuitry. For smaller and medium screen picture tubes, the vertical output amplifier also forms part of an IC and hence, no separate transistor amplifier stage is necessary.

Horizontal Deflection: The horizontal or line deflection circuit performs a similar function of deflecting the beam from left to right across the screen. The horizontal oscillator forms part of the line deflection IC but uses AFC technique to compare generated frequency with the incoming line sync pulses to keep it synchronized with horizontal scanning in the camera at the transmitting end. The line drive and output stages are discrete and employ high voltage, high wattage transistors because most of power consumed by the receiver is spent in horizontal cum dc supply section of the receiver. Several auxillary and extra high voltages needed for the picture tube are obtained by rectifying line trace and retrace pulses. The current demand by the picture tube is very small, around 1.5 mA, and hence does not load very much the line output stage.

11.13 EHT Generation

The inner side of a picture tube screen is coated with a phosphor chemical which when excited (delivered energy) emits light. In order to excite the atoms of phosphor coating, electrons that are emitted at the cathode of picture tube are directed at it with very high velocity. In order to give very high velocity to the electron beam, it is necessary to apply 12 to 16 KV voltage at the final anode of a B&W picture tube. For colour tubes even higher voltages up to 25 kV become necessary. To obtain such high voltages called EHT, by stepping up mains voltage with a transformer and then rectifying it is almost impossible and definitely prohibitive in cost. The novel method of generating this from the line output circuit is shown in Fig. 11.21. During the extremely fast retrace intervals of horizontal scanning, high voltage pulses of amplitude between 6 to 9 kV are developed

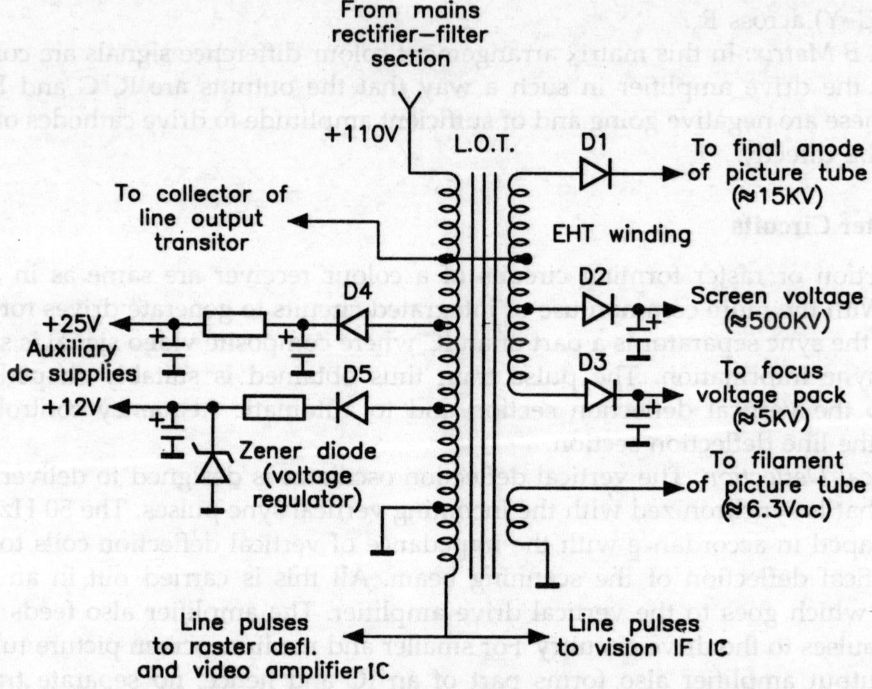

Fig. 11.21. EHT and auxiliary supplies obtained from the line output circuit of a TV receiver.

due to the very high rate of change of flux $\left(\frac{d\phi}{dt}\right)$ in the primary winding of line (horizontal) output transformer. As shown in the figure, these are stepped up by an auto-transformer winding to about 10 to 15 KV and then fed to a high voltage rectifier diode (D_1). The output of rectifier is filtered to provide required dc voltage.

While this method is alright for B&W picture tubes, but not satisfactory for colour picture tubes where EHT around 25 kV is necessary. For this, a 'diode split addition' technique (not shown in the figure) is used where voltages induced across three sections of a multilayer winding on the core of line output transformer (L.O.T.) are separately rectified and then added to obtain EHT.

Auxiliary dc Supplies: The use of bridge rectifier is now standard practice for obtaining dc voltages from ac mains. The transformer steps down the mains voltage as necessary and provides isolation from the supply line thus preventing shock hazards. A large electrolytic capacitor is used to filter out 100 Hz ac ripple without the need of earlier 'L' and 'π' type filters. Both series and shunt type voltage regulators are used to ensure steady dc supply to sound section IC, video output, line deflection IC and picture tube circuitry. Another technique called, Switched Mode Power Supply (SMPS) is getting increasingly popular because of its higher efficiency. High voltage supply for the screen grid and first anode of picture tube are also obtained from the line output stage as shown in Fig. 11.24. One or two low voltage auxiliary dc supplies at 25V and 12V are also usually obtained from the line output stage. The necessary circuits are shown in the figure.

11.14 Receiver Picture Tubes

The TV receiver picture tube [see Fig. 11.22] is a special form of cathode ray tube, the face plate of which serves as screen of the receiver. A monochrome (B&W) picture tube has a single electron gun. The cathode in the gun, when heated to a high temperature through its filament, emits electrons which are focussed to form a narrow beam by

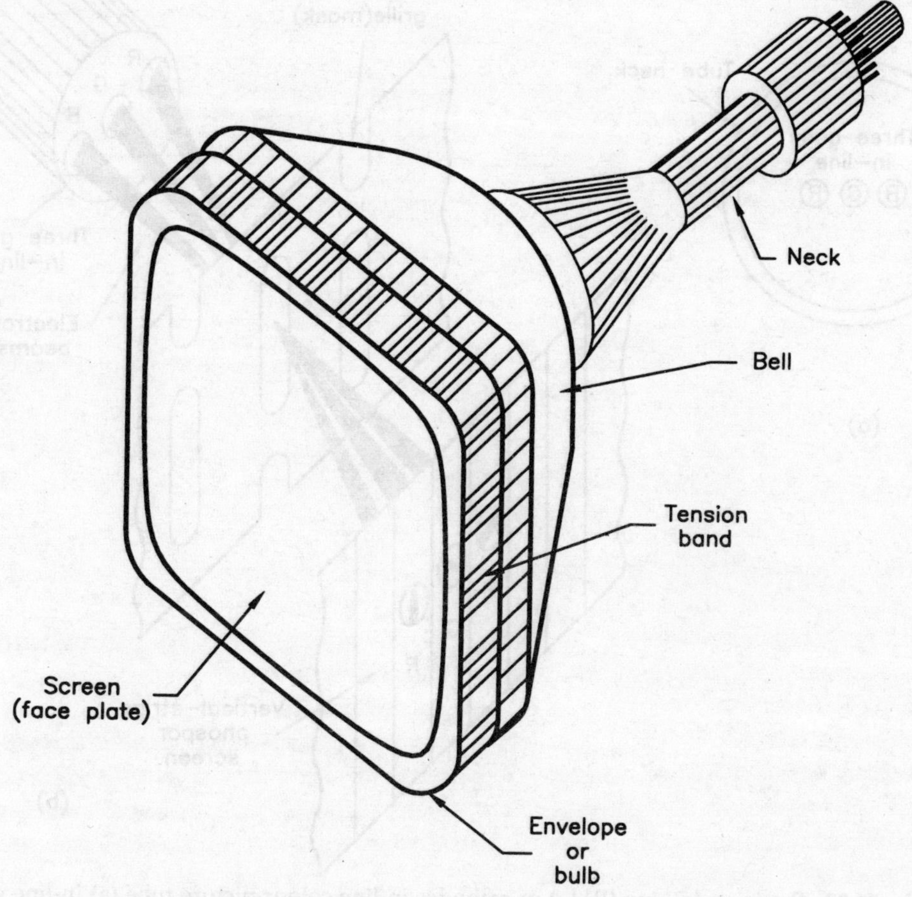

Fig. 11.22. Pictorial view of a television receiver picture tube.

electrostatic deflection action provided in the gun structure. The beam is accelerated to attain a very high velocity by the high positive voltage connected at the final anode of picture tube. The inside of wide glass faceplate that forms the screen is coated with a special material called phosphor which produces white light when excited by the electrons in the beam that on reaching the screen hit it with a great force. Magnetic deflection is used with deflecting coils mounted in the form of a yoke around the neck of picture tube. The vertical and horizontal deflecting currents that flow into corresponding deflection windings simultaneously deflect the electron beam at a linear rate from top to bottom and left to right to form the raster that fills the entire picture area. The received video signal drives the electron gun *i.e.* changes dc bias between the cathode and control grid in the gun assembly in accordance with variations in its magnitude and thus controls the intensity of electron beam at any instant. The resulting variations in brightness of different areas on the raster generate black and white pictures as the beam scans it at the rate of 25 frames (50 fields) per second. The picture seen on the

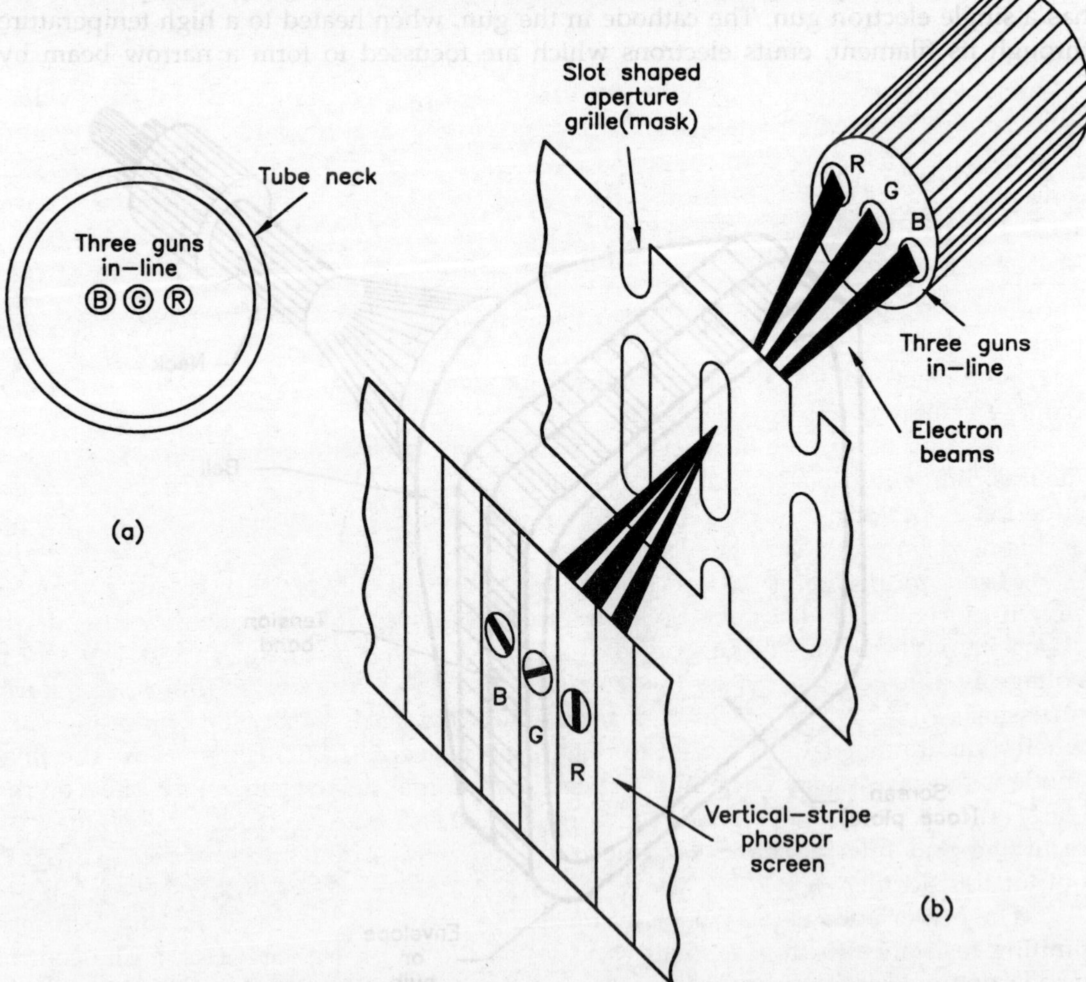

Fig. 11.23. Precision In-Line (P.I.L.) or cathodes in-line colour picture tube (a) in-line guns, (b) electron beams, aperture grille, and striped colour phosphor screen.

other side of screen at any instant is then the same as being televised in the studio at the transmitting end.

Colour Picture Tubes: [Fig. 11.23] In a colour picture tube there are three guns in the same glass envelope. The inside of screen area is coated with very small stripes of red, green and blue light emitting phosphors. These triades *i.e.* groups of R, G and B phosphors are repeated along the breadth of tube faceplate to cover its entire surface area. The breadth of each stripe is so small that there are approximately 333000 triades on the screen of a 51 cm colour picture tube. A thin metal sheet mask [see Fig. 11.23] called aperture mask that has numerous openings to allow the electron beams to pass through them, is mounted close to and at the back of screen. As shown in Fig. 11.23 (b) each vertical line of slots or openings is for one triade or group of red, green and blue phosphor stripes. Since all the three electron beams are in the same plane as shown in Fig. 11.23 (a), the beam in the centre that corresponds to green phosphor stripes moves along axis of the tube. However, because of inward tilt of right and left guns the electron beams that finally hit the blue and red phosphor stripes, travel at an angle and meet the central beam at the aperture grille mask. The slots in the mask are so designed that each beam strikes its own phosphor and is thus prevented from landing on other colour phosphors. Thus, depending on the intensity of each colour beam, three coincident rasters get formed when the receiver is on. The eye integrates the three primary colour lights to see the final colour.

P.I.L. Picture Tube

The picture tube described above is called precision-in-line (P.I.L.) colour picture tube because all the cathodes of the three guns are in one line. This tube superseded long back the earlier delta-gun colour picture tube and is now used in almost all modern television receivers. It is manufactured with minor modifications under different brand names in many countries.

The only other tube in use is the 'TRINTRON' where the gun and mask arrangement is different. It was developed by SONY Corporation of Japan which claim it to be superior in performance over P.I.L. Tubes. TRINITRON is covered by SONY patents and is used only in SONY television receivers.

Application Circuit: A typical application circuit of the P.I.L. colour picture is shown in Fig. 11.24. The tube uses high voltage focussing technique and thus needs about 500 volts at the screen grid and over 5 kV at the focussing anode. The screen voltage is adjusted by a potentiometer (R_6) to obtain optimum brightness. Similarly focussing voltage is applied through a resistance divider network called focussing pack where potentiometer R_{11} is varied to obtain sharp overall focus of the beam. The final anode is fed nearly 25 kV through a special socket that fits over the pin located on the glass bulb. As explained in the previous section, all high voltages are obtained by rectifying and filtering 15625 Hz line trace and retrace pulses that occur in the line output transformer.

The heater voltage (6.3V ac) is connected to one of the heater pins through current limiting resistance R_4 from a small winding on the L.O.T. [see Fig. 11.21]. The control grid structure is grounded and similarly the other heater pins are earthed to enable flow of filament (heater) currents. Since the cathodes receive about 120V dc from collectors of corresponding video amplifier transistors, the earthed control grids remain negatively biased as necessary.

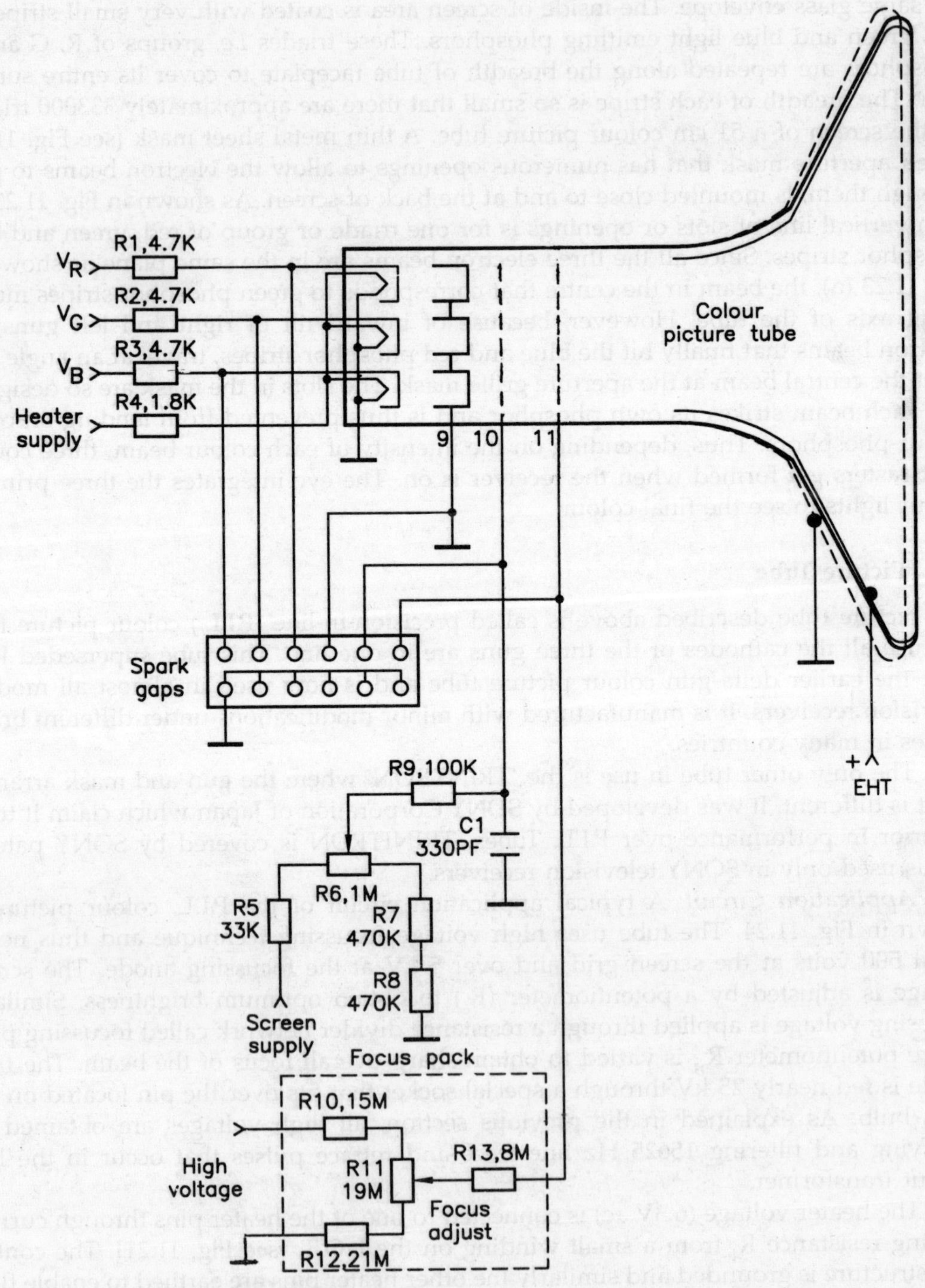

Fig. 11.24. A typical circuit around a P.I.L. colour picture tube.

Colour Pictures: As already explained the three phosphor stripes (R, G and B) though very close to each other are physically separate from one another and each is energized by corresponding video signal amplitude. Thus, three coincident rasters, red, green and blue, are produced on the screen. While seeing from a normal viewing distance, the eye integrates the three colour informations to convey the sensation of hue (colour) at each element or part of the picture. The colour and its brightness as perceived (seen) at any instant depends on the presence and magnitude of colour video signals at that moment.

As an illustration, consider that the camera is scanning a board that is painted red. Thus, the camera output will be only V_R while V_G and V_B will be zero. Also Y will be equal to 0.3 V_R only. As obvious, only V_R will become available at the receiver to drive the red, gun of the picture tube and only a red raster will be produced. The brightness of the red raster will depend on the magnitude of V_R produced at the camera. Similarly, green and blue rasters will be produced and seen when these colour objects are scanned at the transmitting end. When a white area is scanned, all the three *i.e.* R, G and B video signals are present to drive corresponding guns. Since for white, the ratio of $V_R : V_G : V_B$ will be 0.3 : 0.59 : 0.11, the intensities of the three colour rasters on the receiver screen will be such that on integrating the three colour lights the eye will see only white colour while viewing the screen from a normal distance.

Similarly, when an object having yellow hue is scanned, only V_R and V_G will be present to drive the colour picture tube guns and as such the eye will integrate red and green rasters to see yellow. Thus multicolour objects or scenes when scanned by the camera will produce three colour rasters on the screen of picture tube and the eye by additive mixing will see resultant hues as present in the scene at any instant.

11.15 Remote Control of Receiver Functions

The remote control facility if provided enables the television viewer to operate various receiver controls from a distance without having to leave his or her seat. The functions which can be controlled include (i) changing of stations, (ii) change of sound level in a series of small steps, (iii) turning the receiver ON and OFF, (iv) contrast control (v) brightness control, and (vi) colour intensity (saturation). Almost all present-day remote control units rely on infra-red wave transmission with Light Emitting Diodes (LEDS). The circuitry employed is quite complex and integrated circuits are used to process various inputs before infra-red wave transmission. A typical remote control transmitter unit is shown in Fig. 11.25. The various functions it can perform are labelled on the diagram.

Since transmission to the TV receiver is by infrared waves, it is somewhat directional. The surface labelled as such should be pointed towards the receiver. The range of useful operation is around 7 meters with a radius of ± 30° from the remote control receiver on the control panel of TV set.

Remote Transmitter: The transmitter unit shown in Fig. 11.25, needs 24 different commands to carry out all the labelled functions. Thus there is provision to generate 24 different pulse patterns. When the desired button is pressed, a special individual circuitry is activated in the encoder. It produces a distinctive pulse train which modulates a crystal controlled 48 kHz carrier. After due amplification this signal is fed to an LED which emits proportionate radiation in the infrared region to cause its transmission towards the remote controlled television receiver when the unit is aimed at it.

Remote Control Receiver: The pulse modulated signals from the remote infrared transmitter are picked up by the remote infrared senser which is an LED. The picked-up signal is amplified, demodulated and used to generate a pulse pattern corresponding to

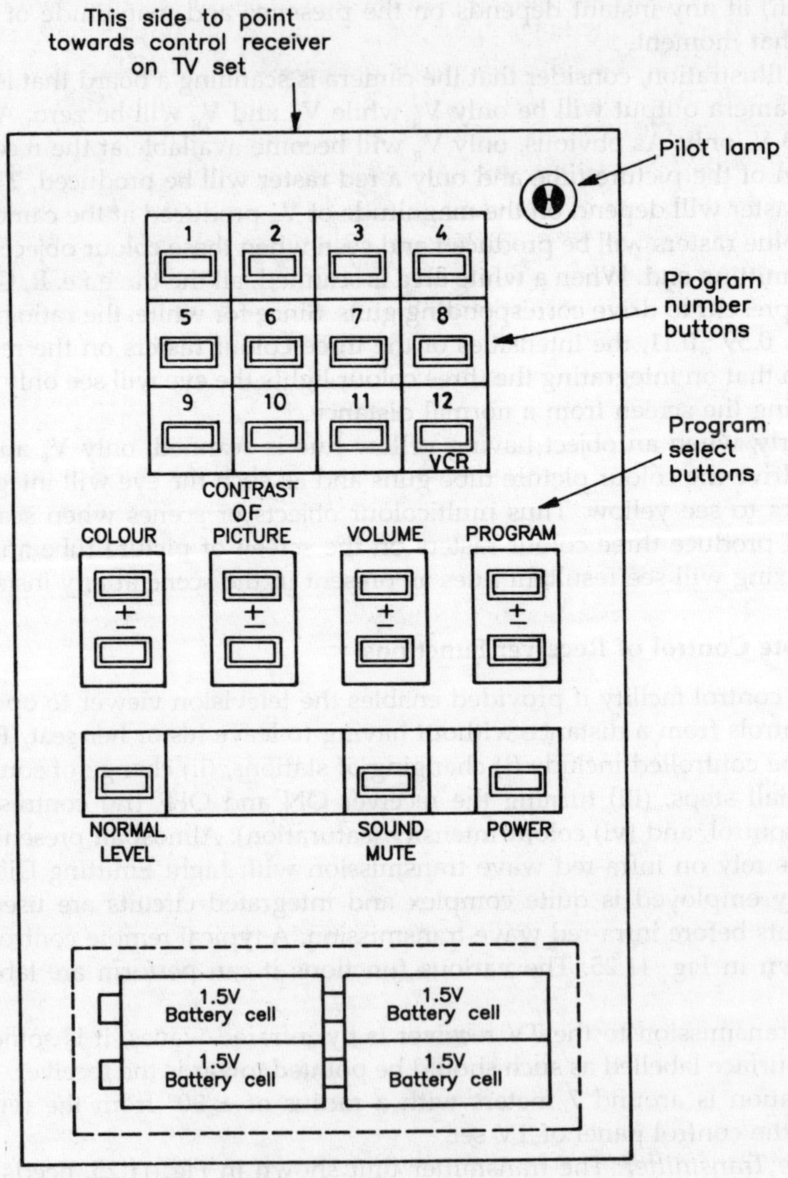

Fig. 11.25. Front view of a typical infrared wave remote control transmitter.

the function button pressed on the remote control unit. The shaped individual pulse pattern is then dc coupled to a microcomputer (mC). The mC on receipt of different inputs produces appropriate outputs to perform the ordered functions.

For example, to turn the TV receiver 'ON', the power button is pressed on the transmitter. For this, the (mC) output at the receiver is a step (+) voltage at one of the pins of its IC. This voltage is fed to a transistor which on conduction lights the lamp of an optical coupler. When the lamp lights, its radiation reduces resistance of the associated Light Dependent Resistance (LDR). The low resistance of the LDR enables application of 220V ac to the mains transformer of the TV receiver through a special gate arrangement and the receiver is turned on.

Note: If necessary the reader is advised to avail of the following books by this author for a detailed study of television transmission and reception:

1. Modern Television Practice (Principles Technology and Servicing)
2. Monochrome and Colour Television
3. Colour Television (Principles and Practice).

Publisher: New Age International (P) Ltd., 4835/24 Ansari Road, New Delhi-2 (earlier Wiley Eastern Ltd., New Delhi).

12

SATELLITE COMMUNICATIONS

Introduction

The geostationary communication satellites orbiting around the earth enable not only national but also international television programmes to be relayed between a number of earth stations around the world. Besides international communication and TV relaying, satellites are being used for distributing national television programmes over extended regions in large countries like U.S.A., China, India and many more. Direct broadcast satellite systems are also in use where a viewer can directly receive TV programmes by installing suitable dish antenna and having a special earth station receiver. This is particularly useful for people living in rural areas not served by many TV stations or cable companies.

Modern cable television owes its existence to satellite television where signals received from various satellites are decoded to obtain baseband (video and audio) signals and then processed in accordance with TV standards adopted by the receiving country and distributed on different VHF channels through coaxial cables to a large number of subscribers in a community area. As such, it is necessary for those interested in cable television to know the essentials of communication satellites. This chapter is therefore, devoted to transmission of TV signals via satellites and their reception at ground stations.

12.1 Geostationary Satellites

The orbit of a satellite is the result of two forces—one the gravitational pull towards the earth which acts radially and the other, the motion of satellite itself, which for an orbit as the goal, must act along a line that forms an angle with the radius, preferably a right angle.

If a satellite is placed in a low orbit, where the gravitational pull is more, it would be necessary for the satellite to have a higher velocity to remain in orbit and will thus rotate around the earth faster than the speed at which earth rotates. This is true of some of the meteorological satellites orbiting our planet that make one complete orbit in about ninety minutes. Conversely, if a satellite is placed in an orbit beyond a certain distance above the equator, it will rotate around the earth slower than the earth rotates, as does our moon, which makes a complete orbit in a little over twenty-eight days.

Knowing this, it is possible to realize that if a satellite is placed above the equator at a suitable distance, it will rotate exactly at the same speed at which the earth revolves. The distance at which this becomes possible is 35887 km (usually referred to as 36000 km), from the earth. If it were possible to observe such a satellite at this distance with naked eye, it would seem as if it was stationary. Hence, the terms 'GEOSTATIONARY' and 'GEOSYNCHRONOUS' are used interchangeably to describe the orbit of a satellite that rotates around the earth at the same speed, the earth itself revolves around its axis.

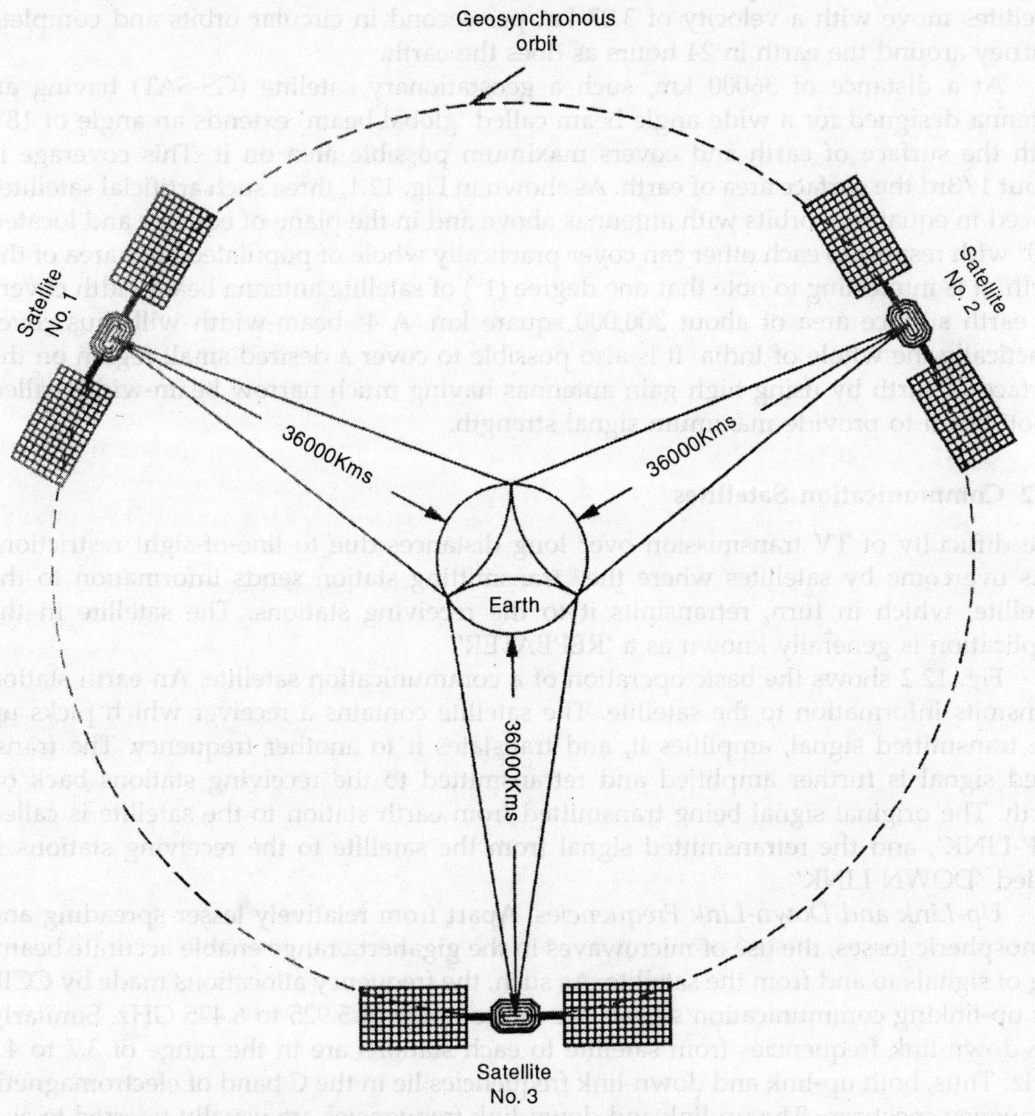

Fig. 12.1. Three geostationary satellites suitably located in circular orbits at 120° apart from each other.

Thus due to apparent stationariness of geosynchronous satellites, the transmitting and receiving antennas between ground stations and satellites do not have to swing around to keep each other in sight to be able to communicate with each other.

While the need for a suitable reflector satellite was felt long back for TV transmission over long distances, but this could not be affected due to practical difficulties of putting it in orbit. It was only around 1950 that necessary technology became available to build huge propelling vehicles that could successfully launch a man-made satellite up at precisely 35887 km, necessary for a geostationary orbit. This not only revolutionised TV transmission round the world but also made possible a very large number of telephone lines and data processing links between any two locations on the earth. Such satellites move with a velocity of 3.07 km per second in circular orbits and complete journey around the earth in 24 hours as does the earth.

At a distance of 36000 km, such a geostationary satellite (GS-SAT) having an antenna designed for a wide angle beam called 'global beam' extends an angle of 18°, with the surface of earth and covers maximum possible area on it. This coverage is about 1/3rd the surface area of earth. As shown in Fig. 12.1, three such artificial satellites placed in equatorial orbits with antennas above and in the plane of equator and located 120° with respect to each other can cover practically whole of populated land area of the earth. It is interesting to note that one degree (1°) of satellite antenna beamwidth covers an earth surface area of about 300,000 square km. A 4° beam-width will thus cover practically the whole of India. It is also possible to cover a desired small region on the surface of earth by using high gain antennas having much narrow beam-width called 'spot' beam to provide maximum signal strength.

12.2 Communication Satellites

The difficulty of TV transmission over long distances due to line-of-sight restrictions was overcome by satellites where the transmitting station sends information to the satellite, which in turn, retransmits it to the receiving stations. The satellite in the application is generally known as a 'REPEATER'.

Fig. 12.2 shows the basic operation of a communication satellite. An earth station transmits information to the satellite. The satellite contains a receiver which picks up the transmitted signal, amplifies it, and translates it to another frequency. The translated signal is further amplified and retransmitted to the receiving stations back on earth. The original signal being transmitted from earth station to the satellite is called 'UP-LINK', and the retransmitted signal from the satellite to the receiving stations is called 'DOWN-LINK'.

Up-Link and Down-Link Frequencies: Apart from relatively lesser spreading and atmospheric losses, the use of microwaves in the gigahertz range enable accurate beaming of signals to and from the satellite. As such, the frequency allocations made by CCIR for up-linking communication signals are in the range of 5.925 to 6.425 GHz. Similarly, the down-link frequencies from satellite to each stations are in the range of 3.7 to 4.2 GHz. Thus, both up-link and down-link frequencies lie in the C band of electromagnetic frequency spectrum. The up-link and down-link frequencies are usually referred to as 6 GHz and 4 GHz respectively.

Although most satellite communications activity takes place in the C band*, there is a steady move towards higher frequencies in the 'Ku' band where up-link frequencies

* Refer band allocation chart in Chapter 2, section 2.4.

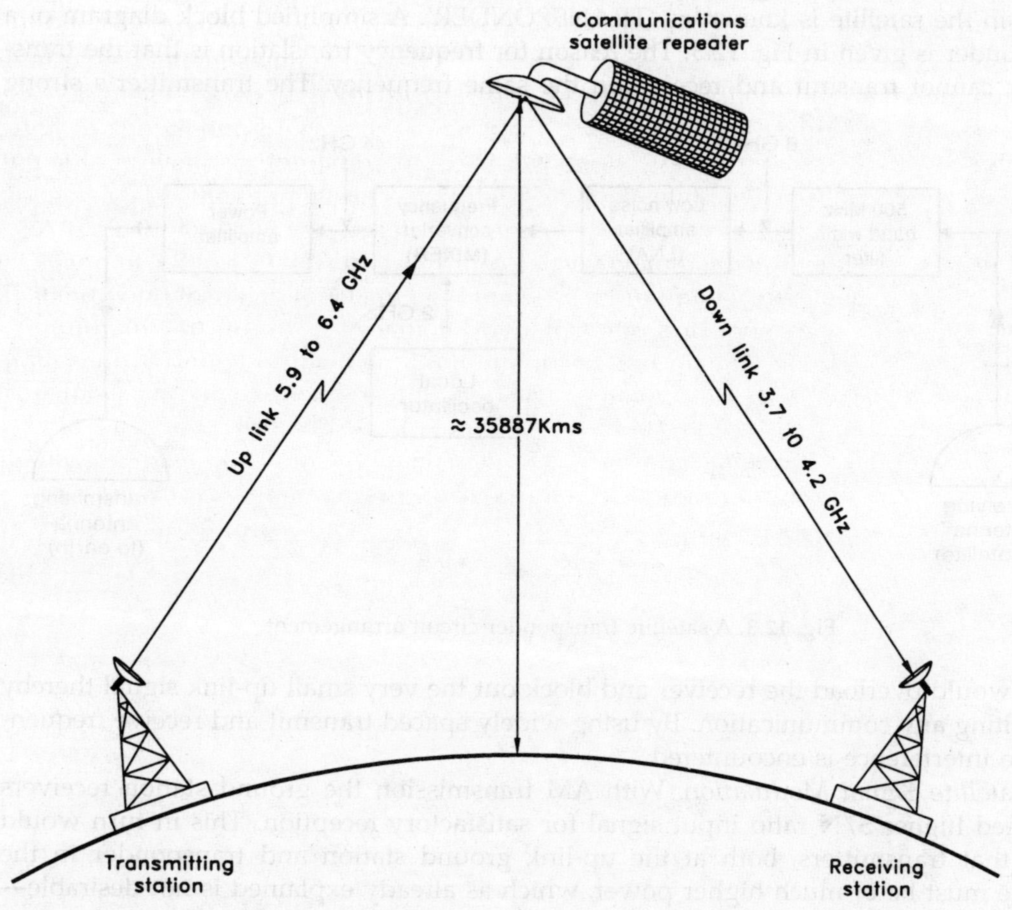

Fig. 12.2. Signal paths in a satellite communication system.

are in 14 to 14.5 GHz range and down-link frequencies lie between 10.95 to 12.75 GHz. The reason for this shift upward in frequency is that the C band is becoming over-crowded with increase in number of satellites operating in this band. The electronic equipment to achieve these higher frequencies is more complex and expensive but there is no other way to accommodate more satellites. However, for a given antenna size, the gain is higher in Ku band than in 'C' band and thus improved communication reliability becomes possible with decreasing antenna size and cost.

Satellite Transponders: Attenuation 'X' of radio waves as they propagate from earth to satellite is around 200 dB. There will be an equal attenuation from the satellite to earth station making total loss between earth transmitter to earth receiver to about 400 dB. An earth station receiver needs atleast an RF input of about 10^{-12} watts for satisfactory operation. Thus, total power radiated by the earth transmitter should be 400 dB above 10^{-12} watts *i.e.* $(-12 + 40) = 10^{28}$ watts, clearly an impossible figure to achieve. This problem is solved by providing a receiver-transmitter combination in the satellite where the received signals are given high amplification and frequency translation before retransmitting at a different frequency. Such a transmitter-receiver combination in the satellite is known as 'TRANSPONDER'. A simplified block diagram of a transponder is given in Fig. 12.3. The reason for frequency translation is that the transponder cannot transmit and receive on the same frequency. The transmitter's strong

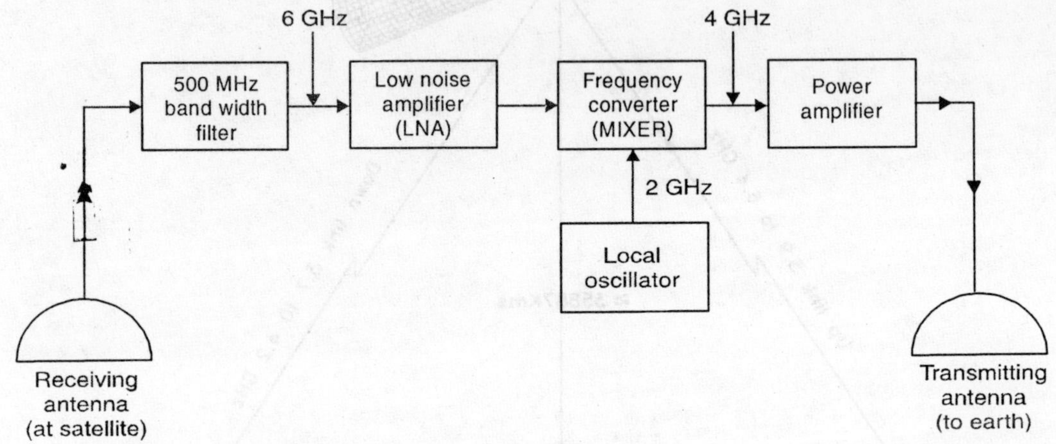

Fig. 12.3. A satellite transponder circuit arrangement.

signal would overload the receiver and block out the very small up-link signal thereby prohibiting any communication. By using widely spaced transmit and receive frequencies, no interference is encountered.

Satellite Signal Modulation: With AM transmission the ground station receivers will need higher S/N ratio input signal for satisfactory reception. This in turn would mean that transmitters, both at the up-link ground station and transponder in the satellite must be of much higher power, which as already explained is not desirable—infact not possible. On the other hand, with FM transmissions a lower S/N ratio of the order of 10 to 15 dB is quite satisfactory for good quality reception. Therefore, FM is the preferred mode of modulation for satellite systems. The wider bandwidth required for FM transmission can be easily accommodated at the GHz frequencies used for satellite communication. Such an arrangement is illustrated in Fig. 12.4.

Transponder Channels: As stated earlier, frequencies designated for the C band up-link and down-link are 5.925 GHz (5925 MHz) to 6.425 GHz (6425 MHz) and 3.7 GHz (3700 MHz) to 4.2 GHz (4200 MHz) respectively. Thus the available bandwidth for up and down transmissions is 500 MHz *i.e.* (6425 MHz – 5925 MHz = 500 MHz) and (4200 MHz – 3700 MHz = 500 MHz). Such a wide band is capable of carrying a large number of signals. The transponder receiver looks up the entire bandwidth and picks up any transmission there. However, the input is 'channelized' because the earth stations operate on selected frequencies and channels. The 500 MHz bandwidth is typically divided into 12 separate transmit channels, each 36 MHz wide. There are 4 MHz guard bands between channels which enable to minimize adjacent channel interference. These communication channels are illustrated in Fig. 12.5. Note the centre frequency for each channel. Remember that the up-link frequencies are translated by frequency conversion to the down-link channels. In both cases, the total bandwidth (500 MHz)

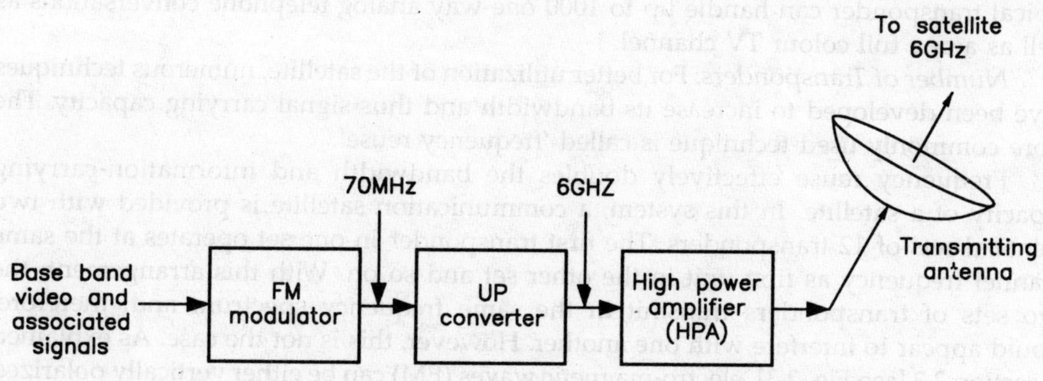

Fig. 12.4. Simplified block diagram of a transmitting earth station.

and channel bandwidth (36 MHz) are the same. On board, the satellite, a separate transponder is allocated to each of the 12 channels.

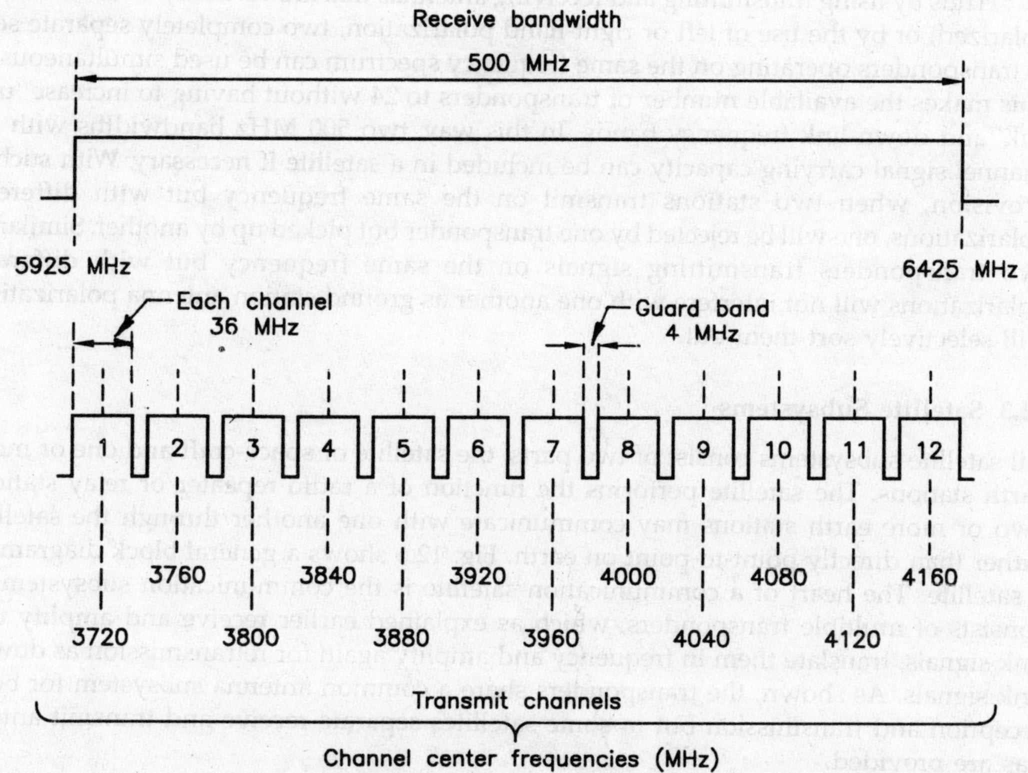

Fig. 12.5. Receive and transmit bandwidths in a 'C' band communication satellite.

Although 36 MHz seems narrow compared to 500 MHz, each transponder bandwidth is still capable of carrying an enormous amount of information. For example, one typical transponder can handle up to 1000 one-way analog telephone conversations as well as a one full colour TV channel.

Number of Transponders: For better utilization of the satellite, numerous techniques have been developed to increase its bandwidth and thus signal carrying capacity. The more commonly used technique is called 'frequency reuse'.

Frequency reuse effectively doubles the bandwidth and information-carrying capacity of a satellite. In this system, a communication satellite is provided with two identical sets of 12 transponders. The first transponder in one set operates at the same channel frequency as first unit in the other set and so on. With this arrangement, the two sets of transponders transmit in the same frequency spectrum and, therefore, would appear to interfere with one another. However, this is not the case. As explained in section 2.3 [see Fig. 2.3], electro-magnetic waves (EM) can be either vertically polarized or horizontally polarized. A signal that is transmitted with vertical polarization will not be received on an antenna that is horizontally polarized. The same is true of a horizontally polarized signal which will not be received by a vertically polarized antenna.

Circular polarization is also used in some types of antenna. There is left-hand and right-hand circulation polarization. An antenna using one cannot work with the other.

Thus by using transmitting and receiving antennas that are vertically or horizontally polarized, or by the use of left or right-hand polarization, two completely separate sets of transponders operating on the same frequency spectrum can be used simultaneously. This makes the available number of transponders to 24 without having to increase 'up-link' and down-link frequency bands. In this way, two 500 MHz bandwidths with 24 channel signal carrying capacity can be included in a satellite if necessary. With such a provision, when two stations transmit on the same frequency but with different polarizations, one will be rejected by one transponder but picked up by another. Similarly, two transponders transmitting signals on the same frequency but with different polarizations will not interfere with one another as ground station antenna polarization will selectively sort them out.

12.3 Satellite Subsystems

All satellite subsystems consist of two parts, the satellite or space-craft and one or more earth stations. The satellite performs the function of a radio repeater or relay station. Two or more earth stations may communicate with one another through the satellite rather than directly point-to-point on earth. Fig. 12.6 shows a general block diagram of a satellite. The heart of a communication satellite is the communication subsystem. It consists of multiple transponders, which as explained earlier receive and amplify up-link signals, translate them in frequency and amplify again for retransmission as down-link signals. As shown, the transponders share a common antenna subsystem for both reception and transmission but in some satellites separate receive and transmit antennas are provided.

Transponder Configurations: The basic circuits of two commonly used transponder configurations are shown in Fig. 12.7. In the single conversion transponder [Fig. 12.7 (a)] only a single frequency translation process takes place from the received signal to the transmitted signal. The up-link signal is picked by the receiving antenna and channelized by a band-pass filter (not shown) before feeding it to the corresponding

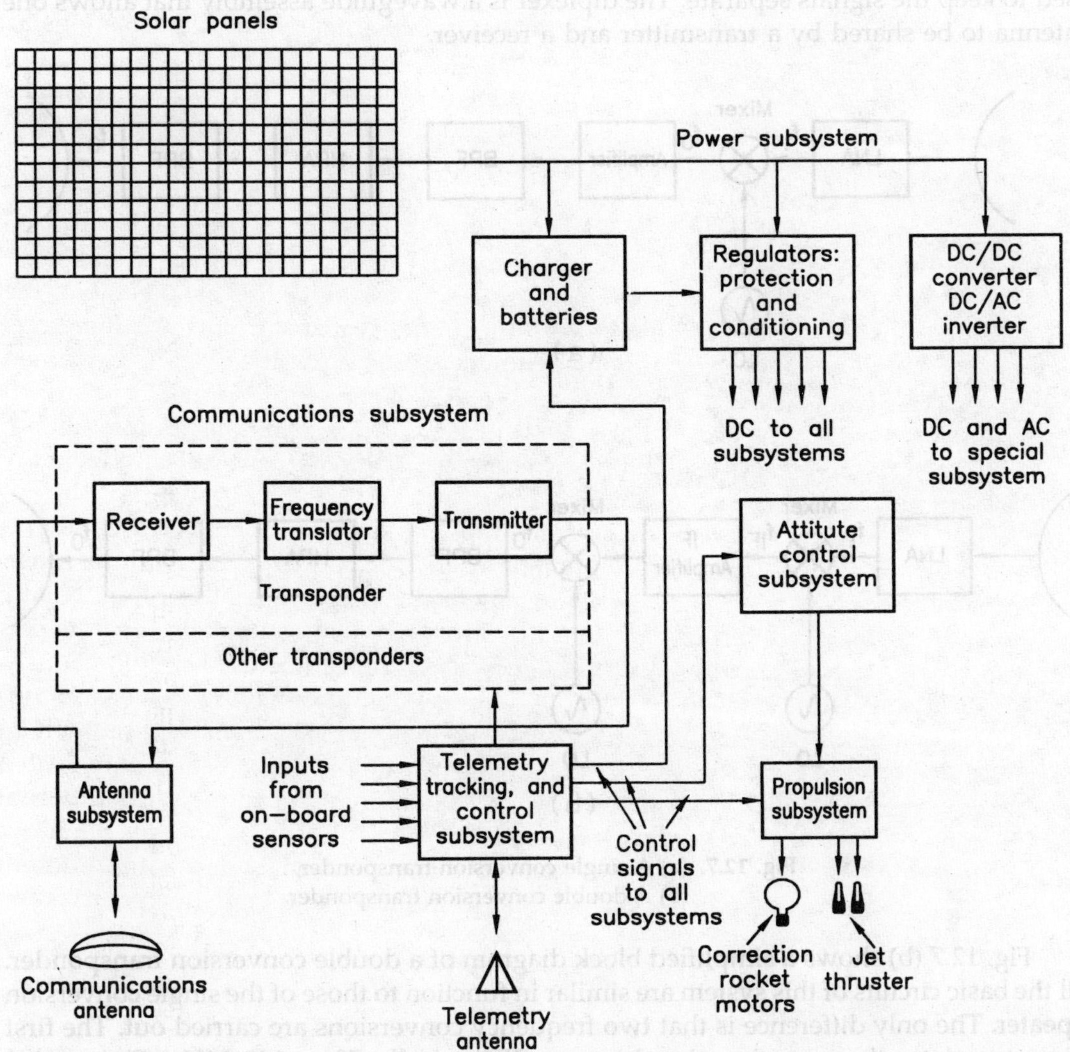

Fig. 12.6. General block diagram of a communication satellite.

transponder. The signal is very weak at this point even though it has been multiplied somewhat by the gain of receiving antenna. The low noise amplifier (LNA) gives it sufficient amplification. It is then translated in frequency. This is done in the mixer circuit where the original modulation is retained during this process. The mixer output is amplified again and fed to a band-pass filter, the purpose of which is to remove all but the desired down-link signal bandwidth. Finally, the down-link signal is amplified by high-power amplifier (HPA) in which usually a travelling wave tube (TWT) is used as the amplifying device. Its output is further filtered to retain only desired channel bandwidth by rejecting harmonies and intermodulation products. The resulting output is combined with outputs of other transponders and fed to a down-link antenna. In some cases, both the receive and transmit antennas are one and the same. A diplexer is used to keep the signals separate. The diplexer is a waveguide assembly that allows one antenna to be shared by a transmitter and a receiver.

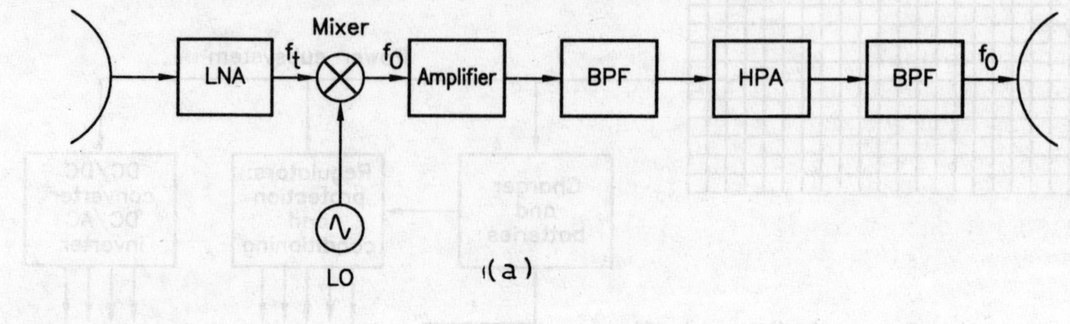

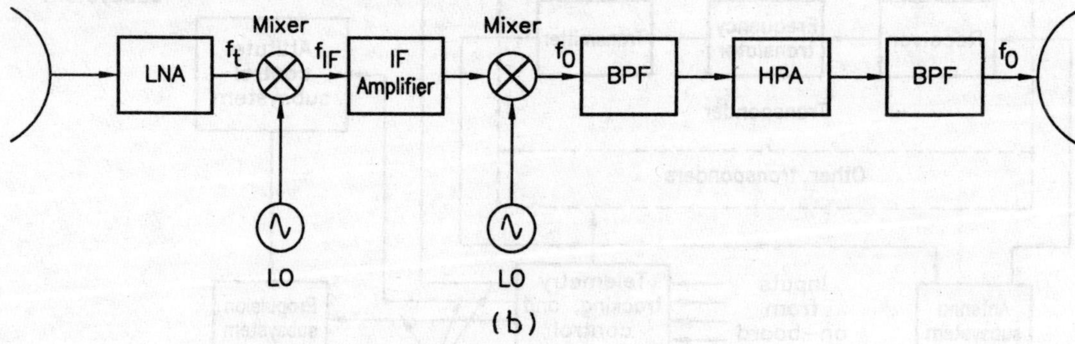

Fig. 12.7. (a) A single conversion transponder,
(b) A double conversion transponder.

Fig. 12.7 (b) shows a simplified block diagram of a double conversion transponder. All the basic circuits of this system are similar in function to those of the single conversion repeater. The only difference is that two frequency conversions are carried out. The first mixer translates the incoming signal into an IF, typically 70 or 140 MHz. The second mixer circuit translates this IF output to the corresponding down link frequency channel. The main advantage of double conversion transponder is that greater flexibility in filtering and amplification can be achieved. Amplification and selectivity at the lower IF level is far easier to attain. In another conversion arrangement, the up-link signal is demodulated to obtain baseband (video and audio in case of TV) signals and then modulated with the necessary carrier frequency. Each configuration has its own merits and is used depending on several other considerations.

Multichannel Architecture: There are two multi-channel architectures in use in communication satellites. One is a broadband system and the other a fully channelized one. The latter transponder system is more common where individual narrow band channels are provided for all the 12 signal bands. This arrangement is illustrated in Fig. 12.8. The receiver antenna feeds a demultiplexer (DEMUX), also called multicoupler which is a wave-guide assembly that separates the single wide-band input into equal feeds for separate channels. As stated earlier, a typical communication satellite spectrum is 500 MHz wide. This is divided into 12 separate channels, each with a bandwidth of 36 MHz. The centre frequency spacing between adjacent channels is 40 MHz, thereby providing a 4 MHz spacing between channels to minimize adjacent channel interference. Each filter selects one designated 40 MHz wide channel and feeds it to the associated LNA which provides desired amount of gain. Each channel has its own mixer driven by a common local oscillator (LO) whose frequency is 2.225 GHz in the C-band. The mixer output feeds into a bandpass filter which selects the desired down-link channel frequency band. For example, for the first channel (5.925 to 5.965 GHz) with centre frequency of 5.945, the filter will pass the difference product only *i.e.* 5.925 – 2.225 = 3.700 GHz to 5.965 – 2.225 = 3.740 GHz, a 40 MHz width with a useful bandwidth of 36 MHz, thus providing translation to the downlink frequency spectrum. The bandpass filter outputs from the channels drive their respective HPAs which usually employ Travelling Wave (TW) tubes. Again B.P. filters are provided at the output of HPAs to prevent passage of harmonics or intermodulation products. The B.P. filter outputs feed into a multiplexer (combiner) unit which creates one feed from 3.7 GHz to 4.2 GHz for the transmitting down-link antenna.

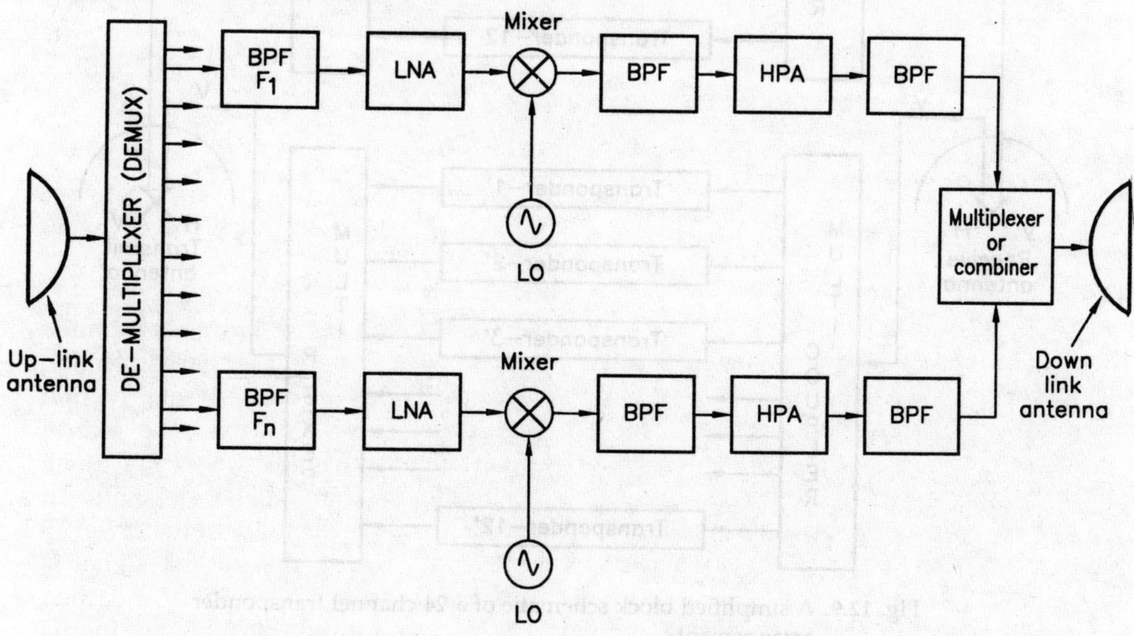

Fig. 12.8. A 12 channel transponder architecture.

24 Channel Transponder Architecture: As explained in section 12.2, by using polarisation discrimination of electro-magnetic (EM) radiations, it is possible to have 24 transponders co-existing in a single satellite and all using the same antenna by multiplexing. A simplified block schematic of such a C band architecture is shown in Fig. 12.9. The centre frequencies of various channels are offset by 20 MHz between the two polarizations and 40 MHz between the same polarization. The H and V polarized signals from the receive antenna are fed to corresponding multicouplers which split the

RF signal into necessary number of outputs, (12 in the case under consideration) to feed all the transponders with equal power. On necessary processing the outputs of both V and H side transponders are combined in separate multiplexers, the outputs of which feed separate transmitting antenna sections. The up-link and down-link centre frequencies of both 'V' and 'H' polarized channels are shown in Fig. 12.10. Note the offset by 20 MHz between the two polarisations and 40 MHz between the same polarisations as stated earlier.

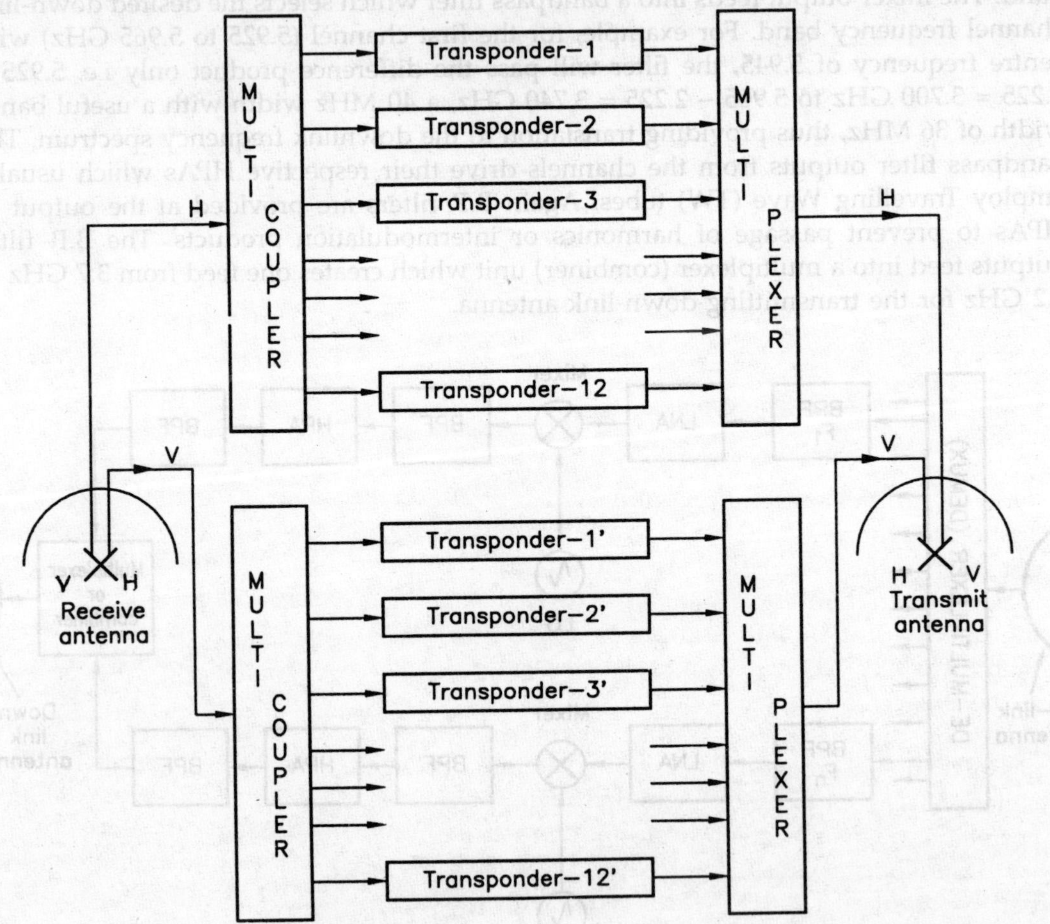

Fig. 12.9. A simplified block schematic of a 24 channel transponder arrangement.

Satellite Orbit Control: A geostationary satellite remains almost stationary with respect to earth because it has a tendency to move a bit due to gravitational forces of the sun, the moon and because of deviations of earth from true spherical shape. Thus, if it is not prevented the satellite may drift more than the permissible 0.1° of location assigned to it by the International Controlling Authority (I.C.A.). A deviation of 0.1° work out to about 0.75 km drift at the geosynchronous orbit. Therefore, it is necessary to constantly monitor its location and apply necessary corrections. For this all satellites have a telemetry tracking and control (TTC) subsystem [see Fig. 12.6] that

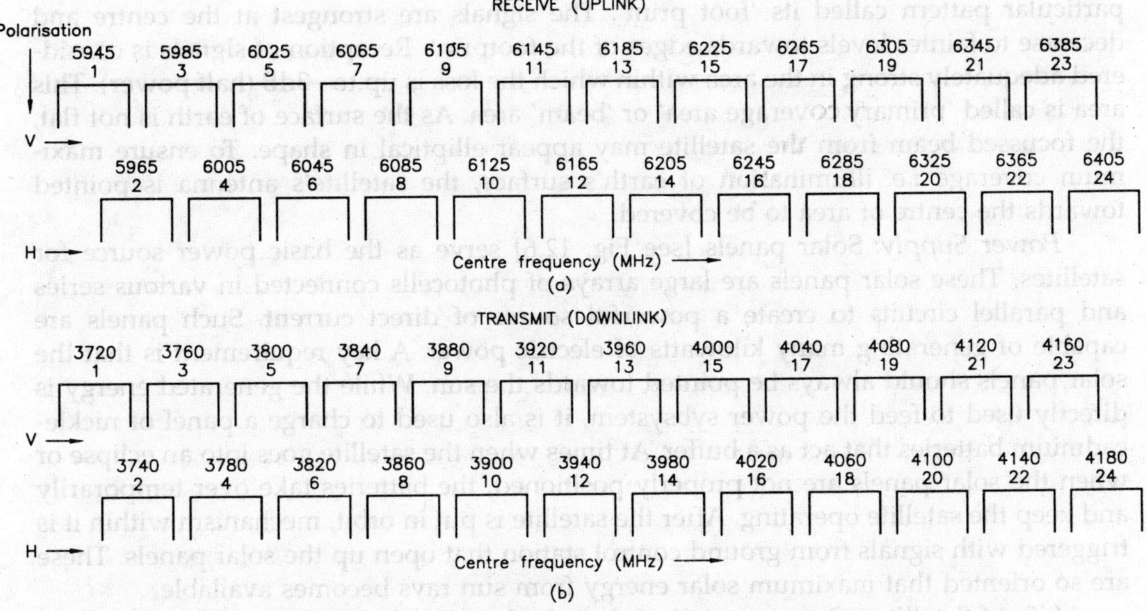

Fig. 12.10. A typical OFFSET Centre frequency plan for V and H channels of a 24 channel satellite system.

allows a ground station to monitor and control conditions in the satellite. The computers located at the land-based control centres keep on tracking the satellite and on detection of any deviation send command signals to the TTC unit to fire correction rocket motors periodically to keep the satellite into its correct course and orbit. When the satellite drifts away from its position in the orbit with respect to earth, the movement is called 'APOGEE' and when comes closer to earth it is called 'PERIGEE'. Thus, as necessary, two rocker motors, called apogee and perigee with sufficient fuel are placed in the satellite for carrying out necessary corrections on receipt of command signals. Such a provision of keeping satellite drift within a specified acceptable limit is called 'STATION KEEPING'. Initially when the satellite is launched, several rounds of corrective pushes are given to bring the satellite in its assigned parking slot.

Satellite Parking: The satellites were earlier parked (located) in the geostationary belt by giant rockets. However, now with the advent of 'space shuttle', satellites are released from the shuttle at proper altitude above the equator in a dormant state. After the satellite is released, sensers within it are triggered to activate it and deploy its antenna system and solar panels.

Earlier satellites were positioned with a spacing of 4° on the geostationary orbit but with increase in number of satellites, the spacing has now been reduced to 1° *i.e.* the longitudes of adjacent satellites differ by one degree. Thus an earth station may be within the footprints (signal available area) of several satellites and a slight error in adjusting the 'Look Angle' of the dish antenna may lead to getting hooked to another satellite than the one to be acquired. Therefore, extreme care is necessary while tracking different satellites.

Foot Print Area: The satellite's antenna transmits signals towards the earth in a particular pattern called its 'foot print'. The signals are strongest at the centre and decrease to fainter levels towards edges of the footprint. Reception of signals is considered adequately strong in the area within which the loss is up to –3dB (half power). This area is called 'primary coverage area' or 'beam' area. As the surface of earth is not flat, the focussed beam from the satellite may appear elliptical in shape. To ensure maximum coverage *i.e.* illumination of earth's surface, the satellite's antenna is pointed towards the centre of area to be covered.

Power Supply: Solar panels [see Fig. 12.6] serve as the basic power source for satellites. These solar panels are large arrays of photocells connected in various series and parallel circuits to create a powerful source of direct current. Such panels are capable of generating many kilowatts of electric power. A key requirement is that the solar panels should always be pointed towards the sun. While the generated energy is directly used to feed the power sybsystem, it is also used to charge a panel of nickle-cadmium batteries that act as a buffer. At times when the satellite goes into an eclipse or when the solar panels are not properly positioned, the batteries take over temporarily and keep the satellite operating. After the satellite is put in orbit, mechanism within it is triggered with signals from ground control station that open up the solar panels. These are so oriented that maximum solar energy from sun rays becomes available.

Life of Satellites: Over a period of time the batteries begin to deteriorate and this puts a limit to the life expectancy of the satellite. In addition, with passage of time, various electronic packages in the satellite also tend to wear out and cause malfunctioning. Another factor that limits the life of a satellite is the fuel provided in it for the rocket motors that are fired periodically to keep the satellite into its correct course of orbit. All this limits the useful life of a satellite between eight to ten years. When the satellite fails due to any of the above reasons, it is moved out of orbit to make room for a replacement.

If a satellite malfunctions for any system failure, it is now possible to service it by the space shuttle personnel or load it into the shuttle cargo bay for return to earth.

12.4 Earth Stations

The earth or ground station is the terrestrial base of the satellite system. It communicates with the satellite to carry out designated missions. The earth station may be located at the end-user's facilities or in a remote open space with ground-based intercommunication links between the earth station and end-users.

An earth station consists of six major subsystems. These are: (i) antenna subsystem, (ii) receiver subsystem, (iii) receive 'ground communication equipment' (GCE), (iv) transmit 'ground communication equipment' (GCE), (v) transmit subsystem, and (vi) power subsystem. In addition, common to all, are the telemetry, control and instrumentation subsystems. Many earth stations have multiple circuits for redundancy and reliability. These are switched on automatically if the main unit fails. A general block diagram of an earth station is shown in Fig. 12.11 and with its help, all the subsystems are described.

(i) *Antenna Subsystem:* It consists of parabolic reflector (dish antenna), waveguide transmission lines, mount with steering and tracking mechanisms and a diplexer. To save on cost and space, in most installations a single antenna is used, both to transmit

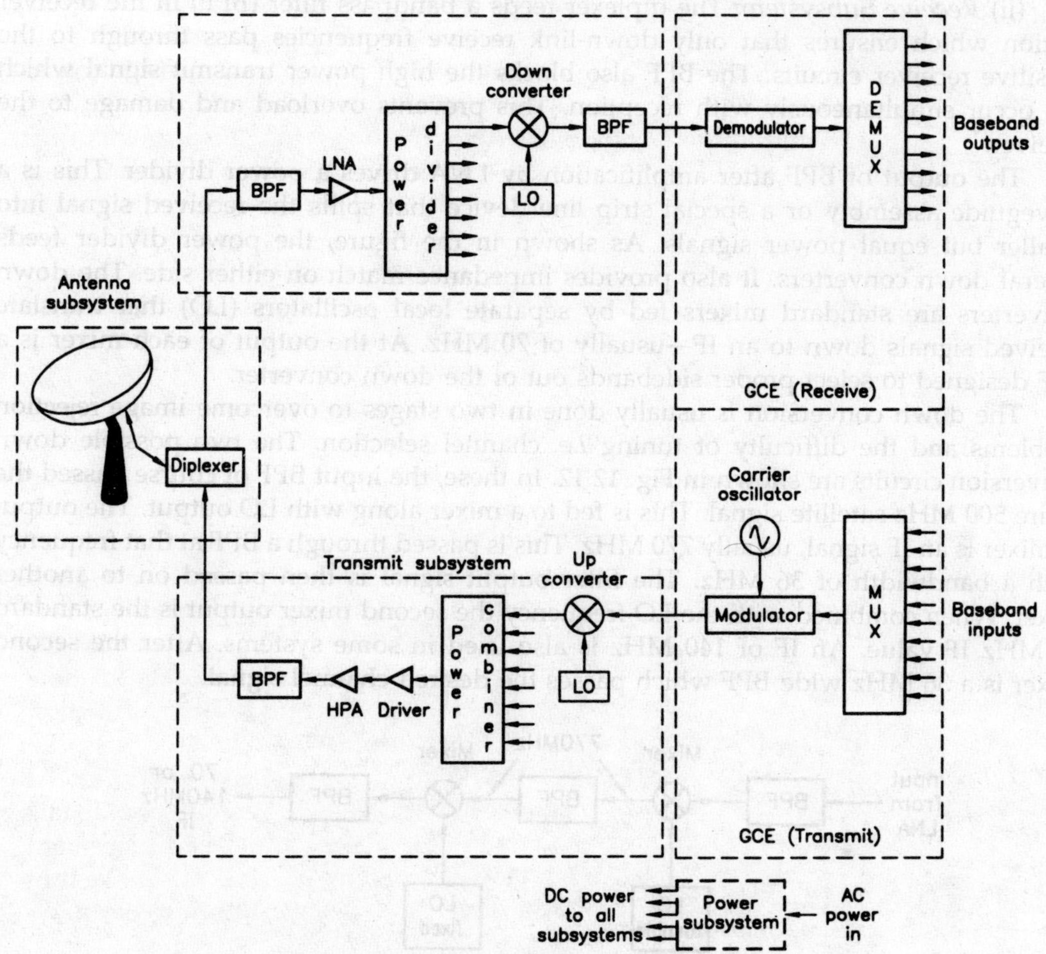

Fig. 12.11. General block diagram of a satellite earth station.

and receive, as the frequencies are far apart, 6 GHz and 4 GHz respectively. Since transmission to the satellite is involved, the antenna size would vary from 5 meters upwards to give tight antenna beam characteristics. The antenna mount is designed and located to meet the line of sight requirements between the satellite and earth station transmit—receive antennas. An earth station antenna has provision to steer it to initially pinpoint the satellite and to permit minor adjustments later if necessary. Remote controlled motor driven automatic tracking facilities are also provided to ensure correct acquisition at all times. Such a provision is also helpful if the earth station is to acquire more than one geostationary satellites during different hours of the day.

The diplexer (not to be confused with multiplexer) is a waveguide assembly designed to have three RF connecting points—say A, B and C. 'A' is connected to antenna, 'B' to the transmitter and 'C' to the receiver such that transmitter output will go only to antenna and RF signal picked up by the antenna will go only to receive side. Additional isolation is provided for transmitting and receiving 'V' and 'H' polarised signals as found in many satellites.

(ii) *Receive Subsystem:* The diplexer feeds a bandpass filter (BPF) in the receiver section which ensures that only down-link receive frequencies pass through to the sensitive receiver circuits. The BPF also blocks the high power transmit signal which can occur simultaneously with reception. This prevents overload and damage to the receiver.

The output of BPF after amplification by LNA drives a power divider. This is a waveguide assembly or a special strip line device that splits the received signal into smaller but equal power signals. As shown in the figure, the power divider feeds several down converters. It also provides impedance match on either side. The down converters are standard mixers fed by separate local oscillators (LO) that translate received signals down to an IF—usually of 70 MHz. At the output of each mixer is a BPF designed to select proper sidebands out of the down converter.

The down conversion is usually done in two stages to overcome image rejection problems and the difficulty of tuning *i.e.* channel selection. The two possible down conversion circuits are shown in Fig. 12.12. In these, the input BPF of course passed the entire 500 MHz satellite signal. This is fed to a mixer along with LO output. The output of mixer is an IF signal, usually 770 MHz. This is passed through a BPF at that frequency with a bandwidth of 36 MHz. The filter output signal is then passed on to another mixer. When combined with the LO frequency the second mixer output is the standard 70 MHz IF value. An IF of 140 MHz is also used in some systems. After the second mixer is a 36 MHz wide BPF which passes the desired channel signal.

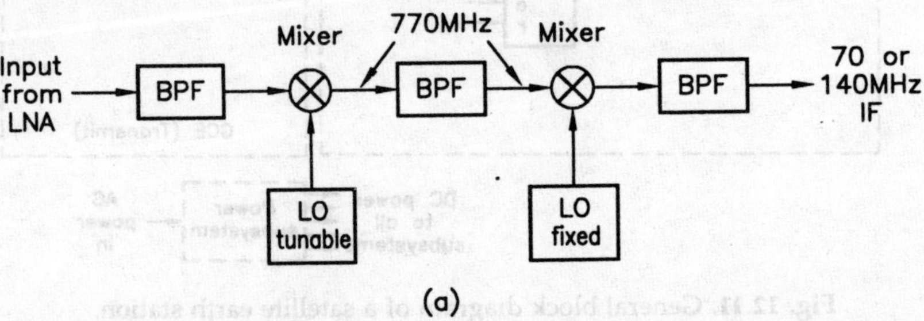

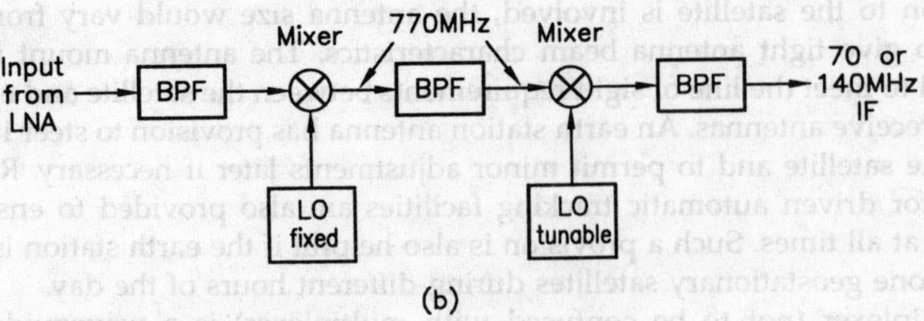

Fig. 12.12. Dual conversion down circuits (a) RF tuning, (b) IF tuning.

In one double conversion circuit [Fig. 12.12 (a)], the first LO is made adjustable and the second LO has a fixed frequency. This is called RF tuning. In the second circuit

[Fig. 12.12 (b)], called IF tuning, the first LO frequency is fixed and that of second is made adjustable, the end result in both cases bring the same *i.e.* 70 MHz IF with a bandwidth equal to 36 MHz.

Fig. 12.13 shows a full set of down converters as used in a standard earth station receiver. The power divider, which is infact a branching network feeds equal signals of 500 MHz bandwidth to each of the 12 dual-converters. A frequency synthesizer is used in this application. It generates a highly stable sinusoidal signal with input from a frequency standard crystal oscillator. Each synthesizer is set to a frequency necessary for each channel circuit. For example in the circuit of Fig. 12.12 (b), for channel one (1) that has a frequency range from 3700 MHz to 3740 MHz, with a centre frequency equal to 3720 MHz will need a LO (synthesizer) frequency setting in the first converter at 4490 MHz (3720 + 770 = 4490 MHz) to obtain an IF output at 770 MHz ± 18 MHz. Similarly, the second channel LO (synthesizer) setting will be at a carrier frequency = 3760 + 770 = 4530 MHz and that for the third channel at 4570 MHz and so on.

However, for all the channels, the second mixer local oscillator (LO) frequency will be 770 + 70 = 840 MHz to obtain the final IF output at 70 MHz. As labelled in Fig. 12.12 (b) the second LO frequency is variable for exact tuning which is indicated by the correctness of the final output. For a television channel, a well defined picture and good quality sound on a TV receiver will indicate correct fine tuning.

(iii) *Receive GCE:* The 70 MHz IF signals from the IF patch panel [see Fig. 12.13] are sent to the receiver GCE where they are demodulated and later fed to demultiplexing circuits where the original signals are finally obtained. In actual practice, depending on the nature of final signals, several levels of demodulation and demultiplexing may have to take place. Thus, TV programmes, multiple telephone calls and digital data for a computer network need somewhat different treatment before emerging in the final form. For TV signals the demodulator is an FM discriminator (detector). The most commonly used type is the PLL discriminator. Equalization and de-emphasis are also taken care of in the demodulators. The demodulators are typically packaged in a thin narrow vertical module that plugs into chassis in a rack. Many such identical demodulators are provided and can be interconnected as desired.

(iv) *Transmit GEC:* In this section, the baseband signals, such as telephone conversations are applied to a multiplexer which permits multiple signals to be carried on a single channel. The multiplexed signals are then applied to a modulator along with a carrier oscillator (LO). In case TV signals are the baseband inputs that need a full channel, such multiplexing is not necessary and are thus directly applied to corresponding modulator circuits. Each modulator output is a 70 MHz IF having a channel width = 36 MHz.

(v) *Transmit Subsystem:* Each IF output from the transmit GCE section is fed to a up-converter circuit in the transmit subsystem where it is translated to the final transmit frequency range. The up-converter circuits are of the same format as used in the receive subsystem. The first up-conversion is the same for all the channel signals to an IF = 770 MHz. However, the second up-converter section has synthesizers that provide a separate LO to each conversions circuit. This enables the final outputs to occupy a separate frequency spectrum in the UP-LINK RF range of 5.925 to 6.425 MHz in the C-band transmission. Note that each of the 12 channel signals continue to occupy a channel width = 36 MHz with ± 2 MHz guard bands.

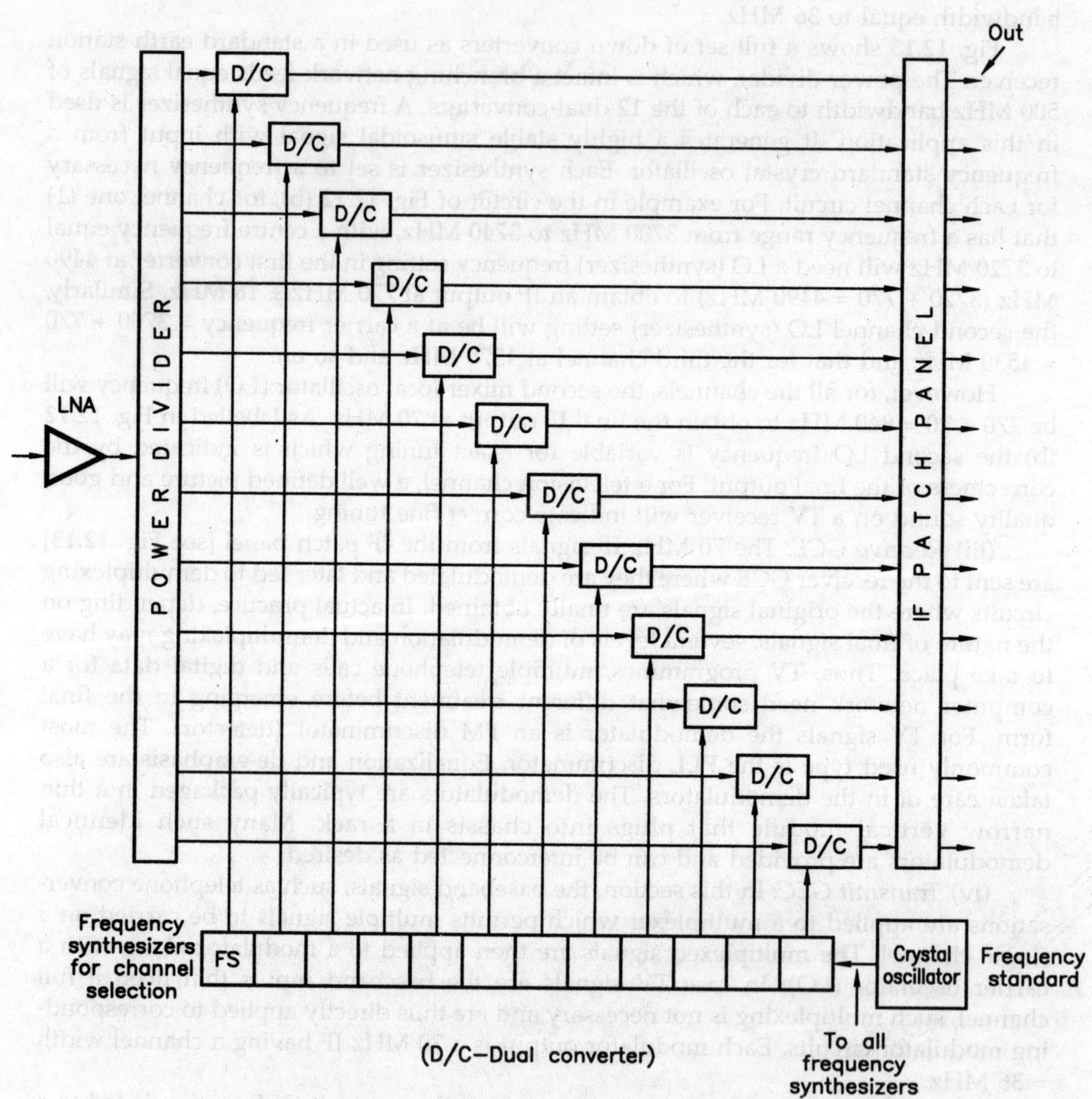

Fig. 12.13. A typical 12 channel down converter arrangement employing frequency synthesizers.

All the up-converted signals are then applied to a power combiner [see Fig. 12.11]. This waveguide like device combines equally all the RF signals into the final signal to be transmitted. The combined signal feeds into a driver circuit which produces sufficient power to operate the final high power amplifier (HPA). This is either a TWT (travelling wave tube) or a Klystron. Such amplifiers generally generate many hundreds of watts of output power. A Klystron amplifier can develop power to about 0.5 kW. The amplifier output is applied to a BPF and then to the common diplexer and antenna as shown in the figure.

(vi) *Power Subsystem:* Most earth stations receive their power from normal ac sources. Standard power supplies convert ac to dc voltages required to operate all the subsystems. However, many earth stations have backup power systems which may consist of diesel engine run ac generator sets which automatically start when mains supply fails for any reason. Smaller systems may use uninterruptable power supplies (UPS) which derive their main power from batteries. The need for uninterrupted power supply is essential for reliable communication of telephone circuits, TV programmes, computer data, etc.

Telemetry Equipment: The telemetry equipment consists of a receiver and recorders and indicators that display telemetry signals. These signals may be received by the main antenna or by a separate telemetry antenna. A special receiver on a frequency different from communication channels is used for telemetry purposes. The telemetry signals from various sensors and transducers in the satellite are multiplexed on to a single carrier and sent to the earth station. The earth station receiver demodulates and demultiplexes these into individual outputs. These are then recorded and also sent to various indicators, such as strip chart recorders, meters and digital displays. The signals can also be sent to a computer where these can be further processed and stored. Mathematical operations on the signals may be necessary. For example, the number received from the satellite representing temperature may have to be processed by some mathematical formula to obtain actual output temperature in degree calsius.

The control subsystem permits the ground stations to control the satellite. This system usually contains a computer for entering commands which modulate a carrier that is amplified and fed to the main antenna. The command signals can make adjustments in the satellite's attitude, turn on transponders and so on. These can also be used to actuate redundant circuits if the circuits they are backing-up fail.

Not all earth stations have the telemetry control and instrumentation subsystems; only those designated as control stations do. Most earth stations just send or receive.

12.5 International and Regional Satellites

There are a number of International Telecommunication Satellites over the Atlantic, Pacific and Indian Oceans operating as relay stations to hundreds of ground stations around the world. This has enabled not only national but also international coverage of TV programmes, data transmission and thousands of telephone lines all over the world. For this, high power highly directive land based transmitters beam signals to high power geostationary satellites which in turn provide large power flux densities enabling very good reception over vast areas even with smaller diameter dish antennas.

Regional Service Satellites: Some important Geostationary Satellites which beam programmes to Middle-East, Asia and South-East Asia are as under:

(1) *ASIA SAT-1:* It is located at 105.5° E in the orbit with foot prints covering Middle-East, Indian Ocean region and Asia excluding Japan. It carries all STAR TV programmes and many others including Zee TV.

(2) *ASIA SAT-2:* It is located at 100.5° E and transmits NBC, TVS, MCM, TVSN and many other channels.

(3) *Pan Ams (PAS)-4:* Its location in the orbit is at 68.5° E and carries channels of great interest including SONY-ENT, BBC World, Home TV, ABN, CNN and DD Intl.

(4) *INTELSET-703:* With its location at 57.0° E it beams programmes of TVI, NEPC, Money TV and most Tamil channels.

(5) *INTELSET-704:* It is located at 66.0° E and carries programmes of YES TV, Fashion TV, and Eenadu TV (Telugu) channels.

(6) *INSAT 1D:* With its location at 83.0° E it beams DD-1 and DD-16 channels.

(7) *INSAT 2B:* It is located at 93.5° E and being a domestic satellite carries some DOORDARSHAN channels inclusive of CNN International.

(8) *INSAT 2C*:* With its location at 93.5° E and as the most important domestic satellite of India it carries most of the Doordarshan channels.

(9) *THAI COM 3:* It is located at 78.5° E and besides other language programmes beams for India RAJ TV and PUNJAB TV channels.

(10) *PALPA C 2:* With location at 113.0° E it carries Australia STAR TV, CNBC and TMT programmes.

(11) *ARABSAT (location 31° E):* It serves Russian countries but has foot prints over part of Asia.

(12) *Stationar (location 80° E):* It serves mostly middle-east countries.

12.6 Domestic Satellites

There are a number of domestic satellite systems meant mainly for television programmes and other services within the territory of owner countries. This enables economic coverage of a large well defined geographical area with mountains and remote areas where terrestrial systems would require transmitters for relatively few users.

INDIA'S DOORDARSHAN is one of the largest domestic broadcasting organisations in the world having over 800 transmitting stations spread all over the country. It has 18 channels using separate transponders on the Indian National Satellites (INSAT), namely, INSAT ID (83° E), INSAT 2B (93.5° E), and INSAT 2C (93.5° E) only 15 km away from INSAT 2B which carries 24 transponders both in the C band and Ku band frequency spectrums.

CCIR has allotted the frequency band—1.55 GHz to 2.7 GHz in the S-band for domestic rebroadcast purposes. Doordarshan uses the range 2.5 to 2.7 GHz for its TVRO (TV Receive only) sets provided at various television stations all over India to receive programmes from its own satellites and rebroadcast them over Low Power (LPT) and Very Low Power (VLPT) transmitters located in different districts of many states.

TVRO Earth Station: The receive subsystem of a TVRO earth station designed to receive down-link signals from two transponders of India's domestic satellite INSAT-ID is shown in Fig. 12.14. The two channels, each of bandwidth 40 MHz (36 MHz ± 2 MHz

* India's Satellite INSAT-3B was launched on 22nd March 2000. Its communication payload will provide 12 extended C-band channels and 3 channels in the Ku band thus enabling more TV channels and introduction of Direct-To-Home (DTH) service.

guard bands) are between 2.555 to 2.635 GHz with centre frequencies of 2.575 and 2.615 GHz respectively. A double conversion circuit is provided to translate down-link signals to an IF of 70 MHz. After the S-band LNA is a BPF that passes the desired frequency spectrum of 80 MHz (2.635 – 2.555 = 80 MHz) which feeds into the first mixer with LO = 1.5 GHz. A crystal oscillator is used to generate 60 MHz sinusoidal signal and its output goes to a multi-harmonic generator which yields a CW output at 1.5 GHz (60 MHz × 2 = 1.5 GHz). The converter output with frequency band between 1.055 to 1.135 GHz is selected by a BPF that follows the mixer. This is amplified through a three stage IF amplifier to give the signal a boost of about 40 dB gain. The amplified signal is fed to the second mixer to obtain IF output of either of the two incoming channels. For this, one voltage controlled oscillator (VCO) is tuned at 1.185 GHz and the other at 1.145 GHz. A change-over switch enables selection of one of these depending on the downlink channel to be selected. The first channel signal lies between 1.055 to 1.095 GHz and that of the second from 1.095 to 1.135 GHz with centre frequencies at 1.075 GHz and 1.115 GHz. With local oscillator switch at LO (L) *i.e.* 1.145 GHz, the difference IF output is 1.145 GHz – 1.095 GH = 50 MHz to 1.145 GHz – 1.055 GHz = 90 MHz with centre IF frequency = 70 MHz. Similarly, when the LO switch is at LO (H) *i.e.* 1.185 GHz, the output is again between 50 to 90 MHz with Centre IF = 70 MHz but with modulation of the second channel. The B.P.F. after the 2nd mixer selects this IF signal (50 to 90 MHz) for further processing.

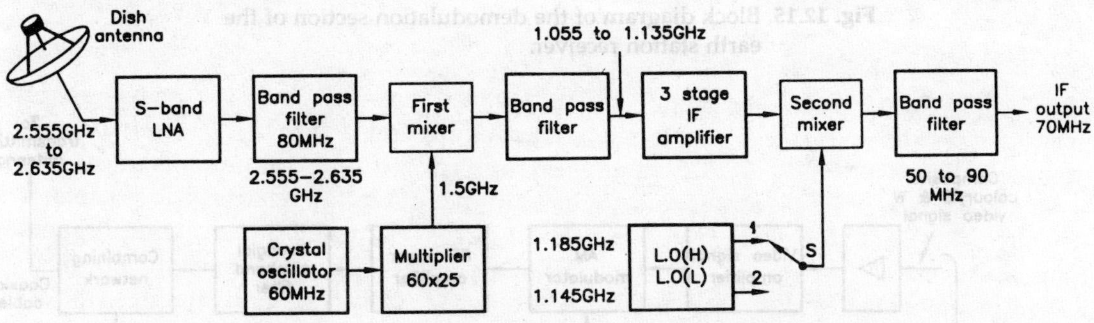

Fig. 12.14. Receive Subsystem of a TVRO earth station.

Demodulation to Baseband Signals: Fig. 12.15 shows block diagram of the earth station receiver's demodulation equipment that converts the FM modulated IF signal at 70 MHz ± 18 MHz to the baseband range of video and audio signals. A two stage bandwidth limited IF amplifier followed by a band-pass filter enables IV (P–P) output at 70 MHz ± 18 MHz. This signal level is necessary for effective detection by the FM demodulator. A limiter precedes the detector to ensure equal amplitude of all the sidebands of frequency modulated signal. A PLL (Phase Lock Loop) type FM demodulator is commonly used to obtain video and intercarrier sound IF baseband range that extends from dc to 6.5 MHz. A de-emphasis circuit at the output of demodulator is for obtaining original amplitude of the modulating signal. The video and intercarrier sound outputs are separated by suitable bandpass filters. The video signal is amplified to provide an output signal which is the same as developed at the transmitting end before modulation.

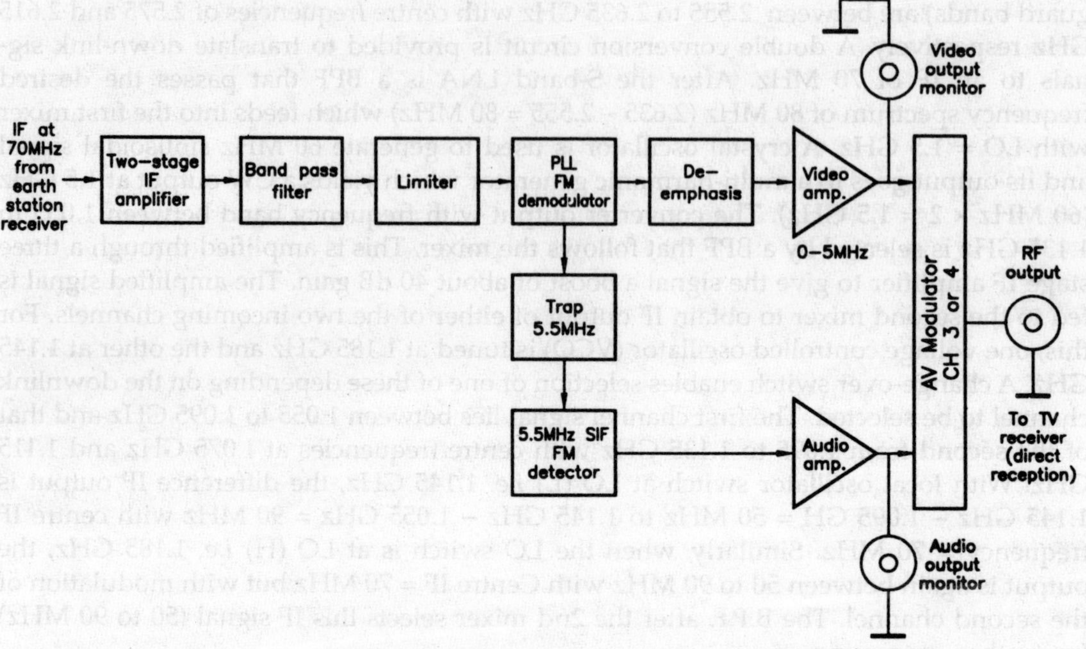

Fig. 12.15. Block diagram of the demodulation section of the earth station receiver.

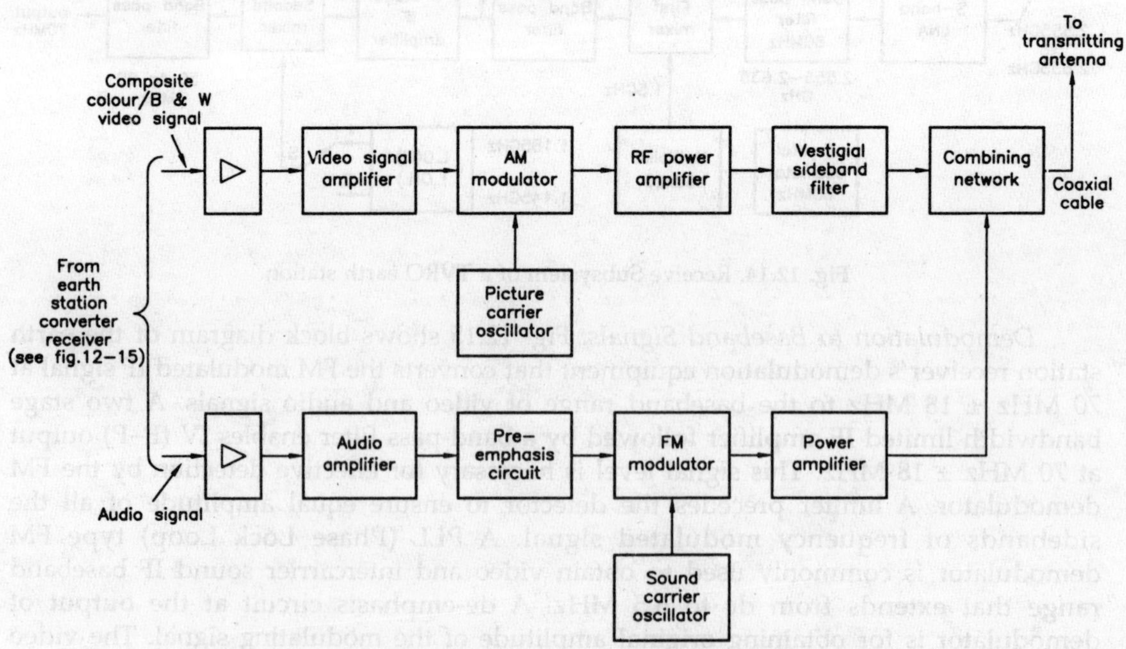

Fig. 12.16. General block diagram of a HPT/LPT transmitter for rebroadcast of TV programmes all over the country.

The INSAT-1D has provision for accommodating two audio channels, one centred around 5.5 MHz and the other on either side of 6 MHz. One of these is for carrying audio in a regional language and the other as usual for Hindi or English. Depending on the desired output, the FM demodulator oscillator is set for either 5.5 or 6 MHz. The demodulated output is amplified to obtain audio output that extends up to 15 KHz.

The video and audio outputs thus obtained are suitable for driving a monitor receiver. However, for direct reception on a conventional TV receiver as in the case of direct reception TAVRO sets, these outputs are modulated on either TV channel 3 or channel 4 carrier frequencies. The circuit block which accomplishes this at low level outputs is called AV (Audio-Video) modulator. Its RF output is fed to the input socket of a TV receiver to obtain picture and sound outputs in the same way as from signals picked up by an antenna.

Rebroadcast over H.P. and L.P. Transmitters: A simplified block diagram of a transmitter for rebroadcast of signals obtained from INSAT 1-D by the earth station receiver is shown in Fig. 12.16.

The earth station receiver outputs [see Fig. 12.15] obtained at the video and audio monitor points serve the same purpose as that obtained from a TV studio. The carrier frequencies, both for amplitude modulation of video signal and frequency modulation of sound signal are chosen depending on TV channel allotted to the station.

It may be noted that for down-link TV signals obtained from an INTEL satellite which originate from a country that employs different standards like NTSC or SECAM, a special converter-receiver becomes necessary along with the earth station receiver for conversion of baseband signals to 625 line PAL standards.

13

CABLE TELEVISION CHANNELS

Introduction

While shielded cables are used for CATV signal distribution, some radiation can still occur due to accidental open or short circuits and also due to partial mismatch at their terminations. Therefore, care is exercised in selecting cable frequencies so that there is no co-channel interference with local TV broadcasts. For this, in some cable systems, the frequencies are kept slightly off the frequencies allotted for television broadcasts, but close enough to allow tuners of TV receivers and converters to tune them in with AFT control at the correct values. One such choice is harmonically related channels (HRCs) where all the picture carrier frequencies are an integral multiple of the channel width. The added advantage of using HRC system is simplification of frequency synthesis circuits used for tuning at the Head-End section of CATV systems and also in converter equipments. However, in India and many other countries, where terrestrial broadcasts in any location are not more than two or three, the need to off-set channel frequencies is not considered necessary and hence, not practised.

Cable channels are generally restricted to VHF bands because of higher cable losses at UHF frequencies. But the problem of adjacent-channel interference arises when all the channels are used. This can lead to wind-shield wiper or venetion-blind effects in the picture. However, this problem is solved by improved IF selectivity in conversion circuits by using 'SAW' filters. In addition, interference in cable systems is minimized by balancing the signals for all channels at a common level. Also, compared with TV broadcast, signals, the sound carrier level is usually kept much lower than the picture carrier level.

Furthermore, there is no restriction in CATV channel allocation between TV channels received via satellites and local TV broadcasts picked up by conventional antennas. This is so because each received channel is processed to obtain base-band video and audio signals which can then be modulated on any CATV channel. Based on the factors described above, cable channel allocations in use at present are detailed in this Chapter.

13.1 Cable Channels

In earlier cable systems 6 or 11 channels were considered enough and thus easily provided in VHF bands. However, with rapid increase in available TV channels from communication satellites, the need for more cable channels was felt. It was then decided to use the frequency spectrum, earlier left unused, that exists between VHF bands I and III and frequency spectrum available after VHF band III and before UHF TV channels. Depending on location, the cable channel allocations are divided into MID, SUPER and HYPER band channels.

The tuners of earlier TV receivers do not have provision to select mid-band, super-band and hyper-band channels. Therefore, a converter has to be used with such receivers for translating all cable channel frequencies to a designated VHF channel—usually 3/4 or a higher UHF channel. The user subscriber keeps his TV receiver tuned to this specified channel and cable channel selection is done at the converter. However, recent models of television receivers have a tuner that can select midband, super band and hyper band channels directly along with VHF and UHF TV channels and thus do not need a converter. Such receivers are called 'cable-ready' television receivers.

Cable channel allocations done by CCIR are different for the three television system. For the American NTSC system, where channel width is 6 MHz, more cable channels can be obtained within a given RF spectrum as compared to European PAL system where channel width is 7 MHz. Here again in the PAL variant 'I', adopted by U.K. where the channel width is 8 MHz, still lesser cable channels can be accommodated in the same frequency spectrum. In the SECAM television system, cable channel allocation is still different because of some other constraints. Therefore, caution must be used while procuring any CATV equipment to ensure that its specifications conform to the TV and cable standards adopted by the user country. We will confine our discussion to cable channel allocations applicable to CCIR 625 line B&W and PAL B&G standards adopted by Continental Europe, Middle-East and most Asian countries. As detailed in Chapter 11, India and many adjoining countries use PAL B&G standards for TV transmission with effective channel width = 7 MHz.

13.2 Cable Channel Allocations

In order to meet growing need for more cable channels, CCIR raised the allocation to 44 channels for the 625 line B&W and PAL B&G TV standards. These are detailed in the chart drawn next. TV channel allocations for this system are also listed along with cable channels to indicate their relative location in VHF and UHF bands.

TV channels available in VHF I and III bands but not used for terrestrial broadcasts are availed of by cable operators to keep the used frequency spectrum as low as possible and thus keep cable losses to a minimum. In addition, FM band that extends from 88-108 MHz is normally not fully used for FM radio broadcasts. Thus another two channels (Z + 1) and (Z + 2), (see chart) are often used for CATV channels. This raises the available cable channels to 46.

As necessary, IF frequencies, both picture and sound remain, the same for TV and cable channels. These are:

Picture (Vision) IF = 38.9 MHz and

Sound (Audio) IF = 33.4 MHz.

Cable and Television Channel Allocation Chart for CCIR 625 Line PAL B&G Colour TV Standards Vision (Picture IF = 38.9 MHz, Sound IF = 33.4 MHz

Channel No. Cable	Channel No. TV	Frequency Band	Channel Designation	Channel Bandwidth (MHz)	Channel's Frequency Range (MHz)
	2	VHF	E_2	7	47-54
		Band-I i.e.			
	3	Lower VHF	E_3	7	54-61
		Band			
	4		E_4	7	61-68
1		Vacant	X/SO1	7	68-75
2		Bands	Y/SO2	7	75-82
3		(68-88MHz)	Z/SO3	7	82-89
4		and FM Band	Z+1	7	89-96
		(88-108MHz)			
5			Z+2	7	96-103
6		M	S_1	7	104-111
7		I	S_2	7	111-118
8		D	S_3	7	118-125
9			S_4	7	125-132
10		B	S_5	7	132-139
11		A	S_6	7	139-146
12		N	S_7	7	146-153
13		D	S_8	7	153-160
14			S_9	7	160-167
15			S_{10}	7	167-174
	5	VHF Band-III	E_5	7	174-181
	6	*i.e.* Upper	E_6	7	181-188
	7	VHF Band	E_7	7	188-195
	8	Range	E_8	7	195-202
	9	174 to	E_9	7	202-209
	10	230	E_{10}	7	209-216
	11	MHz	E_{11}	7	216-223
	12		E_{12}	7	223-230

Channel No.		Frequency Band	Channel Designation	Channel Bandwidth (MHz)	Channel's Frequency Range (MHz)
Cable	TV				
16		S	S_{11}	7	230-237
17		U	S_{12}	7	237-244
18		P	S_{13}	7	244-251
19		E	S_{14}	7	251-258
20		R	S_{15}	7	258-265
21			S_{16}	7	265-272
22		B	S_{17}	7	272-279
23		A	S_{18}	7	279-286
24		N	S_{19}	7	286-293
25		D	S_{20}	7	293-300
(300 MHz to 302 MHz Band is not used)					
26		H	S_{21}	8	302-310
27		Y	S_{22}	8	310-318
28		P	S_{23}	8	318-326
29		E	S_{24}	8	326-334
30		R	S_{25}	8	334-342
31			S_{26}	8	342-350
32		B	S_{27}	8	350-358
33		A	S_{28}	8	358-364
34		N	S_{29}	8	364-374
35		D	S_{30}	8	374-382
36			S_{31}	8	382-390
37			S_{32}	8	390-398
38			S_{33}	8	398-406
39			S_{34}	8	406-414
40			S_{35}	8	414-422
41			S_{36}	8	422-430
42			S_{37}	8	430-438
43			S_{38}	8	438-446
44			S_{39}	8	446-454
45			S_{40}	8	454-462
46			S_{41}	8	462-470

Note: Cable channels with designation 'S' (S_1, S_2 etc.) are also referred to a 'S' band channels.

TV Channels in UHF Band

(i) *UHF (U) Band IV (470 to 582)*
It contains 16 channels between 21 to 36 (channel width = 7 MHz)

(ii) *UHF (U) Band V (660 to 870 MHz)*
It accommodates 33 channels between 37 to 69 with allowed channel width = 8 MHz. However, effective channel width is the same = 7 MHz.

Notes: (1) For cable channels in the HYPER band and TV channels in UHF band V, while the allotted bandwidth is 8 MHz the effective *i.e.* used channel width is 7 MHz as for channels in other bands. The additional 1 MHz space is guard band to prevent any adjacent channel interference that can occur at very high frequencies.

(2) In any cable or TV channel, if the channel width is say from 'a' to 'b' MHz, then picture carrier location is at (a + 1.25) MHz. For example cable channel S_1 of mid-band, that extends from 104-111 MHz has its picture carrier location at (104 + 1.25) MHz *i.e.* 105.25 MHz.

(3) Similarly, the sound carrier is located at (b-0.25) MHz, if the channel width is 7 MHz and at (b-1.25) MHz if allowed channel width is 8 MHz. Thus for cable channel 'S_1' it is 110.75 MHz and for cable channel S_{21} (302-310) MHz it is 308.75 MHz.

(4) In bilingual satellite transmissions, the intercarrier sound carriers are located somewhat differently as shown in the chart of satellite channels drawn in chapter 16.

13.3 Preferred Cable Channels

The equipment and cable run costs rise with increase in frequency of operation and therefore cable operators prefer to use low frequency channels as far as possible. Thus in a 6 channel CATV systems, commonly used in small towns and villages, the preferred channels are E_2, E_3, E_4 of the VHF I band and E5 to E8 out of VHF III band. One of these channels, depending on the location is left for direct reception of terrestrial local broadcast. Similarly, a cable operator offering 11 channels inclusive of local TV channels will choose all the 3 channels of VHF band I and 8 of VHF band III.

The use of VHF bands I and III channels has the additional advantage that subscribers possessing earlier models of TV receivers and VCRs, the number of which is quite large, can receive programmes of all the 11 channels without the need of a set TOP converter.

However, when the channels offered increase to become say 21 as is common in big cities, the mid-band channels S_1 to S_{10} also come into use along with 11 VHF channels. Here, cable operators allot popular channels in VHF bands for the benefit of subscribers having earlier TV receivers. However, for receiving all the channels a TOP converter or a cable ready TV receiver becomes necessary.

When the number of channels offered exceeds 21 to become 31 or more, as is happening in Metros, all the mid and super band channels along with VHF band channels have to be used. This makes the cable system expensive both in equipment cost and need for expensive low loss cables. To circumvent this to some extent, it is common practice to include channels X, Y, Z (also called S_{O1}, S_{O2}, S_{O3}) and (Z+1), (Z+2) and drop equal number of high frequency channels.

The choice of channel numbers also depends on interference that may occur on certain channels due to their use by other services. Thus there is no fixed rule on channel selection and differs somewhat from operator to operator and location of cable station. In due course of time, cable systems offering 50 channels or even more will come into use. This would need the use of hyper band channels and expensive high frequency modulators, mixers and special high grade coaxial cables. Such systems are economically viable only if the number of subscribers is quite large.

14

CABLE SIGNAL PROCESSING

Introduction

Satellite down-link signals on receipt at a 'TVRO' ground station need conversion from their FM mode into normal vestigial sideband (VSB) format which conforms to television standards adopted by the receiving country. For this, selected SAT-TV signals are first processed to obtain baseband signals which are then modulated on assigned CATV channel carriers to obtain standard RF TV signals. This process involves the use of antenna system, Front-end converter and Head-end sections of the cable system. Since 'reception only' is the aim at any CATV station, signal processing is much simpler. There is no need for a diplexer as no transmission occurs from the earth station. Also, there is no need for transmit-reject filters [see Fig. 12.11] and signal received at the dish antenna can be fed directly to an LNBC and its output sent to satellite receivers in the control room. Signal processing of terrestrial transmissions from local TV stations and those obtained from VCRs is much simpler because these are already in the standard VSB format. The processes involved in both cases are described in this chapter. However, before proceeding to know necessary details, the reader is advised to recapitulate the contents of chapter 1 where the entire gamut of CATV is introduced. It will also be helpful to revise chapters 4 and 12 which are devoted to microwave techniques, dish antennas and satellite communication systems.

14.1 Satellite Acquisition

In antennas designed for receiving TV signals from satellites, a horn is mounted at the focal-point of the dish which collects signals from the parabolic reflector and feeds them to the low-noise block converter (LNBC). Such a location of the horn on the dish, which is quite popular in CATV systems, is called 'Prime Focus Feed'. Maximum antenna gain is along its axis and therefore it is aligned exactly in the direction of 'look angle' of the wanted satellite to obtain maximum signal pick-up.

Longitude and Latitude of Earth Stations: Satellite tracking from any location on the surface of earth cannot be done at random in the absence of any data for correct orientation of dish antenna. For this, it is necessary to first fix position coordinates, for both the receiving station and wanted satellite. Longitude and latitude form a system for locating any given point on the surface of earth. The lines drawn from north pole to south pole on the earth's surface are called longitudes or meridians. The reference line for longitude measurements called prime meridian is the one that passes through Greenwich in England. Similarly, lines of latitude are the ones drawn from east to west on the surface of earth and parallel to the equator. The equator that serves as reference for latitude measurements is the central latitude line that divides the earth into north and south hemispheres. Another reference is the central point of earth called 'geocentre'. All the three references are shown prominently on the earth's globe in Fig. 14.1.

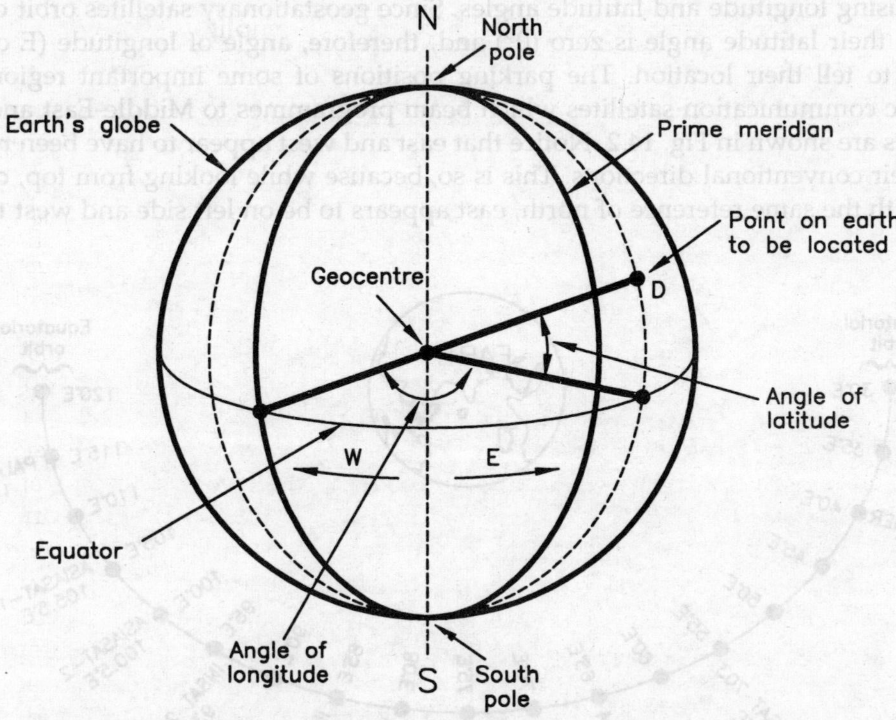

Fig. 14.1. Longitude and latitude lines for locating any point on the surface of earth.

Latitude is defined as the angle between the line drawn from the location (point) of interest ['D' in Fig. 14.1] on the surface of earth to the geocentre and the line between geocentre and intersection of equator and the meredian on which the point of interest lies as illustrated in the figure. As obvious, zero-degree latitude would be at equator whereas 90° latitude will be either at north or south pole. Usually an N or S is added to the latitude angle to indicate northern or southern hemisphere.

Similarly longitude of the same location (D) is the angle between the line connecting geocentre and intersection of prime meredian and equator and the line between geocentre and intersection of equator and meridian on which the point of interest lies. This is also illustrated in Fig. 14.1. The designation east or west (E or W) is usually added to the longitude angle to indicate whether the angle is being measured to the east or west of prime meridian. For example, longitude and latitude angles for Delhi are 77.0° E, and 28.5° N, for Bombay 73.0° E and 19.5° N, for Calcutta 88.0° E and 22.5° N and for Madras 80.5° E and 13.0° N. The location of these cities on the map of India with reference to Fig. 14.1 will enable clear understanding of these angles. All the CATV stations in each of these cities will have nearly the same angles of longitude and latitude. Similarly, the location angles of any other CATV station can be found out from geography map of the country.

Location of Satellites: Since geosynchronous satellites travel at the same speed at which the earth revolves, these are in a relative sense at a fixed point with respect to each other. As such, the location of such communication satellites is specified by a point on the surface of earth directly below each satellite. This point is known as sub-satellite point (SSP) and its location is determined in the same way as for any city or ground station using longitude and latitude angles. Since geostationary satellites orbit over the equator, their latitude angle is zero (0°) and, therefore, angle of longitude (E or W) is enough to tell their location. The parking positions of some important regional and domestic communication satellites which beam programmes to Middle-East and Asian countries are shown in Fig. 14.2. Notice that east and west appear to have been reversed from their conventional directions. This is so, because while looking from top, down at earth with the same reference of north, east appears to be on left side and west towards east.

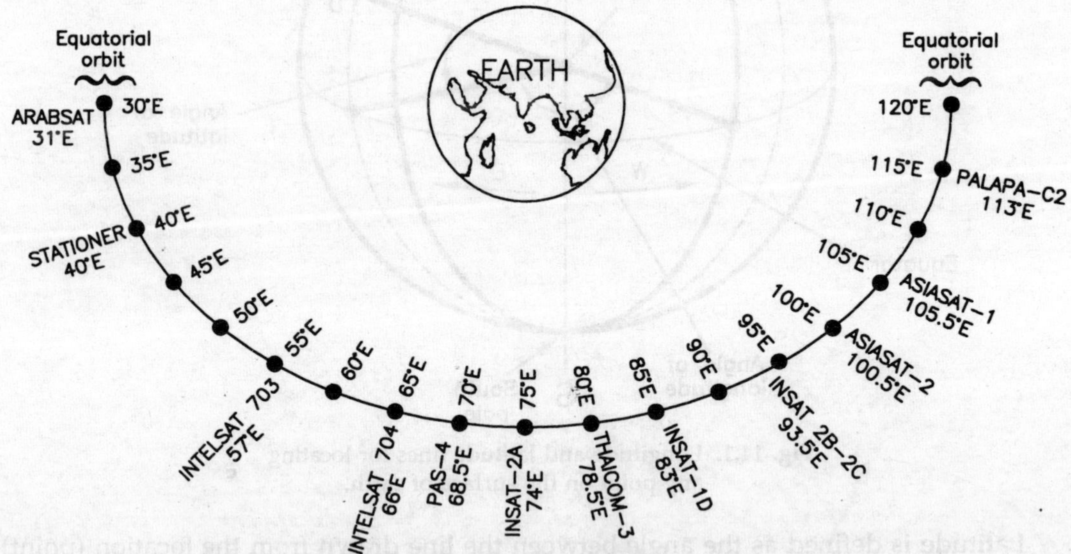

Fig. 14.2. Parking locations of some important satellites having foot-prints over India and adjoining countries.

14.2 Determination of Elevation and Azimuth Angles

The antenna for down-link beaming from a satellite is inclined at a particular angle which depends on its parking location and chosen footprint area on the earth. Therefore, to acquire any satellite, the earth station receiving antenna must point at the 'look angle' of that satellite. For this, two angles, called ELEVATION and AZIMUTH are specified. The angle of elevation is that angle which appears between the line from earth station antenna to the satellite and the line between the antenna and earth's horizon. This is pictured in Fig. 14.3 (a). The azimuth angle is the angle of direction. It

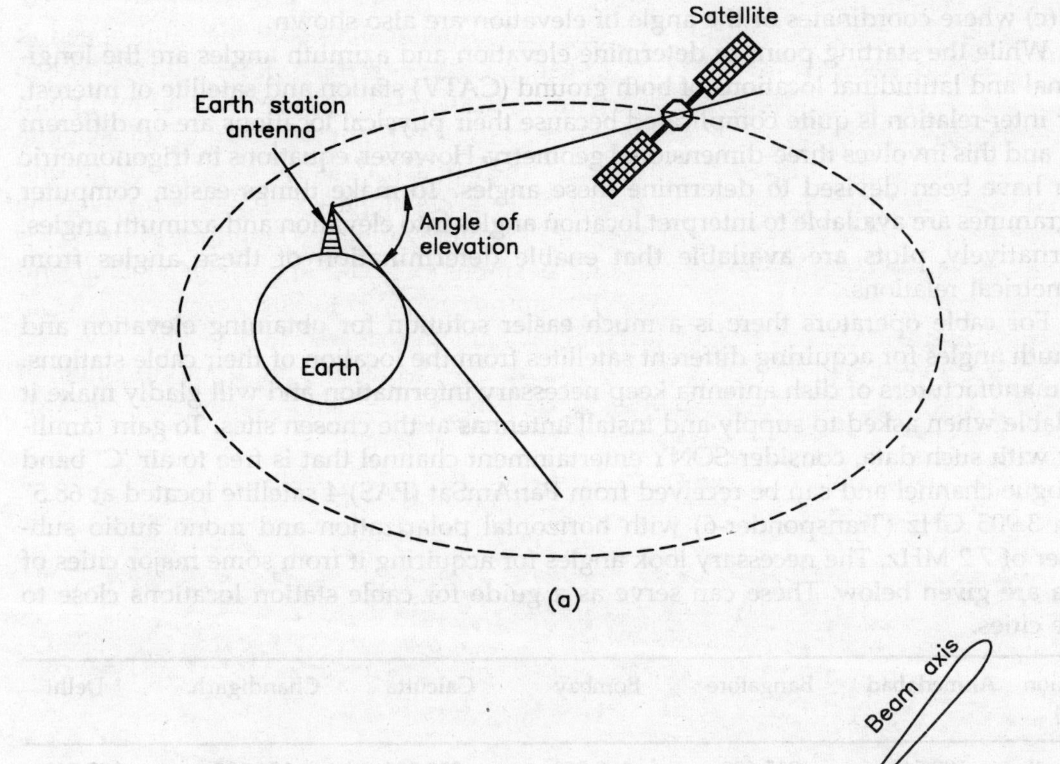

Fig. 14.3. Angles for acquiring any satellite (a) angle of elevation, (b) angle of azimuth, and (c) schematic illustration.

is, therefore, measured clockwise from north pole taken as zero angle point and the plane containing earth station and satellite antennas. This is depicted in Fig. 14.3 (b) and (c) where coordinates of the angle of elevation are also shown.

While the starting point to determine elevation and azimuth angles are the longitudinal and latitudinal locations of both ground (CATV) station and satellite of interest, their inter-relation is quite complicated because their physical locations are on different axis, and this involves three-dimensional geometry. However, equations in trigonometric form have been devised to determine these angles. To make things easier, computer programmes are available to interpret location angles into elevation and azimuth angles. Alternatively, plots are available that enable determination of these angles from geometrical relations.

For cable operators there is a much easier solution for obtaining elevation and azimuth angles for acquiring different satellites from the location of their cable stations. The manufacturers of dish antenna keep necessary information and will gladly make it available when asked to supply and install antennas at the chosen sites. To gain familiarity with such data, consider SONY entertainment channel that is free to air 'C' band analogue channel and can be received from PanAmSat (PAS)-4 satellite located at 68.5° E on 3.905 GHz (Transponder-6) with horizontal polarization and mono audio subcarrier of 7.2 MHz. The necessary look angles for acquiring it from some major cities of India are given below. These can serve as a guide for cable station locations close to these cities.

Location (City)	Ahmedabad	Bangalore	Bombay	Calcutta	Chandigarh	Delhi
Azimuth Angle	190.54°	215.62°	193.35°	222.33°	196.13°	197.76°
Elevation Angle	62.60°	71.42°	67.24°	55.32°	53.06°	55.18°

Notice that the given centre frequency of SONY channel is 3905 MHz instead of the standard centre frequency of 3900 MHz. It seems that only half of the transponder band width is allotted for the TV channel and the remaining width for carrying some other signals. This does not cause any serious problem in tuning at satellite receivers because deviation from standard value for channel 10 [see Fig. 12.10 (b)] to which the receiver tuner is pre-set is only 5 MHz (3905 – 3900 = 5) *i.e.* 0.1% away and the AFT circuit around the tuner will adjust the LO at the correct value to tune-in the incoming TV channel.

Orientation of Dish Antenna:* For acquiring any satellite from a given location, the dish antenna is first set in the horizontal plane at the corresponding azimuth angle with the help of a magnetic compass. Since the magnetic needle always points in North-South direction, the angles marked below it on the dial enable setting the dish at the desired angle. Next, the dish is vertically oriented to set it at the known elevation angle.

* More details on setting dish antenna to acquire different satellites are given in chapter 17 that is devoted to cable station set-up.

For this an angle protector in the form of ANGLE METER can be used. Final setting is done by slightly moving the dish both horizontally and vertically to firmly locate the satellite for maximum signal pick-up.

In general, it may be noted that angles of elevation will be large for cable stations located close to the equator. For example, to acquire PAS-4 satellite at Bangalore, the angle of elevation is 71.42° and will become nearly 90° at Kanyakumari close to which the equator line passes. This means that the dish will point almost upwards towards the sky. Furthermore, for CATV stations in 'SRI-LANKA' the elevation angle for acquiring the same satellite will become more than 90°, meaning the dish will face towards north *i.e.* towards India for receiving SONY CHANNEL programmes.

Similarly, for ground stations located far-away from the equator towards north, say Srinagar (J&K), the angle of elevation will become much less than 90° for receiving PAS-4 programmes. Lower the angle of elevation, larger the slant range of the satellite and hence, greater the path of attenuation due to spreading and atmospheric losses. Attenuation increases rapidly when the elevation angle becomes less than 10°. The reception will thus be poor unless dish diameter is greatly increased. Even otherwise, for stations located at the outer edge of footprint area of any satellite, a large diameter dish becomes necessary as compared to locations in the prime focus area of down-link beam from the satellite.

14.3 Dish Antenna System

The dish antenna system is a vital part of any ground station. It consists of dish mount, dish itself, LNBC and feeder connections to the satellite receiver. For reception from international satellites, as at India's Doordarshan, dish antenna of diameter up to 30 meters become necessary. Since such dishes are very heavy, special concrete foundation becomes necessary for supporting the mount on which the dish system is installed. Usually a motorised feed system is provided for automatic selection of necessary polarization, setting of azimuth and elevation angles for acquiring the satellite and also for fine tuning, all with a computerised feedback system. Some dishes have multiple feed-horn and LNBC sets mounted on the same dish for receiving more than one down-link signals from two different satellites. Once the wanted satellite is acquired, it is seldom necessary to readjust the dish. With the 0.1° station keeping accuracy of the satellite, a ground station with an antenna bandwidth of about one degree does not have to be constantly adjusting its antenna to accurately point it at the satellite after once acquiring it. The need to adjust the antenna again comes only when another satellite is to be acquired. Such a need can also arise if the satellite's original orbit changes or if the antenna is not firmly anchored (fixed) and shifts its position due to strong winds or storm.

In contrast at cable TV stations, where only TV signals are to be received from regional and domestic satellites (DOM-SATS), the dish antenna set-up is much simpler. The dish diameter seldom exceeds 6 meters and therefore elaborate mount and foundations are not necessary. The azimuth and elevation angles are usually set manually with the help of a wheel and screw assembly and on acquiring the satellite the dish is locked in this position. Some cable operators prefer motor driven units for faster and convenient acquisition of any satellite.

Receive Only Set Up: As explained in section 4.4 (microwave antennas), a horn in conjunction with a parabolic reflector forms the so-called dish antenna. The parabolic nature of the dish enables high gain directivity to the very weak signals it receives from the satellite. The dish is made from a large number of either solid, expanded metal or mesh type sections, usually out of a light metal like aluminium. These sections are bolted together firmly on suitably shaped extruding trusses to give the dish a parabolic shape. While a solid plate dish will gather more signal as no beam loss will occur in the absence of any holes, it would need a very strong foundation and mount support to withstand heavy pressure due to occasional strong winds. As such, for CATV station antennas, a mesh type dish construction is preferred. For reception in India and adjoining countries, in the C-band frequency range of 3.7 to 4.2 GHz, a dish diameter from 2.5 to 5 meters serves the purpose. In fact, dish dia also depends on the location of cable

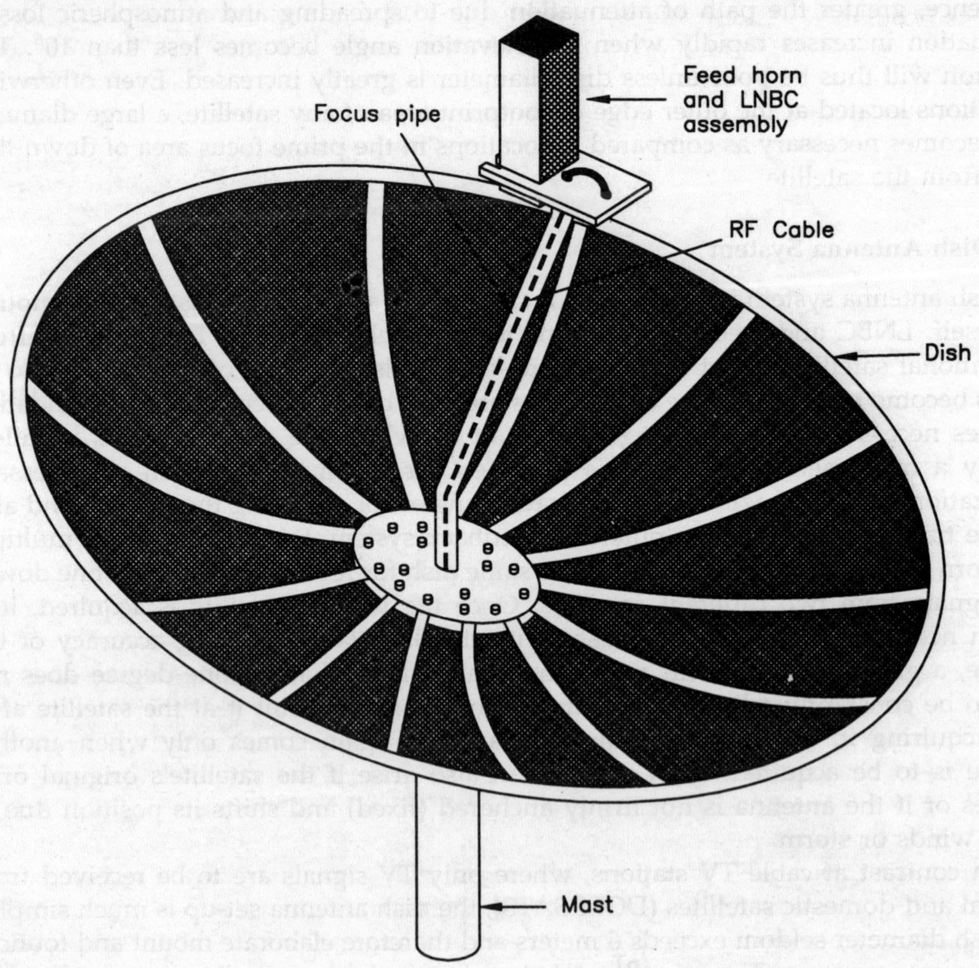

Fig. 14.4. A high gain chicekn-mesh construction dish antenna suitable for reception in the C-band. (Courtesy: 'SHYAM' communication systems).

station. As already explained, in weaker reception areas, a relatively large dish dia is required as compared to good reception areas because of difference in angles of elevation at which antennas must be set. For pick-up from domestic satellites, a dish dia of about 2.5 meters is enough for satisfactory reception. Furthermore, if the antenna is located in snow-bound locations, some type of de-icing facility must be added to the antenna to prevent accumulation of ice and water on the surface of dish. Such a provision becomes all the more necessary if the dish is of solid construction.

A typical parabolic wire-mesh dish antenna is shown in Fig. 14.4 for reception in the C-band of 3.7 to 4.2 GHz. In its 3.7 meter dish version, it has a gain of 42 dB and can withstand wind loads up to 80 km per hour. The mounting facility enables 360° rotation in azimuth and 10° to 90° in elevation. Actuators and positioners are also available for remote control setting of the dish. Both motorised and manual versions of the antenna are available. As shown in the figure, the feed element (horn) together with associated LNBC is positioned at the focus of parabolic dish on a central support structure which is made out of a non-conducting material to keep losses as low as possible. The support has provision for selecting desired polarization position of the feed element.

Feed Configurations: As already explained, it is the prime focus feed that is

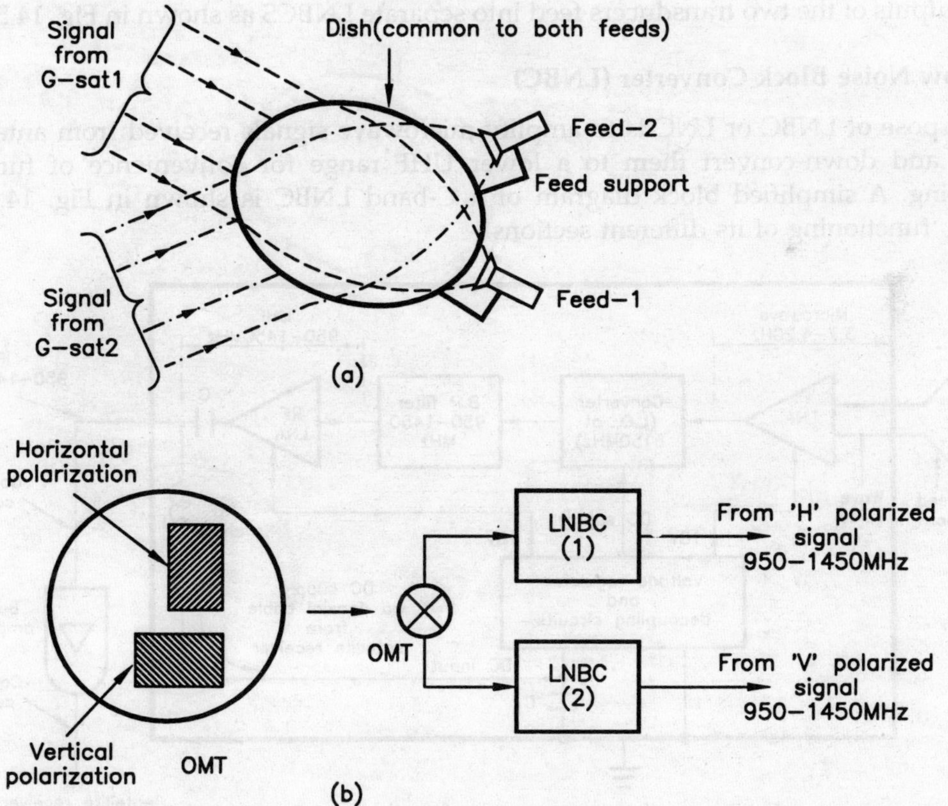

Fig. 14.5. (a) Offset feed configuration, (b) Orthogonal feed arrangement.

employed in most dish antennas where the horn is positioned at the focus of parabolic dish.

In another arrangement called 'OFFSET', the feed horn is mounted towards the edge of dish away from its aperture to prevent any obstruction by the support structure. In practice, this arrangement enables the use of two feeds sharing the same dish so that two satellites can be acquired if available within a reasonable arc area of the dish. Such an arrangement is shown in Fig. 14.5 (a).

It is also possible to locate two horns-cum LNBC assemblies side-by-side, slightly offset from the focal point, on separate supports to obtain two similar signal streams when more than four channels are to be availed from the same transponder of the acquired satellite at the same time. However, this is possible in locations close to the central area of the satellite's foot-print where the signal strength is adequate and thus a small loss of signal due to location of horns slightly away from the focal point will not affect the performance of the thus picked up channels.

Simultaneous Reception of 'V' and 'H' Polarized Signals: Most satellites transmit in both polarizations thereby achieving a 24 channel (TV) capacity in the 500 MHz bandwidth available that extends from 3.7 to 4.2 GHz. For this, orthogonal mode transducer (OMT) is used where there are two feed horns, one oriented to receive 'V' polarized signals and the other for 'H' polarized signals. There is no mix-up of signals since outputs of the two transducers feed into separate LNBCS as shown in Fig. 14.5 (b).

14.4 Low Noise Block Converter (LNBC)

The purpose of LNBC or LNC is to amplify microwave signals received from antenna system and down-convert them to a lower UHF range for convenience of further processing. A simplified block diagram of a C-band LNBC is shown in Fig. 14.6 to explain functioning of its different sections.

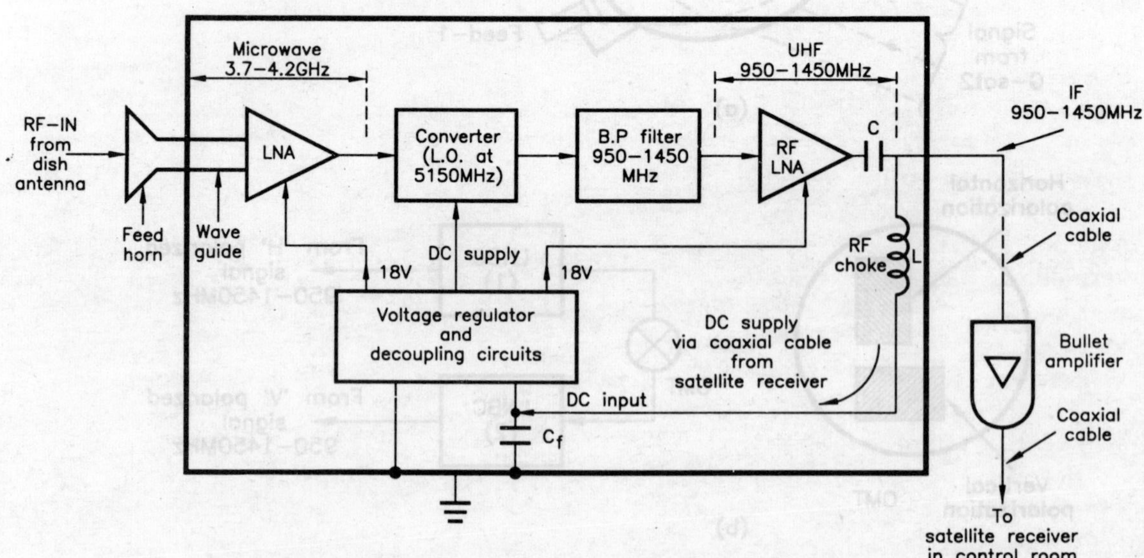

Fig. 14.6. A simplified block diagram of a typical C-band LNBC.

(i) *Feed Horn:* As detailed in section 4.4, the feed horn is a tubular metalic device that is located at the focal point of the parabolic dish. It captures signals reflected from antenna dish and directs them to the low noise amplifier (LNA). Since the feed horn is actually a tubular waveguide, it not only conducts electromagnetic waves to the LNA with impedance match on either side but also acts as a high-pass filter (HPF) for the incoming signal. As explained in section 4.2 [see Fig. 4.11] the dimensions of guide section of the horn can be so chosen that it does not pass signals below a specified cut-off frequency, which for C-band is 3.7 GHz. After C-band, the next higher transmission band, called KU band, is far-away from the C-band range and hence, does not cause any interference.

The feed horn is suitably oriented while fixing it at the focal point of dish to match polarization of electromagnetic wave signals to be received from the satellite. In some feed horn designs, an isolator made out of a dielectric material, fixed at its input side enables selection of H or V polarization. A probe placed at the other end of feed horn converts E.M. wave signals to a tiny voltage which becomes input to the low noise amplifier.

(ii) *Low Noise Amplifier* (LNA): The low noise amplifier as the name suggests, is a specially designed two-stage amplifier which introduces least possible noise in the very weak signal it amplifies. For this, noise temperature rating of the amplifier must be less than 35° K *i.e.* Noise Factor less than 0.5 dB. The active devices used in the amplifier are special field-effect transistors (FETs) made out of gallium-arsenide (Gn-As) semi-conductor materials. These transistors have least transit time and shunting capacitance effects. Besides, special microwave components and coupling techniques, as described in chapter 4, are used to achieve desired performance of the LNA. Its bandwidth is kept equal to 500 MHz (4.2 GHz – 3.7 GHz = 500 MHz) to provide gain to all the 12 channels of the transponder. Because of such a large bandwidth and other constraints the gain of LNA is limited to about 20 dB.

(iii) *Down Converter:* As explained in chapters 1 and 12, microwave signals are down converted to a lower UHF range before feeding them to the satellite receiver. This is done to avoid losses that would otherwise occur in the coaxial cable connecting LNA output to the receiver located in the control room. For cable television, the IF frequency range is chosen to lie between 950 to 1450 MHz for C-band to maintain the same bandwidth of 500 MHz. For this, as shown in the figure, the local oscillator (LO) frequency is kept at 5150 MHz to convert the C-band 3.7 to 4.2 GHz signal to this intermediate frequency range.

The converter circuit [see Fig. 9.2] employs a point-contact diode, preferably Schottky type, which on application of input signal and LO output operates as a square-law detector to generate sum and difference frequency components. The difference components (950 to 1450 MHz) are selected by an LC tuned band-pass amplifier having the same bandwidth of 500 MHz.

(iv) *IF Amplifier:* The IF output from the converter is amplified by a 3-stage UHF amplifier called IF amplifier or simply RF LNA as in Fig. 14.6, to give it a boost of about 40 dB. The amplifier configuration is nearly the same as that of preamplifiers in UHF tuners of television receivers. Automatic gain control (AGC) is employed to obtain steady output despite variations in the received microwave signals.

(v) *Power Supply:* The 18V. dc supply to both the LNA and IF amplifier section comes from the associated satellite receiver's power supply through the coaxial cable that connects them. It is further regulated and decoupled before feeding to each amplifier. As shown in Fig. 14.6, the capacitor 'C' blocks dc supply directly towards LNBC but permits RF signals to freely flow to the output feeder cable that goes to the control room. The choke (inductor L) blocks flow of RF signals towards the dc voltage regulator but allows easy path to the dc supply. A by-pass capacitor C_f is usually connected across the dc supply line to short circuit any RF that may leak through the choke.

The feed horn, LNA, converter and IF amplifiers are interconnected as a PCB assembly and the unit thus formed is called LNBC. It is suitably enclosed in a box so that no rain water gets into it. It is also protected from excessive heat and direct sunlight. The antenna system and LNBC unit put together is called Front-End Converter (FEC) or outdoor equipment as shown in Fig. 1.2.

(vi) *Line Drive Amplifier* (LDA): The signals arriving at the dish are extremely weak being of the order of .001 μμw power due to heavy attenuation in the atmosphere. An often used 3.7 meter dish antenna provides a gain of about 40 dB (10^4) to bring the signal level to nearly 1 μμw necessary for the LNA to maintain desired S/N ratio. The power gain from the LNA is close to 20 dB and that of the IF amplifier about 40 dB. This put together raises the signal level to nearly 1 μw at the output of LNBC. This power level is considered adequate at the input of satellite receiver. However, if the cable run from the FEC to the control room is long, a signal loss of about 6 to 10 dB can occur. This is taken care of by providing an amplifier in between the cable length. It is of the same type as one section of IF amplifier in the LNBC. The dc voltage to this amplifier also comes from the satellite receiver power supply through the same cable that carries the IF signal. It may be noted that flow of dc current through the cable does not affect transmission of RF signal through it. Necessary isolation is, however, provided as for the LNBC. The line drive amplifier is also called bullet amplifier because of its small size and shape.

14.5 Power Divider (Splitter)

A power divider divides the input RF power, received from LNBC, equally into separate parts to feed several satellite receivers. The signal splitting is done by passive elements, usually microwave strips that constitute the power divider. Some loss occurs in dividing circuits but major reduction in signal strength is due to sharing of input power amongst a number of outputs. A 4 port divider, that is commonly used is shown in Fig. 14.7. In this, the power will get divided into four equal parts and if each receiver is to receive nearby 1 μw, the input to the divider must be about 4 μw. In practice it must be more than 6 μw when divider losses are taken into account. Therefore, the divider is preceded by an amplifier to ensure 1 μw signal level to all the satellites received. The amplifier is similar to the bullet amplifier with a gain of about 8 dB. Usually dc supply to the divider unit comes from the satellite receiver but in some divider circuit designs a separate power supply is provided.

It may be noted that all the outputs [see Fig. 14.7] have a blocking capacitor to keep dc supply away from the circuits while coupling RF energy to the output side (O/P) *i.e.* to satellite receivers. As shown in the figure, one output port, labelled O/P_4 has a coupling arrangement of RF choke (L_2) and decoupling capacitor C_f to carry dc supply to the voltage regulator that feeds the amplifier. A similar arrangement is shown on the input side to feed dc supply to the LNBC and if needed to the bullet amplifier.

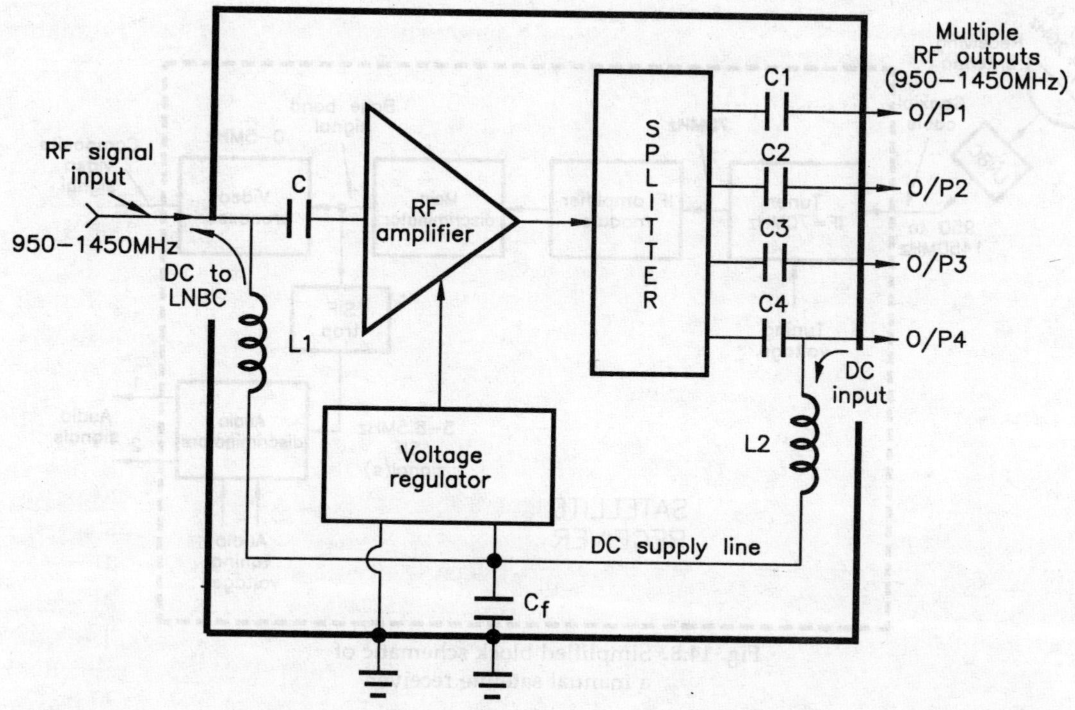

Fig. 14.7. Simplified block diagram of a power divider.

Power dividers with more than 4 port output are also available for receiving more than four channels from the same satellite. The divider circuit impedances not only divide the signal as necessary but while doing so also isolate one output from the other. This ensures that any feedback or load variation in one output circuit does not affect the performance of other output connections.

14.6 Satellite Receivers–Manual and Remote Controlled

Signal processing receivers designed for transmit-receive at earth stations are called VIDEO receivers while those meant for TV receive only (TVRO) at cable stations are called SATELLITE receivers. An over-simplified block schematic of a manual satellite receiver is shown in Fig. 14.8. While its building blocks are nearby the same as that of a TV receiver but channel width is not restricted to 7 MHz. It extends up to nearly 10 MHz to accommodate two audio channels which usually lie between 5-8.5 MHz. Thus two inter-carrier sound IFs are present and as shown in the figure, two audio discriminators are provided to obtain two audio signals. This enables stereo sound and multi-lingual audio receptions.

Detailed Receiver Circuit: The block diagram of a complete C-band satellite receiver is shown in Fig. 14.9. While it is designed to process two audio signals, only one audio channel has been shown as an illustration of its processing. The receiver accepts

950-1450 MHz signal from the power divider and heterodynes it in a converter to select any one channel out of the possible 12 present in the input signal. Each section of the receiver is briefly described.

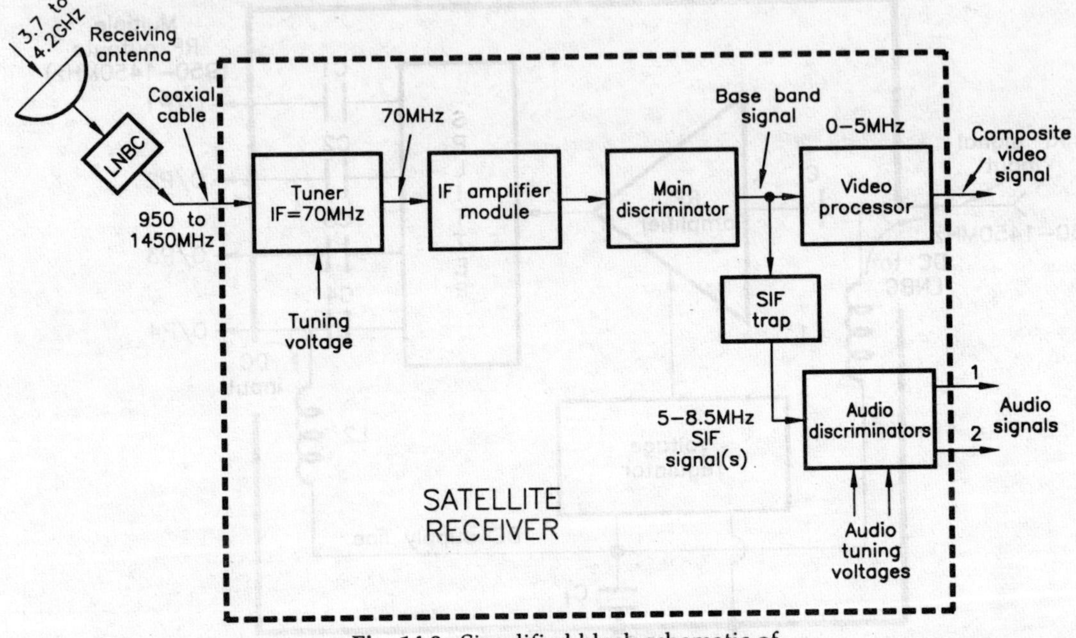

Fig. 14.8. Simplified block schematic of a manual satellite receiver.

(1) *Converter:* The input signal is amplified by an RF amplifier before feeding it to the converter. The gain of this 500 MHz wideband amplifier is stabilized by delayed AGC action as explained in section 11.8 [Fig. 11.18]. The receiver has an IF frequency of 70 MHz and therefore LO frequency of the converter is kept 70 MHz above the centre frequency of the wanted channel so that on heterodyning, the difference products lie in the IF band of 52-88 MHz assuming effective bandwidth of the channel equal to 36 MHz out of the allotted width of 40 MHz. This is separated by a B.P. filter and made input to the IF subsystem.

(2) *Channel Selection:* The frequency of LO in the converter is set by varactor control at correct value necessary for tuning-in the desired channel. As explained in chapter 11 [see Fig. 11.15 (a)], the varactor is a capacitive diode, the capacitance of which decreases with increase in reverse-bias voltage across it. Since the diode forms part of the tuned circuit of the oscillator, its frequency changes accordingly. As the oscillator frequency is varied by changing a voltage, the circuit is called voltage controlled oscillator (VCO). As shown in the block diagram, a separate dc voltage is applied to the varactor diode for each channel through a potentiometer. Though the potentiometer is preset to obtain correct voltage but can be varied slightly when necessary. A clear picture on the associated monitor TV receiver indicates correct tuning. Besides, as explained in chapter 11 [Fig. 11.17], AFT control is employed to keep LO frequency locked at the correct value.

(3) *IF Subsystem:* It is a multistage tuned voltage amplifier configuration assembled

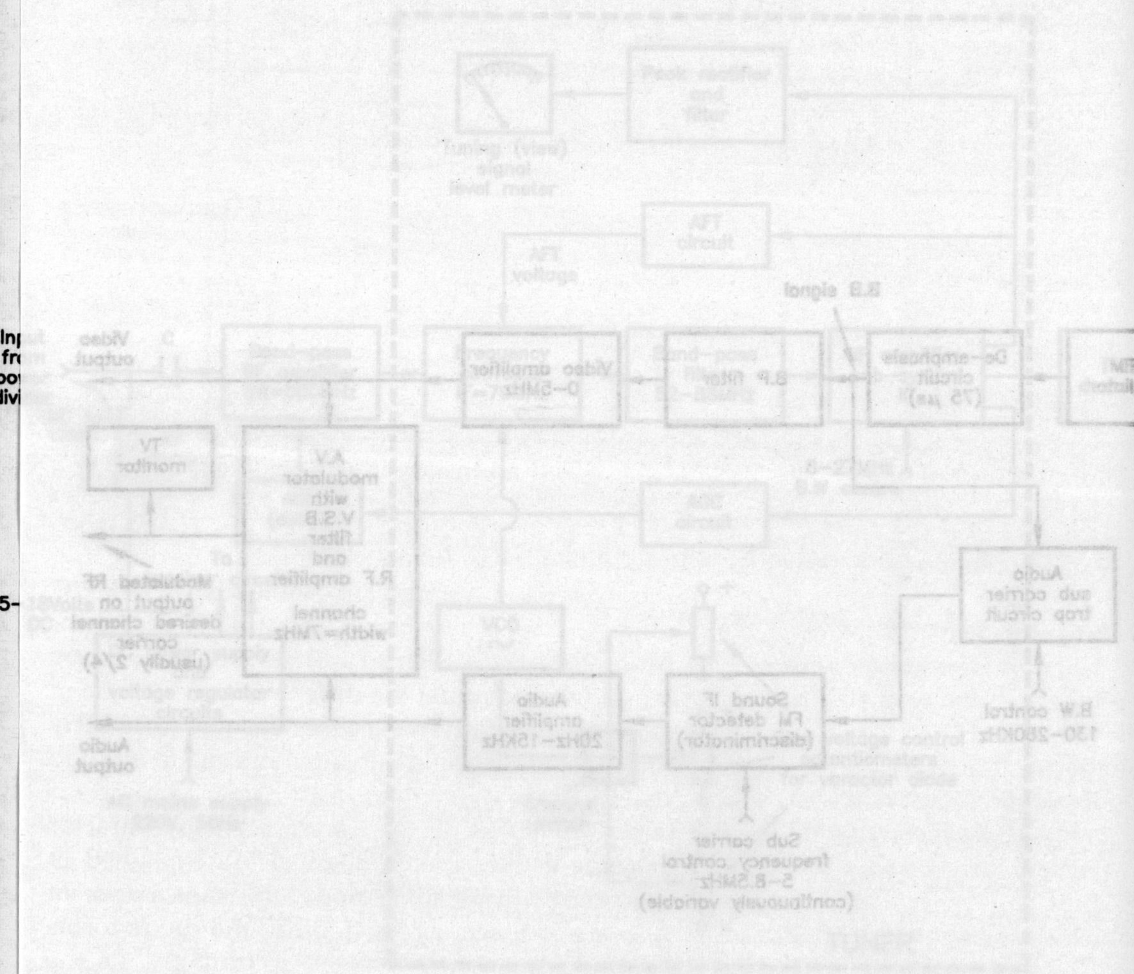

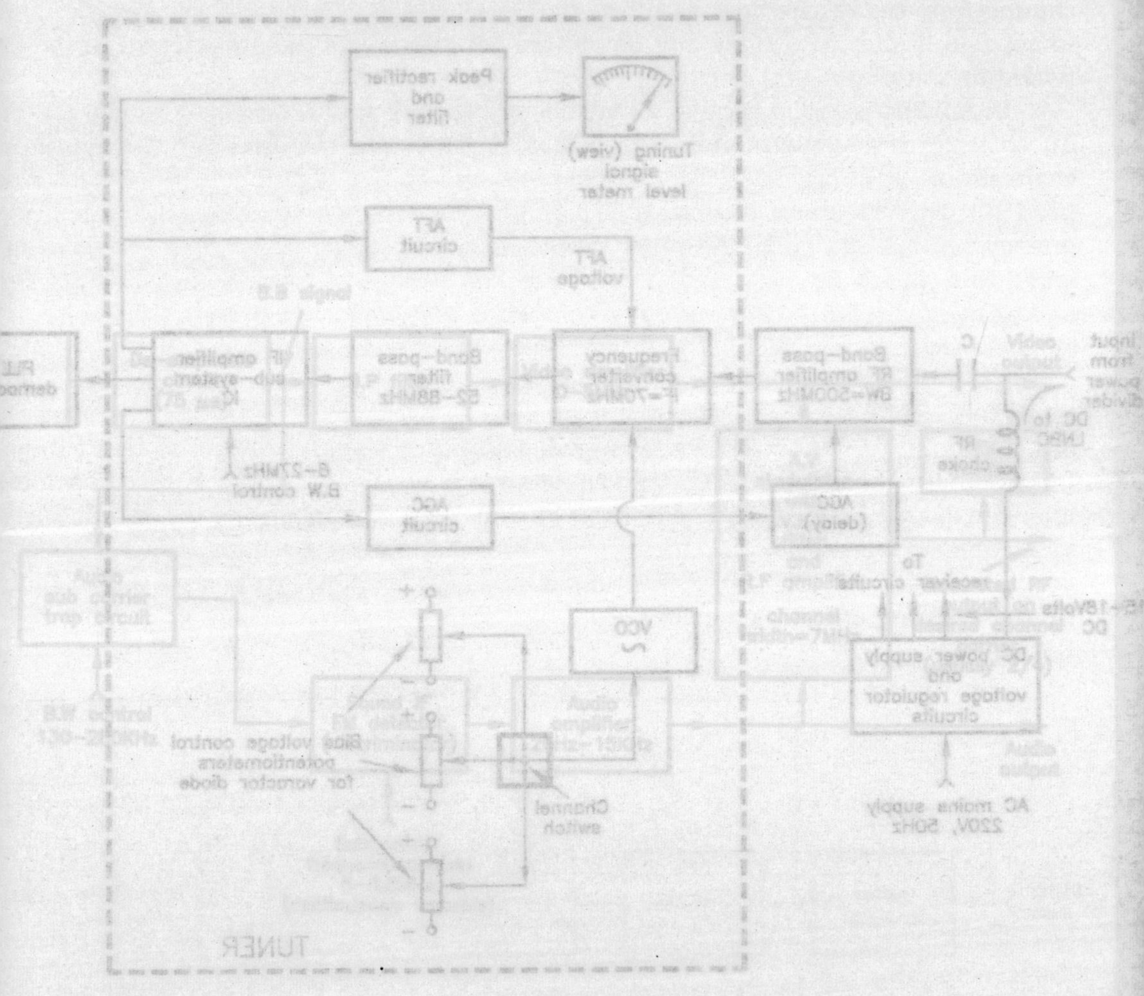

on a separate module unit. The amplifier is AGC controlled and provides stability and high gain to the IF signal. As shown in the figure, delayed AGC as obtained from this section is applied to the band-pass RF amplifier to keep noise level as low as possible. From the output of IF subsystem, a sample of IF signal is also fed to the AFT circuit which enables automatic frequency control of the converter circuit. The IF signal is also rectified and filtered to become input to the VIEW METER (tuning indicator) as shown in the figure. Another important feature in the IF section is a provision for varying effective bandwidth of IF amplifiers to accommodate signals having different formats coming from many satellites. In addition, the last IF stage of the subsystem is designed to act as a limiter to remove any amplitude variations that may be present in the FM (video and audio) signal.

In modern satellite receivers the subsystems 1, 2 and 3 described above are produced to form a single module which then becomes the TUNER as shown by dotted chain lin block in Fig. 14.9. Such compact tuners are pre-aligned and thus there is no need for expensive and sophisticated equipments for aligning critical RF and IF frequency settings. As such, the tuner provides correct signal output from which video and audio signals are extracted as shown in Fig. 14.9.

(4) *Demodulation:* Various circuit blocks of a demodulator used in satellite receivers are shown in Fig. 14.10. The IF amplifier output with centre frequency at 70 MHz enters a buffer amplifier which provides additional amplification. The limiter circuit, limits the amplitude of IF signal to a fixed level. A filter circuit follows it to remove any 3rd harmonic content from the IF signal. Its output feeds into the FM demodulator (discriminator) which is of PLL type as detailed in section 8.4 [Fig. 8.12]. The amplifier after the discriminator and de-emphasis circuits is for obtaining original amplitudes of video and SIF signals as components of the baseband signal. A feedback circuit from the output of baseband (BB) amplifier to input side is to keep the signal level within limits, thus preventing any noise generation.

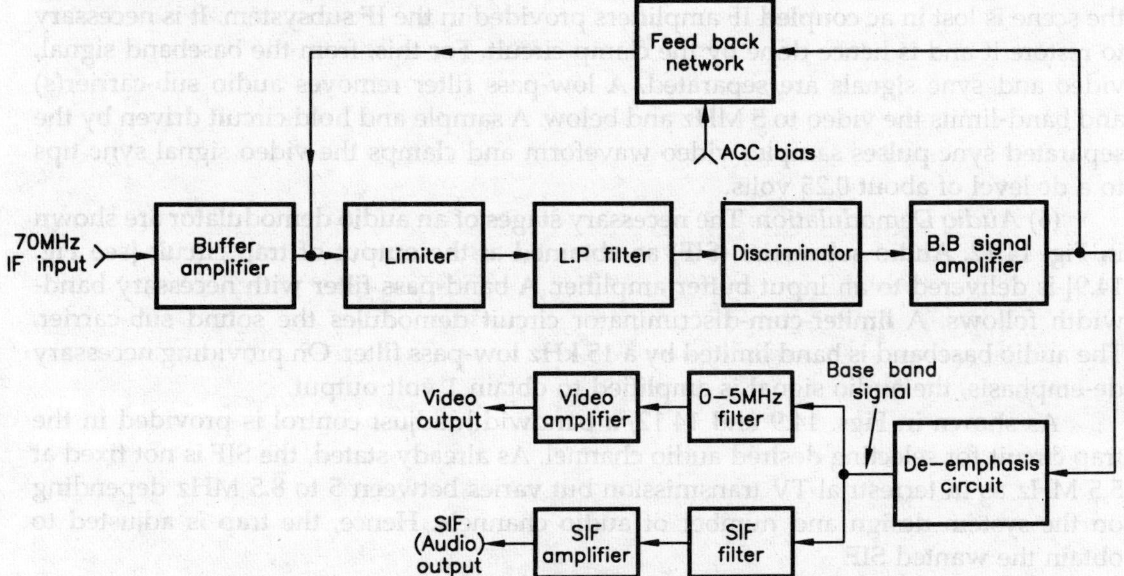

Fig. 14.10. Schematic diagram of a satellite receiver demodulator.

In satellite transmission, frequency modulation (FM) is employed to process both video and audio signals with a pre-emphasis of 75 μs. As such the de-emphasis circuit in the demodulator is of 75 μs time constant. However, in some satellite transmission systems a pre-emphasis of 50 μs is used. For this a switch is provided in satellite receivers to choose the time-constant of de-emphasis circuit between 75 μs and 50 μs. At the output of demodulator the two components, video and audio of the BB signal are separated and processed (as shown in the figure) to obtain video and audio signals separately.

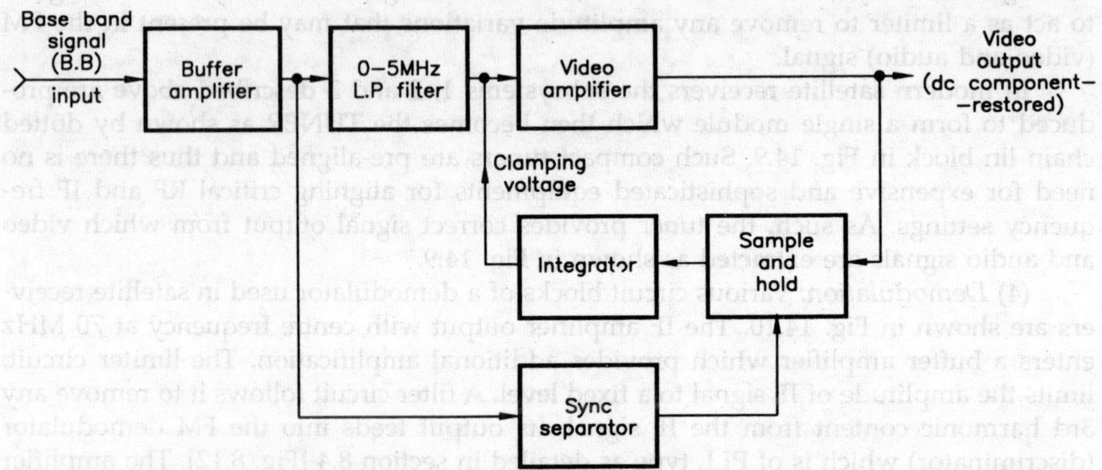

Fig. 14.11. Block diagram of a video clamp circuit in a satellite receiver.

(5) *Video Processing:* Video clamp circuit not included in Fig. 14.9 is shown separately in Fig. 14.11. The dc component of video signal which determines brightness of the scene is lost in ac coupled IF amplifiers provided in the IF subsystem. It is necessary to restore it and is hence done by the clamp circuit. For this, from the baseband signal, video and sync signals are separated. A low-pass filter removes audio sub-carrier(s) and band-limits the video to 5 MHz and below. A sample and hold circuit driven by the separated sync pulses samples video waveform and clamps the video signal sync tips to a dc level of about 0.25 volts.

(6) *Audio Demodulation:* The necessary stages of an audio demodulator are shown in Fig. 14.12. Audio subcarrier (SIF) as obtained at the output of trap circuit [see Fig. 14.9] is delivered to an input buffer amplifier. A band-pass filter with necessary bandwidth follows. A limiter-cum-discriminator circuit demodules the sound sub-carrier. The audio baseband is band limited by a 15 kHz low-pass filter. On providing necessary de-emphasis, the audio signal is amplified to obtain 1 volt output.

As shown in Figs. 14.9 and 14.12, a bandwidth adjust control is provided in the trap circuit for selecting desired audio channel. As already stated, the SIF is not fixed at 5.5 MHz as in terrestrial TV transmission but varies between 5 to 8.5 MHz depending on the system design and number of audio channels. Hence, the trap is adjusted to obtain the wanted SIF.

(7) *A.V. Modulator:* The last section of the satellite receiver [see Fig. 14.9] is the

A-V modulator with associated circuits to provide RF output, as at conventional TV transmitters. The chosen channel carrier can be that of any desired channel. The A-V modulator output is sampled by a monitor receiver to keep a check on the quality of both picture and sound reproduction.

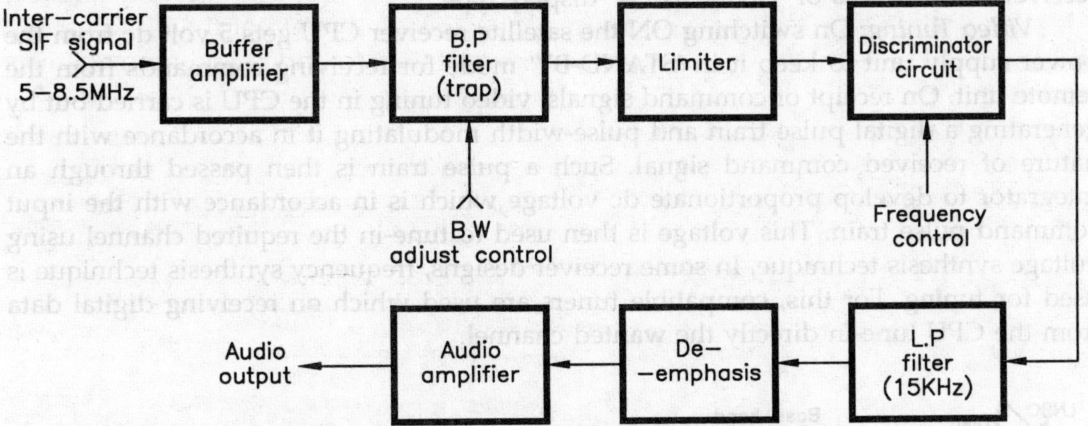

Fig. 14.12. Block diagram of the audio demodulator.

REMOTE CONTROLLED SATELLITE RECEIVER: The main difference between a manual satellite receiver [Fig. 14.9] and a remote controlled receiver is that in the former, video and audio tuning is carried out by voltage synthesis technique through potentiometers whereas in the latter, these functions are affected by a Central Processing Unit (CPU) commonly called Microprocessor (µp). As in a digital computer, CPU consists of control unit, logic gates and memory. It is suitably programmed through RAMS and ROMS located in the memory to carry out desired functions with commands *i.e.* Instructions fed to it from outside through an input interface unit. The CPU uses pulse-width modulation or frequency synthesis technique to carry out various functions assigned to it.

The block schematic of a typical remote controlled satellite receiver is shown in Fig. 14.13. The remote control hand-held unit sends out different commands for executing all the functions labelled on it. For each function, it generates a distinctive pulse which after modulation on an assigned carrier is sent out in the form of varying infrared radiation from a light emitting diode (LED). A senser which is located on front panel of the receiver picks up these signals when remote unit is aimed at it. The senser unit, which is also a LED, converts the received pulses, which are unique for each function, into electrical pulses for feeding into the CPU. The CPU in turn decodes these pulses and performs respective functions for which it is pre-programmed.

The CPU also receives commands from control switches mounted on front panel of the receiver. These are called KEY-SCAN commands and their number varies with receiver make and model. The three basic functions provided on all receivers are:

(i) *Stand-by ON/OFF:* This enables switching ON or OFF a 5 volt dc supply to the CPU.

(ii) *Channel (+):* It is for advancing channel numbers in incrimanting order to select desired channel.

(iii) *Channel (–):* It performs a similar function but for changing channel numbers in decremanting order.

Most receivers are designed for receiving 50 or more channels. The channel number that gets tuned on selection is LED displayed on front panel of the receiver or on TV receiver screen if it is of 'ON-SCREEN' display type.

Video Tuning: On switching ON the satellite receiver CPU gets 5 volt dc from the power supply unit to keep it in 'STAND-BY' mode for receiving commands from the remote unit. On receipt of command signals, video tuning in the CPU is carried out by generating a digital pulse train and pulse-width modulating it in accordance with the nature of received command signal. Such a pulse train is then passed through an integrator to develop proportionate dc voltage which is in accordance with the input command pulse train. This voltage is then used to tune-in the required channel using voltage synthesis technique. In some receiver designs, frequency synthesis technique is used for tuning. For this, compatible tuners are used which on receiving digital data from the CPU tune-in directly the wanted channel.

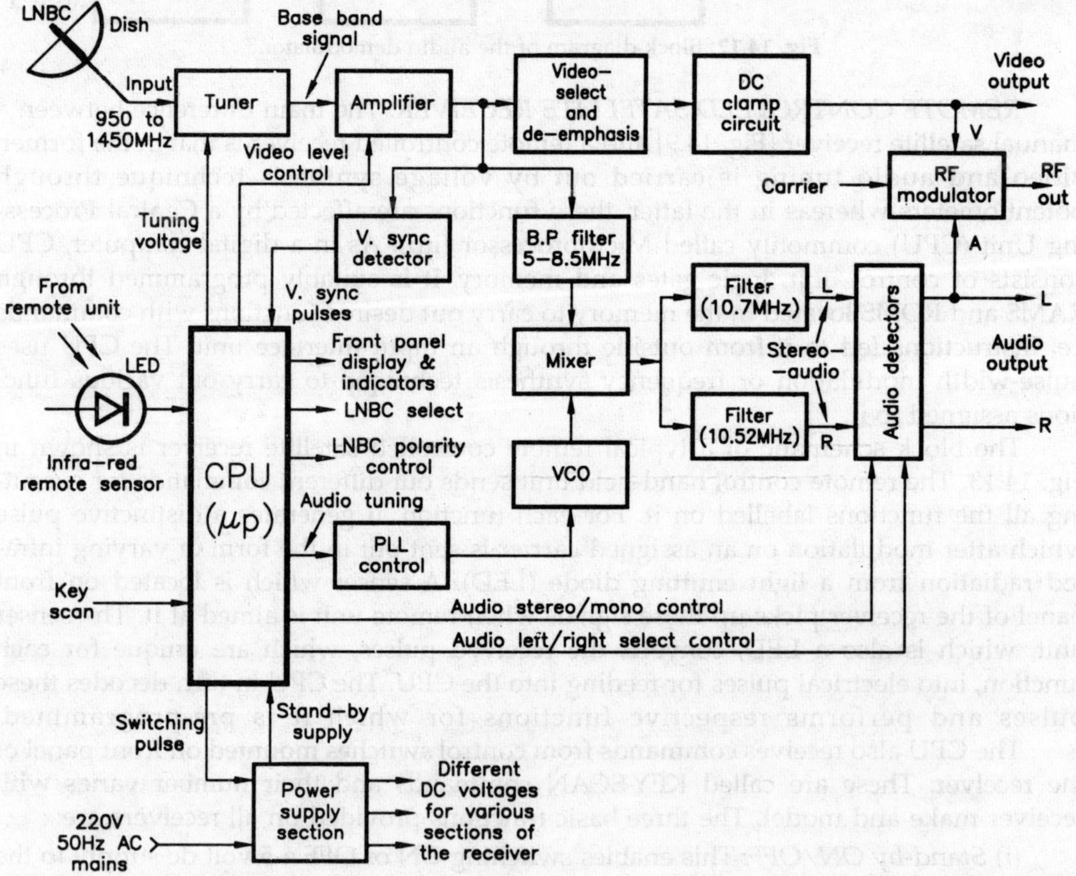

Fig. 14.13. Functional block diagram of a remote controlled satellite receiver.

Audio Signal Separation and Processing: The audio subcarrier signal (SIF) is separated from the baseband signal by a B.P. filter designed to pass signals between 5 to 8.5 MHz. It is heterodyned in a mixer to convert it to 10 MHz range for which the local oscillator is of VCO type. This is set at the required frequency by an audio tuning voltage developed by the CPU in a similar way as for video tuning. A PLL circuit is provided to lock VCO output at the exact frequency determined by the tuning voltage.

Remote controlled satellite receivers are invariably designed to produce two audio outputs for stereophonic sound output. For this, the two audio signals are obtained by separating mixer output by two filter units centred around 10.7 MHz and 10.52 MHz respectively. The two inter-carrier SIF thus obtained are demodulated in the usual way by two separate discriminator (detector) circuits. The two audio outputs (L&R) thus obtained are made available on separate jacks located on the receiver to enable stereophonic sound. However, as shown in the block diagram [Fig. 14.13], RF output from the receiver is in mono mode only with left side (L) audio signal.

Auto Search: In receivers which have facility for auto-tuning, a vertical sync detector inputs vertical sync pulses from the baseband signal into the CPU. When the sync amplitude gets stabilized it is an indication to the CPU that tuning search is complete.

CPU Inputs and Outputs: The main inputs to the CPU are:

(i) Signals from the remote sensor,
(ii) Key-scan signals,
(iii) Vertical sync pulses, and
(iv) Stand-by 5 volt dc supply.

Similarly CPU outputs which perform and control various functions are:

(i) Video tuning voltage.
(ii) Audio tuning voltage.
(iii) Video signal level control voltage for the B.B. amplifier.
(iv) Power switching 'ON' pulse to the power supply unit which in turn generates various dc voltages.
(v) Multiple LNBC selection signal to facilitate switching between 2 or 4 LNBCS mounted on one or more dish antennas.
(vi) LNBC polarity control signal to enable reception of satellite signals having different polarizations.
(vii) Audio L/R channel select control signal to enable selection of either audio channel. This facility is required for incoming transmission having multilingual audio signals for the same visual programme. One language audio is transmitted on the left channel and the second on the right channel on either side of the SIF centre frequency. The bilingual provision is necessary because the foot prints of most satellites cover a large area where different languages are spoken. This L/R channel select facility thus enables subscribers to watch the same picture programme with audio language of their choice.

Choice of Satellite Receiver Type: Another aspect of satellite receivers which needs attention is that manually controlled receivers are preferred at cable stations and remote controlled versions find applications where a 'STAND-ALONE' set-up is desired as in houses for personal use. Remote controlled receivers are not recommended for cable operation as commands from one remote control unit though aimed at one particular receiver can effect the functioning of other similar receivers located close-by.

14.7 Channel Modulators—Single Frequency and Agile

A CATV channel modulator has all the building blocks of a similar unit in a TV transmitter though at a much lower power level. As per 625 line B&W and PAL B&G colour TV standards adopted by India and many other countries, the video signal is amplitude modulated and audio signal frequency modulated*. The modulation is usually affected in two stages, both for better stability and ease of filter circuit designs.

Channel modulators are of two types *i.e.* (1) fixed channel carrier frequency, and (2) agile, where the channel carrier can be chosen as necessary. As obvious, the number of channel modulators needed at a cable station is equal to the number of channels for which the system is designed. Fixed frequency modulators are preferred at cable stations because there is no need to change their carrier frequency once set for a particular channel number. It is common practice to procure channel modulators set for different channels depending on the programmes to be offered at the cable station. However, one or two agile modulators are kept in stock to replace any defective modulator so as to prevent interruption of any channel programme. Both types are described with suitable diagrams.

(1) *Fixed Frequency Channel Modulator:* The block diagram of a fixed frequency channel modulator is shown in Fig. 14.14. The circuit can be broadly divided into four sections as labelled in the figure.

Each section is described separately:

(i) *Video and Audio Signal Processing:* The video signal is successively passed through (i) a video filter to suppress any unwanted frequency components, (ii) a level control to adjust signal amplitude for effective modulation, and (iii) a video amplifier with matching impedance on either side. Similarly, the audio signal is amplified, accorded pre-emphasis of 50 μs and passed through a signal level control before making it input to the frequency modulator. A 5.5 MHz carrier, as in TV transmitters is used for modulation so as to place the FM sound signal ahead of the 5 MHz video signal. A 5.5 MHz ± 75 KHz filter selects the desired FM output which is then mixed with the 0 to 5 MHz video signal in a combiner network as shown in the figure. The composite TV signal (0 to ≈ 5.5 MHz) thus becomes available at the output of combiner for amplitude modulation (A.M).

(ii) *IF Section:* As already mentioned, amplitude modulation of the composite TV signal is carried out in two stages. The local oscillator of 1st modulator is chosen to be 38.9 MHz which is the same as the vision IF of the adopted TV system. With f_c = 38.9 MHz, the L.S.B. of modulated output will extend from 33.4 to 38.9 MHz and USB from 38.9 to 44.4 MHz. The following band-pass amplifier is designed to pass signals from 33.4 to 44.4 MHz but suppress the high amplitude of IF carrier of 38.9 MHz. A SAW FILTER that follows the IF filter and amplifier is designed to fix relative amplitudes of vision IF and sound IF besides shaping response of side-bands as shown by the wave-

* All aspects of amplitude and frequency modulation are explained in chapter 8.

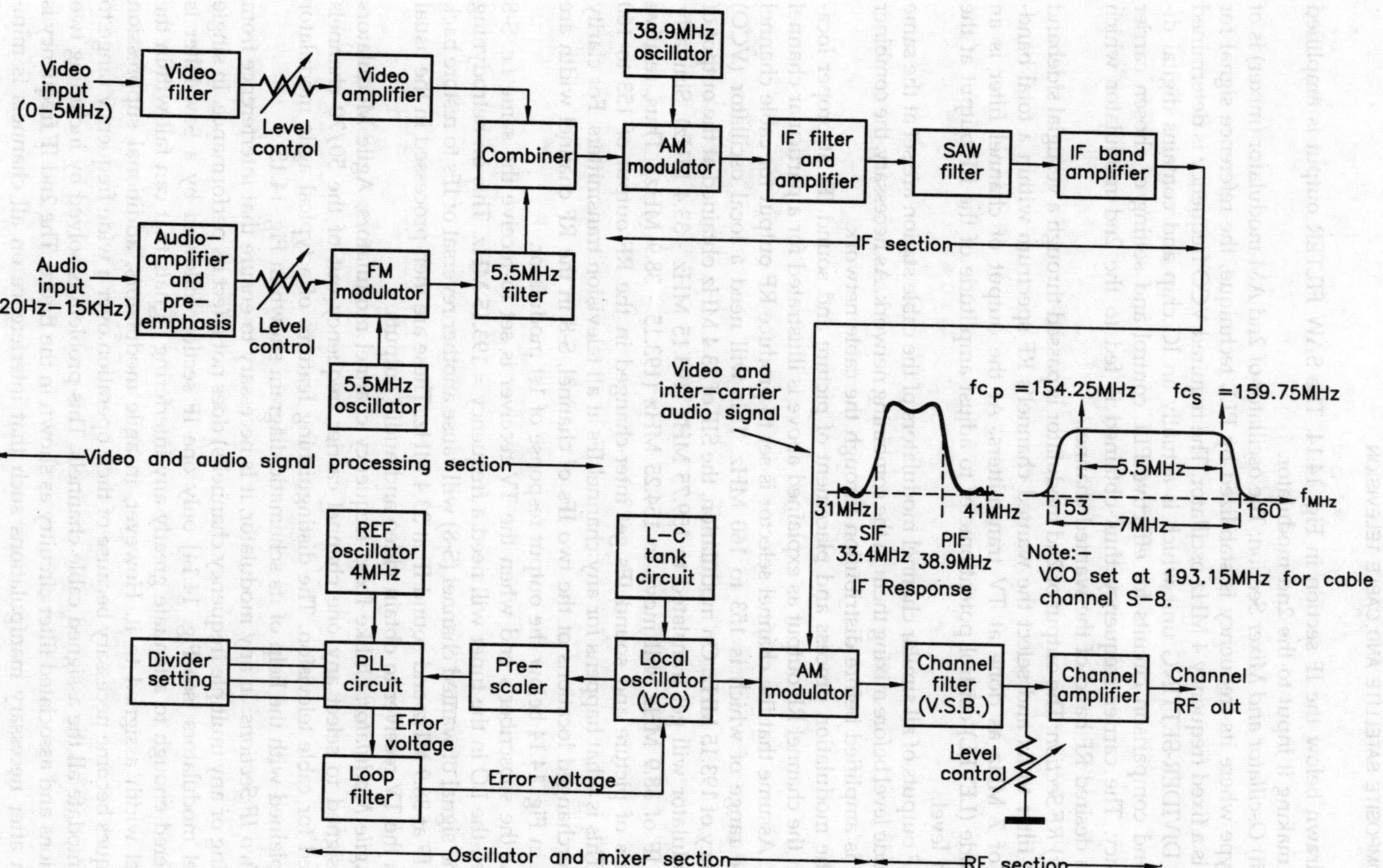

Fig. 14.14. Block diagram of a fixed frequency channel modulator.

form drawn below the IF section in Fig. 14.14. The SAW FILTER output is amplified before making it input to the 2nd modulator.

(iii) *Oscillator and Mixer Section:* The oscillator of 2nd AM modulator (mixer) is of VCO type where its frequency is stabilized by PLL technique, the reference signal for which is a fixed frequency 4 MHz oscillator. The required VCO frequency is determined by the DIVIDER-SETTING unit which is actually an IC chip and contains digital divider and comparison circuits for effective PLL control and setting of chosen carrier frequency. The carrier frequency thus obtained is fed to the 2nd modulator which enables desired RF band of the wanted channel.

(iv) *RF Section:* The output of 2nd modulator is passed through a vestigial sideband (V.S.B.) filter to fix and select the wanted channel's RF spectrum within a total bandwidth of 7 MHz as done at TV transmitters. At the output of channel filter is an amplitude (LEVEL) control potentiometer to adjust amplitude of the RF output at the desired level.

RF outputs of all similar channel modulators of the cable station are set at the same amplitude level before mixing them in the combining network. As necessary, the combiner output is amplified before distribution through the cable network.

The modulation process and placement of picture and sound IFs at proper locations in the channel RF output as explained above is illustrated for a particular channel setting. Assume that the channel selector is set to produce RF output for cable channel S-8, the range of which is 153 to 160 MHz. This will need a local oscillator (VCO) frequency of 193.15 MHz. On modulation, the SIF of 33.4 MHz obtained at the output of 1st modulator will get translated to 159.75 MHz (193.15 MHz - 33.4 MHz). Similarly, vision IF of 38.9 MHz will move to 154.25 MHz (193.15 - 38.9 MHz). Thus, relative locations of picture and sound IFs get inter-changed in the RF output of 153 to 160 MHz. This is what happens for any channel IFs at all television transmitters. For clarity the interchanged locations of the two IFs of channel, S-8 in the RF channel width are shown in Fig. 14.14 below the output response of 1st modulator.

At the subscriber's end when the TV receiver is set to receive the same *i.e.* S-8 channel, the LO in the tuner will need a frequency = 193.15 MHz. This on hetrodyning with the signal of wanted channel (S-8) will cause another reversal of IFs to restore back vision IF at 38.9 MHz and sound IF at 33.4 MHz. These are then processed in the usual way in the TV receiver to obtain video and audio outputs.

Agile Modulator: Unlike fixed frequency channel modulators, Agile Modulators are designed to select any one channel carrier frequency out of the 50/70 channels assigned for cable television. The distinguishing features of a typical agile modulator are explained with the help of its schematic diagram shown in Fig. 14.15.

Two IF Sections: In any modulator, it is necessary to ensure that interference from adjoining or any multiple frequency channel(s) does not affect its performance. In single channel modulators [see Fig. 14.14] only one IF section followed by a Saw-Filter is considered enough to attenuate greatly any interfering signals that can fall within the channel width assigned to it. However, in agile modulators, additional suppression techniques become necessary because of their operation over a wide frequency range to accommodate all the assigned cable channels. This problem is solved by including two IF sections and associated filter circuitry as shown in the figure. The 2nd IF frequency is chosen after necessary manipulations such that interference on all channels is minimized. Furthermore, present-day agile modulators are designed to enable baseband scrambling without external clamping interface. These have in addition visual or composite IF loop option to ensure compatibility with most RF scrambling systems. There is also provision for a mixer circuit [see Fig. 14.15] to insert output from a character generator (C-G).

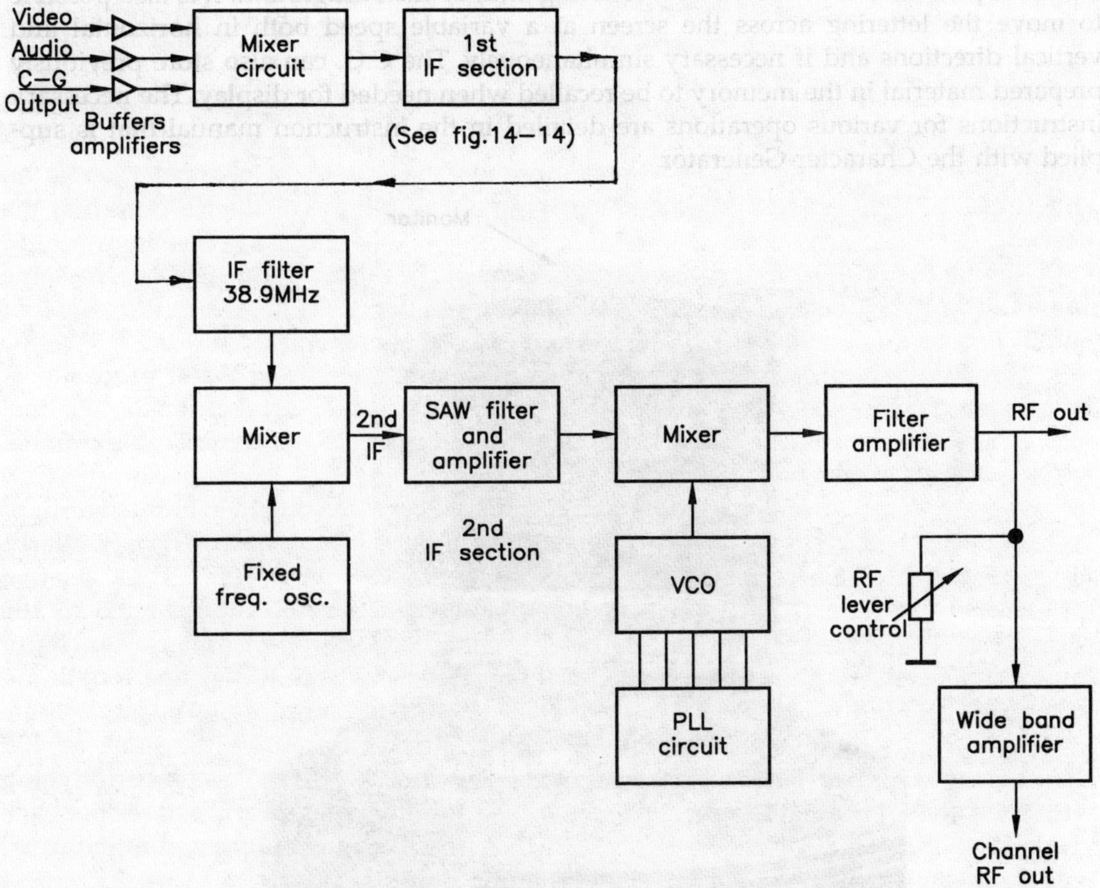

Fig. 14.15. Block schematic of an Agile Cable TV modulator.

Channel Selection: The RF output of most Vestigial Side Band (VSB) AM modulators designed for PAL systems is agile from 50 to 470 MHz. The unique PLL synthesized tuning used in these modulators enables ultra-stable RF output frequencies that are virtually spurious free. The channel selection is via Direct THUMB-DIAL method with the result that the operator can select directly the output frequency for the desired channel without having to look at charts and set associated dip switches as provided for in earlier similar modulators.

Character Generator: It is often necessary to superimpose spot news, messages or advertisements on pictures received via satellite or from local transmissions. This is done at the input of IF stage of the modulator as shown in Fig. 14.15. The information to be inserted is provided by a micro-processor controlled equipment called Character Generator [see Fig. 14.16]. It has a typewriter type keyboard which is used to create lettering as necessary and to control the position of characters on the screen. A position or cursor key enables positioning of typed message anywhere on the screen. Once the message gets typed on the screen, it can be manipulated in a number of ways. The character Generator (C.G.) has provision through separate keys to flash individual words or phrases and to colour both lettering and/or the background. It is also possible to move the lettering across the screen at a variable speed both in horizontal and vertical directions and if necessary simultaneously. The C.G. can also store previously prepared material in the memory to be recalled when needed for display. The necessary instructions for various operations are detailed in the instruction manual that is supplied with the Character Generator.

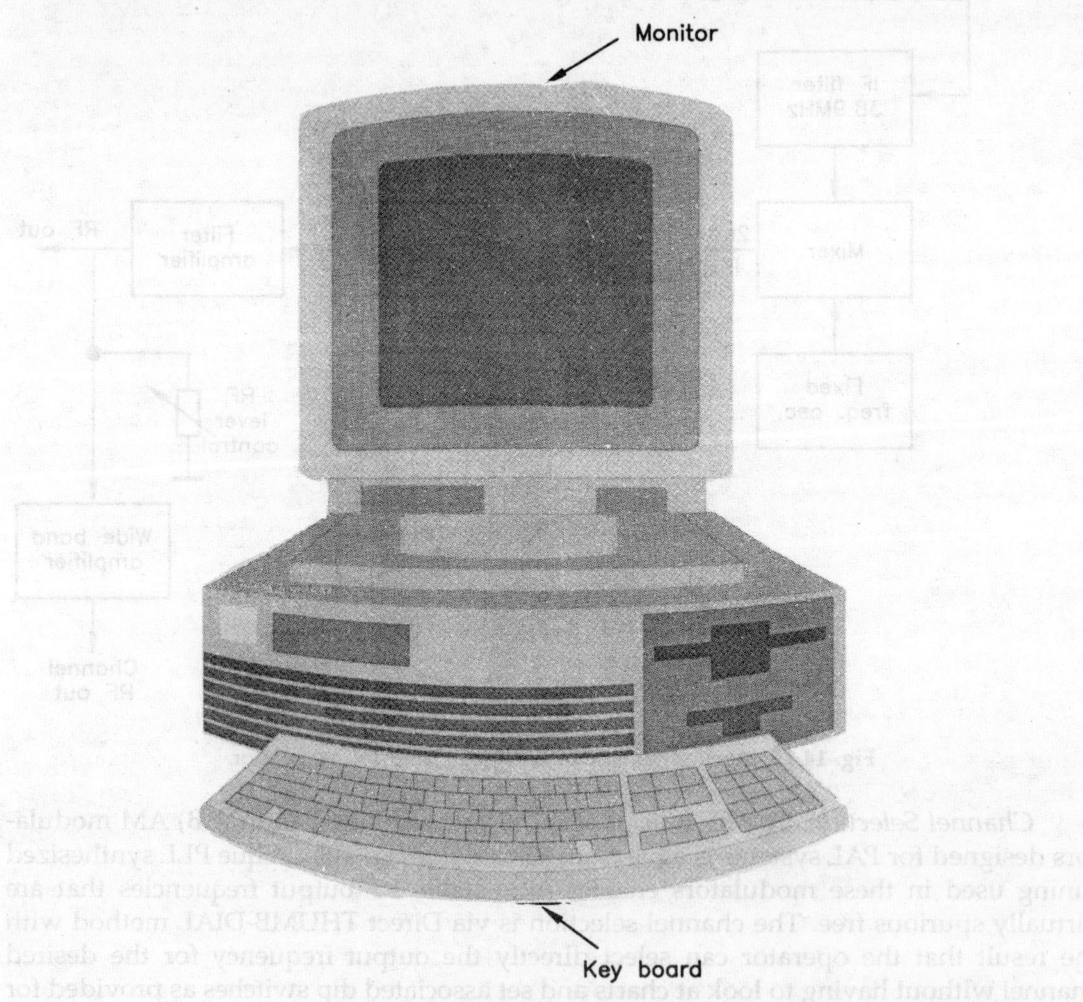

Fig. 14.16. Pictorial view of a basic character generator.

Professional Class character generators that are used in TV and Cable Satellite production centres have in addition two monitors. The first, called 'Programme' or

'Line' displays the message which is currently available to be placed on 'air'. The second monitor called 'Previous' or 'Set-up' displays another message which the C.G. operator is preparing. A cursor or position marker shows the locations of each character as it is typed on the screen. The use of a 'Paint Box' along with the C.G. eliminates the need for an artist to use paper, pencil, inks or photographic techniques in the production of what is called television graphics. This enables the display of messages along with desired illustrations. The electronics needed to incorporate all these functions in the C.G. is quite complex but essentially digital in nature and controlled by a set of microprocessors.

14.8 Signal Processor

The local 'OFF-AIR' telecasts are often included in the cable network for distribution to customers. Similarly outputs from VCRs and/or C.D. players are also carried on cable channels. Usually it is not possible to assign same channels on which these are received. As such, it is common practice to transfer such programmes to other convenient channels. It is not necessary to obtain baseband signals for doing so, but instead, the modulated RF signals are down converted to the IF band with standard TV IF frequencies of 38.9 MHz for video and 33.4 MHz for audio. This IF band is then up-converted to the assigned cable channel band of 7 MHz. This is the purpose of a Signal Processor as shown in block diagram form in Fig. 14.17. Its various sections are:

(i) *Input Down Converter:* At the input of this section is a channel filter to prevent entry of adjacent channel signals if present. It is followed by a BP. RF amplifier whose gain is stabilized by delayed AGC action. The amplified RF signal is down converted by setting LO frequency of the converter at the correct value necessary to obtain IF band as the difference output product. The IF band signal thus obtained feeds into a variable sound IF, trap network to set the sound IF frequency level at least 20 dB below the picture IF signal to ensure proper inter-carrier audio signal detection in the TV receiver.

(ii) *IF Amplifier:* The IF amplifier section is of the same type as in television receivers with a 'SAW FILTER' at its output for necessary waveshaping of the television signal. As shown in the figure, an IF detector and filter circuit after the IF amplifier is used to develop video and audio AGC circuits. The potentiometer in the audio AGC circuit is used to adjust sound IF attenuation in the trap circuit.

(iii) *Output up Converter:* This section is designed to up-convert the IF signal band to the assigned video and audio carriers of the allotted cable channel. For this, the output converter's LO frequency is chosen accordingly. A potentiometer along with a meter is provided on the front panel of signal processor to set the RF carrier level at the input of up-converter. The converter output is amplified by an RF amplifier and then passed through a channel filter, the B-W of which is set for the allotted cable channel. The modulated RF output, thus obtained is combined with other modulator outputs in the combiner network.

14.9 Video Cassette Recorder (VCR)

Video Cassette Recorders are used in cable stations to show programmes earlier recorded on magnetic tapes. Out of the three available transport systems, Video Home System (VHS) is commonly used for recording and playback purposes. In this system, the tape

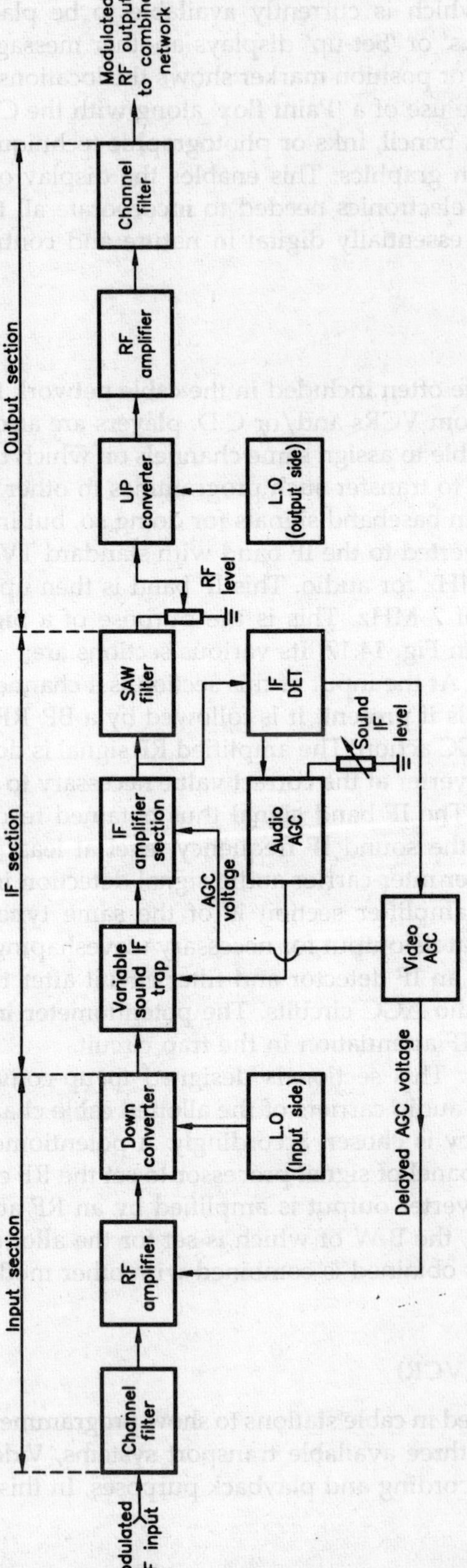

Fig. 14.17. Block schematic of a TV signal processor.

transport speed is 2.339 cm/s and writing speed (tape speed relative to head drum rotation) 4.85 m/s. Though not common, when a free channel on the cable system is not available, any incoming 'OFF-AIR' or satellite transmission can be recorded with the VCR and shown later when a free cable channel becomes available.

Modern VCRs are designed to record programmes 'OFF-AIR' *i.e.* direct from the antenna down-lead and for replay to supply to the antenna socket of the receiver a simulated TV signal which appears to the receiver as if it had come through the air on a normal channel. Any VCR in this way can be used with any TV receiver directly or through the cable system and the machine can be used in the recording mode even when the receiver is switched 'off' or even not there at all. The manner in which the VCR is connected to the antenna on its input side and to the TV receiver or Signal Processor from its output is explained with the help of a skeleton block diagram [Fig. 14.18] of the video cassette recorder. The transmission line (feeder cable) from the antenna does not go to the TV receiver/signal processor directly but is routed via the VCR. As shown, it terminates at the input socket of the VCR and through that to the splitter network located in it. The two-way splitter provides parallel paths for the antenna signal to respective tuners in the VCR and TV receiver. The VCR, which besides tuner has its own IF and detection section, recovers baseband video and audio signals for recording on the tape. When the VCR is operated in playback mode, the video and audio outputs from the tape go to a modulator which is designed to produce RF output whose format is the same as that from a TV transmitter. The carrier frequencies used are those of channel 3/4 or 31-39 in the UHF band depending on channel allocation of local and nearby TV stations to avoid interference. A switch (S_1) on the rear of VCR enables selection of channel for modulated output. Another switch (S_2) provided on the front panel is for selection of input to the signal processor either from the VCR output or from antenna. Thus for playback from the video tape, S_2 is kept at VCR (VTR) enabling modulator output to be fed to the input terminals of TV or signal

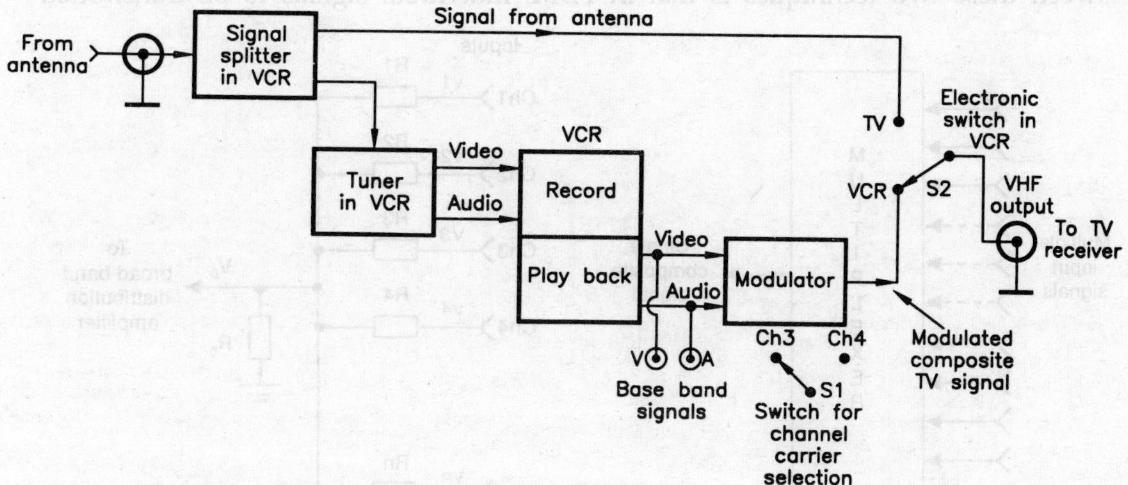

Fig. 14.18. An oversimplified block diagram of the VCR showing connections to the antenna and TV receiver or signal processor.

processor. Usually a balun is provided to enable impedance match between the VCR output and input of TV receiver or signal processor. It is also possible to obtain baseband video and audio signals directly from the VCR as shown in the figure and feed these to a channel modulator either directly or through an RF amplifier, the gain control of which enables setting of correct input amplitudes to the modulator.

Video Camera: The baseband video signals are also available from a video camera when used for direct transmission via the cable system. However, this is normally not done. In practice, several professional TV cameras are used to televise live programmes and their outputs are continuously monitored by the programme director in the control room who then selects the scene to be shown almost frame by frame. The selected signal is either sent directly to the cable station via a coaxial cable or a mini-transmitter. Further processing is the same as for signals from a VCR.

Camcorder: Many big cable companies collect local news and events during the day and show these in the evening on their cable system. For this, camcoders are used. A Camcorder (CAMERA-RECORDER) is essentially a video-cum-sound camera and a VCR combination. The recordings which become available on a tape are edited and transferred on a standard VHS tape for playing on a VCR for distribution to subscribers over the cable system.

14.10 Combining Networks

The combining of several modulator and signal processor outputs is essentially multiplexing of two or more individual signals for simultaneous transmission over a single communication channel. The concept of a simple multiplexer is illustrated in Fig. 14.19 (a), where several input signals are combined by the multiplexer into a single composite signal. The two basic types of multiplexing are: Frequency Division Multiplexing (FDM) and Time Division Multiplexing (TDM). The primary difference between these two techniques is that in FDM, individual signals to be transmitted

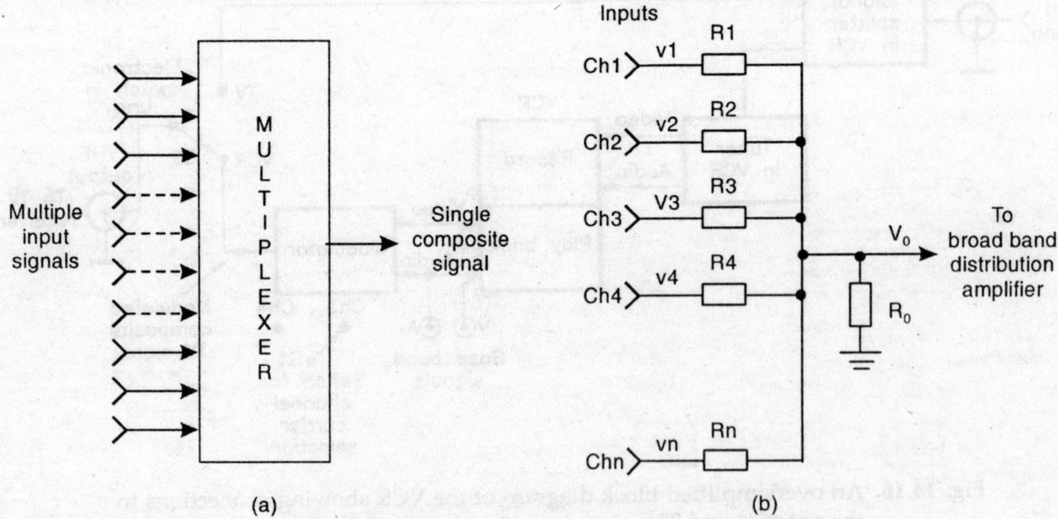

Fig. 14.19. (a) Concept of signal multiplexing
(b) Resistive summing network.

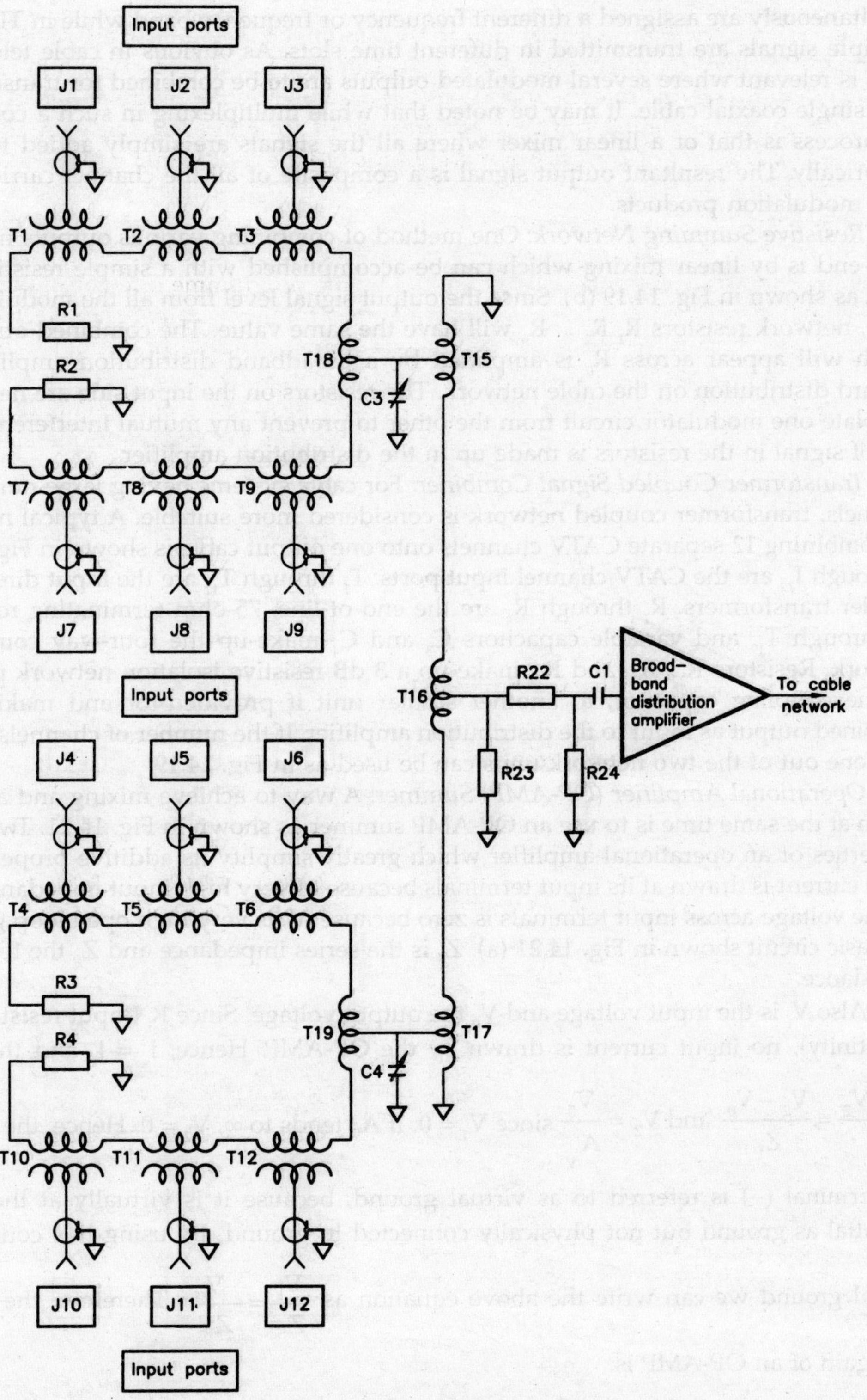

Fig. 14.20. Simplified diagram of a 12 channel network combiner.

simultaneously are assigned a different frequency or frequency band while in TDM, the multiple signals are transmitted in different time slots. As obvious in cable television, FDM is relevant where several modulated outputs are to be combined for transmission on a single coaxial cable. It may be noted that while multiplexing in such a combiner, the process is that of a linear mixer where all the signals are simply added together algebrically. The resultant output signal is a composite of all the channel carriers and their modulation products.

Resistive Summing Network: One method of combining various outputs from the head-end is by linear mixing which can be accomplished with a simple resistive network as shown in Fig. 14.19 (b). Since the output signal level from all the modulators is same, network resistors R_1 R_2 ... R_n will have the same value. The combined output V_0 which will appear across R_0 is amplified by a broadband distribution amplifier for onward distribution on the cable network. The resistors on the input side are necessary to isolate one modulator circuit from the other to prevent any mutual interference. The loss of signal in the resistors is made up in the distribution amplifier.

Transformer Coupled Signal Combiner: For cable systems having large number of channels, transformer coupled network is considered more suitable. A typical network for combining 12 separate CATV channels onto one output cable is shown in Fig. 14.20. J_1 through J_{12} are the CATV channel input ports. T_1 through T_{12} are the input directional coupler transformers. R_1 through R_4 are the end-of-line, 75-ohm terminating resistors. T_{15} through T_{19} and variable capacitors C_3 and C_4 make-up the four-way combining network. Resistors R_{22}, R_{23} and R_{24} make-up a 3 dB resistive isolation network prior to C_1, the coupling capacitor, to another similar unit if provided-for and making the combined output as input to the distribution amplifier. If the number of channels is less, only one out of the two network units can be used as in Fig. 14.19.

Operational Amplifier (OP-AMP) Summer: A way to achieve mixing and amplification at the same time is to use an OP-AMP summer as shown in Fig. 14.21. Two basic properties of an operational amplifier which greatly simplify its additive property are: (i) no current is drawn at its input terminals because of very high input impedance, and (ii) the voltage across input terminals is zero because of its very high open loop gain. In the basic circuit shown in Fig. 14.21 (a). Z_S is the series impedance and Z_F the feedback impedance.

Also V_i is the input voltage and V_0 the output voltage. Since R_i (input resistance) $\approx \infty$ (infinity), no input current is drawn by the OP-AMP. Hence, $i_s = i_F$ and therefore $\frac{V_i - V_2}{Z_s} = \frac{V_2 - V_0}{Z_F}$ and $V_2 = \frac{V_0}{A_V}$ since $V_1 = 0$. If A_V tends to ∞, $V_2 = 0$. Hence, the inverting terminal (–) is referred to as virtual ground, because it is virtually at the same potential as ground but not physically connected to ground. By using this concept of virtual ground we can write the above equation as $\frac{V_i}{Z_S} = \frac{-V_0}{Z_F}$. Therefore, the closed loop gain of an OP-AMP is:

$$AV_f = \frac{V_o}{V_i} = -\frac{Z_F}{Z_S}$$

Thus, the closed loop gain is independent of the open-loop gain A_v as long as A_v is large. The negative sign indicates inversion of signal but with modulated signals as inputs it is of no consequence.

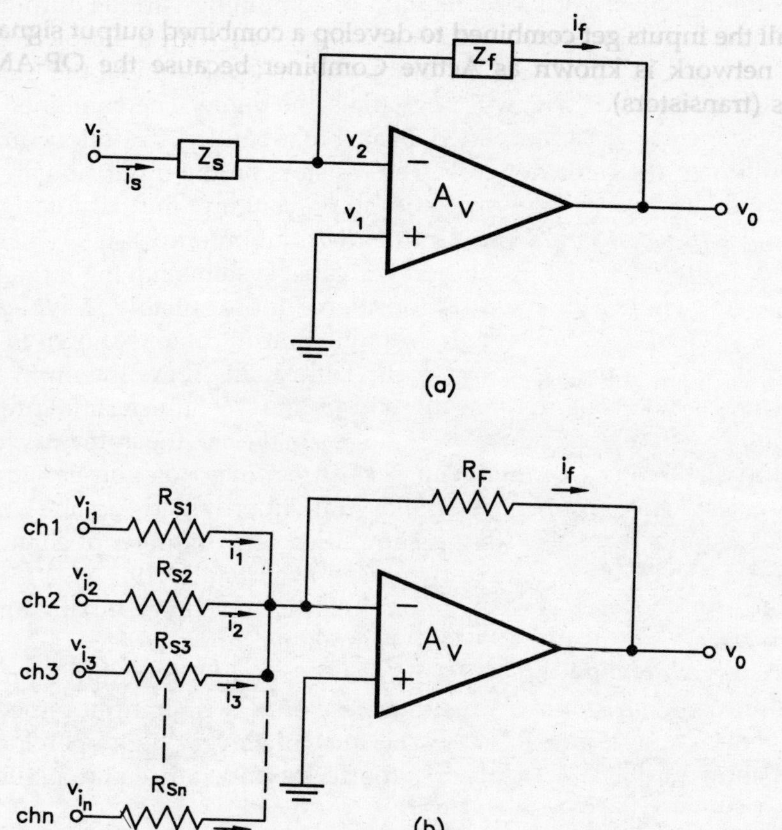

Fig. 14.21. Summing network employing OP-AMP (a) Basic circuit, (b) Practical circuit configuration.

Next consider the additive circuit [Fig. 14.21 (b)] which can be used for mixing (combining) modulator outputs. Since the amplifier input acts as a virtual ground:

$$i_1 = \frac{Vi_1}{Rs_1},\ i_2 = \frac{Vi_2}{Rs_2} \ldots\text{and } i_n = \frac{Vi_n}{Rs_n}$$

Also $\quad i_f = \dfrac{V_0}{R_F}$ and $i_1 + i_2 + i_3 \ldots\ldots\ldots i_n = i_f$

$$\text{Hence, } V_o = -i_f \times R_F = \left(-\frac{R_F}{Rs_1} vi_1 + \frac{R_F}{Rs_2} vi_2 + \frac{R_F}{Rs_3} vi_3 \ldots\ldots + \frac{R_F}{Rs_n} vi_n \right)$$

If $Rs_1 = Rs_2 \ldots\ldots = Rs_n$ is made equal to R_F

Then $V_0 = -(Vi_1 + Vi_2 + Vi_3 \ldots\ldots + Vi_n)$

Hence, all the inputs get combined to develop a combined output signal = V_0. Such a combining network is known as Active Combiner because the OP-AMP contains active devices (transistors).

15

CABLE SIGNAL DISTRIBUTION

Introduction

Cable TV distribution system is essentially a coaxial cable network that feeds multiplexed channel signals from the combiner to subscribers over an area that can be quite widespread in big cable systems. As explained in chapter 1 [see Fig. 1.4], trunk amplifiers are provided at regular intervals along trunk routes to make-up for cable losses. In the distribution area, directional couplers and splitters are used to divide the signal into branch lines and finally tap-off networks are connected to supply cable signal to subscribers through short runs of feeder cables. One millivolt (0 dBmV) is the minimum signal voltage measured across 75 ohms input impedance of the receiver for a noise-free picture. For an input impedance equal to 300 ohms, the minimum required voltage will be 2mV (6 dBmV). In practice, signal voltage amplitude is kept somewhat higher to ensure good picture reception and enable different input connections.

The cable signal can be connected either directly at the input socket of TV receiver or through a VCR. It is also possible to connect the two in parallel. In cable-ready TV receivers and VCRs, all the incoming channels can be tuned-in as for off-air TV station channels. However, earlier models of television receivers need a SET-Top converter for receiving all the channels.

In well planned towns and society complexes, coaxial cables are laid during construction period in well protected ducts. In older cities, where ducting is not easily possible, cables are spread in the open on specially erected wooden poles or where permitted, hung from electricity and telephone poles. Such a layout looks quite messy, more so when feeder cables are run for giving connections to new subscribers.

The solution to laying and maintaining cables over wide areas of operation lies in Microwave Multichannel Distribution System (MMDS). It is an over-the-air means of transmitting up to about 30 channels of TV programmes in the 2 GHz band after translating the channel mixer output at the head-end to this frequency spectrum. At the receiving end, each subscriber will need a small roof-top antenna, preferably a small dish, and a converter to translate back microwave signals to the VHF band necessary for usual reception. However, this makes MMDS quite expensive unless the distribution system is large as it would be in big cities. We will confine our discussion to the existing practice of signal distribution by coaxial cables.

15.1 Signal Distribution Network

There is no fixed network plan for CATV signal distribution because it depends on the number of subscribers and nature of area to be covered. For example, a 100,000 customer cable system will be very widespread having nearly 2600 km of cable length, needing over 1750 amplifiers (trunk and distribution) and thousands of splitting and distribution tap-off networks. On the other hand, a small cable operator serving around 500 subscribers in a small locality will have a much smaller layout spread over 3 to 5 km needing 2 or 3 trunk amplifiers, 5 to 10 distribution amplifiers, a few splitters and nearly 100 to 150 tap-off networks. The plan shown in Fig. 15.1 visualises a medium network drawn to include different amplifiers, splitters and tap-offs with a view to introduce all types of amplifiers and networks that form part of CATV systems.

The combiner in the head-end equipment is followed by a broad band amplifier to raise the signal level to about 1V (p-p) *i.e.* 60 dBmV or 120 dBμV. This amplifier feeds to several trunk amplifiers on cable lines which carry signals to distant locations. Trunk and distribution amplifiers are located at several points along cable runs to overcome progressive signal attenuation which occurs due to cable losses. The signal distribution is done through splitters and directional couplers. Such networks divide signal power equally in all the branches and also provide impedance match to keep losses to a minimum. The signal distribution from splitters to tap-off points is done through coaxial cables from where feeder cables carry cable signals to customers. Where necessary, distribution amplifiers are provided between splitters and tap-offs. Tap-off networks may have 2, 3 or four output ports for signal distribution to an equal number of subscribers. Tap-off networks divide the signal equally and also maintain continuity of signal flow along the distribution cable. Since improperly terminated lines will develop standing waves, the end of each distribution cable is terminated with a 75 ohms resistor called a terminator. More details about CATV amplifiers, splitters and tap-off networks are given in subsequent sections.

15.2 Cable Losses and Signal Distortion

Signal attenuation occurs mostly in lengthy cables and distribution networks. Power loss due to dissipation at the input of TV receivers is quite small in comparison to cable losses. For example, with signal voltage amplitude equal to 1.5 mV, only 20 μA current will flow in the input impedance (75 Ω) of each receiver.

1. *Cable Losses:* The energy dissipated in coaxial cables is due to:

(i) *I^2R Losses:* The characteristic impedance which is resistive (R) of most cables is around 75 ohms. Hence, sufficient I^2R loss occurs when signal current 'I' flow through cable conductors.

(ii) *Dielectric Losses:* A capacitor is formed whenever an insulator *i.e.* dielectric separates two conductors between which a difference of potential can exist. A small loss of energy occurs when the capacitor formed due to insulation between cable conductors charges and discharges rapidly due to reversal of polarity of the voltage that exists across them. Such a loss is called dielectric loss and rises with increase in frequency. Since cable currents are in the VHF and lower UHF band, dielectric loss in coaxial cables becomes significant.

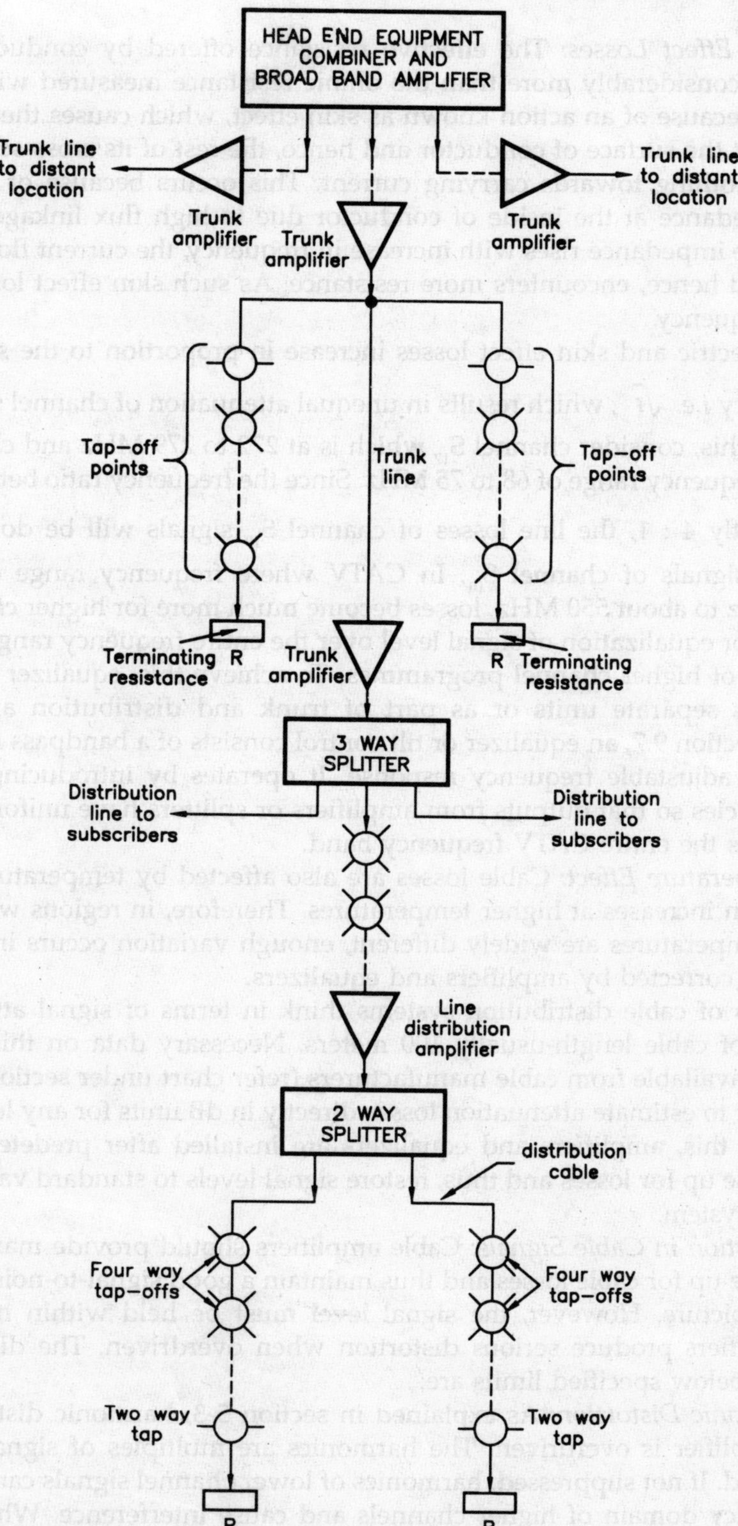

Fig. 15.1. A typical signal distribution plan for a CATV system.

(iii) *Skin Effect Losses:* The effective resistance offered by conductors to radio frequencies is considerably more than the ohmic resistance measured with direct currents. This is because of an action known as skin effect, which causes the current to be concentrated at the surface of conductor and hence, the rest of its cross-section contributes little or nothing towards carrying current. This occurs because of much higher inductive impedance at the inside of conductor due to high flux linkages over there. Since inductive impedance rises with increase in frequency, the current flows just along the surface and hence, encounters more resistance. As such skin effect losses rise with increase in frequency.

Both dielectric and skin effect losses increase in proportion to the square-root of signal frequency *i.e.* $\sqrt{f}$, which results in unequal attenuation of channel signals. As an illustration of this, consider channel S_{17} which is at 272 to 279 MHz and channel S_{01} (X) which has a frequency range of 68 to 75 MHz. Since the frequency ratio between the two is almost exactly 4 : 1, the line losses of channel S_{17} signals will be double $(\sqrt{4})$ as compared to signals of channel S_{01}. In CATV where frequency range extends from around 45 MHz to about 550 MHz, losses become much more for higher channels. Thus there is need for equalization of signal level over the entire frequency range for satisfactory reception of higher channel programmes. To achieve this, equalizer networks are used either as separate units or as part of trunk and distribution amplifiers. As explained in section 9.7, an equalizer or tilt control consists of a bandpass filter arrangement with an adjustable frequency response. It operates by introducing less loss to higher frequencies so that outputs from amplifiers or splitters have uniform amplitude response across the entire CATV frequency band.

(iv) *Temperature Effect:* Cable losses are also affected by temperature variations. The attenuation increases at higher temperatures. Therefore, in regions where summer and winter temperatures are widely different, enough variation occurs in signal level. These are also corrected by amplifiers and equalizers.

Designers of cable distribution systems think in terms of signal attenuation per unit distance of cable length-usually 100 meters. Necessary data on this at specified frequencies is available from cable manufacturers (refer chart under section 3-7). It then becomes easier to estimate attenuation losses directly in dB units for any length of cable run. Based on this, amplifiers and equalizers are installed after predetermined cable lengths to make up for losses and thus, restore signal levels to standard values throughout the cable system.

(2) *Distortion in Cable Signals:* Cable amplifiers should provide maximum signal output to make up for cable losses and thus maintain a good signal-to-noise ratio for no snow in the picture. However, the signal level must be held within narrow limits, because amplifiers produce serious distortion when overdriven. The distortions that must be kept below specified limits are:

(i) *Harmonic Distortion:* As explained in section 5-3, harmonic distortion occurs when the amplifier is overdriven. The harmonics are multiples of signal frequencies being amplified. If not suppressed, harmonics of lower channel signals can become part of the frequency domain of higher channels and cause interference. While restricting amplifier operation to its linear region will keep distortion at its minimum, harmonic distortion if present can be suppressed more effectively by using at least one stage of push-pull amplification. This will also enable necessary power gain. As shown in section 5.3, even harmonics are cancelled in the output circuit of a push-pull amplifier and its output voltage has very small content due to harmonic distortion. In particular 2nd harmonic voltage which has highest amplitude is eliminated and consequent interference very much reduced.

(ii) *Cross Modulation:* As detailed in section 9.1, non-linear amplitude characteristics also cause an overloaded amplifier to operate as a mixer. This results in detection and remodulation of some cable signals on carriers of several other channels. Such a transfer of modulation can cause appearance of video (picture) of one channel on another. The interfering picture can be a negative one. The sync signal which is of maximum amplitude in the video signal will cause bars on the picture of another channel due to cross-modulation. Overloading of amplifiers is prevented by using automatic gain control (AGC) in amplifier circuits. This prevents non-linear operation and hence, cross-modulation is nearly eliminated.

(iii) *Spurious Signals:* Another effect of amplifier overload is the production of random frequency components spread over a wide frequency range. This is caused by mixer action in the amplifier which in turn generates sum and difference frequency components called spurious signals. These on combining with picture signals of different channels cause an annoying disturbance on the reproduced picture. As far cross-modulation, generation of spurious signals can also be prevented by restricting amplifier operation in its linear region. This again is made possible by the application of AGC as explained in the next section.

15.3 Cable Amplifier Requirements

Cable amplifiers are basically low power, medium gain, wideband RF amplifiers. While power ratings of line, trunk and distribution amplifiers are different, their performance characteristics are nearly same. These are:

(i) *Bandwidth:* The bandwidth of cable amplifiers is kept between 45 to 300 MHz and if necessary up to 550 MHz to accommodate the provided range of CATV channels inclusive of some TV channels that are not in use for local off-air broadcasts and can be used as cable channels. Double tuned circuits with windings on special low-loss ferrite cores are used for coupling one amplifier stage to another. Stagger tuning is preferred for obtaining desired bandwidth. In stagger tuning, each amplifier stage is tuned to a frequency slightly off the centre frequency to increase overall bandwidth of the composite amplifier.

(ii) *Gain:* High frequency Bipolar Junction Transistors (BJTs) are used as active devices in cable amplifiers. Because of large bandwidth requirement, two or three amplifier stages are cascaded to produce overall gain of about 30-40 dB meaning a voltage gain of up to 100. Both pre- and post-amplifier stages employ push-pull configuration to obtain higher output voltage, more power gain and to eliminate harmonic distortion effects.

The line amplifier that follows the combiner of head-end equipment is designed to raise the signal amplitude to about 120 dB μV (60 dBmV) meaning 1.0 V P-P. On trunk lines, the signal amplitude is not allowed to fall much below 100 dB μV (40 dBmV) *i.e.*

0. IV. However, on long trunk routes where several trunk amplifiers operate in tandom, a lower signal amplitude is preferred to prevent overall distortion exceeding preset limits. In any case, signal amplitude on trunk lines in-between two amplifiers is not allowed to fall below 70 dBμV (10 dBmV).

Losses in distribution cables and splitters are made up by distribution amplifiers which maintain a signal amplitude level close to 80 dB μV *i.e.* 10 mV P-P at their output terminals. Additional losses occur in tap-off networks and individual subscriber feeder cables with the result that a signal level of about 60 to 65 dB μV (1 mV to 1.7 mV) becomes available at the input of each customer's TV receiver. This is enough to produce a good quality picture.

Gain Control: The gain control compensates for differences in flat loss and overall signal attenuation caused by coaxial trunk cabies. Trunk amplifiers employ both manual and automatic gain controls to adjust output signal level as necessary.

Attenuators: When input signal level is higher than the optimum value, it can be attenuated to the correct value by plugging in appropriate value of attenuation pad at the input of trunk amplifiers. Variable attenuators in small steps of 5 dB are available together with a 0 dB jumper wire for selecting required value of attenuation.

Automatic Gain Control (AGC): To compensate for inherent gain variations in amplifiers and those due to temperature effects, trunk amplifiers have AGC circuits. The principle of AGC is explained in section 11.8 [see Fig. 11.18] and a typical arrangement as used in trunk amplifiers is shown in Fig. 15.2. The system operates with a

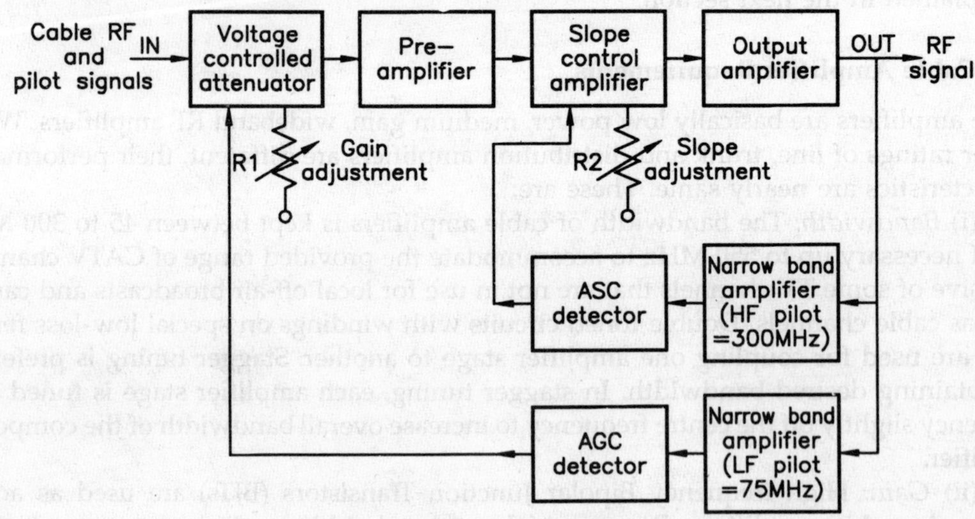

Fig. 15.2. Simplified block diagram of a trunk amplifier to explain AGC and ASC operations.

continuous wave (CW) pilot signal that is inserted at the input of amplifier. In the figure, a pilot frequency of 75 MHz is indicated but higher frequencies around 150 MHz are also used in some trunk amplifiers. A narrow-band amplifier tuned to select 75 MHz signal is connected at the output of amplifier. On selection and amplification, the pilot signal is detected by the AGC detector to produce a proportionate dc control voltage. It is applied to the voltage controlled attenuator which varies signal input to the preamplifier for a steady RF out at the output amplifier. A manual control R_1 is also provided for initial adjustment of gain for a specified output voltage.

(iii) *Frequency Response:* As already explained, signal attenuation increases at higher frequencies and thus, affects performance of higher channels. To compensate for this down tilt with increase in frequency, the gain of trunk and distribution amplifiers is made to slope upwards for higher channel frequencies.

Slope Control: The trunk amplifiers usually employ both manual and automatic slope control methods. The manual slope control's output changes with any variation in input signal level but automatic slope control maintains originally set output slope regardless of changes in input levels.

Equalizers: As explained in section 9.7, equalizers are used to correct changes in gain with frequency. These are made of L-C components that provide shunt paths to signals whose gain is to be reduced. Both fixed and variable equalizers are inserted in series with the signal path to keep gain for all the cable frequencies to be nearly same.

Automatic Slope Control (ASC): The automatic slope control (ASC) is also a gain control system to equalize gain over the entire frequency range of cable signals. While series and shunt peaking techniques can provide some boost to the down sloping frequency response towards the higher end but major slope tilt correction is done by reducing gain of lower channel signals. For this, another pilot carrier is injected at the input of slope control amplifier. Its frequency is chosen to be in the higher frequency range, typically 300 MHz. As shown in Fig. 15.2, the corresponding narrow band amplifier and ASC detector are tuned to this frequency. The detector output is a dc control voltage proportionate to the difference in gain between upper and lower ranges of signal frequencies. The control voltage is applied to the equalizer that forms part of the input circuit of slope control amplifier. Depending on the magnitude of control voltage, the equalizer connects shunt paths to input signal components for which the gain must be reduced. A manual control (R_2) [see Fig. 15.2] is also provided for initial adjustment of gain for a flat response over the entire cable frequency range.

15.4 Trunk and Distribution Amplifiers

The trunk and distribution amplifiers have similar performance characteristics but their gain and output voltage and power ratings are somewhat different. To gain familiarity with the nature of circuitry of such amplifiers two common types of trunk amplifiers are briefly described.

(1) *Push Pull Trunk Amplifier:* The functional block diagram of a two-stage push-pull trunk amplifier is shown in Fig. 15.3. The RF input signal feeds into the push-pull preamplifier through an impedance matching network. The gain of preamplifier is about 20 dB. Its output becomes input to the flatness control board section where frequency response of the entire forward signal bandwidth (45 to 300/450/550 MHz) is adjusted manually. The signal then passes to the variable attenuator section. In it, gain control potentiometer R_1 sets gain of the composite amplifier. As shown in the figure, dc control voltage is amplified before supplying to the variable attenuator. The output of attenuator section feeds into the variable equalizer section. The dc voltage obtained

from slope control circuit formed by potentiometer R_2 and resistor R_3 is amplified and fed to the variable equalizer section for frequency slope correction. The control voltage is automatically adjusted by the temperature sensitive thermistor T_1. The equalizer output is made input to the post-amplifier stage which is a push-pull amplifier. It provides a gain of 16 to 18 dB and suppresses even harmonic distortion if present. Its output is connected to the on-going trunk line cable through an impedance matching network.

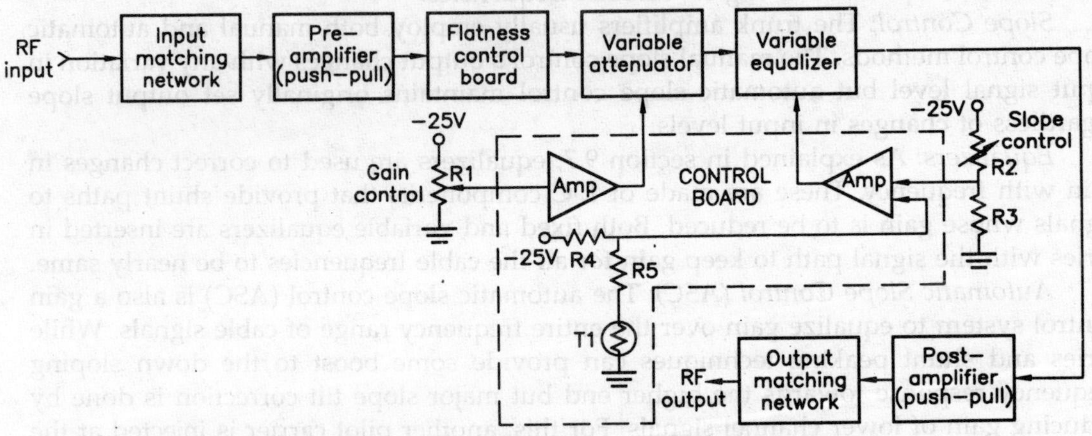

Fig. 15.3. Simplified functional block diagram of a push-pull trunk amplifier.

(2) *Parallel Hybrid or Power Doubling Trunk Amplifier*: A simplified block diagram of a parallel hybrid also called power doubler trunk amplifier is shown in Fig. 15.4. Trunk line input jack J_1 provides entry point for the forward RF signal into the trunk amplifier module from where it is routed to the optional plug-in input attenuator. This is needed if the amplifier stations are short-spaced and require a specific value of attenuation or padding. If not required, it is shorted by a 0 dB jumper wire. Signal out from the attenuator feeds into the plug-in equalizer. The equalizer attenuates lower frequencies (40 MHz and up) more than upper frequencies (300/450/550 MHz and down) to compensate for unequal signal attenuation through the preceding coaxial trunk cable. As such, output signal from the plug-in equalizer has a flat gain-frequency response. It is applied to the 1st stage (preamp) push-pull amplifier whose gain is about 17 dB. The amplified RF signal feeds into the control board via an optional interstage attenuator. The control board has necessary circuitry for manual and automatic gain controls and a general purpose equalizer besides manual and auto slope controls for optimum slope adjustment. The output of interstage control board is applied to the 2nd stage (post-amp) power doubler amplifier. As shown in the figure, the signal is split into two equal halves for feeding to the two post amplifiers. The two outputs are combined and sent to the trunk line through an impedance matching network. A directional coupler (splitter) [Fig. 15.4] diverts some of the amplifier output to a distribution cable which is later amplified by the distribution amplifier before distribution to area subscribers.

Amplifier Power Supply: All aspects of power supply to trunk and distribution amplifiers are detailed in Chapter 10. As explained there, cable powering is an

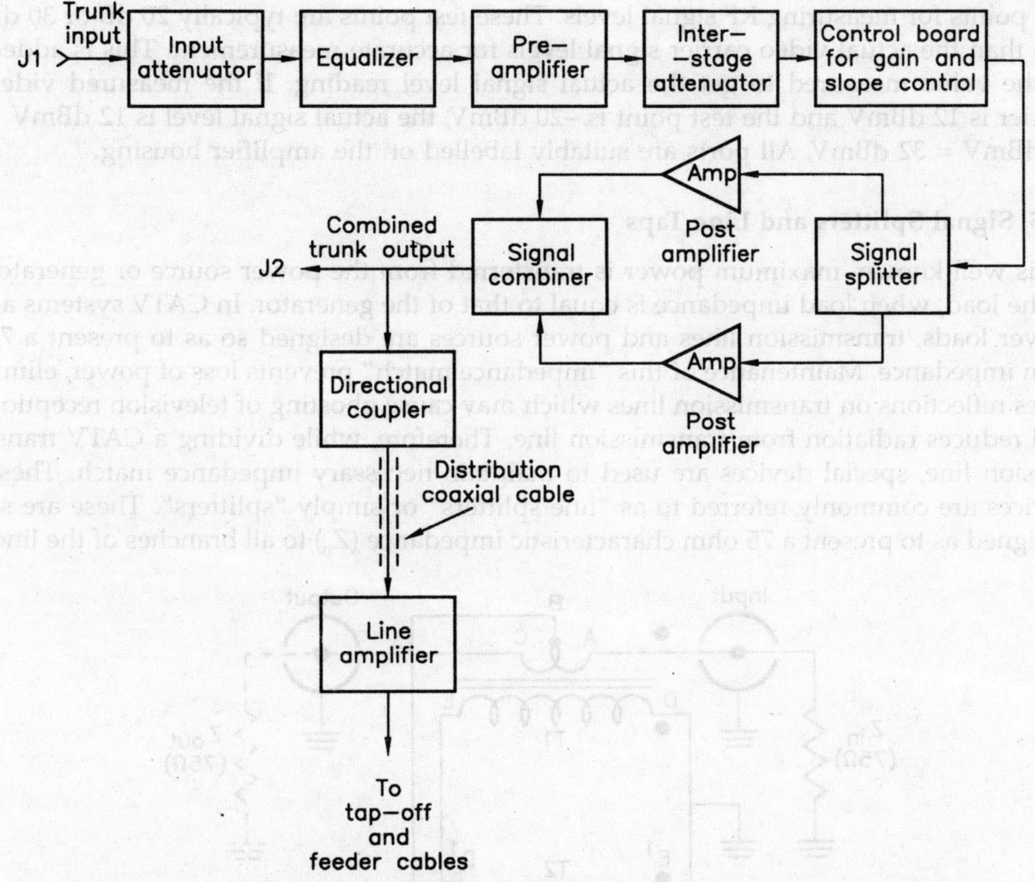

Fig. 15.4. Block diagram of a power doubling trunk amplifier.

important aspect of a cable television system. For large cable system, 60V ac is carried by cables along with RF signals to trunk and distribution amplifiers. At the amplifier it is bypassed for feeding other amplifiers on trunk routes. Such amplifiers are called POWER PASS TRUNK/DISTRIBUTION amplifiers. For smaller CATV systems, each amplifier is fed ac separately to be rectified, filtered and regulated for use in the amplifier. For more details refer to sections 10.4 and 10.5.

Amplifier Housing: The housing that contains CATV amplifiers is made of die-cast aluminium alloy or mild steel and painted for protection against harsh environment. In-between the main housing and its cover, there are two gaskets. One is a silicon-rubber gasket that provides a hermatic seal to prevent entry of water, dust and any other pollutants. The other gasket is of metal mesh to prevent ingress (induction or pick-up) and egress (radiation) of RF signals. The housing also has cooling fins to permit dissipation of heat created by the amplifier and power supply module.

At the base or sides of amplifier housing, there are ports (openings) for input and output connections of RF signals. Similarly, 60V ac is fed to the amplifier through a separate port and taken out through another port for feeding to other trunk line amplifiers.

Test Points: The trunk and distribution amplifiers also have input and/or output test points for measuring RF signal levels. These test points are typically 20 dB or 30 dB less than the actual video carrier signal levels for accurate measurement. This is added to the value measured to get the actual signal level reading. If the measured video carrier is 12 dBmV and the test point is –20 dBmV, the actual signal level is 12 dBmV + 20 dBmV = 32 dBmV. All ports are suitably labelled on the amplifier housing.

15.5 Signal Splitters and Line Taps

As is well known, maximum power is transferred from the power source or generator to the load, when load impedance is equal to that of the generator. In CATV systems all power loads, transmission lines and power sources are designed so as to present a 75 ohm impedance. Maintenance of this "impedance match" prevents loss of power, eliminates reflections on transmission lines which may cause ghosting of television reception and reduces radiation from transmission line. Therefore, while dividing a CATV transmission line, special devices are used to maintain necessary impedance match. These devices are commonly referred to as "line splitters" or simply "splitters". These are so designed as to present a 75 ohm characteristic impedance (Z_0) to all branches of the line.

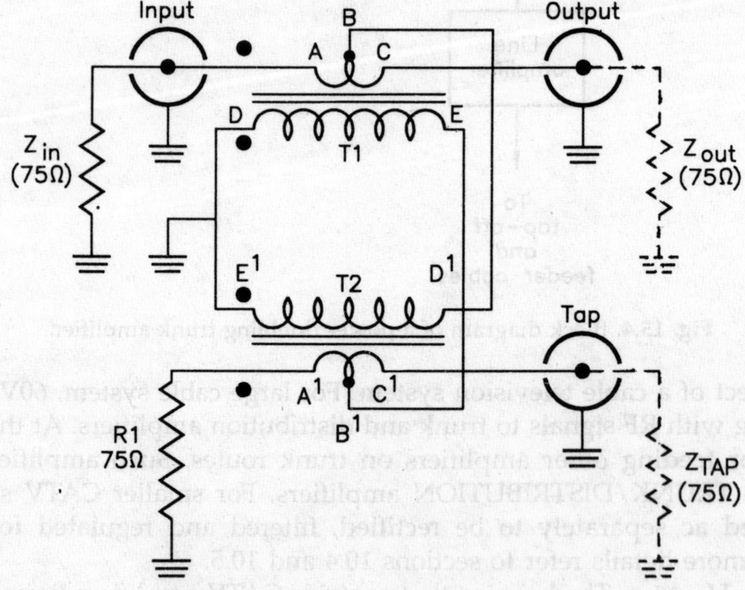

Fig. 15.5. Simplified circuit of a two-way splitter.

A two-way line splitter is constructed with tapped transformers in a circuit configuration designed to equally divide power between the two branches while maintaining 75 ohm characteristic impedance in all branches. A simplified circuit diagram of such a splitter is shown in Fig. 15.5. The input signal is equally divided between the two output taps and each 'looks into' a 75 ohm impedance.

There are three- or four-way splitters also which are actually combinations of two-way line splitters. For example, a three-way line splitter consists of a two-way line splitter with one of the branches being split again. Thus, the branches of a three-way line splitter will have outputs that are 6 dB down on two of the branches and 3 dB down on one branch. A four-way splitter will have both branches split and hence will have all the four output branches down by 6 dB. In actual practice, losses, will be somewhat higher depending on 'Q' of coils and efficiency of transformer circuitry.

The actual circuit used for two-way line splitters is more complex as shown in Fig. 15.6. Additional circuitry is added to provide facility to bypass flow of ac power supply for feeding successive amplifiers and preventing it from entering RF signal transformers.

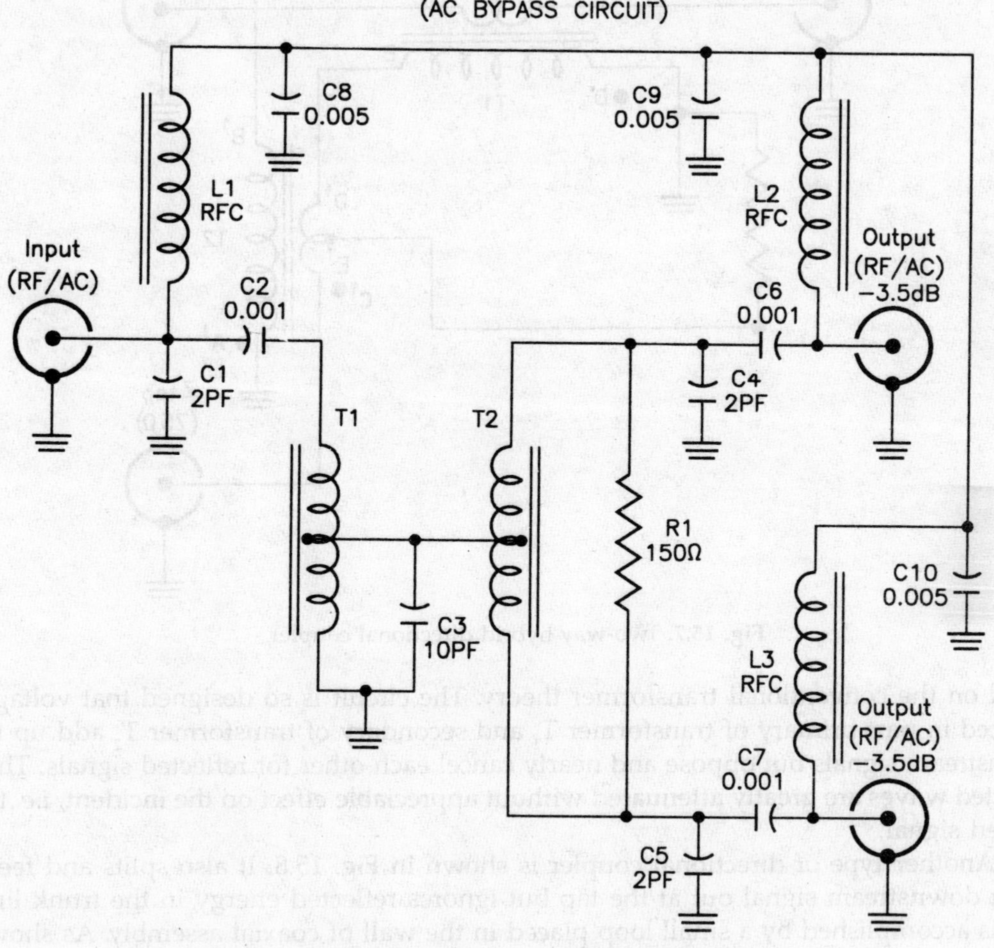

Fig. 15.6. AC pass two-way splitter.

AC power is bypassed through L_1, L_2 and L_3 to the two output ports while RF signal from input port enters transformer. T_1 through capacitor C_2. Capacitors C_1, C_4 and C_5 improve impedance match and frequency response. The signal applied to transformer T_1 is divided by transformer T_2 into two equal 3 dB down parts, which reach the two output points via capacitors C_6 and C_7. The circuit otherwise operates to split the signal in the same way as for a non-bypass splitter described earlier.

Directional Couplers: Despite using all means to keep correct impedance match on trunk lines, some mismatch persists and this causes reflections from far-end of the trunk line. If reflected signals are allowed to enter feeder lines, ghosting and interference can occur on television pictures. To prevent this, special splitters called directional couplers are used. A hybrid type of directional coupler is shown in Fig. 15.7. Its action is also

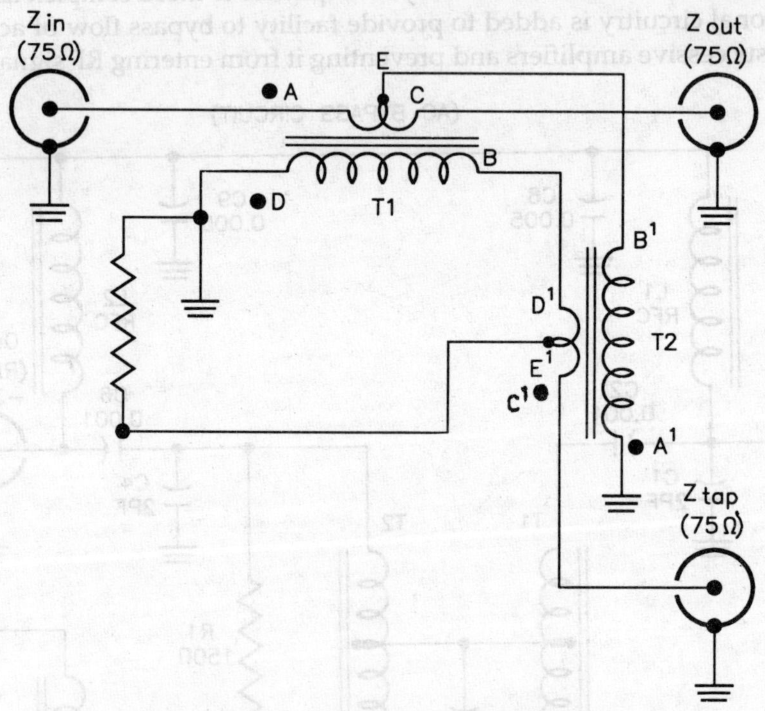

Fig. 15.7. Two-way hybrid directional coupler.

based on the conventional transformer theory. The circuit is so designed that voltages induced in part primary of transformer T_1 and secondary of transformer T_2 add up for downstream signals but oppose and nearly cancel each other for reflected signals. Thus reflected waves are greatly attenuated without appreciable effect on the incident, *i.e.* the desired signal.

Another type of directional coupler is shown in Fig. 15.8. It also splits and feeds direct downstream signal out at the tap but ignores reflected energy in the trunk line. This is accomplished by a small loop placed in the wall of coaxial assembly. As shown in the figure, this loop is terminated with a 75-ohm resistor. The loop acts both as a capacitor and an inductor. Its capacitance charges to the potential difference between the inner and outer conductors at that point on the line. As a one turn coil, the loop is magnetically coupled to the centre conductor to tap-off the signal. Directional couplers are used in big CATV systems where trunk lines are very long. These are usually followed by bridging amplifiers just close to the split-tap to boost the signal to line levels. For medium and small CATV set-ups, splitters serve the purpose of dividing the signal.

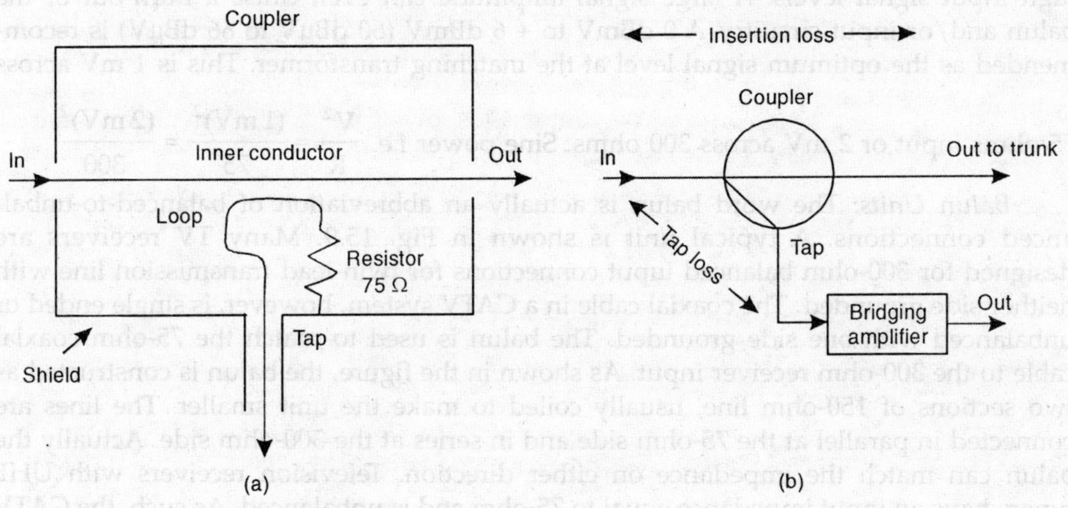

Fig. 15.8. Inductive—capacitive coupled directional coupler (a) schematic, (b) symbol.

Line Taps: Line taps or tap-offs are also passive devices like signal splitters and are connected along distribution cables at suitable locations to tap-off signal for customer receivers. The distribution cable is cut to interpose tap-off network still providing continuity to signal flow along the cable.

Line taps can be one-way, two-way, three-way or four-way (see Fig. 15.1) for connecting equal number of TV receivers. Each tap-off has input and output points (connecting points) for the distribution cable input and output connections besides tap-off terminals for feeder cables.

As in splitters, passive elements are used in line taps for dividing the signal while, maintaining impedance match on either side for maximum transfer of power. However, line taps are designed to have a high tap-loss because isolation is necessary so that a mis-terminated cable at the subscriber's TV receiver does not set up reflections in the cable system.

Two possible terminations that can cause reflections are a cable that is not connected or leads that are short-circuited. Tap units are available with many values of tap loss so that signal levels can be balanced for different subscriber drop lines along the branch cable. In general, a tap loss of 20 dB is enough to prevent echos or ghosts entering feeder cables from main distribution line and to stop mutual interference between TV receivers. As stated earlier, gain of distribution amplifiers is kept high to counter losses in tap-offs. As for splitters, line-tap units are also housed in sealed boxes to prevent entry of moisture or rain water.

Before concluding distribution of cable signals in a CATV system, it may be mentioned that some non-technical system operators express a desire for as much as + 10 dBmV (+ 70 dBμV) *i.e.* a signal amplitude equal to 3.1 mV at the customer's TV set. Such a high signal level is not only undesirable but also very expensive to achieve. Many TV receivers are not capable of suppressing adjacent channel interference with high input signal levels. A large signal amplitude can even cause a burn-out of the balun and/or input circuitry. A 0 dBmV to + 6 dBmV (60 dBμV to 66 dBμV) is recommended as the optimum signal level at the matching transformer. This is 1 mV across 75-ohms input or 2 mV across 300 ohms. Sine power *i.e.* $\frac{V^2}{R} = \frac{(1\,mV)^2}{75} = \frac{(2\,mV)^2}{300}$

Balun Units: The word balun is actually an abbreviation of balanced-to-unbalanced connections. A typical unit is shown in Fig. 15.9. Many TV receivers are designed for 300-ohm balanced input connections for twin-lead transmission line with neither side grounded. The coaxial cable in a CATV system, however, is single ended or unbalanced with one side grounded. The balun is used to match the 75-ohm coaxial cable to the 300-ohm receiver input. As shown in the figure, the balun is constructed as two sections of 150-ohm line, usually coiled to make the unit smaller. The lines are connected in parallel at the 75-ohm side and in series at the 300-ohm side. Actually the balun can match the impedance on either direction. Television receivers with UHF tuners have an input impedance equal to 75-ohm and is unbalanced. As such, the CATV feeder cable can be directly connected at the input socket of the receiver.

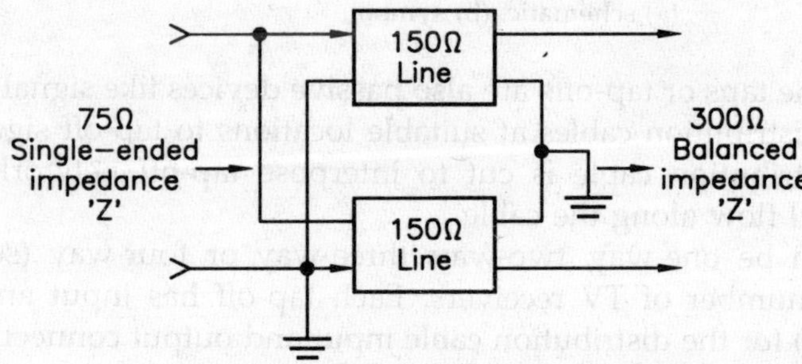

Fig. 15.9. Schematic diagram of a balun.

15.6 Cable Signal Converters

The converter placed on top of a TV set or close to it is an RF tuner used to select any desired channel. All cable channels are heterodyned to a specific frequency band chosen to be one of the low band VHF channels. Usually channel 3 or 4 is the choice depending on which is not an active broadcast channel in the area.

However, a simple frequency conversion to channel 3 or 4 presents some problems. For channel 3 (54 to 61 MHz.) conversion, the local oscillator (LO) in the converter will operate 54 MHz above the upper end of the selected channel. Suppose cable channel S_{01} (68-75 MHz) is selected by the converter. Then the LO operates at 54 + 75 = 129 MHz. This local oscillator frequency is inside the cable mid-band channel S_4 which is between 125 to 132 MHz. In addition, there are several other frequency combination in which LO signal can be a source of interference, especially since the broadband preselector in the 'converter' is not very selective for RF tuning owing to wide range of cable channels. The solution to these problems is to use a double super-heterodyne circuit for the cable converter with a UHF intermediate IF frequency.

Up-Down Cable Converter: The block diagram of the up-down converter is shown in Fig. 15.10. In this, double superheterodyne circuit, first the cable channels are heterodyned up to the IF range of 608 to 615 MHz in the UHF band. The oscillator frequencies for conversion to this IF band do not fall in any band allotted to cable channels and hence do not cause any interference. As necessary, a varactor tuned oscillator provides a specific local oscillator (LO) frequency for each cable channel. A bandpass filter in the output circuit of up-converter selects only the IF signal (608-615 MHz) which contains modulation of the selected cable channel.

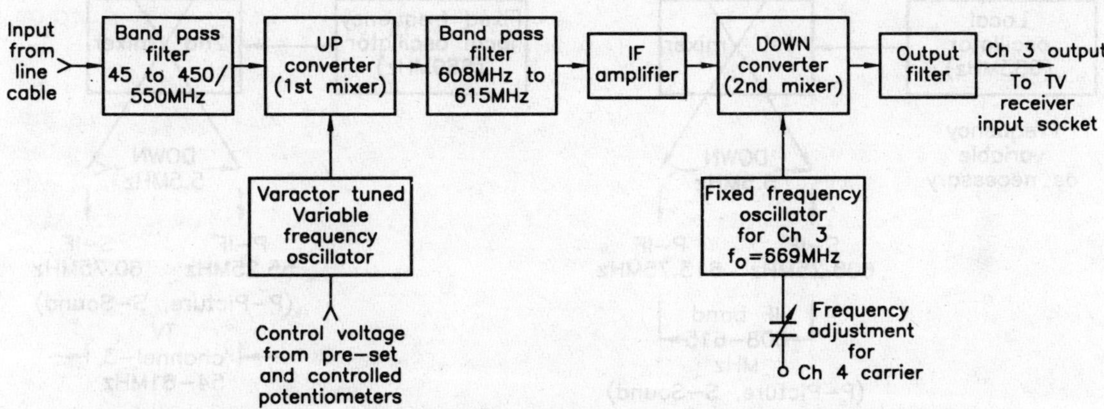

Fig. 15.10. Basic block diagram of UP-DOWN cable TV converter.

The second mixer in the down converter has a fixed frequency oscillator because both input and output signal bands have fixed values. It heterodynes the IF signal down to either channel 3 or 4 for TV receivers. An adjustment in the local oscillator is set for either output channel.

As an illustration of the process of double conversion, consider that cable channel S_{01} (68-75 MHz) with picture carrier (P) = 69.25 MHz and sound carrier(s) = 74.75 MHz is to be converted to TV channel 3 (54-61 MHz) with picture carrier = 55.25 MHz and sound carrier = 60.75 MHz. Since the IF range for up-conversion is chosen to be 608 to 615 MHz, the local oscillator frequency of the 1st converter must be set at 608 + 75 = 683 MHz. One heterodyning, 'P' carrier at 69.25 MHz is converted 683 – 69.25 = 613.75 MHz at the upper end of IF passband. Similarly, 'S'-carrier at 74.75 MHz is converted to 683 – 74.75 = 608.25 MHz at the lower end of IF band. This is illustrated in Fig. 15.11(a) for clarity. The reason for picture IF becoming higher than the sound IF is that any heterodyning that occurs when the oscillator beats above the mixer input frequencies produces inversion of the IF signal frequencies. However, this inversion is corrected in the 2nd mixer and its output has correct locations of picture and sound carriers as shown in Fig. 15.11 (b). Note that local oscillator frequency of the 2nd mixer remains fixed at 669 MHz (608 + 61 = 669 MHz).

Channel Selection: Each channel is tuned by varactor control of the local oscillator frequency in the up-converter. As explained in a previous chapter, the varactor is a capacitive diode. Its capacitance decreases with increase in reverse bias voltage across it. Since frequency is varied by dc bias, the circuit is a voltage controlled oscillator (VCO).

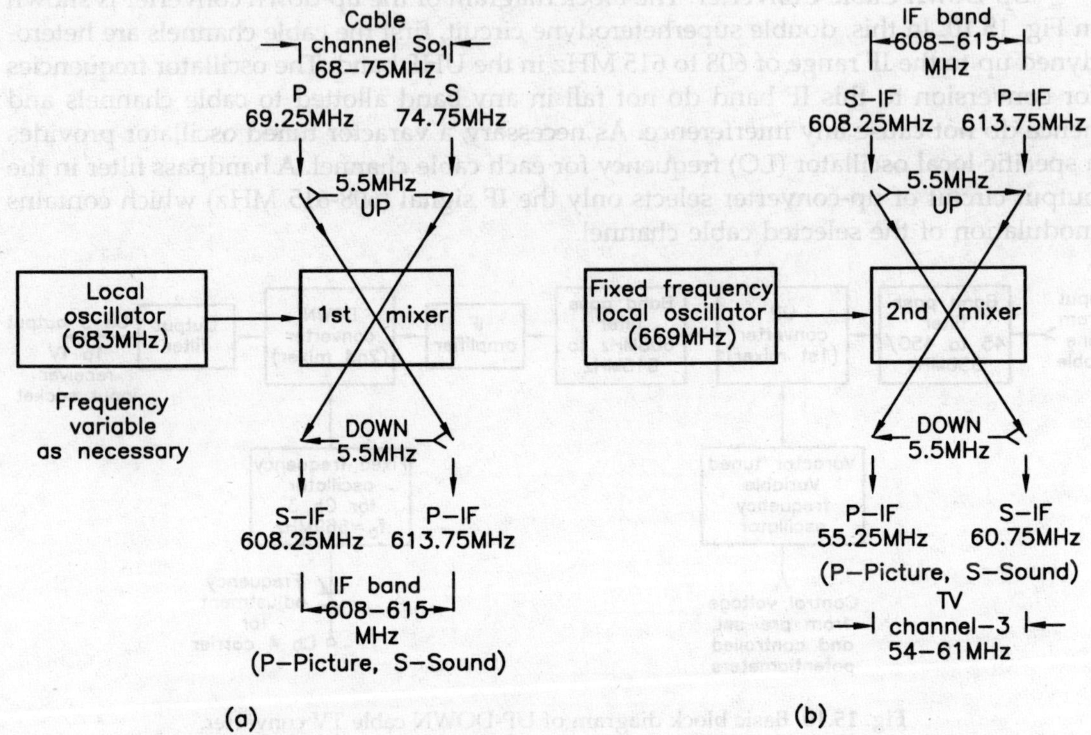

Fig. 15.11. Illustration of double heterodyning process: (a) up conversion to IF band (608-615 MHz), (b) down conversion to TV channel 3 (54 to 61 MHz).

Frequency Synthesizer: Since the number of available cable channels is quite large, it is cumbersome and time consuming to select and change from one channel to another manually. Therefore, modern top convertors use a frequency synthesizer for instant channel selection with a remote control. The advantage of frequency synthesis is that many oscillator frequencies can be supplied with the accuracy of crystal control. Frequency synthesizers are available in the form of IC chips and are designed to suit different television systems as for satellite receivers.

The block diagram in Fig. 15.12 shows the complete circuit for an up-down conversion using frequency synthesis. The synthesizer section illustrates how a specific frequency is set for the VCO which is the first local oscillator for up-conversion. A crystal oscillator is made as the reference. Its frequency is between 8 to 10 MHz depending on the frequencies to be generated in each TV system. The crystal oscillator output is divided by a suitable number (around 1000) to obtain a frequency close to 8 KHz. This becomes one input to the phase detector or comparator circuit. The other input to the comparator is a sample from the output of VCO in the up-converter. The sample is first divided by a fixed number and made input to the synthesizer chip. It is further divided by a factor 'N' which can take different values depending on the channel to be selected. The value of number 'N' is determined by the programmable 8-bit counter actuated by clock pulses generated with input from converter control circuit. The idea is to divide

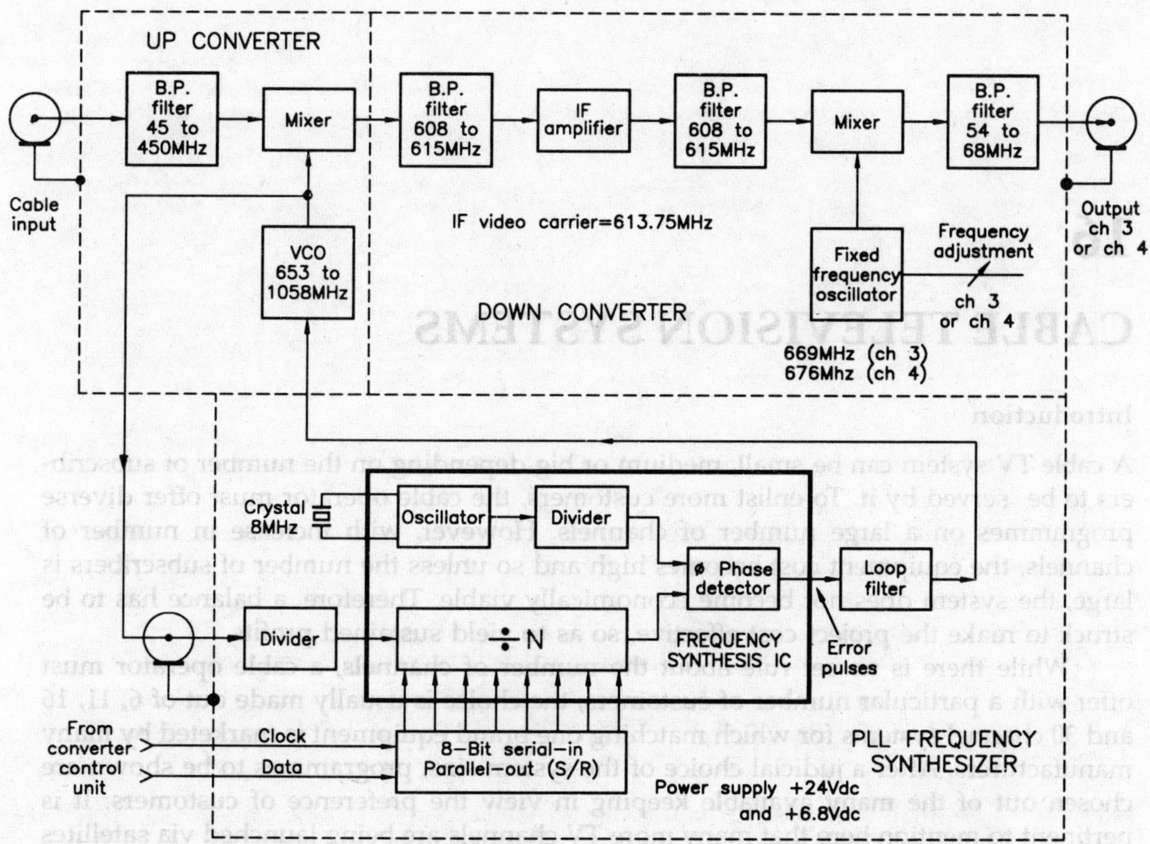

Fig. 15.12. Block diagram of a PLL frequency synthesizer for control of VCO of the up-converter.

by whatever 'N' is needed to obtain the same frequency signal as obtained by dividing the crystal oscillator output. The two are compared in the comparator that develops a control voltage proportional to any difference between the two inputs. This is applied to the VCO for keeping it at the correct frequency. As a result, the VCO in the up-converter has the same accuracy as the standard crystal oscillator for any selected channel. The factor 'N' can take any value between around 330 to 475 depending on the channel to be selected. This is determined by an 8-bit binary code set with push-button switches on either the counter or a remote control unit.

16

CABLE TELEVISION SYSTEMS

Introduction

A cable TV system can be small, medium or big depending on the number of subscribers to be served by it. To enlist more customers, the cable operator must offer diverse programmes on a large number of channels. However, with increase in number of channels, the equipment cost becomes high and so unless the number of subscribers is large, the system does not become economically viable. Therefore, a balance has to be struck to make the project cost effective, so as to yield sustained profits.

While there is no set rule about the number of channels, a cable operator must offer with a particular number of customers, the choice is usually made out of 6, 11, 16 and 30 channel systems for which matching one brand equipment is marketed by many manufacturers. After a judicial choice of the system size, programmes to be shown are chosen out of the many available keeping in view the preference of customers. It is pertinent to mention here that many more TV channels are being launched via satellites and at the same time some get terminated for one reason or the other. Similarly, sometimes a software owner company changes over from one satellite to another similar satellite for different reasons. As such, there is no fixed chart on the availability of programmes at any given time. Even then, the programmes that are currently available from both regional and domestic satellites are detailed in a tabular form in the following pages.

Satellite Channel Guide for India and Adjoining Countries

Channel by Satellite	Channel Carrier Frequency (GHz)	Polarization	Audio sub-carrier (MHz)
PAS-4, 68.5°E			
SONY ENT TV	3.905	H	7.2
BBC WORLD	3.825	V	6.6/7.2
DISCOVERY (INDIA)	3.785	V	6.8
DD (INT)	4.035	V	6.8
DD (MOVIE CLUB)	4.030	V	6.8
HOME TV	3.840	V	6.6

(Contd.)

Channel by Satellite	Channel Carrier Frequency (GHz)	Polarization	Audio sub-carrier (MHz)
ESPN	3.865	H	7.2
MTV (ASIA)	4.185	H	6.8/7.5
ABNI	3.785	H	6.8
TNT (CARTOON NET)	4.115	H	6.8
ATN	4.180	V	7.55
CNN (INDIA)	4.093	H	6.8
MUSIC ASIA	4.178	V	6.6
ASIASAT-1 105.5°E			
CHANNEL (V)	3.900	V	6.8
STAR SPORTS	3.860	V	6.8
PAKISTAN TV(PRE-EMP = 50μs)	4.100	V	6.6
STAR PLUS	4.020	V	6.8
STAR MOVIES	4.180	V	6.8
ZEE CINEMA	4.060	V	6.8
ZEE (INDIA) (EL)	3.940	V	6.8
INSAT-2B 93.5°E			
DD 13 (ASSAMESE)	2.618	H	5.5
DD 10 (MARATHI)	4.140	H	5.5
DD (CNN INT)	4.170	H	5.5
DD 6 (ORYA)	4.650	H	5.5
DD 11 (GUJARATI)	4.690	H	5.5
DD 17 (BIHAR)	4.730	H	5.5
INSAT-2C 93.5°E			
DD1 (NATIONAL)	4.125	V	5.5
DD2 (METRO)	4.165	V	5.5
DD3 (HINDI-ENGLISH)	3.805	V	5.5
DD4 (MALAYALAM)	3.765	V	5.5
DD5 (TAMIL)	3.885	V	5.5
DD7 (BENGALI)	3.845	V	5.5
DD8 (TELGU)	4.085	V	5.5
DD9 (KANNARA)	3.965	V	5.5
DD12 (KASHMIRI) PLUS MOVIE CLUB	4.045	V	5.5
DD14 (RAJASTHANI)	4.690	V	5.5
DD15 (HINDI M.P.)	4.730	V	5.5
INSAT-1D 83.0°			
DD1 (NATIONAL)	2.575	V	5.5
DD16 (HINDI U.P.)	2.615	V	5.5

(Contd.)

Channel by Satellite	Channel Carrier Frequency (GHz)	Polarization	Audio sub-carrier (MHz)
INTELSAT 703, 57.0°E			
SUN TV (TAMIL)	3.705	RHC	6.3/6.6
TVI (HINDI-ENGLISH)	4.140	LHC	6.6
NEPC (HINDI)	4.058	LHC	6.6
MONEY (E/H)	3.805	RHC	6.3/6.6
ASIA NET	3.980	RHC	6.3/6.6
UDAYA TV (T.S.)*	3.805	LHC	6.3/6.6
GEMNI TV (T.S.)*	3.805	LHC	6.3/6.6
(*TS-TIME SHARING)			
THAI COM-3 78.5°E			
RAJ TV (TAMIL)	3.640	V	6.6
PUNJAB TV (PUNJABI)	3.545	V	6.6
VIJAY TV (TAMIL)	3.620	V	6.6
PALA PAC-2 113.0°E			
AUSTRALIA TVI	3.880	H	5.95
STAR TV (AUS)	3.580	H	6.6
CNBC	3.620	H	6.6
DISCOVERY (ENGLISH)	3.720	H	6.6
ESPN (INT)	4.110	V	6.6
TNT (CARTOON)	3.760	H	6.8
MTV (ENGLISH)	4.120	H	6.8
HBO	4.000	H	6.8

Notes:

1. For all satellite channels PRE-EMPHASIS =75μs unless specified otherwise.
2. Each satellite channel has an effective BW=36 MHz.
3. In many cases, channel carrier is not at the exact centre of allotted BW for various reasons. This does not cause any serious problem because AFT circuit in the TUNER of satellite receivers tunes-in the channel correctly.
4. All the above listed satellites have their down-link frequencies in the C-BAND except INSAT 1-D which operates in S-BAND.
5. Cable operators/individual users are advised to check the above given data from current issues of CABLE TV magazines or equipment suppliers.

16.1 Six Channel Cable TV System

Block representation of an economical 6-channel cable system is shown in Fig. 16.1. It is quite suitable for small towns and low tariff lodges. The dish antenna (1) can be set to receive ASISASAT-1/PAS-4 downlink signals. The associated satellite receivers can

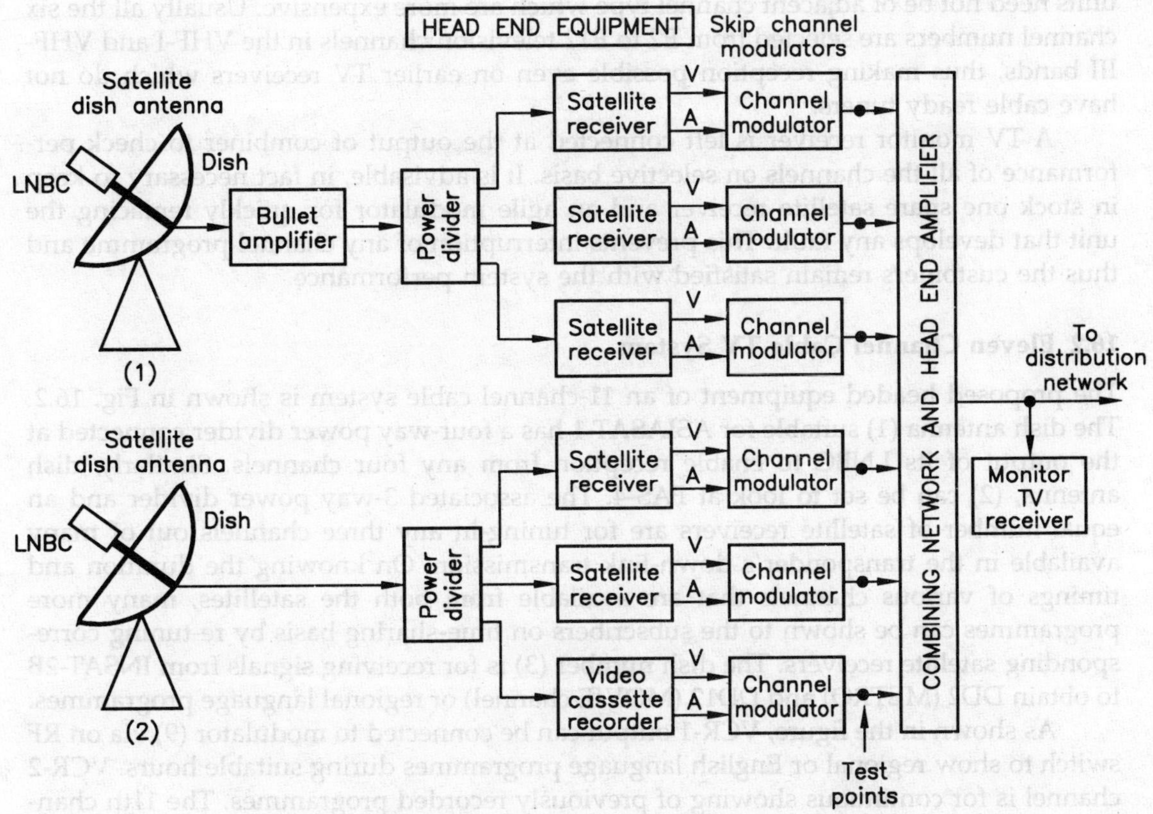

Fig. 16.1. Block schematic of a 6-channel CATV system.

select any three TV channels out of the many available from the transponder. Unless the customers are very particular on viewing specific programmes, many channels can be shown on time sharing basis from the two satellites. In fact, this is the normal practice for which the dish is oriented as necessary and satellite receivers tuned accordingly. The second dish antenna, (2) is suitable for looking at INSAT-2C for receiving METRO (DD2) and MOVIE club (DD12) signals. During some hours, one or two regional language programmes can also be selected. Since downlink signals from INSAT-2C are quite strong, the use of a bullet amplifier after the LNBC is not necessary.

To attract more customers, a VCR channel forms part of this cable system. It can be used to show popular movies, request songs, plays and devotional programmes from video tapes. Some programmes can be recorded on the VCR to be shown later. The subscribers are expected to receive any one local transmission directly by installing a suitable antenna at their respective sites. This will need disconnecting the cable feed and connecting antenna lead-in transmission line to the input socket of the TV receiver for which a RF changeover switch is used.

The baseband video (V) and audio (A) signals from each satellite receiver and VCR are fed to connected modulators for modulation on carrier frequencies of assigned channels. Since the number of channels is only six, it is possible to skip one channel number in between while allotting them to the modulators. Therefore, the modulator units need not be of adjacent channel type which are more expensive. Usually all the six channel numbers are selected from E2 to E12 television channels in the VHF-I and VHF-III bands, thus making reception possible even on earlier TV receivers which do not have cable ready tuners.

A TV monitor receiver is left connected at the output of combiner to check performance of all the channels on selective basis. It is advisable, in fact necessary to keep in stock one spare satellite receiver and an agile modulator for quickly replacing the unit that develops any fault. This prevents interruption of any channel programme and thus the customers remain satisfied with the system performance.

16.2 Eleven Channel Cable TV System

The proposed headed equipment of an 11-channel cable system is shown in Fig. 16.2. The dish antenna (1) suitable for ASIASAT-1 has a four-way power divider connected at the output of its LNBC to enable reception from any four channels. Similarly, dish antenna, (2) can be set to look at PAS-4. The associated 3-way power divider and an equal number of satellite receivers are for tuning-in any three channels out of many available in the transponder's down-link transmission. On knowing the duration and timings of various channels that are available from both the satellites, many more programmes can be shown to the subscribers on time-sharing basis by re-tuning corresponding satellite receivers. The dish number (3) is for receiving signals from INSAT-2B to obtain DD2 (METRO) and DD12 (MOVIE channel) or regional language programmes.

As shown in the figure, VCR-1 output can be connected to modulator (9) via on RF switch to show regional or English language programmes during suitable hours. VCR-2 channel is for continuous showing of previously recorded programmes. The 11th channel is for including local terrestrial transmission via an RF amplifier or signal processor.

The modulators in this *i.e.* 11-channel system are of adjacent channel type where SAW filters are used to achieve desired selectivity. Thus, there is no need to skip channels as suggested for the 6-channel system. It is both desirable and economical to restrict channel allotment to the lower H.F. bands as far as possible. It is suggested that two channels out of VHF band I and six out of VHF band III be chosen. The remaining three can be S_{01}, S_{02}, S_{03} (X, Y, Z) or S_1, S_2 and S_3 channels as detailed in section 13.2, Furthermore, popular programmes should be carried on channels selected from VHF-1 and III TV bands, so that subscribers not having cable ready receivers or top converters can also receive them.

As shown in Fig. 16.2, it is convenient to mix modulated RF output from one modulator to the next and so on and finally mix combined signals from all the four receiver-modulator blocks in an active combiner to obtain wide-band RF signal for distribution on the cable network.

Finally, it is suggested that the cable operator instals facility for easy change of orientation of dish antenna (2). He can then quickly re-set the dish to receive signals from other satellites during selected hours and thus satisfy more customers. The 11-channel cable system is suitable for district level towns and society complexes.

16.3 Sixteen Channel Cable TV System

The proposed plan of a 16 channel CATV system shown in Fig. 16.3 has four dish antennas.

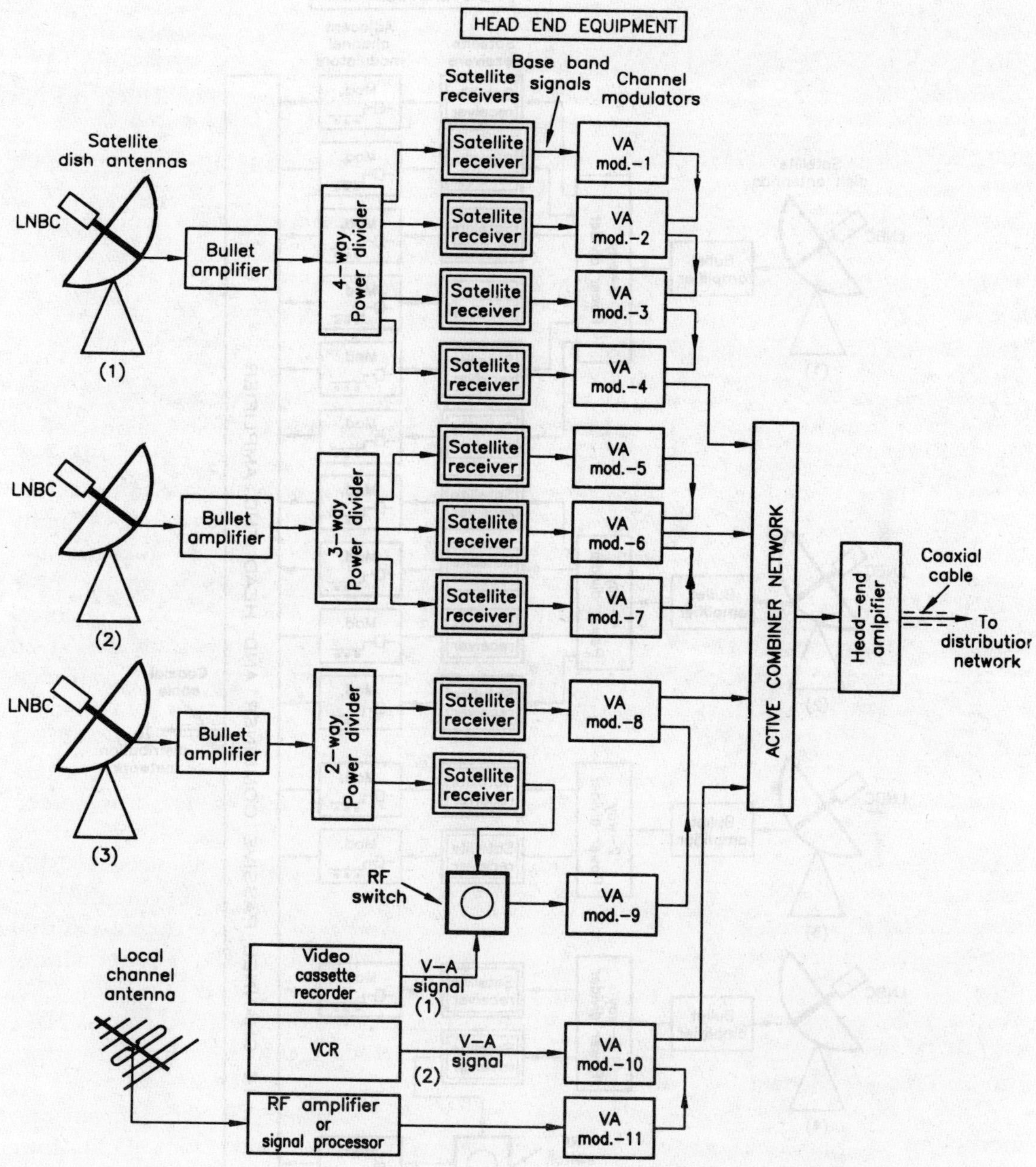

Fig. 16.2. Headed equipment of a 11-channel CATV system.

The dishes 1, 2 and 3 are suitable for receiving signals from regional satellites while dish antenna labelled number 4 is more suited to look at INSAT 2-B or INSAT 2-C.

Notice that power divider connections to associated satellite receivers and that of receivers to modulators have been simplified to save space in the diagram. The modulator outputs are combined in the passive network of the type shown in Fig. 14.20. VCR-1 channel is on time sharing basis with one of the receivers associated with satellite

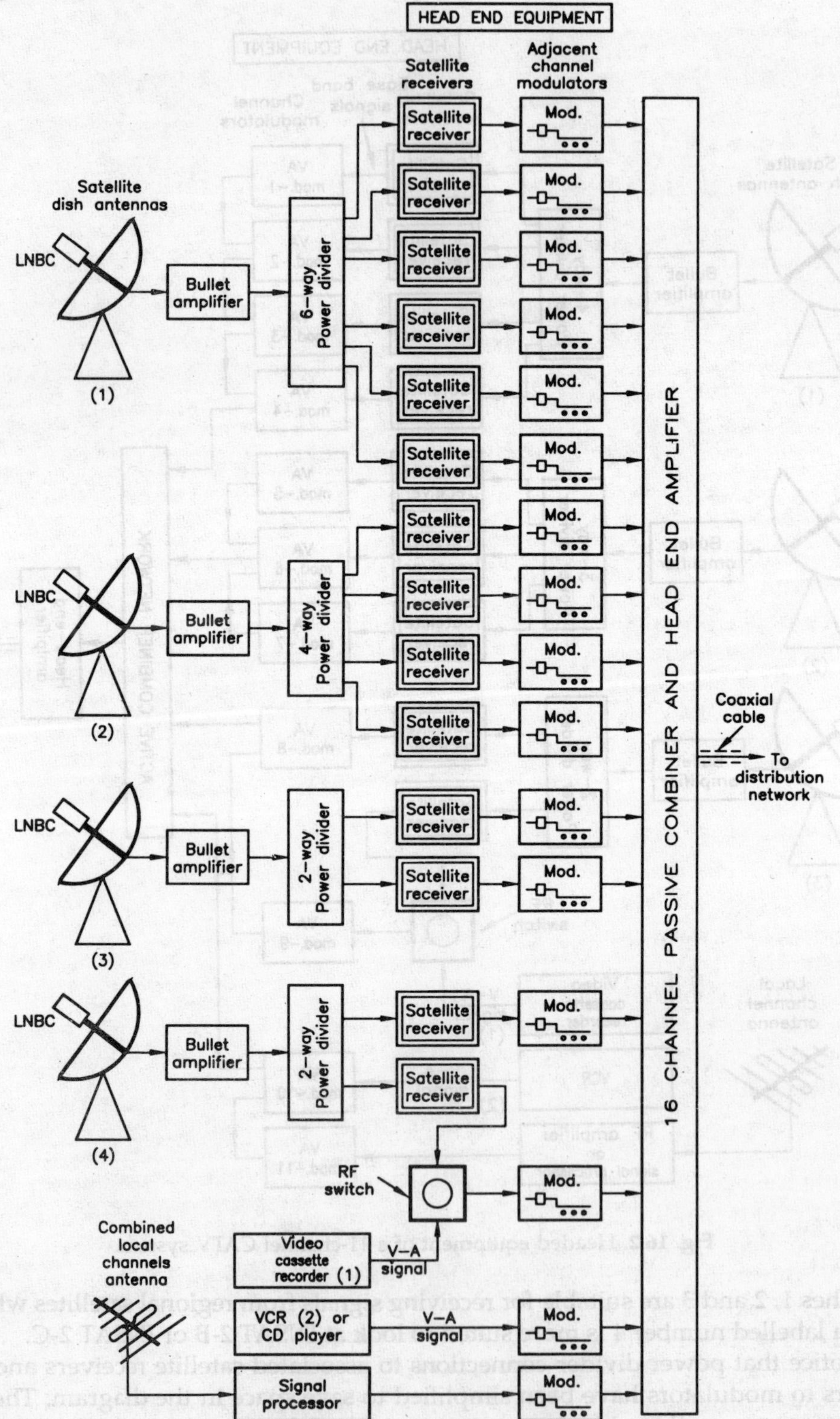

Fig. 16.3. Proposed plan of a 16 channel cable TV system.

INSAT-2-B or 2-C. VCR-2/CD PLAYER channel is an independent channel for showing popular programmes recorded earlier. The antenna for local terrestrial TV stations is a combination antenna to receive DD1, DD2 and DD3 transmissions. As necessary, the input (tuner) section of the signal processor is tuned to receive desired transmission. Such a 16 channel cable system is quite suitable for big towns and different locations in cities where there are enough subscribers to make the system economically viable.

16.4 Thirty and Higher Channel CATV Systems

Thirty channel systems are now becoming common in metros but it is difficult to decide on a standard plan for such a system because of a large number of variables that influence the decision. To begin with, it is necessary to ascertain preferences of potential customers about programmes they would like to receive during different hours of the day. The planning will then begin with preparing a list of available satellite and local channels and note against each the duration for which these are available during 24 hours of the day. This will enable making another list to find out which programmes can be shown on time sharing basis on different cable channels. From this exercise will emerge a final schedule acceptable to most customers which can vary from area to area.

Front-End Equipment: Normally 6 to 8 dish antenna will be required to receive at least 24 channels simultaneously from available satellites during prime time hours. Since many channels are not available round the clock, at least 27 channel programmes can be shown on time sharing basis with equipment installed for showing 24 channels simultaneously. Besides, as in the 16-channel system two VCR/CD channels will be necessary to distribute special programmes including major events of the day collected by special camera teams. With one or two channels for showing local TV telecasts, the system will become a 30-channel cable network.

While deciding on the number of dish antennas to be installed, it should be kept in mind that if 6 channels are to be received simultaneously from the same satellite, it will be necessary to use two dish antennas each feeding a 3-way power divider because a 6-way power divider, unless the downlink signal is very strong, is likely to reduce signal strength at its output ports to a level below the minimum for satisfactory operation of associated satellite receivers. Alternatively, two LNBCs can be mounted side-by-side for feeding two sets of receivers through separate 3-way power dividers. In some situations, it may be possible to use two LNBCs in an 'off-set' configuration to acquire two satellites with one dish. It may as well turn out to be economical to install one or two dishes with provision for motor driven remote control units for setting azimuth and elevation angles thus needing lesser number of dish-antenna set-ups.

Another aspect that will need consideration while installing dish antennas is the consideration of down-link signals to be received for which it will be necessary to orient each horn as necessary. As explained in a previous chapter, the polarization can be vertical (V), horizontal (H) or circular with either Right Hand Circular (RHC) or Left Head Circular (LHC) field directions. As explained in section 14.3 [see Fig. 14.5(b)] when a satellite transmits in both polarizations *i.e.* 'V' and 'H' to obtain higher channel capacity it becomes necessary to install an orthogonal mode transducer (OMT) to avail of all the down link channels. Assuming that 4 channels from each mode of polarization are of interest at any given time, the cable station set-up for the acquired satellite will be as shown in Fig. 16.4. The schematic design shown is of another type as used by many manufacturers of CATV equipment.

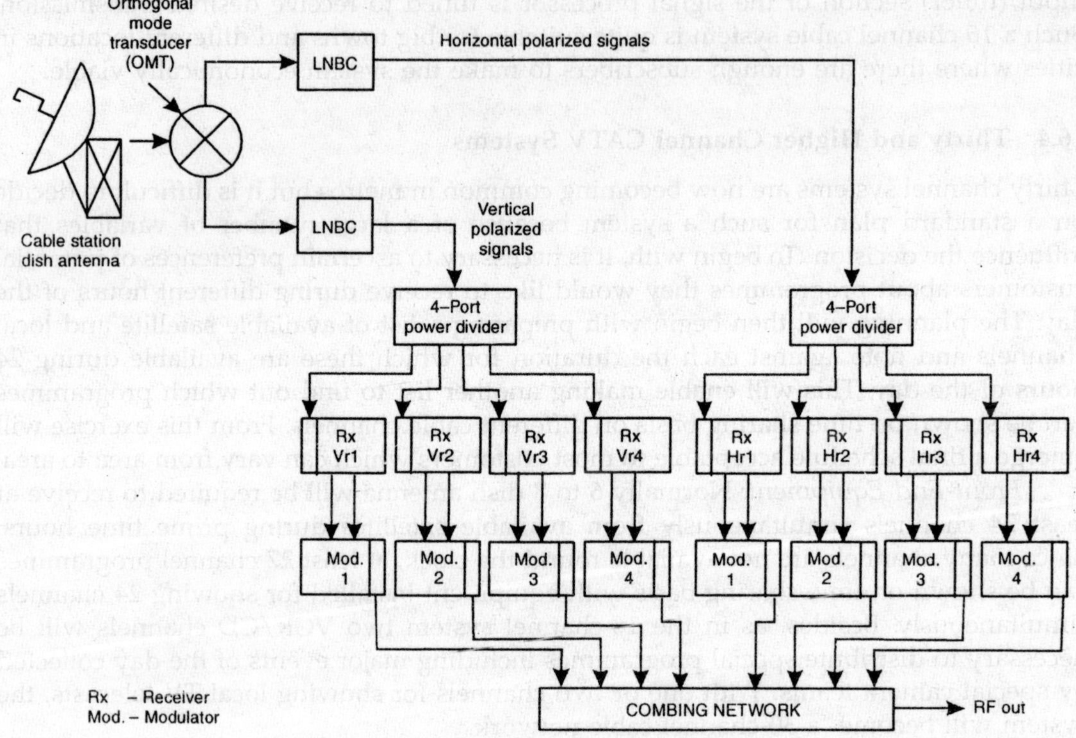

Fig. 16.4. An 8 channel layout with OMT installed at the dish antenna.

Headend Equipment: In order to process the estimated 24 channels at the same time, an equal number (24) of satellite receivers and modulators will be needed. Another two modulators will be required for the two VCR channels besides one or two signal processors to include local TV broadcasts. In fact, it will be advisable to install two sets of 16-channel headend units on separate racks as independent blocks. This will not only leave enough spare equipment for replacing faulty receivers and modulators but also enable easy addition of a few more channels that may become available later on.

The broadband RF outputs from the combiners of two 16 channel units will need further mixing in a suitable 32 channel passive or active combiner. The final RF output on mixing can then be fed to a headend amplifier before connecting it to the distribution cable network.

Channel allocations: The need to restrict system operation below 300 MHz as far as possible was explained in section 13-3. As such it will be prudent to choose 2 channels from VHF band (1) leaving one for local terrestrial broadcast and allot another 13 channels that are available in the vacant and mid-S band *i.e.* S_{01}, S_{02}, S_{03} (also called X, Y, Z) and S_1 to S_{10} thus covering a frequency range from 68 to 174 MHz. Another 6 channels can be chosen from VHF band III (174 to 230 MHz) leaving two channels—one for terrestrial DD2 and another for DD3 if broadcast locally. This will add up to 21 (2+13+6) channels and the remaining 6 channels can be allotted S_{11} to S_{16} from super band thus bringing the RF range to 272 MHz. When it becomes necessary to add a few more channels, the remaining 4 super band channels (S_{17} to S_{20}) can be used still keeping the cable system operation within desired upper limit of 300 MHz.

Higher Channel Systems: Initially the 30 channel systems grew very fast in big cities and with that the existing 11 and 16 channel stations merged with them and thus disappeared. However, the process did not stop there because of keen competition and consequent reduction in subscription rates. At present only big cable companies are surviving and the smaller units are becoming only distribution centres on commission basis. In due course of time, when say 50 channels will become available the best course would be to install two 24-channel systems and run two cables, one for each with provision for a changeover switch to each subscriber.

16.5 Direct Reception System (DRS)

This is the simplest satellite cable system meant for exclusive use in individual homes. A parabolic dish antenna of diameter 8′ to 12′ depending on the location in the foot print area of the satellite(s) of interest is usually mounted on roof top of the building or ground nearby if a location with a clear view to the satellite(s) is available. The dish should have a mechanised system for changing its position for acquiring required satellite. In case it is desired to be able to watch programmes from many satellites, it would be necessary to install a still large dish antenna along with an OMT with change over switch from V to H polarization and remote controlled automatic tracking mechanism. This will enable the viewer to remotely adjust the dish to acquire wanted satellite and tune-in any of its channels from the comfort of his or her seat in front of the television receiver.

Direct Reception: A possible arrangement of connections to direct reception satellite receiver is shown in Fig. 16.5. The amplified RF output from the LNBC is brought inside the house through a low-loss coaxial cable and connected to the input of a special satellite receiver that is located close to the TV receiver. The special receiver delivers two types of outputs, one the baseband signal which is connected to the V and I input sockets of a VCR if available. The second output is through a modulator which gives RF output suitable for direct viewing on the television receiver. This output is generally on channel 3 or 4 and hence the TV receiver channel selector is set accordingly. The SAT-receiver is then tuned, antenna oriented and OMT set if installed to receive desired channel programmes. It will also be necessary to set on the SAT-receiver the sound sub-carrier frequency which is between 5 to 8 MHz depending on the satellite from which the programme is to be received. Since only a single TV receiver is connected in the house, no additional equipment, as in multichannel cable system, would be required. The satellite receiver can, however, feed 2 or 3 television sets in parallel that are positioned in different rooms in the house but all would offer the same programme at one time. If the programme is to be viewed later on, the VCR can be switched on for recording it.

Terrestrial Reception: As shown in the figure, the special direct reception receivers normally have an additional RF input for processing local terrestrial broadcasts. The outputs are in the same form as for receiving satellite channels. This enables direct viewing of local broadcasts.

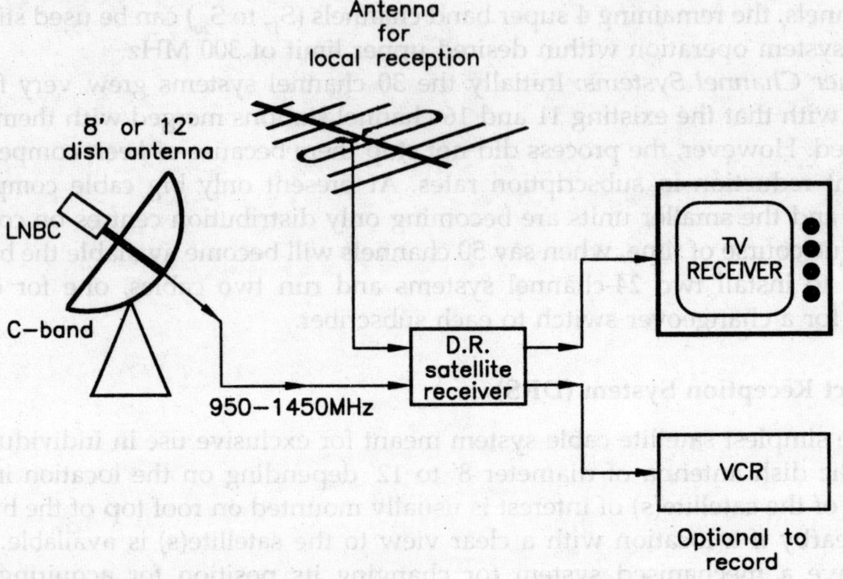

Fig. 16.5. Simple home satellite direct reception cable TV system.

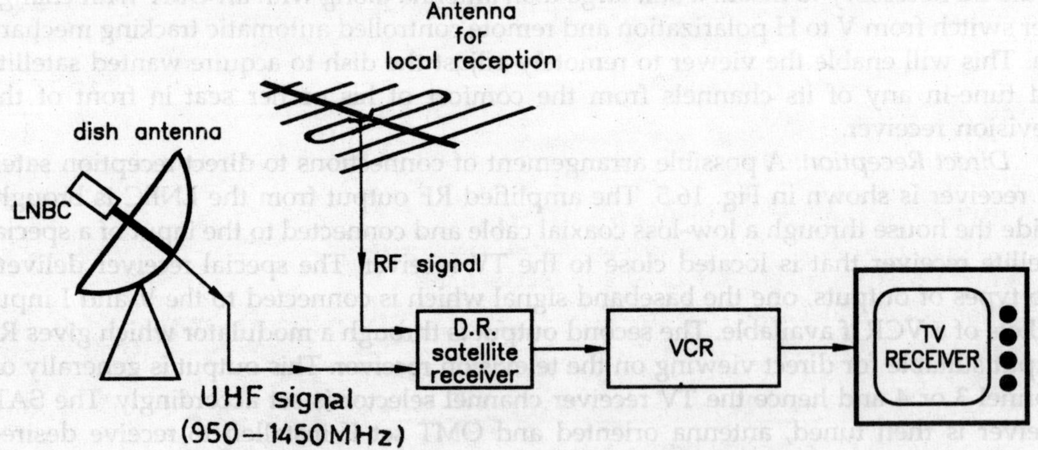

Fig. 16.6. Direct home receiver TV connections through a VCR.

Another arrangement of connecting the satellite receiver, VCR and home TV set is shown in Fig. 16.6 where the VCR comes in-between the satellite receiver and television set. The VCR is kept 'ON' though not set on recording mode unless necessary. This mode of connections is preferred if recording of programmes if often required.

17

CABLE SYSTEM PLANNING AND INSTALLATION

Introduction

Cable systems which distribute signals of 80 to 100 channels to over 100,000 subscribers are common in big cities of USA, Japan and many European countries. Such units have elaborate control rooms to process as many channels and many have more than 2000 km of cable network lines with over 2000 trunk and line amplifiers and several hundred power supplies for cable powering. In order to reach such a large number of customers spread over a wide area, cables are laid underground in trenches dug along roads, lanes and bridges. Regulatory authorities in these countries make it mandatory for cable operators to monitor programmes of their set-ups on a day-to-day basis to ensure that good quality signal is distributed to all the end-users. This needs lot of investment on costly instruments and staff salaries for necessary measurements and documentation. All this makes the system quite complex with the result that big cable corporations prefer to hire professionals for planning, design and installation of the entire system on a turn-key basis and later manage it with their own staff.

However, the situation in India and many other countries where cable TV started not many years ago, is quite different. No firm laws have yet been formulated and it is still considered illegal to dig public roads and lanes to reach the subscribers. As such, the operation of cable companies is restricted to small areas and hence, to a limited number of customers. The number of channels offered by these cable units is around 25 because of economic considerations and constraints to keep the monthly rental within affordable limit. Even otherwise, not too many channels with good programmes in Indian languages are available at present via satellites.

Some big Indian companies like SITI-CABLE, INDUS-INDIA and BITV have overcome the problem of signal distribution over large areas by installing 10 or more identical and independent cable stations in different parts of big cities like Mumbai and Delhi. Each unit offers about 30 channel out of which 8 to 10 carry English language programmes which are very popular with elite section of the society. Such innovations, though welcome, make cable systems quite expensive, both in terms of equipment cost and staff, with the result that customers are made to pay higher rentals.

It is estimated that besides big cable companies operating in metropolis, there are about 60,000 medium and small units spread all over India in cities, towns and big villages. Such set-ups offer 6, 11, 16 or 25 channels depending on the number of customers which subscribe to the system. For such units, no elaborate planning or design is considered necessary and cable entrepreneurs prefer to collect necessary information from cable equipment manufacturers and suppliers. Such systems are then installed with the help of technicians who are conversant with cable television. This chapter is, therefore, written to benefit such enterprising cable operators who wish to install a CATV system or enlarge their existing set-up.

17.1 Cable System Economics

The first and most important aspect which needs consideration while planning CATV business is to find out expected number of customers in the area chosen for the venture. For this, it is necessary to make a reasonable assessment of firm and undecided customers who may join later, by a door-to-door survey in the area, preceded preferably by distribution of hand-bills giving main features of the system. This exercise will enable estimation of expected annual revenue based on monthly rental being charged by nearby cable operators. This must not only balance with annual expenses which account for interest on expected investment, control room and antenna farm rentals, payment for channels which are not 'free-to-air', insurance, depreciation, electricity and telephone bills, security, entertainment, staff salaries and other sundry expenses but also leave a reasonable profit margin.

To be on safe side, it is advisable to start business by offering 11 or 16 channels in cities and towns where the estimated clientele is reasonably large and 6 or 11 channels in small towns and villages where growth prospects are not very bright. In any case, it must be remembered that key to success lies in showing educative and popular programmes and distributing undistorted signal through the cable network. Besides, as far as possible, any interruption in the functioning of cable system should be prevented and prompt fault service provided to all the customers.

17.2 Control Room and Antenna Farm Sites

Two suitable sites located close to each other are required—one for the control room and the other, usually called antenna farm, to fix dish antenna system. The control room location should be at a prominent place and as far as possible in the centre of signal distribution area. This will need smaller cable runs and hence, saving in cable plant cost. The control room should have two sections separated by a panel with door to install head-end equipment in one section and locate auxiliary components like spares, monitors, etc. besides telephone and furniture for staff on the other side. The control room should also be well lighted, dust-proof and maintained cool by installing an air-conditioner or at least ceiling fans. A desert-cooler is not to be used because it will bring in moisture which can affect proper operation of the installed equipment.

The antenna farm site may be at ground level or on roof top but it must provide a clear and unobstructed view in the look angles of dish antenna to the satellites to be acquired. It should also be ensured that no high-rise buildings or structures are likely to come up near the chosen site because these may obstruct the line of sight path between dish antennas and satellites. If this happens, the cable system will fail and the only alternative would be to shift antennas to an another location nearby.

It may be stated again that distance between antenna farm and control room should be as short as possible. If not, excessive, signal attenuation will occur in RF cables that link the two, making it necessary to install bullet amplifiers in-between.

17.3 Cable System Equipments

On deciding the number of channels, the proposed cable station will offer, the next step is to prepare a list of equipments and accessories which will be needed. These are described for each section separately.

(1) *Front End Equipment:* There can be some variation in the front-end equipment. A dish antenna can either be set to receive signals continuously from one particular satellite or from two to three satellites on a time sharing basis. For the latter, there must be provision at the mount to swing the dish around as necessary. While it is possible to do so manually, a motor driven unit is convenient to choose different azimuth and elevation angles. Big cable operators can opt for satellite receivers having remote facility which enable reception from different satellites with push-button controls provided on their front panels.

Polarization: Another aspect that needs consideration while deciding on the number of dish antennas and LNBCs is polarization of down-link signals to be received. When programmes are to be received from two transponders of the same satellite, one employing horizontal (H) and the other vertical (V) polarization, the receiving dish antenna will need two LNBCs with their feed horns oriented accordingly. For such applications, the two LNBCs are constructed in one block for convenience of mounting at the focal point of same parabolic dish. However, if signals from the two transponders are not to be received at the same time, only one LNBC having an adjustable scaler ring to orient the receiving wave guide probe as necessary will serve the purpose.

Some satellites are designed to operate on circular polarization, with right-hand circular (RHC) mode in one transponder and left-hand circular (LHC) polarization in the adjoining transponder. The electric field 'E' with such radiations is either clockwise or anti-clockwise. As such, two separate feed elements each feeding its own LNBC will be needed. Though not desirable, it may be mentioned that the linear H-V probe can as well be used to receive circular polarized signals but with a loss of a few dB_S of signal strength which can be made-up by either using a bigger diameter dish or a better (high gain) LNBC.

Thus, the number of front-end components required at a cable station depends on many factors and need careful consideration, while preparing the list of equipments to be ordered. To begin with, it is desirable to manage with two dish antennas for a 6-channel cable system, 3 for 11-channel and 4 for 16-channel system. A 25/30 or more channel system will, however, need 6 to 8 frond-end units for satisfactory operation.

(2) *Head-End Equipment:* There cannot be much variation in equipment required in the control room. The required number of satellite receivers will naturally be equal to the down-link channels to be shown simultaneously plus at least one for immediate replacement on failure of any one of the operating receivers. In addition, it will be necessary to buy one VCR for smaller cable system and at least two for large channel systems. One or two signal processors with equal number of conventional YAGI antennas will be needed to obtain signals of local TV station(s). The required number of

modulator units will be equal to the number of channels to be distributed plus at least one agile unit as stand-by to avoid interruption in any channel. One signal combiner will be enough for lesser channel systems and a multi-combiner network will be needed for higher channel cable systems.

Monitor TV Receivers: It is desirable that each satellite receiver-cum-modulator units has its own monitor as provided for in TV broadcasting stations. If such a provision is considered to be expensive, at least one monitor TV receiver must be installed on each rack with provision to switch to receive different modulator outputs. A remote controlled television receiver will also be necessary to check sequentially, picture and sound quality of all the outgoing channels with input from the output of signal combining network. It is to be placed close to the staff position for his/her convenience.

Keeping in view wide fluctuations in mains supply, it is imperative to install an automatic voltage stabilizer to feed steady ac voltage to all equipments of the cable station. The mains switch board must have proper fuses (MCBs) to cut-off supply in case of any accidental short in wiring, any plug point or input of any equipment.

(3) *Coaxial Cables:* For distributing cable signal, routes should be so chosen that load on each cable-run is nearly the same. This is essential, because unbalanced distribution of signal would entail additional cost due to longer cable runs. An estimate should then be made of the total length of distribution cable inclusive of provision for prospective customers on each route. It will also be necessary to work out the length of high grade low-loss RF cable required for connecting LNBC outputs to inputs of satellite receivers. A certain length of flat twin lead-in cable will also be needed for connecting local transmission antennas to corresponding signal processors. Necessary length of each type of cable can then be chosen out of the many available in the market. It is advisable to buy good quality cables to keep losses at a minimum and prevent undesired radiation.

(4) *Trunk and Line Amplifiers:* Cable manufacturers provide necessary data on attenuation per unit length for different types and grades of cables. Based on that and lengths of cable runs required on different routes, it should then be possible to decide on the number of trunk and line amplifiers not forgetting that the ultimate aim is to provide 80 to 90 dB(μv) level signal at tap-off points and about 55 to 60 dB(μv) at the input of customer TV receivers. At this stage it should also be possible to find out required number of distribution and tap-off networks together with line terminating resistors.

(5) *Power Supplier for Amplifiers:* In big cable systems, elaborate cable powering methods have been devised as explained in Chapter 10 (sections 10.4 and 10.5), but medium and small cable operators prefer to feed regulated ac supply to amplifiers from the control room through a separate insulated wire that is run along the signal cable on all trunk and line routes. The cable sheath serves as the second wire.

17.4 Equipment Specifications

There are not many manufacturers of good quality cable TV equipments in India. Some produce total system requirements while others manufacture specific items like dish antennas, amplifiers and other small accessories. Though there may not be much difference in their equipment specifications, but there can be large variation in product quality and reliability of items marketed by small manufacturers. It is, therefore, advisable to procure full system needs from a single well-established and reputed manufacturer. Equipments of excellent quality from USA, Taiwan, Korea, and some European countries

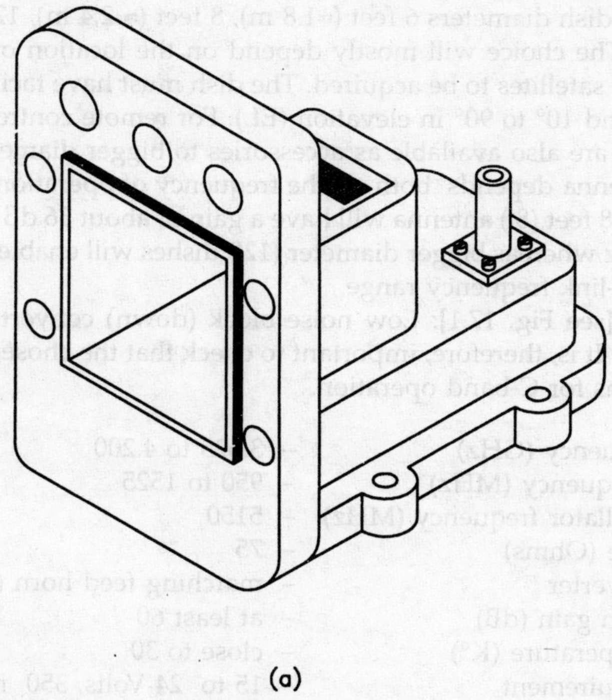

(a)

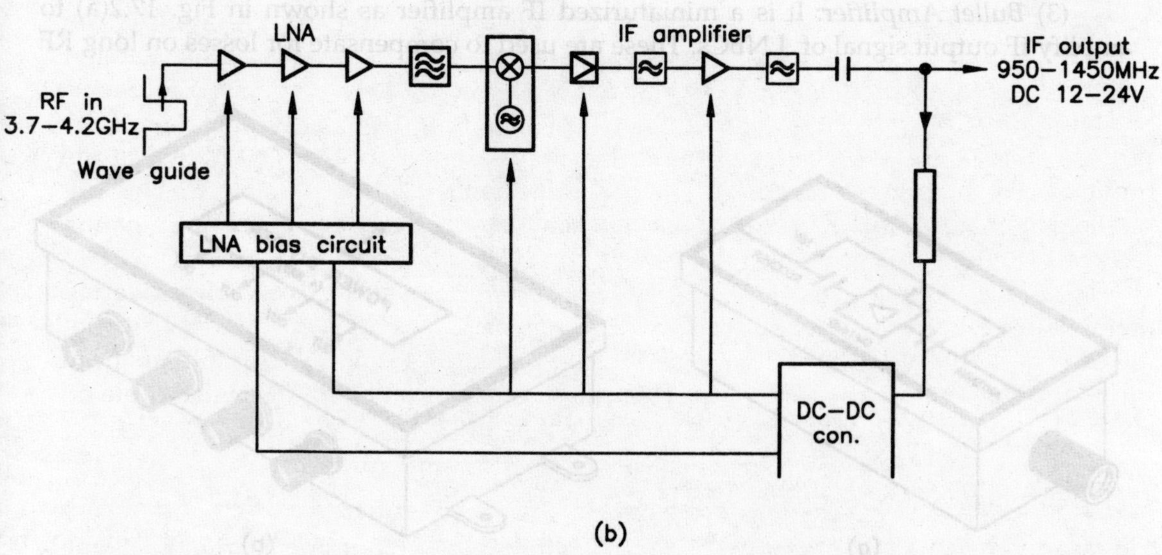

(b)

Fig. 17.1. A typical C-Band LNBC unit
(a) pictorial view, (b) block diagram.

are also available in India but these are expensive. In any case, cable system equipments that are selected must conform to television standards adopted by the user country.

(1) *Parabolic C-Band Dish Antennas:* Dish antennas are available in various models having different diameter and structural arrangement. The antennas commonly available are of dish diameters 6 feet (≈1.8 m), 8 feet (≈ 2.4 m), 12 feet (≈ 3.7 m), and 16 feet (≈ 4.8 m). The choice will mostly depend on the location of cable station in footprint area of the satellites to be acquired. The dish must have facility for 360° rotation in azimuth (AZ) and 10° to 90° in elevation (EL). For remote control operation, actuators and positioners are also available as accessories to bigger diameter dish antennas. The gain of any antenna depends both on the frequency of operation and diameter of dish. For example an 8 feet (8′) antenna will have a gain of about 36 dB in the frequency range of 3.7 to 4.2 GHz whereas bigger diameter (12′) dishes will enable gains around 42 dB in the same down-link frequency range.

(2) *LNBC* [see Fig. 17.1]: Low noise block (down) converters are imported from many countries. It is, therefore, important to check that the chosen make has the following specifications for C-band operation.

Input frequency (GHz)	– 3.625 to 4.200
Output frequency (MHz)	– 950 to 1525
Local Oscillator frequency (MHz)	– 5150
Impedance (Ohms)	– 75
Input converter	– matching feed horn (wave guide)
Conversion gain (dB)	– at least 60
Noise temperature (K°)	– close to 30
Power requirement	– 15 to 24 Volts, 350 mA

(3) *Bullet Amplifier:* It is a miniaturized IF amplifier as shown in Fig. 17.2(a) to amplify IF output signal of LNBCs. These are used to compensate for losses on long RF

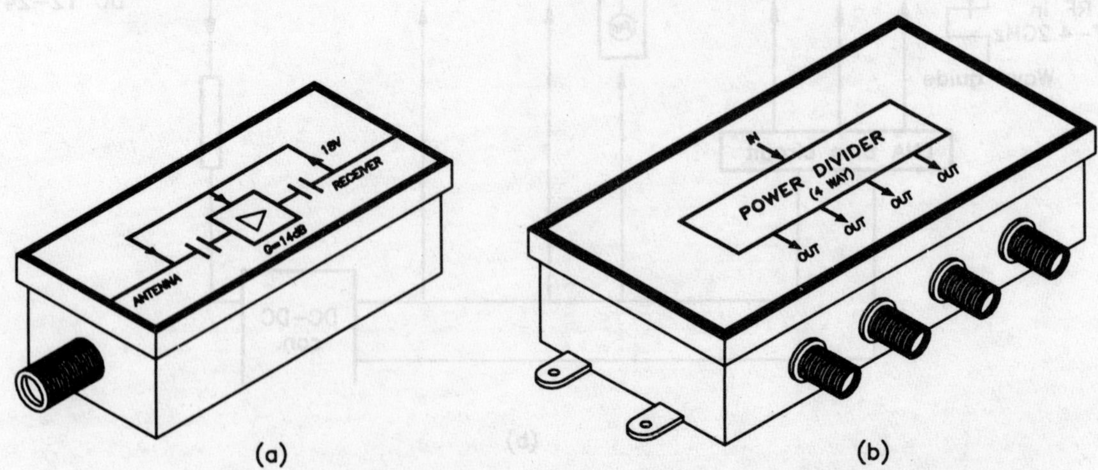

Fig. 17.2. (a) Pictorial view of a bullet amplifier. (b) A 4-way power divider. (*Courtesy:* Shyam Communication Systems).

cable runs and also in power dividers. A bullet amplifier must provide a gain around 20 dB, have flat frequency response up to 1525 MHz, low noise figure and small power consumption.

(4) *Power Divider:* A power divider is specified by its output ports. A typical 4-way power divider is shown in Fig. 17.2(b).

It has necessary RF network to provide desired (75Ω) input and output impedances. When the number of output ports exceed four, an HF bullet amplifier is included to keep insertion loss less than 5 dB at each output terminal.

(5) *Satellite Receiver:* It is a vital piece of equipment and no compromise should

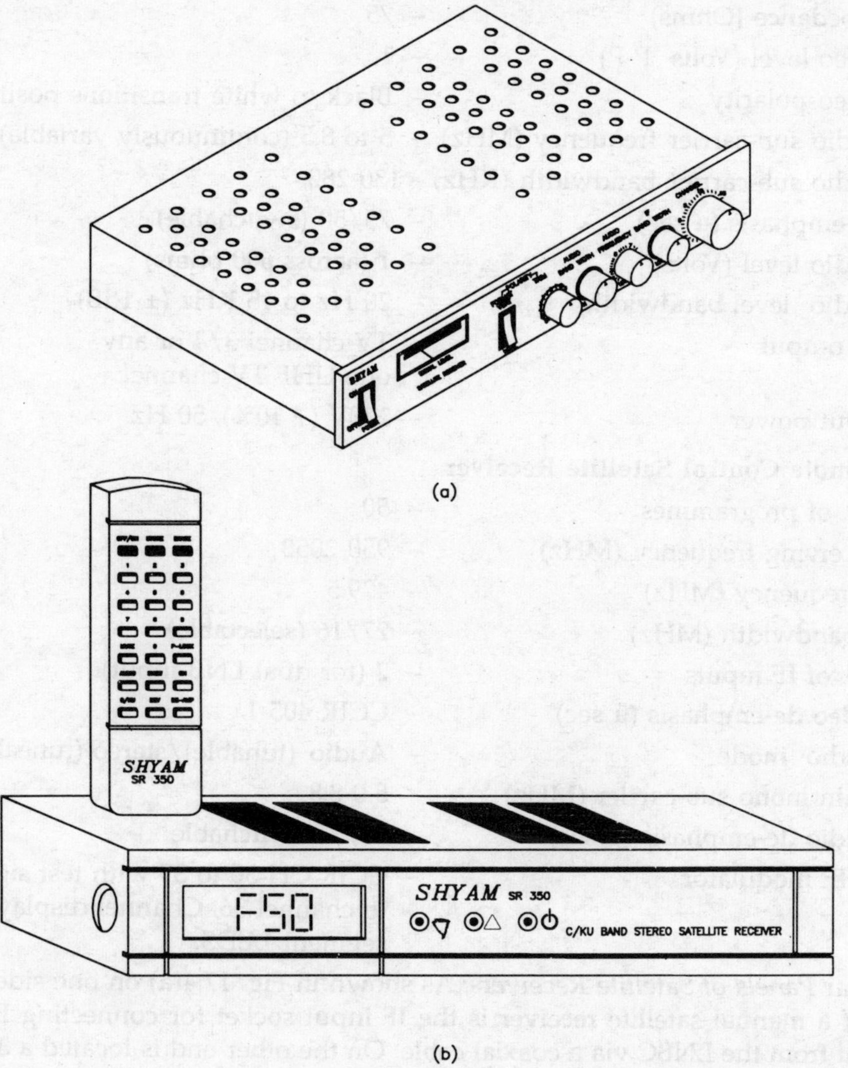

Fig. 17.3. Typical Satellite Receivers (a) with manual control, (b) with remote control
(*Courtesy:* Shyam Communication Systems)

be made on its quality and reliability against cost. Remote controlled receivers with facility to tune up to 50 or more programmable channels and digital display of channel numbers are available [see Fig. 17.3].

Main specifications of both types are as under:

Manually Operated Satellite Receiver

Input frequency (MHz)	– 950-1750
If frequency (MHz)	– 70
If bandwidth (MHz)	– 8 to 27 (continuously variable)
Output (video) frequency response	– 10 Hz to 5 MHz
Impedance (Ohms)	– 75
Video level (Volts P-P)	– 1
Video polarity	– Black to white transitions positive going.
Audio sub-carrier frequency (MHz)	– 5 to 8.5 (continuously variable)
Audio sub-carrier bandwidth (KHz)	– 130-280
De-emphasis (μ sec)	– 75/50 (switchable)
Audio level (Volts)	– 1 (across 600 ohms)
Audio level bandwidth	– 20 Hz to 15 KHz (± 1dB)
RF output	– TV channel 3/4 or any one UHF TV channel
Input power	– 220V (± 10%), 50 Hz.

Remote Control Satellite Receiver

No. of programmes	– 50
Receiving frequency (MHz)	– 950-2050
If frequency (MHz)	– 479.5
If bandwidth (MHz)	– 27/16 (selectable)
No. of IF inputs	– 2 (for dual LNB input)
Video de-emphasis (μ sec)	– CCIR 405-1
Audio mode	– Audio (tunable)/stereo-(tuneable)
Main mono sub-carrier (MHz)	– 5.0-8.8
Audio de-emphasis (μ sec)	– 50/75 switchable
UHF modulator	– CCIR CH 30 to 39 with test signal. Preset to channel 36. Channel display 3-7 segment LEDS.

Rear Panels of Satellite Receivers: As shown in Fig. 17.4(a) on one side of the rear panel of a manual satellite receiver is the IF input socket for connecting IF output as obtained from the LNBC via a coaxial cable. On the other end is located a 300 mA fuse unit and close to it is the input point for 230/240V AC supply to the receiver. In the middle, are output sockets for BB (base-band) signal output and video and audio outputs separately. On another socket 'MAC' (multiplexed analog components) signal becomes available when transmitted. Since such as output is 'ENCRYPTED', meaning it is in code form to prevent unauthorised access, this signal is fed to a special decoder for obtaining normal B-B outputs. The TV 'OUT' socket located close to the middle section of the panel enables modulated output on channel 3/4 as desired. It can be fed directly to a TV receiver for receiving tuned-in channel.

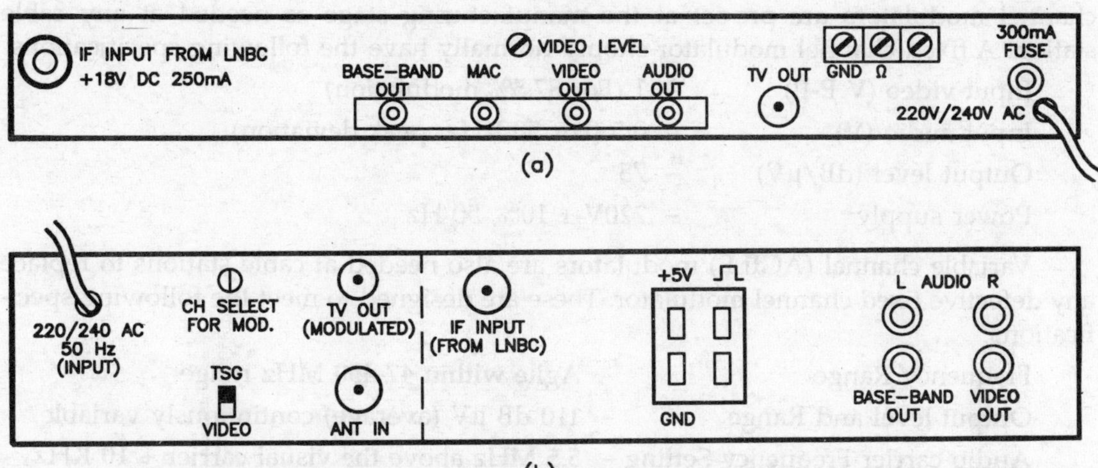

Fig. 17.4. (a) Rear panel of one type of manual satellite receiver, (b) Rear panel of a typical remote control receiver.

The rear panel of a typical remote control receiver is shown in Fig. 17.4(b). The left side of panel is essentially the same as that of VCR. The two-position switch 'TSG/VIDEO' when thrown on TSG connects a standard test signal to the receiver for testing and if necessary adjusting its performance. When the switch is thrown on 'VIDEO' position, the receiver performs in the normal mode. For receiving terrestrial broadcasts, signal from the receiving antenna is made input at the ANT-IN socket. The TV-OUT socket provides modulated output of the tuned-in channel on either channel 3 or 4 adjustable with a 'slot' switch as shown in the figure. On the other side of dotted-chain line is the usual 'IF-INPUT' socket for connecting satellite signals as obtained from the output of LNBCs. On extreme right side of the panel are located sockets for obtaining B-B signal out, video-out and L and R audio signals for stereo operation. It may be noted that this facility is for individual reception setups because, as of now, CATV operation is in MONO mode only. In the middle of rear panel a 5V dc supply is available for feeding LNBC and bullet amplifier as necessary.

(6) *Video Cassette Recorder:* A general purpose VHS video cassette recorder will serve the purpose because it is to be used mostly in 'PLAY' mode. It should meet the following specifications:

Video Recording System	– 2 rotary heads with helical scanning
Tape speed	– 23.39 m/sec
Tape format	– VHS tape
Television system	– CCIR-625 lines, 50 fields, PAL B&G colour
Receiving channels	– VHF 2-12, UHF 21-69, Mid and Superband S_1 to S_{20}, and Hyper band S_{21} to S_{41}.

RF Modulated output – VHF TV channel 3/4 or any one UHF TV channel.
Input power – 110-240 AC (Auto Voltage), 50 Hz

(7) *Channel Modulators:* As for satellite receivers, it is equally important to select good quality modulators to ensure stability of output and reliability of service. Fixed channel modulators are pre-set at the manufacturing stage as needed at any cable station. A fixed channel modulator should normally have the following specifications:

Input video (V P-P) – 1 (For 87.5% modulation)
Input audio (V) – 0.5 (For 50 KHz peak deviation)
Output level (dB/μV) – 75
Power supply – 220V ± 10%, 50 Hz

Variable channel (AGILE) modulators are also needed at cable stations to replace any defective fixed channel modulator. These are designed to meet the following specifications:

Frequency Range – Agile within 47-450 MHz range
Output level and Range – 110 dB μV (average) continuously variable.
Audio carrier Frequency Setting – 5.5 MHz above the visual carrier + 10 KHz
Video input for 87.5% depth of modulation – 0.7V P-P (typical)
Vestigial sideband response – Optimised for adjacent channel operation.

Pictorial view of a typical Agile modulator is shown in Fig. 17.5.

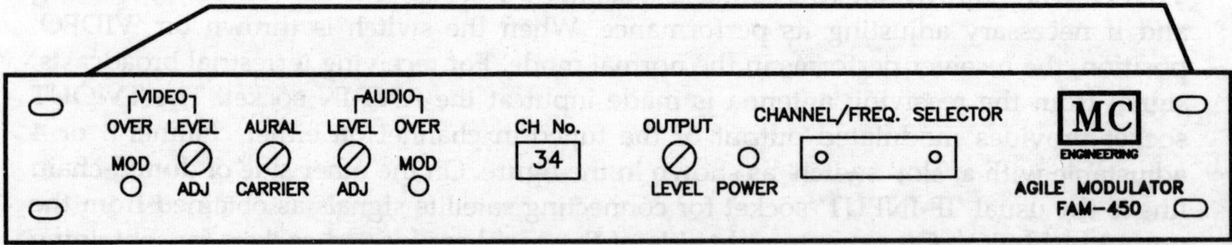

Fig. 17.5. Pictorial view of an Agile modulator (*Courtesy:* MC Engineering).

(8) *Channel Combiner (Mixer):* A variety of combiners are available as detailed in Chapter 14 (section 14-10) for simple as well as complex combining applications in CATV headends. It is common practice [see Fig. 16.2] to mix modulated RF output from one modulator to the next and then finally combine the output of such units. Such a technique is pictured in Fig. 17.6.

Depending on the size of cable system, both active and passive combiners are manufactured to suit different cable channel systems. It is advisable to buy a matching unit to avoid any mismatch problems at the time of commissioning the plant. In general a good quality mixer must have a frequency response flat up to 550 MHz, insertion loss less than 3 dB and an isolation of about 35 dB.

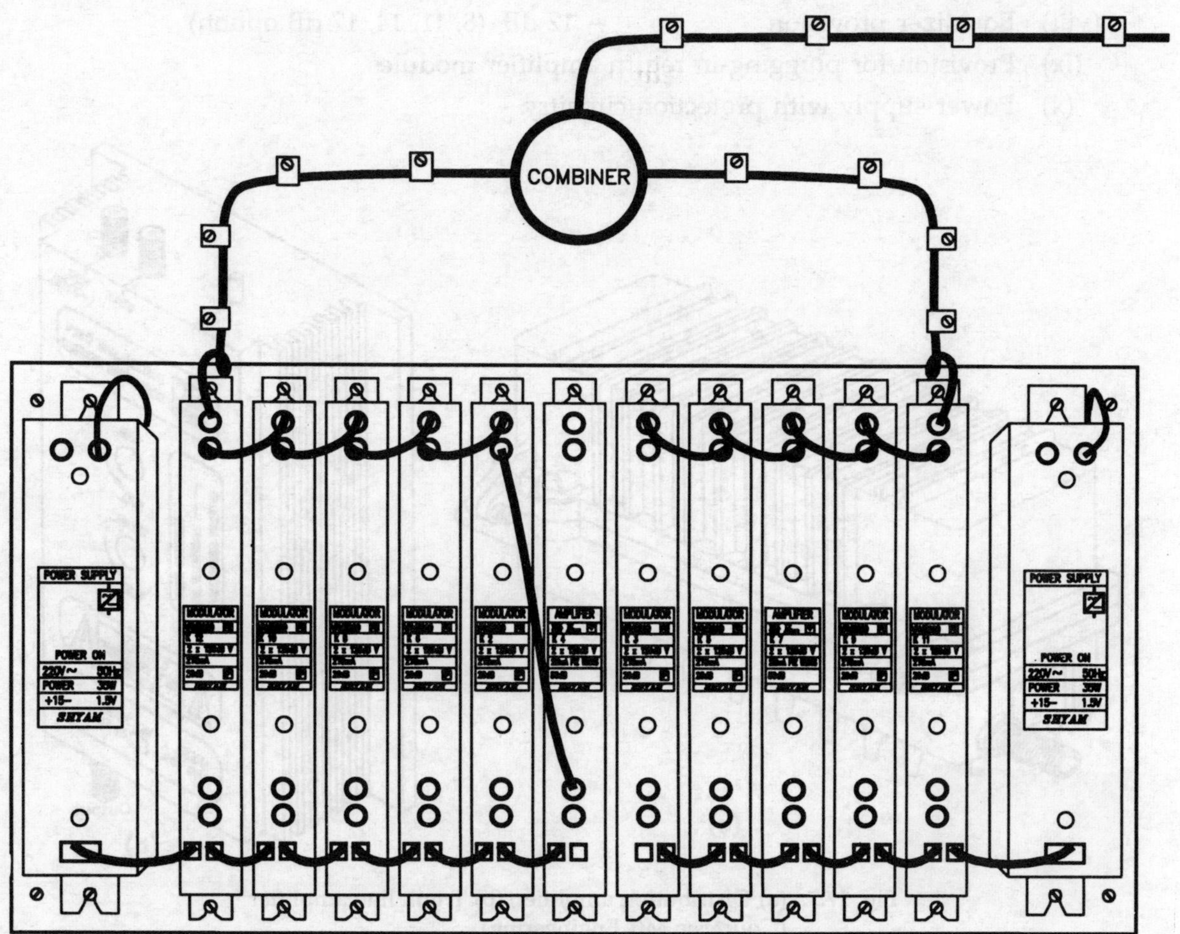

Fig. 17.6. A compact channel modulator unit.
(*Courtesy:* Shyam Communication Systems).

(9) *Signal Distribution Amplifier:* A wide variety of trunk, distribution and line extender amplifiers with or without powerpass provision are available. The choice depends on system needs and location of the amplifier. A typical power pass distribution amplifier and a small trunk line amplifier are pictured in Fig. 17.7.

The main specifications of a power-pass distribution amplifier are:

(i)	Frequency range (forward)	– 45 to 550 MHz
(ii)	Gain (forward)	– 25 dB/30 dB/40 dB (as needed)
(iii)	Gain control provision	– 0-20 dB
(iv)	Slope control provision	– 0-16 dB
(v)	Noise figure	– 7 dB (typical)
(vi)	Input, output impedance	– 75 ohms
(vii)	Voltage range	– 30/40/50/60V AC

(viii) Equalizer provision – 12 dB (8, 11, 14, 17 dB option)

(ix) Provision for plugging-in return amplifier modules

(x) Power supply with protection circuitry.

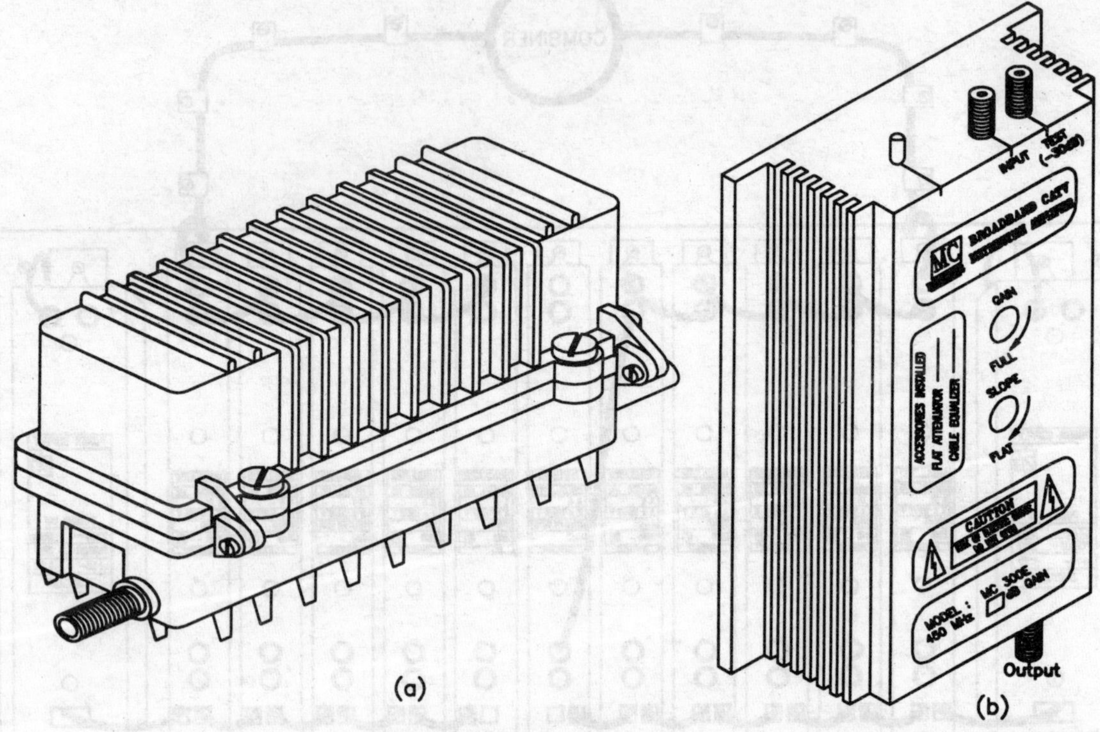

Fig. 17.7. (a) Distribution amplifier, (b) Trunk line amplifier (*Courtesy:* MC Engineering).

(10) *Distribution Network Accessories:* Good quality TAP-OFFS and Splitters having different output ports in protective die-cast housing are manufactured. In general these are designed to have high frequency response, low insertion loss and all-port power-pass facility. Some typical units are shown in Fig. 17.8 and their specifications are as under:

Tap-Offs – Tap loss = 15 dB, Through loss 0.5-1.5 dB Isolation > 40 dB and return loss > 20 dB

Splitters – Through loss ≈ 7 dB, Isolation > 20 dB and return loss > 15 dB.

Sockets – Through loss ≈ 2 dB and Isolation > 15 dB.

As necessary, one-way to four-way tap-offs with loss not more than 15 dB and frequency range up to 850 MHz can be chosen. Similarly, 3-way to 5-way splitters having frequency range up to 850 MHz are available and can be selected depending on the nature of distribution network.

17.5 Equipment Ordering and Preparation at Sites

In order to keep the plant cost low, it is possible to build a cable system by obtaining

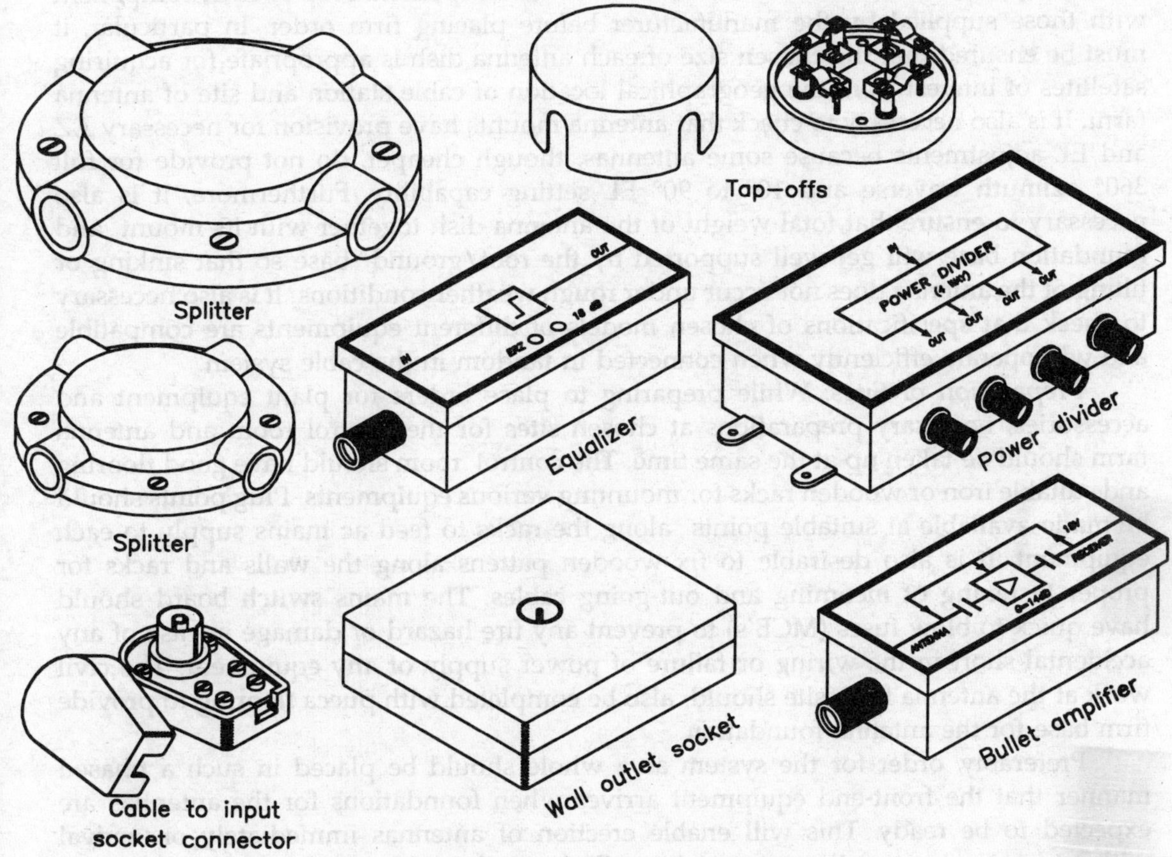

Fig. 17.8. Typical network distribution accessories. (*Courtesy*: Shyam Communication Systems).

major pieces of equipment from different manufacturers on price competitive basis. However, it is not advisable to do so for many reasons. Mainly it is quite possible that putting together components of different makes may not result in satisfactory operation of the cable system. If this happens, it will be difficult to hold any one manufacturer/ supplier responsible for this because it is not easy to check the performance of each equipment against given specification. For example, the dish antenna, the exactness of which is essential for efficient performance of the system, may not give specified signal output due to its incorrect parabolic shape or misalignment with its focal point because of not too stringent manufacturing standards or because of damage while packing or during transport. It is not easy to pin-point the defect because antenna testing needs special facilities and expensive instruments. Similarly, testing of satellite receives and other equipments is equally difficult because of the need of sophisticated instruments and skilled staff to operate them. Hence, it is prudent to buy entire front and head-end equipments from a single and reliable manufacturer and hold him responsible for any deficiency in overall performance of the plant.

The next step after opting for a dependable supplier is to prepare a detailed list of all the requirements of cable system and cross-check specifications of each component with those supplied by the manufacturer before placing firm order. In particular, it must be ensured that the chosen size of each antenna dish is appropriate for acquiring satellites of interest from the geographical location of cable station and site of antenna farm. It is also necessary to check that antenna mounts have provision for necessary AZ and EL adjustments because some antennas, though cheaper, do not provide for full 360° azimuth traverse and 10° to 90° EL setting capability. Furthermore, it is also necessary to ensure that total weight of the antenna dish together with its mount and foundation base will get well supported by the roof/ground base so that sinking or tilting of the antenna does not occur under rough weather conditions. It is also necessary to check that specifications of chosen models of different equipments are compatible and will operate efficiently when connected in tandom in the cable system.

Preparation of Sites: While preparing to place orders for plant equipment and accessories, necessary preparations at chosen sites for the control room and antenna farm should be taken up at the same time. The control room should have good flooring and suitable iron or wooden racks for mounting various equipments. Plug points should be made available at suitable points along the racks to feed ac mains supply to each equipment. It is also desirable to fix wooden pattens along the walls and racks for proper fastening of incoming and out-going cables. The mains switch board should have quick to blow fuses (MCB's) to prevent any fire hazard or damage in case of any accidental short in the wiring or failure of power supply of any equipment. The civil work at the antenna farm site should also be completed with pucca flooring to provide firm base for the antenna foundation.

Preferably, order for the system as a whole should be placed in such a phased manner that the front-end equipment arrives when foundations for the antennas are expected to be ready. This will enable erection of antennas immediately on arrival without having to store them somewhere till the civil work is over. Later when rest of the equipment arrives, it should be checked with the order list so that no undue delay occurs during its installation and commissioning.

17.6 Choice and Laying of Cables

Correct choice of coaxial cable types for signals from LNBCs and those for the distribution system is essential for fault-free performance of the cable system. Main data of commonly used cables CATV systems is given in a tabular form at the end of chapter 3. Besides, all aspects of cable signal distribution are fully covered in chapter 15 and may be referred to if necessary. Proper and firm cable connections must be ensured all along cable runs, more so when these are hung along poles as is the current practice. Where possible, cables should be laid underground protected by 'COX-SEAL' and taken through pipes. Otherwise, while anchoring off along electricity and telephone poles, it should be made sure that cables are not close to mains supply lines and telephone wires. If such a precaution is not taken, the maintenance staff of electricity supply and telephone services may cut/detach them from poles causing interruption of cable signal distribution in the area.

The number and location of trunk and line amplifiers, splitters and tap-offs will depend on cable runs on different routes. The distance between two amplifiers can be computed from the cable data where signal attenuation per unit length is specified. It may be necessary to affect some changes while commissioning the plant on measuring signal level along different cable runs.

While making connections between cables or to amplifies, splitters and tap-offs it is very necessary to use good quality weather proof coaxial connectors and seals to

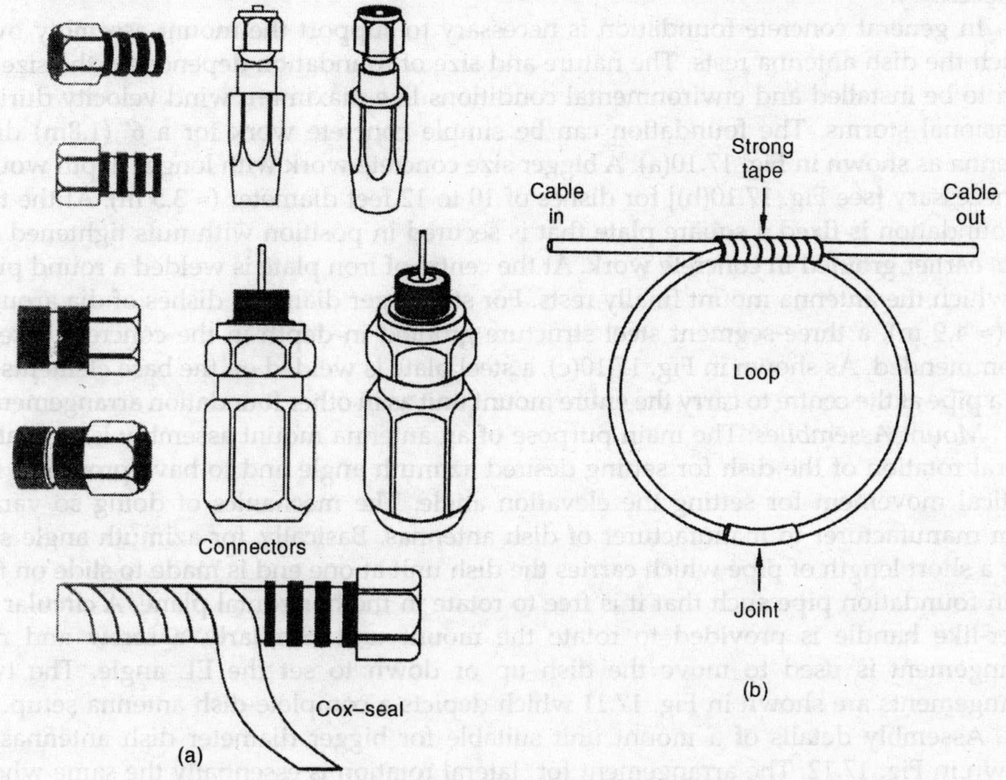

Fig. 17.9. (a) Coaxial seal and cable connectors of various types, (b) Loop at cable joints.

make sure that no signal power is lost due to radiation from junction points. Such a loss of signal not only causes interference in reception on TV sets in many ways but also necessitates installation of more line amplifiers. Some good quality connectors are shown in Fig. 17.9 (a) which also shows how cox-seal is applied where necessary.

Loop: It is often necessary to join two lengths of cables on CATV distribution lines. On fixing the joint, it is common practice to make a loop at the point as shown in Fig. 17.9 (b) and tie the two cables with a strong tape. This prevents any pull at the joint and thus any disconnection is prevented.

17.7 Assembly and Installation of Antenna Subsystems

The first subsystem to be installed at the cable station is the antenna system. It should normally be agreed upon at the time of placing order for the cable plant that front-end equipment will be installed by its manufacturer/supplier. If for any reason this be not possible, full instructions for doing so must be obtained in advance from the supplier. These are usually in the form of a manual, separate for each antenna unit. It gives details of civil work at the site, list of items and their specifications, step-by-step procedure for assembling the mount and dish structure and their erection in position. The manual also contains detailed instructions for adjusting azimuth and elevation angles as necessary.

In general concrete foundation is necessary to support the mount assembly over which the dish antenna rests. The nature and size of foundation depends on the size of dish to be installed and environmental conditions like maximum wind velocity during occasional storms. The foundation can be simple concrete work for a 6′ (1.8m) dish antenna as shown in Fig. 17.10(a). A bigger size concrete work with longer depth would be necessary [see Fig. 17.10(b)] for dishes of 10 to 12 feet diameter (≈ 3.5 m). At the top of foundation is fixed a square plate that is secured in position with nuts tightened on bolts earlier grouted in concrete work. At the centre of iron plate is welded a round pipe on which the antenna mount finally rests. For still larger diameter dishes of dia around 16′ (≈ 4.9 m), a three-segment steel structure grouted in depth in the concrete base is recommended. As shown in Fig. 17.10(c), a steel plate is welded on the base elements. It has a pipe at the centre to carry the entire mount unit as in other foundation arrangements.

Mount Assemblies: The main purpose of an antenna mount assembly is to enable lateral rotation of the dish for setting desired azimuth angle and to have provision for vertical movement for setting the elevation angle. The mechanics of doing so varies from manufacturer to manufacturer of dish antennas. Basically, for azimuth angle setting a short length of pipe which carries the dish unit at one end is made to slide on the main foundation pipe such that it is free to rotate in the horizontal plane. A circular or lever-like handle is provided to rotate the mount unit. Similarly, a screw and nut arrangement is used to move the dish up or down to set the EL angle. The two arrangements are shown in Fig. 17.11 which depicts a complete dish antenna setup.

Assembly details of a mount unit suitable for bigger diameter dish antennas is shown in Fig. 17.12. The arrangement for lateral rotation is essentially the same where on loosening the bolts, the lever handle is rotated to set the AZ angle [Fig. 17.12(a)]. For setting EL angle, open end of a long screw [see Figs. 17.12(b) & (c)] is fixed in a pivot joint arrangement on the bracket which holds the dish unit. This screw is held straight by passing it through a hole in a bracket which is clamped to the main antenna mount as shown in the figure. For moving the dish at the desired EL angle, the screw is moved up or down as necessary with the help of nuts on the screw, provided on either side of the bracket [see Fig. 17.12(b) and (c)]. One nut is loosened and other tightened depending on the direction in which the dish is to be moved in the vertical axis. On obtaining desired location of the dish, both the nuts are tightened on to the bracket to prevent any further movement of the screw and hence that of dish in the vertical plane.

All the components needed for erecting a mount assembly are provided by the dish manufacturer. Thus, before proceeding to erect the unit, make sure that all components and tools taken to the site check with the supplied list. Also inspect to make sure that the round pipe at the centre of foundation plate is straight and firm in position. A stout string with a small weight tied at its one end (plumb-line) can be used to check that it is vertical at 90° on the foundation.

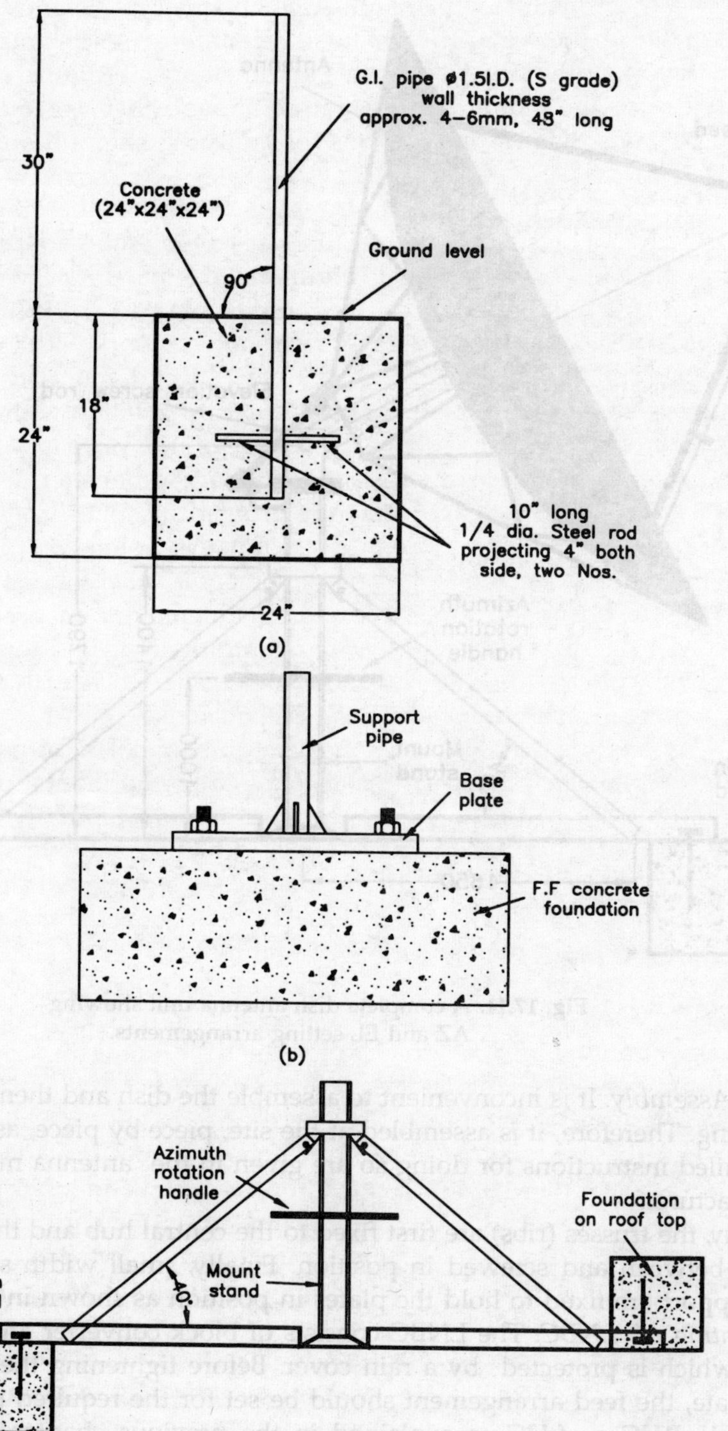

Fig. 17.10. Foundations for erecting dish antennas (a) Foundation suitable for a 6 feet (≈ 1.8m) dish antenna. (b) Foundation for bigger size dish antennas. (c) Foundation for 16' feet (≈ 5m) and above dish antenna.

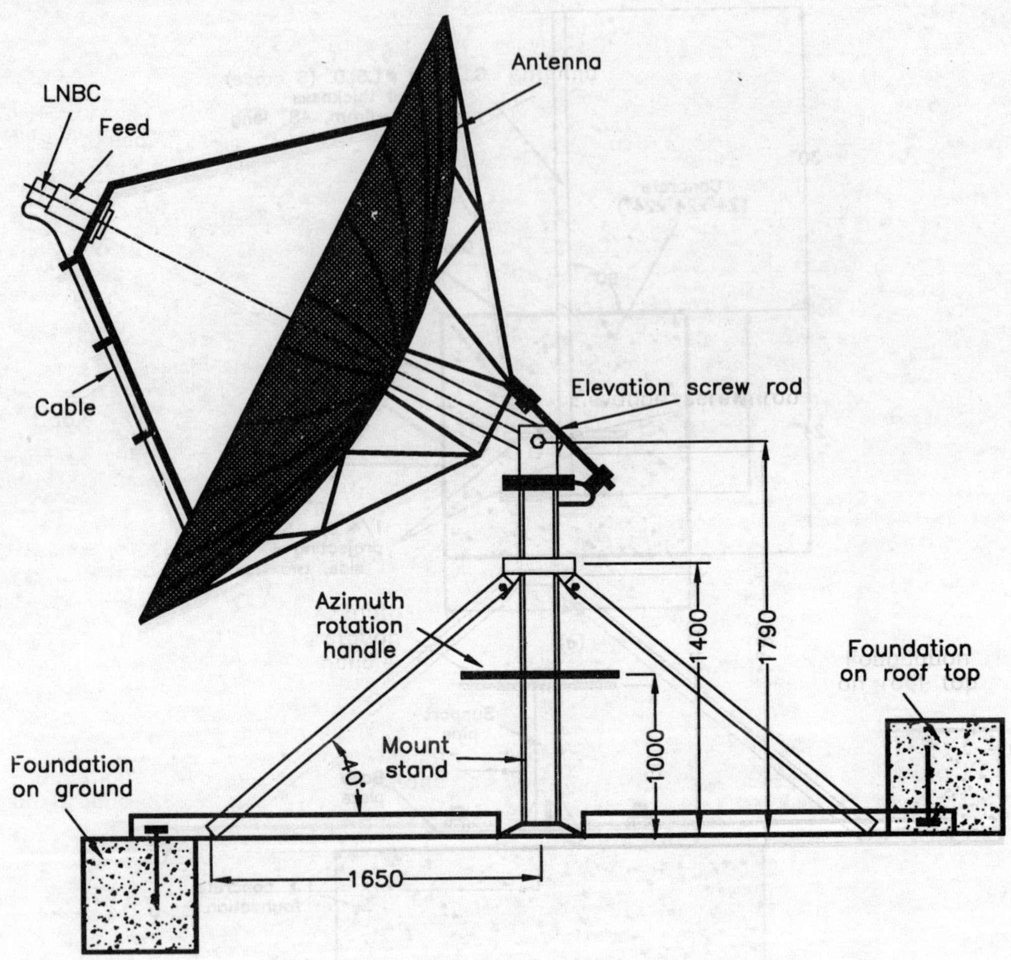

Fig. 17.11. A complete dish antenna unit showing AZ and EL setting arrangements.

Dish Assembly: It is inconvenient to assemble the dish and then carry it to the site for mounting. Therefore, it is assembled at the site, piece by piece, as illustrated in Fig. 17.13. Detailed instructions for doing so are given in the antenna manual supplied by the manufacturer.

Mainly, the trusses (ribs) are first fixed to the central hub and then segment plates slipped in-between and screwed in position. Finally, small width segments for outer circular support are fixed to hold the plates in position as shown in the figure.

Mounting of LNBC: The LNBC consists of block converter and feed unit as one assembly which is protected by a rain cover. Before tightening the LNBC unit on its support plate, the feed arrangement should be set for the required polarization which may be H,V, RHC or LHC as explained in the previous chapter. On doing so, the assembly is fixed on the mounting plate side of the feed pipe with provided screws as shown in Fig. 17.14 (a). One end of the main RF cable that takes IF signal to the satellite receiver in the control room is then slided in the feed pipe and the other provided with

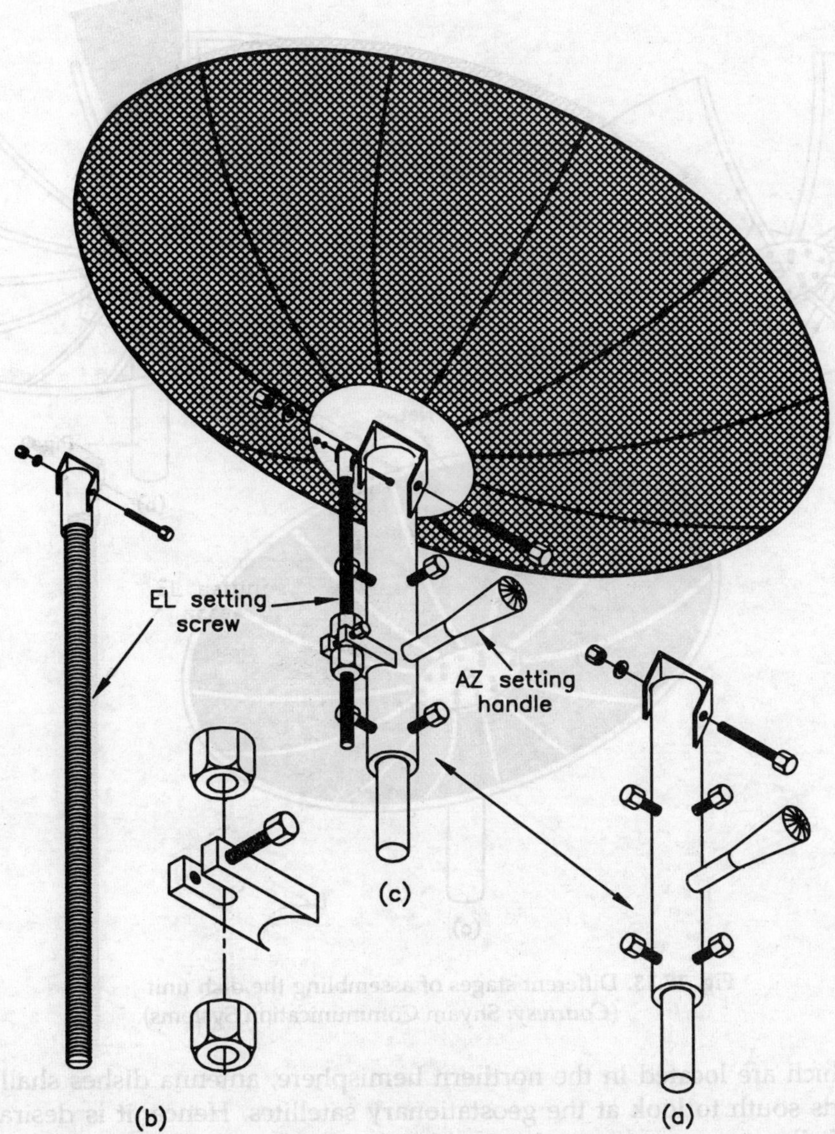

Fig. 17.12. AZ and EL setting arrangements for bigger dish antennas. (a) AZ setting, (b) EL setting, (c) complete unit.

a male connector. It is then screwed on the corresponding socket at the output of LNBC. The other (lower) end of feed pipe is firmly mounted in a bush located at the bottom plate of the mount unit which has provision for the RF cable to pass through. It finally goes to the control room. Component and mounting details of another make LNBC are shown in Fig. 17.14 (b).

The details given above pertain to an antenna subsystem to be operated manually. For antennas that have motorised tracking facility, the mount assembly is more elaborate to suit remote control setting of look-angles. In any case, for India and other

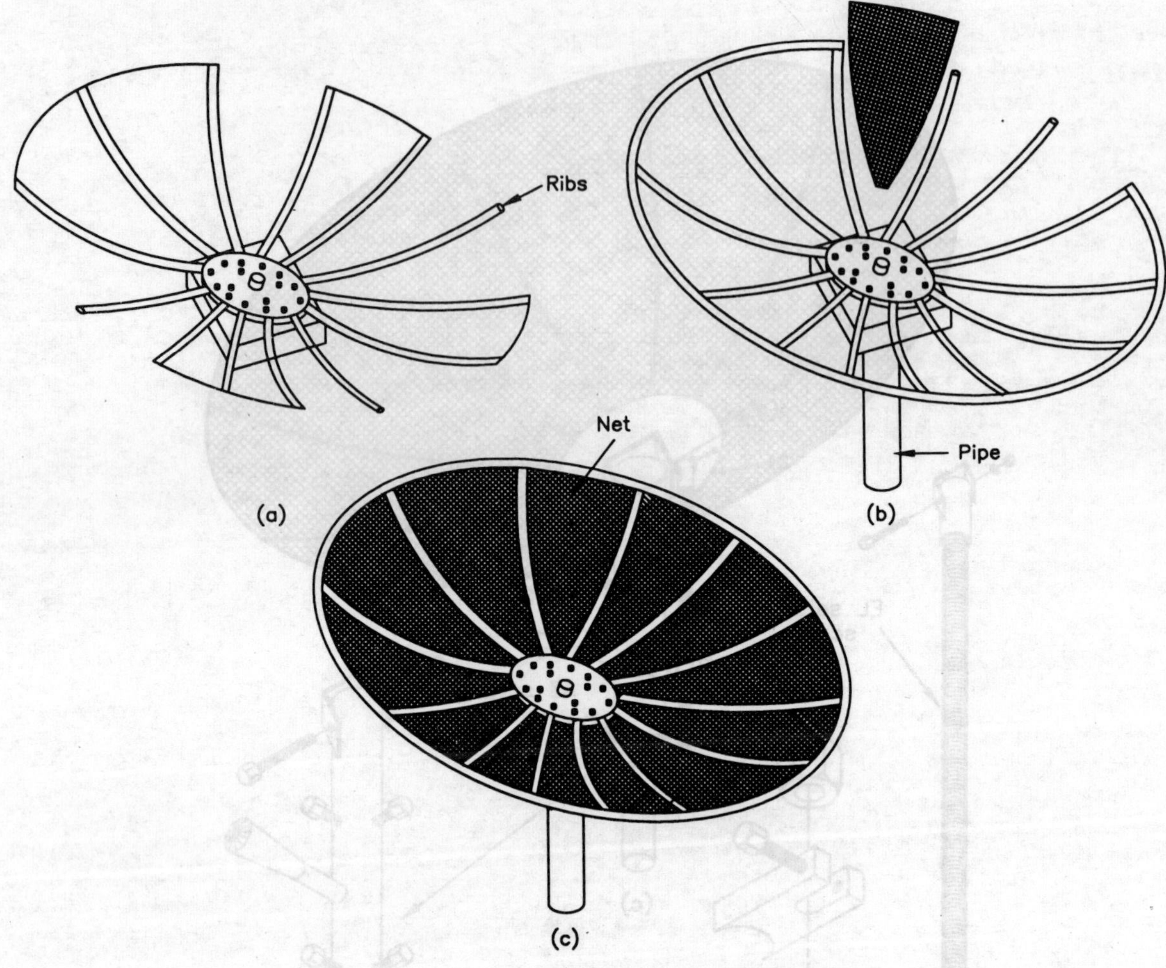

Fig. 17.13. Different stages of assembling the dish unit (*Courtesy:* Shyam Communication Systems).

countries which are located in the northern hemisphere, antenna dishes shall have to point towards south to look at the geostationary satellites. Hence, it is desirable, that dish faces of all antennas are turned in south-west direction while fixing them on their mounts. This will enable easy manoeuvrability at the time of setting AZ and EL angles. Furthermore, it may be noted that if the dish is to be aimed at a satellite where transmission is both horizontally and vertically polarized, as from PAS-4, a dual feed unit is to be provided and suitably oriented. As obvious, two LNBCs will be needed and their output cables will go to two different satellite receivers.

Antennas for Receiving Local Telecasts: The receiving antennas for picking up signals of local TV stations should correspond to their channels and be of good design with sturdy elements. These should be fixed on roof-top for a clear line-of-sight view of transmitting antennas. As for direct reception on TV receivers, these should be mounted horizontally for maximum signal pick-up and at a height of about 10 metres from ground (roof) level to prevent any interference. The twin lead-in transmission cable, one end of which is fixed at the closing end of folded dipole is taken to the control room to feed received signals to the amplifier, signal processor or VCR as the case may be.

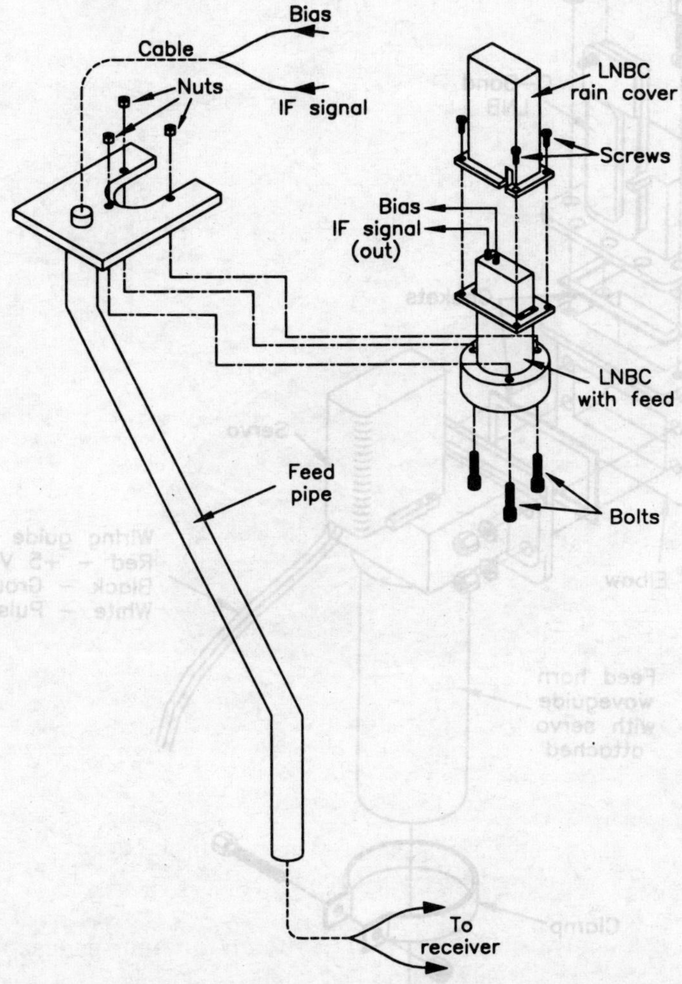

Fig. 17.14. (a) Assembly details of one type of LNBC unit.

A necessary precaution while mounting such antennas is that these should be at least 2 metres away from each other and any large metal objects. If the reproduced pictures are of not good quality, changing height of the antenna with respect to ground level (roof) will enable increased signal strength. Similarly, rotating the antenna a bit from its normal position which is in the direction of transmitting antenna, will eliminate ghost images if seen on the TV receiver screen.

In most cases, local telecasts are also available via satellites. In that case, it is advisable to obtain signals from the satellite instead of through a YAGI antenna. This will enable better reception but will need another satellite receiver-modulator unit for each channel for processing received signals as for other satellite channels.

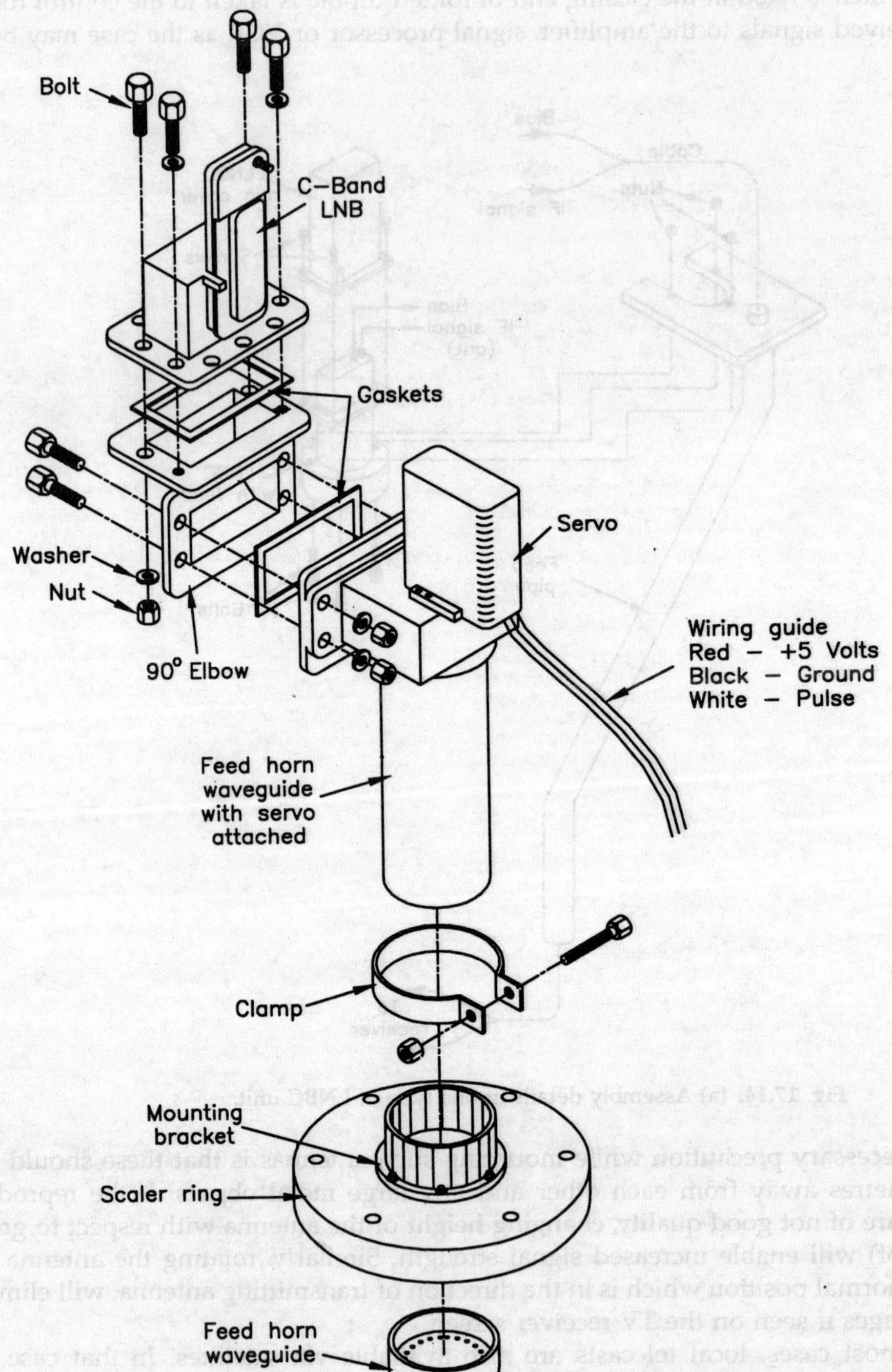

Fig. 17.14. (b) Assembly and mount details of LNBC of another make.

17.8 Dish Antenna Alignment

Since AZ and EL angles vary from place-to-place for the same satellite, it is necessary to obtain this information, beforehand for the location of proposed cable station. This can be had from the supplier of front-end equipment or collected from magazines devoted to cable television. It is helpful to tabulate such data for all the satellites of interest and display it at a prominent place in the control room for ready reference.

Preparation for Dish Alignments: Before proceeding to set dish antennas at the required look angles the following preparations are necessary:

(1) Make sure that in the southern direction of the antenna* farm *i.e.* location of dish antennas, there is no obstruction. This is to ensure that line-of-sight from dishes to satellites is not obstructed.

(2) Arrange for mains supply at a suitable location very close to dish antennas. This is necessary for feeding ac mains to equipments needed for acquiring satellites of interest.

(3) Bring at the antenna site, one C-band satellite receiver, a colour TV receiver and a VCR. The VCR will be needed if the colour receiver does not have provision to accept base-band (video and audio) signals directly.

(4) Normally, at the time of equipment installation in the control room, RF coaxial cables are provided for connecting LNBC outputs to satellite receivers. If not, the same should be done at this stage. Each cable should be terminate at both ends with male connectors—usually called 'F' connectors. The cable lengths will depend on the distance between each dish and control room. However, if this exceeds 25 meters, a bullet amplifier will be needed in-between the total length to make-up for excessive signal loss.

(5) Arrange for a magnetic compass [Fig. 17.16 (a)] and an angle meter [Fig. 17.18 (a)] for locating desired AZ and EL angles respectively. A 'D' angle protector commonly found in geometry boxes can also be used for setting EL angle.

(6) Place feed horns and associated LNBCs in the rubber gasket that comes with each unit and then fix the two with the supplied nuts and bolts on top of the focus (support) pipes that rise from the centre/sides of dish assembly.

(7) On dishes which are meant for receiving signals from satellites which transmit in both horizontal and vertical polarized modes, there is a provision for fixing a dual feed horn unit and two associated LNBCs. The manual gives necessary details for fixing such dual units. As necessary, two RF cables will connect the outputs of these LNBCs to two different satellite receivers in the control room.

(8) Connect one end of each RF cable to the corresponding LNBC by screwing the 'F' connector on the matching female socket available at its output side.

(9) While fixing outgoing cables along the focus (support) pipes, make sure that no pressure is exerted on LNBC connections. Leave other ends of all the RF cables close to the dish antennas.

---

* In India dishes face south-westwards for having line of sight link with the satellites located along the equator. However, the direction can be different for other countries. For example, CATV dishes in SRI LANKA must face north-wards for receiving signals from the same satellites.

(10) Recheck the above enumerated preparations and if found in order, the set-up can be taken as ready for aligning each dish at the desired look angle. It is advisable to take help of another person for setting the dishes as necessary.

Dish Alignment Procedure: The procedure for setting the antenna dish at the required look angle is as follows:

Azimuth Angle Setting: Zero reference for measuring azimuth angle is taken to be north. Thus AZ ∠90° is towards east, AZ ∠180° towards south and AZ ∠270° is in the west. These locations are illustrated in Fig. 17.15. As explained next, a magnetic compass is used to locate the desired AZ angle.

Procedure: (i) Loosen the bolts that hold horizontal rotation pipe [see Fig. 17.12 (a)] in position to enable free movement of dish in the horizontal plane.

(ii) Place the magnetic compass, [shown in Fig. 17.16 (a)] close to the dish unit, either on a high wooden stool or at any other suitable location but at a small distance away from the dish assembly to prevent any effect on deflection of magnetic needle caused by the magnetic material of dish unit components. In case a stool is not available and any other location is not convenient, the compass can be held on the palm of one's hand kept steady.

(iii) Rotate the compass dial to bring N-S directions as labelled on it, in line with the rest position of magnetic needle [see Fig. 17.16 (a)].

(iv) Locate the desired AZ angle on the dial and rotate the dish with the provided handle till its axis appears parallel to the line (direction) that would form (obtain) from the centre of dial which is the needle's pivot point, to the wanted angle mark, as shown in Fig. 17.16 (b).

(v) Tighten back AZ setting bolts to prevent any further movement of the dish in the horizontal plane.

Azimuth angle needed for receiving PAS-4 signals at Delhi is 197.6°≈198°. As an illustration, position of the dish when set at this AZ angle and as viewed from top is shown in Fig. 17.16(c). The line (X-X) shown in Fig. 17.16(b) on the compass dial corresponds to AZ ∠198° and thus the dish axis must lie parallel to this as indicated by the line (X-X) on the dish.

Elevation Angle Setting: Elevation angle setting is raising the dish upwards towards sky from its location earlier set in the horizontal plane. The elevation angle EL can vary from ∠ 0° to ∠ 90°. As an illustration of this, the two extreme locations are depicted in Fig. 17.17(a) and (b) for a dish earlier set at an AZ angle of 270°. In Fig. 17.17(c), though the dish is still facing southwards, it is shown raised upwards to acquire an EL angle = ∠55°, which is the required elevation angle for receiving PAS-4 signals at Delhi.

Procedure: (i) Set the angle meter at the desired EL angle. Such a setting for EL ∠55° is shown in Fig. 17.18 (b). Notice inclination of the lever when set at this angle.

(ii) Loosen the nuts that hold the EL setting screw and hence the dish in the vertical plane. [see Figs. 17.11 and 17.12].

(iii) Hold the angle meter set earlier at the desired EL angle (55°) along the vertical mount pipe such that flat side of its 'D' (X-X) lies parallel to the pipe axis. Depending on the initial horizontal position of dish (270°) the angle meter when so placed will appear inclined as seen in Fig. 17.18 (b).

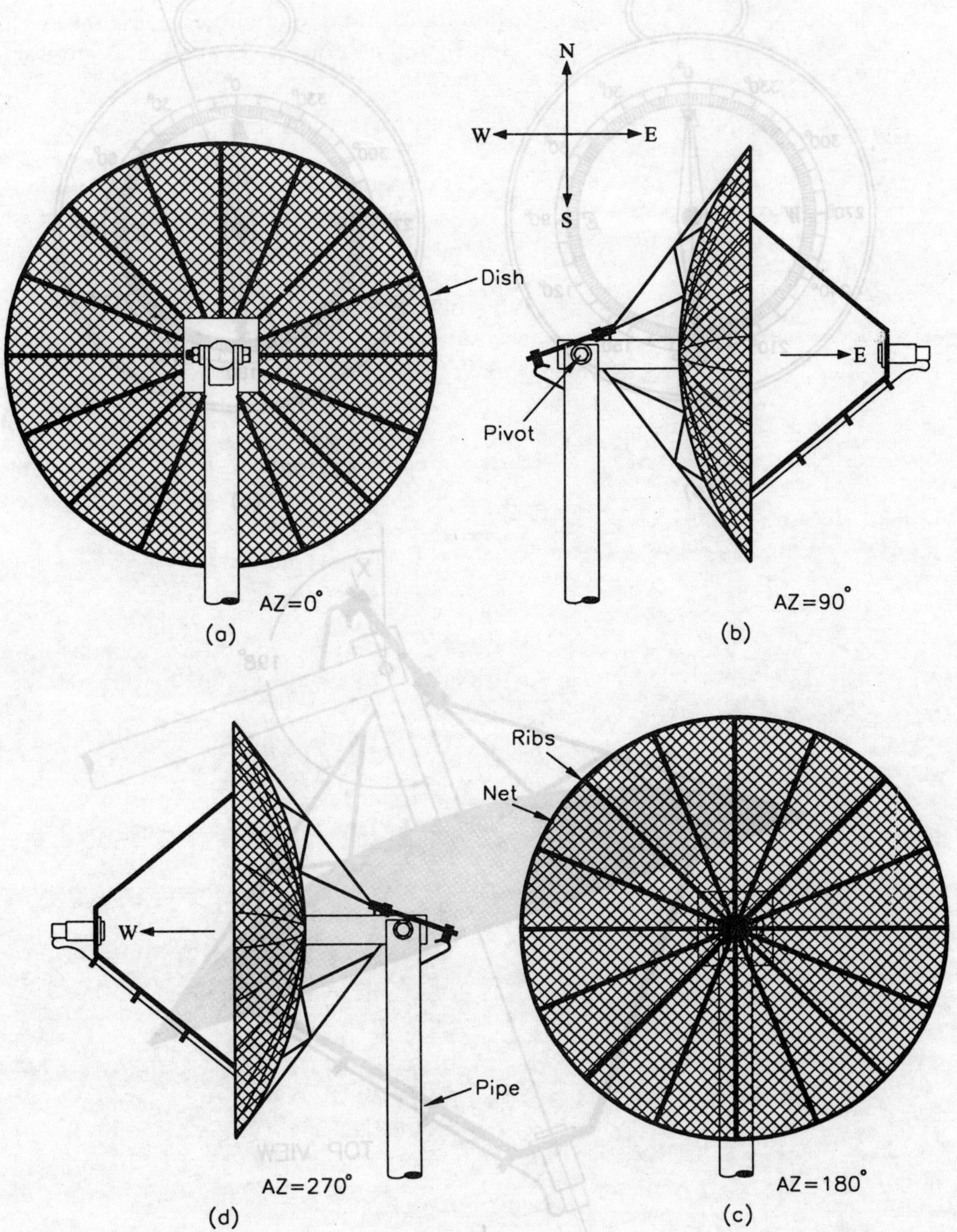

Fig. 17.15. Pictorial views of an antenna dish when oriented in four different directions. (a) AZ ∠0°, (b) AZ ∠90°, (c) AZ ∠180°, and (d) AZ ∠270°.

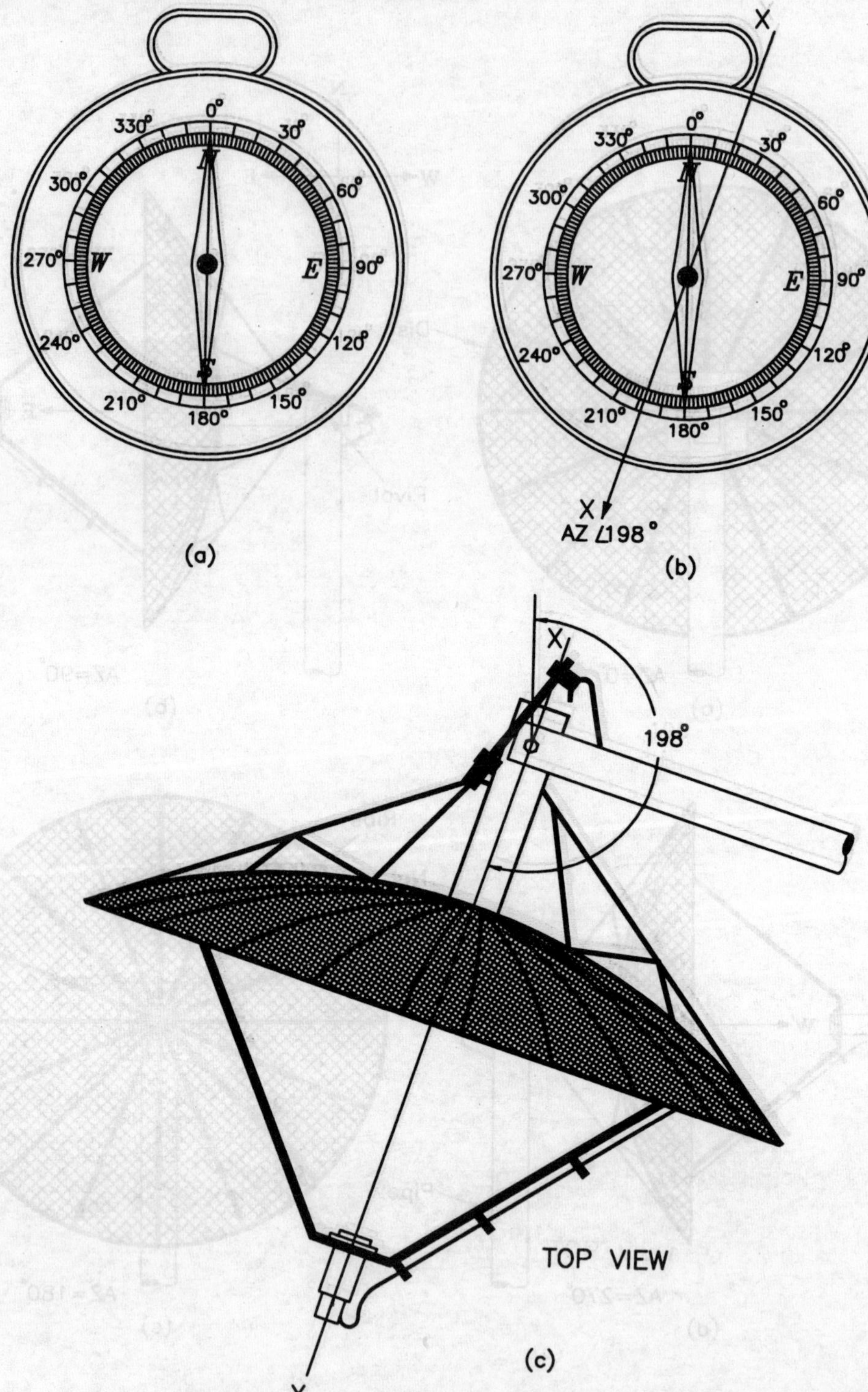

Fig. 17.16. Azimuth angle setting, (a) magnetic compass,
(b) Compass dial showing the direction of AZ ∠198°,
(c) Pictorial top view of the dish when set at this AZ angle.

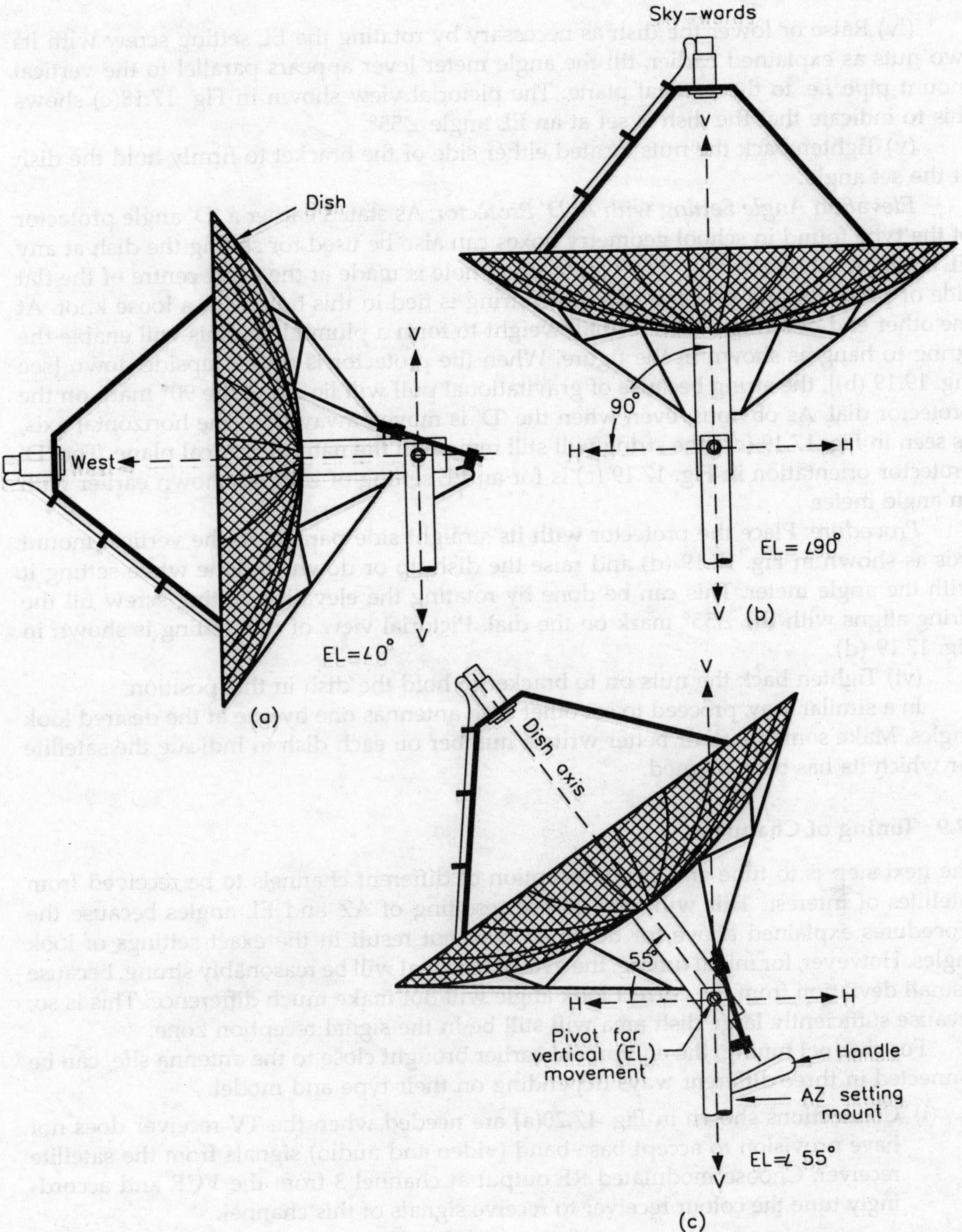

Fig. 17.17. Location of antenna at different elevation angles with dish initially oriented for an AZ angle = 270°, (a) EL ∠0° (b) EL ∠90°, and (c) EL ∠55°.

(iv) Raise or lower the dish as necessary by rotating the EL setting screw with its two nuts as explained earlier, till the angle meter lever appears parallel to the vertical mount pipe *i.e.* to the vertical plane. The pictorial view shown in Fig. 17.18(c) shows this to indicate that the dish is set at an EL angle ∠55°.

(v) Tighten back the nuts located either side of the bracket to firmly hold the dish at the set angle.

Elevation Angle Setting with A 'D' Protector: As stated earlier a 'D' angle protector of the type found in school geometry boxes can also be used for setting the dish at any EL angle. As shown in Fig. 17.19 (a), a small hole is made at the exact centre of the flat side of protector and one end of a stout string is tied in this hole with a loose knot. At the other end of string is tied a small weight to form a plumb line. This will enable the string to hang as shown in the figure. When the protector is turned upside down [see Fig. 19.19 (b)], the string because of gravitational pull will lie along the 90° mark on the protector dial. As obvious, even when the 'D' is moved away from the horizontal axis, as seen in Fig. 17.19 (c) , the string will still remain in the natural vertical plane. The 'D' protector orientation in Fig. 17.19 (c) is for an EL setting of ∠55° as shown earlier with an angle meter.

Procedure: Place the protector with its straight side parallel to the vertical mount axis as shown in Fig. 17.19 (d) and raise the dish up or down as done while setting it with the angle meter. This can be done by rotating the elevation setting screw till the string aligns with the ∠55° mark on the dial. Pictorial view of this setting is shown in Fig. 17.19 (d).

(vi) Tighten back the nuts on to bracket to hold the dish in this position.

In a similar way, proceed to set other dish antennas one by one at the desired look angles. Make some mark or better write a number on each dish to indicate the satellite for which its has been aligned.

17.9 Tuning of Channels

The next step is to tune and check reception of different channels to be received from satellites of interest. This will need minor resetting of AZ and EL angles because the procedures explained above for doing so may not result in the exact settings of look angles. However, for initial tuning, the available signal will be reasonably strong, because a small deviation from the correct look angle will not make much difference. This is so, because sufficiently large dish area will still be in the signal reception zone.

For channel tuning, the equipment earlier brought close to the antenna site, can be connected in three different ways depending on their type and model.

(i) Connections shown in Fig. 17.20(a) are needed when the TV receiver does not have provision to accept base-band (video and audio) signals from the satellite receiver. Choose modulated RF output at channel 3 from the VCR and accordingly tune the colour receiver to receive signals of this channel.

(ii) In the connections shown in Fig. 17.20(b) a VCR is not necessary because the satellite receiver output is connected directly to the base-band (V&A) input sockets which are available on the TV receiver. These signals feed into vision, sound and sync pulse processing circuits in the receiver. The output composite video and audio signals thus obtained are modulated on vision and sound carriers of a

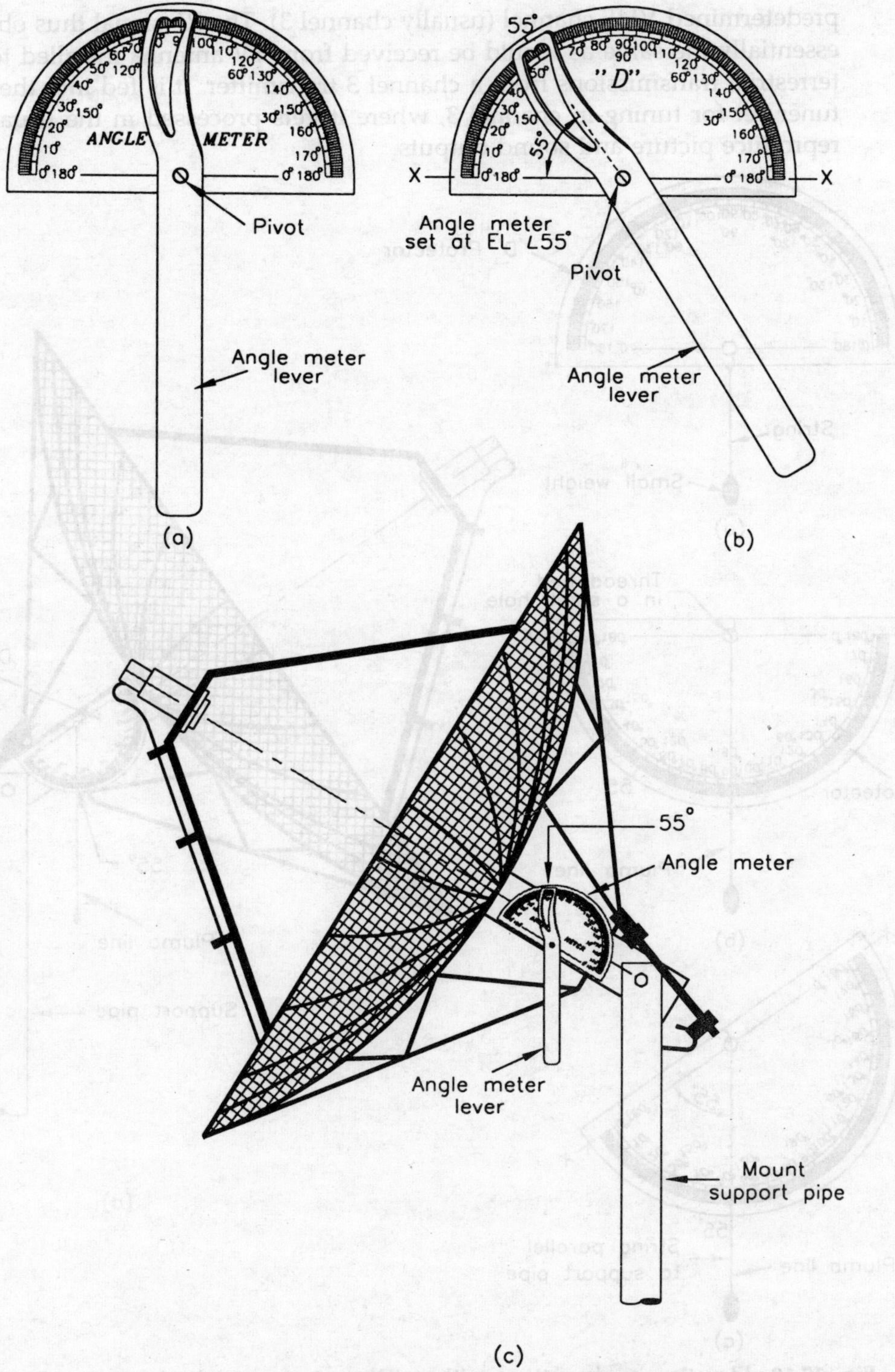

Fig. 17.18. Elevation angle setting with an angle meter
(a) angle meter, (b) angle meter set for EL ∠55°,
(c) pictorial view of the dish unit when set at EL ∠55°.

predetermined VHF channel (usually channel 3). The RF signal thus obtained is essentially the same as would be received from TV antenna installed to receive terrestrial transmissions from a channel 3 transmitter. It is fed into the receiver tuner set for tuning-in channel 3, where it gets processed in the usual way to reproduce picture and sound outputs.

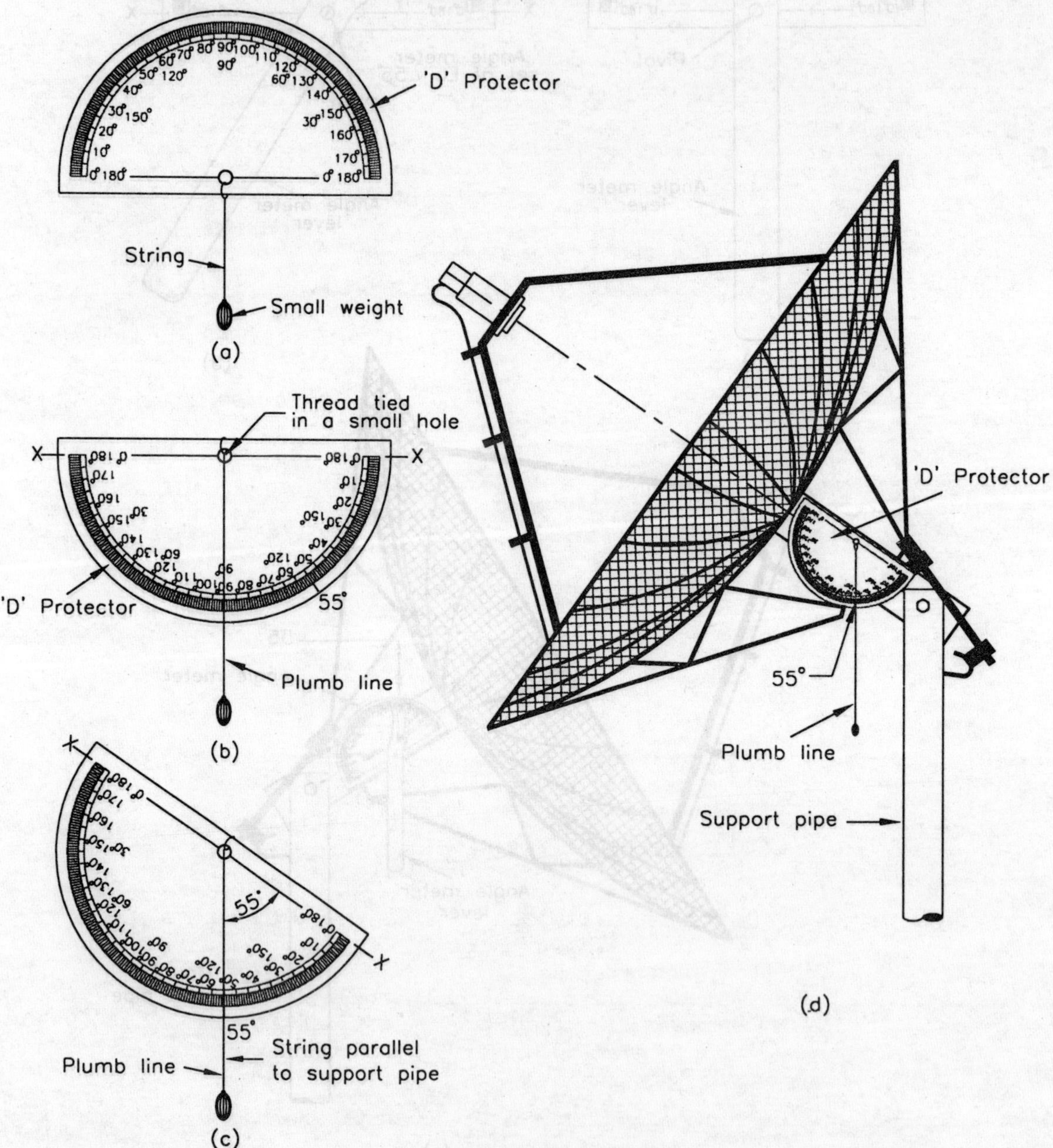

Fig. 17.19. Elevation angle setting with a 'D' protector. (a) basic arrangement, (b) protector 'D' turned upside shown, (c) a ∠55° tilt of protector to show location of string on the dial at ∠55°. (d) Pictorial view of the dish unit when raised to attain an EL angle ∠55°.

(iii) In case the satellite receiver has provision for enabling modulated RF output on channel 2/UHF channel, it can be directly connected to the TV receiver aerial input socket as shown in Fig. 17.20(c) and its tuning set for the chosen RF channel.

In any case make sure that the mains ON-OFF switch of the satellite receiver is at 'OFF' position till all connections on the equipments have been made and RF cable connected at its input socket.

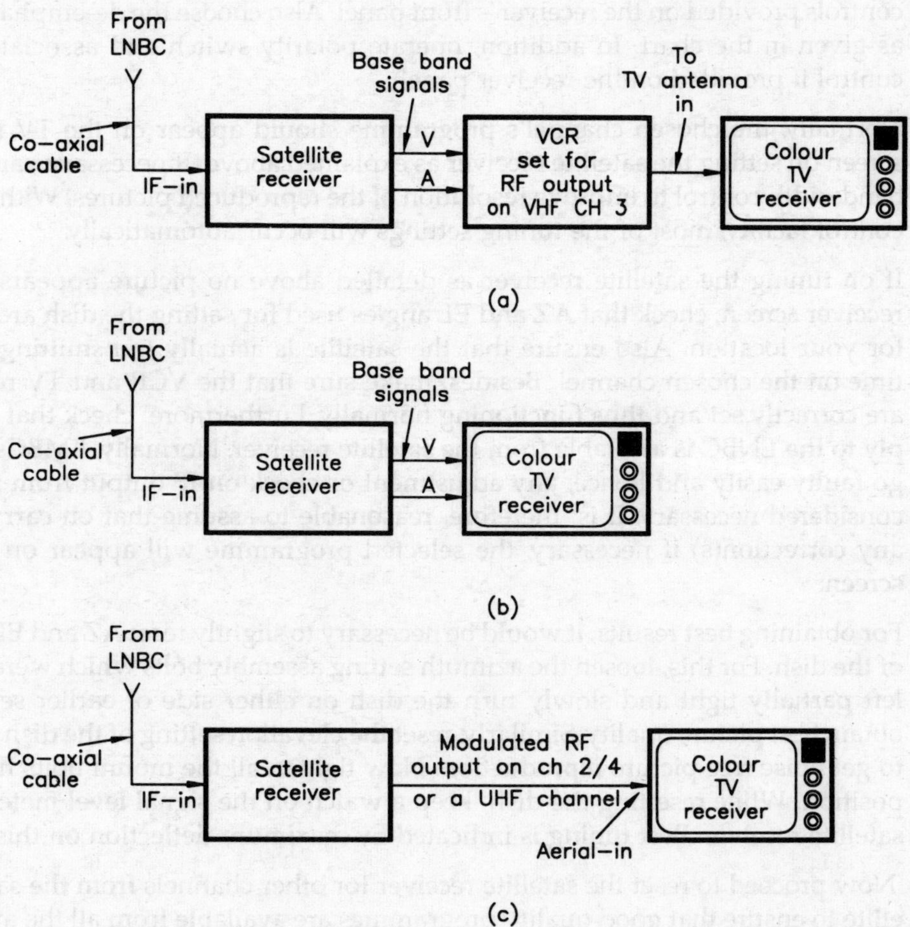

Fig. 17.20. Equipment connections for tuning satellite channels (a) with VCR when the colour receiver does not have input points for Base-Band video and audio signals, (b) without VCR when Base-Band input sockets are available on the colour receiver, (c) when RF modulated output is directly available from the satellite receiver.

Now proceed as under for tuning required channels from different satellites:

(i) Connect other end of the RF cable coming from any one dish to the IF-IN socket of the satellite receiver.

(ii) Make sure that RF cable from the LNBC output to receiver input is firmly connected at both ends and then switch 'ON' the satellite receiver and other equipment(s) depending on the chosen set-up out of the three shown in Fig. 17.20. Note that dc supply for the LNBC is obtained from the satellite receiver through the coaxial (RF) cable itself as explained in an earlier chapter.

(iii) Choose any one channel out of the many available at the time of alignment from the satellite being acquired. Note its centre frequency (channel number) from the chart prepared earlier and tune/set the satellite receiver at this channel. Select the given sound sub-carrier frequency (f_{sc}) and its bandwidth if necessary with controls provided on the receiver's front panel. Also choose the de-emphasis time as given in the chart. In addition, operate polarity switch and associated trim control if provided on the receiver panel.

(iv) Normally, the chosen channel's programme should appear on the TV receiver screen on setting the satellite receiver as explained above. If necessary vary the IF bandwidth control to improve resolution of the reproduced pictures. With remote control facility, most of the tuning settings will occur automatically.

(v) If on tuning the satellite receiver as detailed above no picture appears on the receiver screen, check that AZ and EL angles used for setting the dish are correct for your location. Also ensure that the satellite is actually transmitting at that time on the chosen channel. Besides, make sure that the VCR and TV receivers are correctly set and thus functioning normally. Furthermore, check that dc supply to the LNBC is available from the satellite receiver. Normally, LMBCs do not go faulty easily and hence, any adjustment or check on IF output from it is not considered necessary. It is, therefore, reasonable to assume that on carrying on any correction(s) if necessary, the selected programme will appear on the TV screen.

(vi) For obtaining best results, it would be necessary to slightly reset AZ and EL angles of the dish. For this, loosen the azimuth setting assembly bolts which were earlier left partially tight and slowly turn the dish on either side of earlier setting to obtain best picture quality. Similarly, reset the elevation setting of the dish slightly to get noise free picture reproduction. Now tighten all the mount bolts firmly in position. While resetting the dish, keep a watch on the signal level meter of the satellite receiver. Best tuning is indicated by maximum deflection on this meter.

(vii) Now proceed to reset the satellite receiver for other channels from the same satellite to ensure that good quality programmes are available from all the available channels.

(viii) For acquiring other satellites, repeat the procedure outlined above. Make sure that the IF-IN cable to the satellite receiver pertains to the dish aligned earlier to receive signals from the satellite to be acquired.

(ix) If the transmission to be received is both vertically and horizontally polarized as from PAS-4, a dual feed unit is provided, as explained earlier, on corresponding dish antennas. For proper adjustments such that both transmissions are received correctly, it would be necessary to use two satellite receivers as shown in Fig. 17.21. RF cables from the two LNBCs are made input at IF-IN points of the two satellite receivers. The selector switch enables selection of modulated RF output from any one satellite receiver. Thus H and V channels can be tuned alternately. The procedure for tuning remains the same. However, the channels transmitted in one mode, say horizontal polarization, can be tuned first and then the vertically polarized channels by switching on corresponding satellite receiver.

(x) For fine tuning, loosen the screws holding the scaler ring to the feed assembly. With both satellite receivers switched 'ON' and set on chosen channel, put the selector switch to H position and rotate the feed assembly till best reception is obtained on the TV receiver. Next change the selector switch to position 'V' and if necessary rotate slightly the feed assembly to obtain equally good reception on the receiver screen. Finally, ensure that best quality reception is obtained from transmissions on either polarized mode. Now tighten the screws that hold the scaler ring. A good check of best possible tuning is that programmes coming from horizontally polarized feed are not received via the satellite receiver connected to receive vertically polarized channels and *vice versa*.

After setting all the dish antennas at the required look angles, connect RF lead-in cable from all LNBCs to corresponding satellite receivers in the control room. Also put back the VCR and satellite receiver(s) used for tuning on the racks in the control room from where these were removed and make necessary connections.

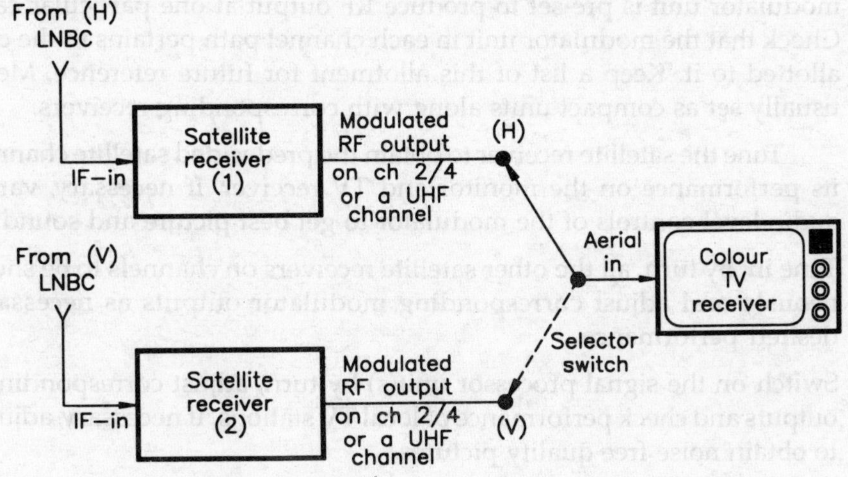

Fig. 17.21. Equipment connections for tunning V and H polarised channel programmes from the same satellite.

(xi) If necessary fine tuning to the feed/LNBC units that receive only V or H polarized signals can be made to get best results. This is done by loosening the LNBC mounting screws slightly and rotating the assembly in either direction to identify position where the received signal (picture) begins to deteriorate. Mark these positions and adjust the assembly midway between the two positions. On doing so, tighten back the holding screws. Such an exercise minimises polarization loss and hence, enables maximum signal pick-up. A signal strength meter will be helpful for fine tuning and may be borrowed from the equipment supplier if available with him.

Such a feed adjustment could have affected the EL angle setting slightly. As such, the setting on the dish may again be checked and corrected, if necessary. Finally note that it may not be possible to receive all the channels with the same high quality. Also remember that if the satellite receiver frequency is not correctly set, the colour sub-carrier frequency is the first to weaken and the TV receiver will not reproduce colours properly. In this case, the pictures could be in B&W only.

17.10 Setting of Control Room Equipments

Now that all dish antennas have been aligned and all the required programmes are available from various satellites, it is time to set control room equipments. Assuming that all the receivers, modulators, signal processors, VCRs and combining network have already been mounted and interconnected as suggested earlier proceed as under:

(i) On checking that the mains supply voltage stabilizer is functioning, switch 'ON' power supply to all the equipments in the control room.

(ii) The procedure for tuning and setting a satellite receiver is detailed in the previous section. Switch 'ON' any one cable channel set-up that includes satellite receiver, modulator, monitor and TV receiver connected at the output of combiner. Each modulator unit is pre-set to produce RF output at one particular cable channel. Check that the modulator unit in each channel path pertains to the cable channel allotted to it. Keep a list of this allotment for future reference. Modulators are usually set as compact units along with corresponding receivers.

Tune the satellite receiver to obtain the predecided satellite channel and check its performance on the monitor and TV receiver. If necessary, vary video and audio level controls of the modulator to get best picture and sound outputs.

(iii) Tune in, by turn, all the other satellite receivers on channels to be shown simultaneously and adjust corresponding modulator outputs as necessary to obtain desired performance.

(iv) Switch on the signal processor unit(s) by turn, adjust corresponding modulator outputs and check performance of local TV stations. If necessary, adjust antenna(s) to obtain noise free quality pictures.

(v) Similarly check VCR channel(s) by playing a pre-recorded video cassette and make adjustments as necessary.

If any equipment appears faulty and does not perform as expected, take help of the supplier and if necessary, insist on its replacement by a good piece as all equipments are under a guarantee period for at least one year.

17.11 Connections to Customers and Commissioning of Cable Plant

Before proceeding to give connections to customers, it should be ensured that all the channels are operative with good signal strength on each trunk line. Switch on ac power supply to trunk amplifiers to enable signal flow on all cable routes. Connections to subscribers should be made one by one, starting from control room side. As explained earlier, cable-run to each customer is made from the nearest tap-off point available on the distribution cable.

On commissioning the cable plant, a technician should visit each customer to tune his/her TV set to enable reception from all the channels available from the cable station. If the receiver is faulty in any manner, the customer cannot expect quality performance from the cable system. Similarly, if it does not have the cable ready version tuner, channels carrying programmes on 'S' and hyper bands cannot be received.

Before making cable connection to the receiver input, a dB meter should be used to check signal level at that point. It should be 55 to 65 dB(μv). Higher signal strength will cause distortion in the displayed images and lower level will produce weak pictures depending on the sensitivity of the TV receiver.

Trouble Shooting: No installation will work perfectly the very first time on commissioning. The problems may be as simple as loose connections somewhere or quite complex like some equipment(s) not performing as per specifications or low IF output from one or more LNBCs. When the antenna installation responsibility is that of the equipments supplier, the cable operator will face minor problems with the rest of set-up. The faults can be improper connections, cable breakage or no satellite transmission for some reason. These can be easily tackled with simple checks without the need of elaborate test instruments. Chapter 19 is devoted to trouble shooting in cable TV systems.

Interference: There can be RF interference to the cable system from ignition system of passing vehicles, radiation from welding sets operating nearby, electrical household gadgets and radio and TV transmitters if located near the cable station location. There is nothing the cable operator can do to eliminate such interferences. However, he can minimise such pick-up by the cables, by running them through metal conduits and providing earth connection to all equipments in the control room. Interference can also be avoided by selecting cable station site away from locations that produce H.F. radiations.

17.12 Set Top Converter

Earlier TV receivers which do not have cable ready tuners need a converter to receive all the cable channels. The operating principle of such a converter is fully explained in chapter 15 (section 15.6). It converts all incoming channels to one RF channel, usually 3/4 as in VCRs. The cable channel selected by the subscriber on the converter panel is demodulated to recover composite TV signal. It is then remodulated on either channel 3 or 4 which remains same for all CATV channels. This output is fed to the TV receiver which is left tuned to receive channel 3/4 programmes. Most converters commonly called 'Set Top Converters' are designed to accept up to 50/60 cable channels. While manual selection and control is available, all functions are usually controlled with a remote control unit that comes with the converter. A typical Set Top Converter is shown in Fig. 17.22.

The remote control has channel selection, fine tuning, favourite channel, recall, delete, memory and channel lock features. Channel 'SCAN UP' button, when pressed

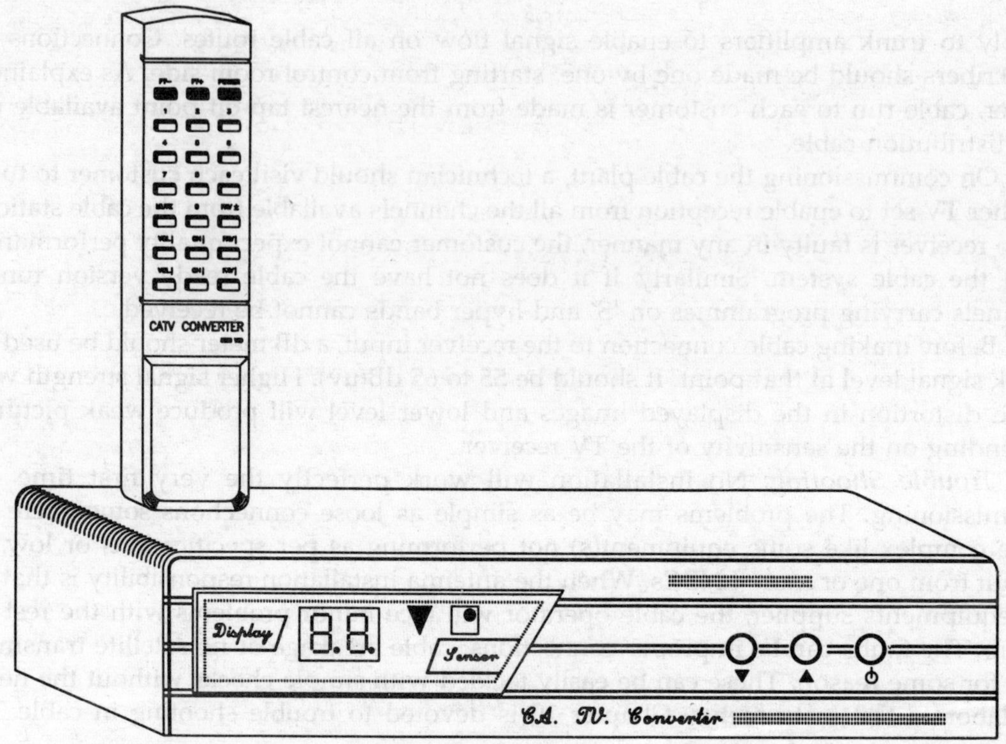

Fig. 17.22. Table Top CATV channel converter.

once, scans up by one channel. However, when the button is held in pressed state, it continuously scans channels upwards till the button is released. Similarly, the SCAN-DOWN button does the same but in the reverse direction. There is provision to sort out and store favourite channels and also to remove or add more channels to the favoured channel list. The chosen channel number is indicated on the display unit. In some models there is provision to set the time usually up to 99 minutes after which the TV receiver is turned off.

18

CABLE TELEVISION PERFORMANCE MEASUREMENTS

Introduction

At any television station, time duration of each constituent of the transmitted RF signal is continuously monitored to ensure that it conforms to the standards of chosen TV system. However, such tests are not necessary in CATV because time durations of various pulses and information signals are not likely to be disturbed during signal processing at the cable plant. But, the visual and aural quality of pictures and sound can suffer due to different types of distortions that can originate at any section of the cable plant. For this, various tests are carried out, both at the cable station and distribution network to make sure that distortion free signal is delivered at the input of all receiving TV sets. A variety of signal level meters, high frequency oscilloscopes, spectrum analyzers and power meters besides 'dB' meter and multimeter are available for carrying out various measurements. As necessary, this chapter is devoted to explaining the working principle of such meters and details of measurements that can be taken with them at various stages of signal flow from dish antenna down to customer locations.

18.1 Front-End Signal Levels

The down-link microwave signals that are picked up by dish antennas are extremely weak and it is not possible to carry out any signal analysis at this stage. Even if a field strength meter is available, knowing how strong is the received signal will not be of much consequence. In fact, output from the LNBC which is quite small in amplitude and still in the UHF range, can only be measured with a very sensitive signal level meter which is quite expensive. Therefore, the best course is that on acquiring the wanted satellite, the associated satellite receivers should be tuned for maximum deflection on their signal level meters. If the deflection for each channel is normal and a well defined picture appears on associated TV monitors, it is reasonable to assume that the received microwave signal is of good quality and has adequate strength.

18.2 Signal Analysis Meters

A signal analysis meter (SAM) or signal level meter (SLM) is basically a frequency tuned voltmeter by which signal level of specific frequencies can be measured. The output magnitudes are expressed on the meter scale directly in dBmV with reference to I mV *i.e.* 0 dBmV which equals 1mV RMS into 75 ohms and power equal to 0.0133 μw. As such, the difference between two measured signal levels will be in dB, for example it will be 12 dB between +5 dBmV and –7dBmV.

SAM Principle: The basic technique of measurement in any make of signal analysis meter (SAM) is essentially the same as shown in Fig. 18.1. The frequency component, the level of which is to be measured is selected out of the composite input signal by heterodyning it in a mixer with the fixed amplitude output of the local oscillator (LO). The mixer output feeds into a highly selective crystal or SAW filter unit having a bandwidth of a few cycles. Thus, it passes only a particular frequency signal and rejects all other sum and difference components that become available at the mixer output. The selected signal is then measured with a dB meter.

As an illustration of this process, if the filter is designed to select 10 MHz output and a signal at 1 MHz is to be measured out of the complex input signal, the LO is set at 9 MHz. The sum product which will be equal to 10 MHz gets selected and then measured. Since the amplitude of carrier (LO) is fixed the meter scale can be suitably calibrated to read the signal level. It may be noted that the dial of LO is marked directly in values of signal frequencies to be measured thus making it convenient to tune-in the wanted signal. Therefore, though the LO frequency will actually be 9 MHz, the dial will indicate 1 MHz.

As it is not possible to have one local oscillator circuit to give equal amplitude output over a wide frequency range, several oscillator circuits are provided with provision to switch from one band to the other to select the desired frequency.

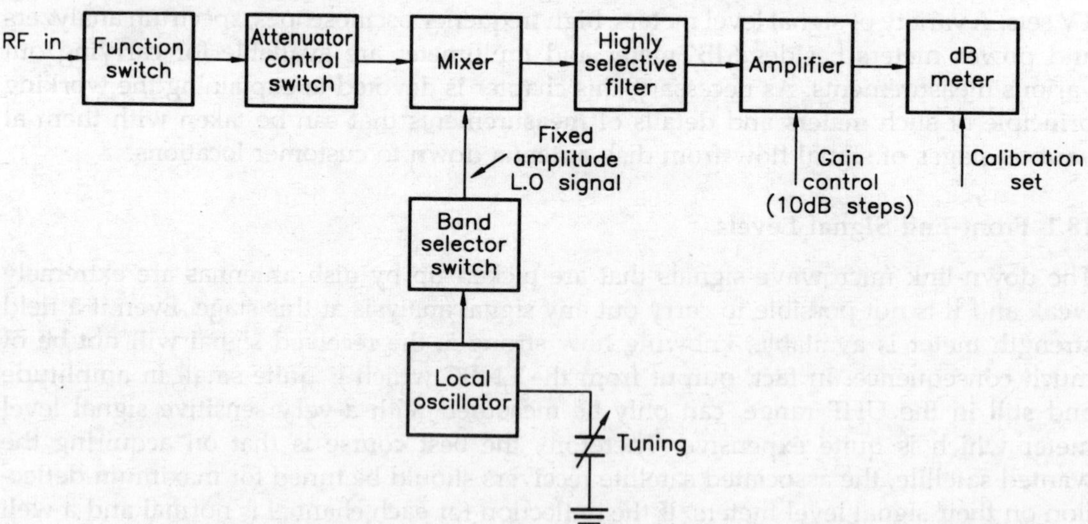

Fig. 18.1. Basic block schematic of a signal analysis meter.

Specifications and Operator Controls: A wide range of signal level meters are available where the least expensive measure signal levels only and the most expensive are digital and microprocessor controlled. These are coupled with spectrum analyzers where besides signal levels, various distortions, can also be measured and cable plant's channel and amplifier frequency responses seen and analyzed. In order to gain familiarity with various CATV measurements, SAM-1 signal analysis meter has been chosen. It is out of a range of analysis meters manufactured by WAVETEK corporation of USA. SAM-1' main specifications are as under:

	Measurement	*Range*	*Accuracy*
(i)	Frequency	4 to 300 MHz (4 to 450 MHz optional)	± 2 MHz
(ii)	Amplitude	– 40 to + 60 dBmV	± 0.5 dB
(iii)	Hum-modulation	0.5 to 5%	± 0.3%
(iv)	Voltage	5 to 100V. AC/DC (5 to 250V. optional)	± 10%

Operator Controls: Front panel view of SAM-1 is shown in Fig. 18.2. All the operator controls which the user can vary for different measurements are labelled on it. Since a spectrum analyzer can also be added to SAM-1 for time versus frequency displays, corresponding controls are also provided on the panel. Various controls and their functions are given in the chart that follows:

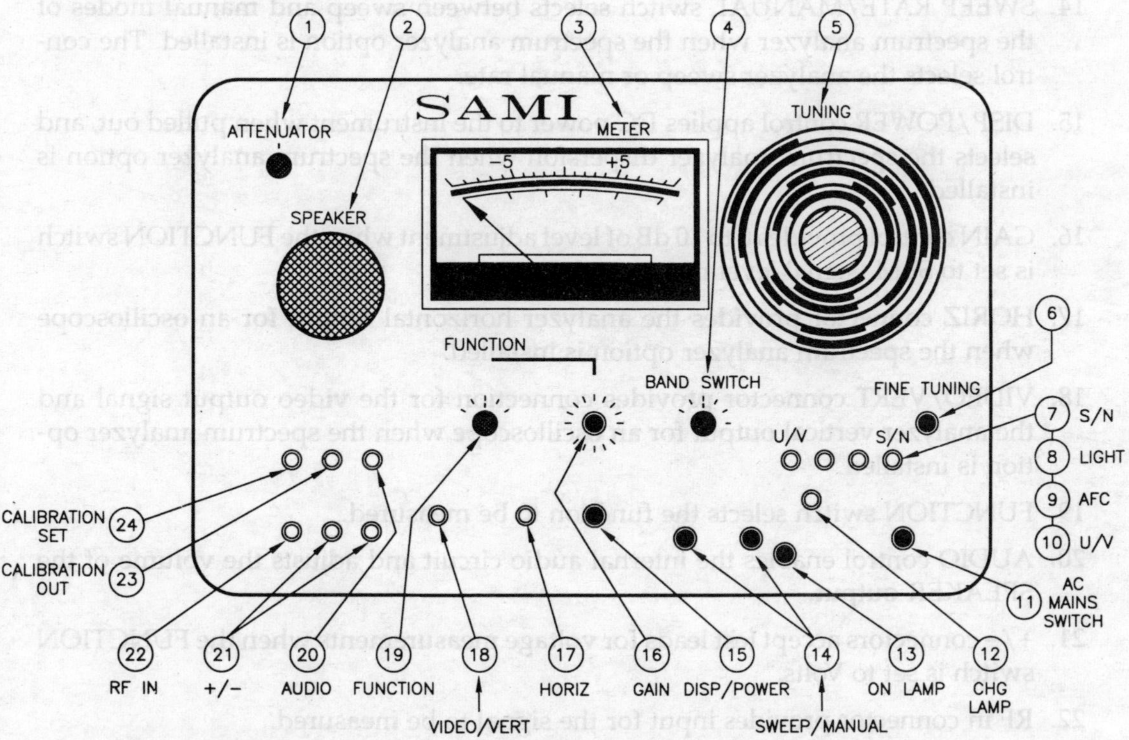

Fig. 18.2. Front panel view of analog signal analysis meter SAM-1.
(*Courtesy:* Wavetek Corporation, U.S.A.).

1. ATTENUATOR control selects the RF input level range and Meter scale to be read.
2. SPEAKER gives audio output of demodulated signal.
3. METER indicates value of function being measured.
4. BAND switch selects TUNING control frequency range.
5. TUNING control selects the frequency to be measured.
6. FINE TUNE control provides fine adjustment for TUNING control setting.
7. S/N switch enables internal compensation for 4 MHz bandwidth noise measurements.
8. LIGHT switch provides backlighting for the Meter.
9. AFC switch enables the internal AFC circuit.
10. U/V switch selects UHF or VHF frequency range when UHF option is installed.
11. AC line connector provides connection to AC mains via three-conductor line cord.
12. CHG lamp indicates that the internal battery is being charged.
13. ON lamp indicates operation of the instrument.
14. SWEEP RATE/MANUAL switch selects between sweep and manual modes of the spectrum analyzer when the spectrum analyzer option is installed. The control selects the analyzer sweep or manual rate.
15. DISP/POWER control applies DC power to the instrument when pulled out, and selects the spectrum analyzer dispersion when the spectrum analyzer option is installed.
16. GAIN control provides 0 to 10 dB of level adjustment when the FUNCTION switch is set to SLM (signal level meter mode).
17. HORIZ connector provides the analyzer horizontal output for an oscilloscope when the spectrum analyzer option is installed.
18. VIDEO/VERT connector provides connection for the video output signal and the analyzer vertical output for an oscilloscope when the spectrum analyzer option is installed.
19. FUNCTION switch selects the function to be measured.
20. AUDIO control enables the internal audio circuit and adjusts the volume of the SPEAKER output.
21. +/– connectors accept test leads for voltage measurements when the FUNCTION switch is set to Volts.
22. RF in connector provides input for the signal to be measured.
23. CALIBRATION OUT connector provides output from the internal calibrator to calibrate the METER for amplitude measurements.
24. CALIBRATION SET control calibrates the METER for amplitude measurements.

18.3 Carrier and Associated Signal Level Measurements

Detailed procedure for each measurement is given in the instruction manual that accompanies the signal level meter and should be followed. However, the procedure in general is essentially the same. A short coaxial cable with suitable connectors on either side is used to connect the output of unit under test to the RF input point of the meter.

(i) *Video Carrier Signal Level:* After initial calibration, the attenuator control switch is set at approximate level of attenuation and band selector switch moved to correspond with the signal frequency of the carrier. The main tuning dial is then brought to correspond with the signal frequency of the carrier as indicated on the dial. This results in the deflection of meter needle. It is peaked with fine tuning control and video signal amplitude read on the corresponding dBmV scale. The desired amplitude levels are ≥ 0 dBmV at subscriber to drop with ≤ 3dB difference between adjacent video carriers and ≤ 12 dB difference between greatest and least-amplitude video carriers in the system. Also there should be 15±2 dB difference between video and audio carriers on each channel.

(ii) *Audio Carrier Level:* The procedure for audio carrier level measurement is essentially the same as for video carrier level. The only difference is that the main tuning dial is set at the audio carrier of the same channel.

(iii) *Hum-modulation Measurement:* Hum-modulation is essentially low frequency amplitude modulation caused by the interaction of system signals with the ac supply source. The most common cause of hum modulation is a defective amplifier power supply. This can cause an ac ripple component on the amplifier's dc supply, thus causing hum modulation. Corroded connections on couplers, taps, splitters etc. can also be a source of hum-modulation, as corrosion acts as a mixer diode for the ac and system signals. Typically hum-modulation is present at frequencies below 250Hz.

Since TV signals are amplitude modulated, the hum-modulation test is performed on an unmodulated (pilot) carrier. Hum-modulation is generally not frequency dependent, therefore, the test must be performed at only one frequency.

Procedure: The meter circuitry for hum measurement is designed to filter out *i.e.* separate hum component from the system signal and heterodyne it with the pilot carrier for its measurement. With the system signal connected at RF 'IN' of the meter, various controls are set as far measuring signal level. The main tuning dial is then set at the chosen pilot carrier frequency and fine tuned to get maximum deflection. Finally, the function switch is set at 'HUM' and hum-modulation amplitude read on the corresponding scale. Maximum acceptable hum content should be less than 5%, typically less than 2% P.P. (peak-to-peak).

(iv) *Signal-to-Noise Ratio:* There are three main sources of noise:– system amplifiers, head-end processors and interference from external sources. Noise due to system amplifiers will generally be quite flat, varying only slightly across the entire spectrum. Headend noise is typically frequency selective, occurring only in the processor's output band. External interference can occur at any frequency, depending on the external source generating the noise.

The meter circuitry is designed to determine difference between video carrier signal level and noise floor. The procedure as outlined in the instrument manual enables this measurement. It should typically be > 36 dB.

(v) *AC/DC Voltage:* The SAM reads ac/dc voltage as in a common multimeter on different scales. The circuit details are given in section 18.7, while explaining the working of a VOM (Voltage-ohms-mill-amps) meter.

18.4 Measurements with Spectrum Analyzer

Cable systems up to 16 channels often distribute TV signals in close-by localities and hence, do not have too many amplifiers on trunk routes. As explained in a previous section, a signal level meter is enough to take necessary measurements at the cable station. Similarly, no special test instruments are considered necessary to monitor the performance of line and trunk amplifiers. A decibel meter is often used to measure signal levels at various points in the distribution network to make sure that all amplifiers and associated passive networks are operating satisfactorily.

However, in big cable systems, it becomes necessary to maintain stringent operating characteristics where the signal passes through several cable spans having many amplifiers before reaching customer receivers. Any excessive change in signal levels and frequency response of amplifiers can lead to distortion in the form of beats and noise on receiver screens. The amplifiers are, therefore, checked regularly and necessary adjustments made to maintain steady gain and correct frequency response. The ideal frequency response should be flat over the entire frequency range of the signal as shown in Fig. 18.3. However, after passing through many amplifiers the response may exhibit peaks and valleys as shown in Fig. 18.3(a). Similarly, the gain may fall-off, both at the lower and upper ends of the frequency bands [Fig. 18.3(b)]. Therefore, special techniques called sweep systems have been devised to enable fast checks at various trunk route stations. On detecting any fault, the response is corrected by adjusting amplifier gain and frequency response controls as provided in the amplifier circuitry. Similarly, passive components on distribution network are checked and replacements made where necessary.

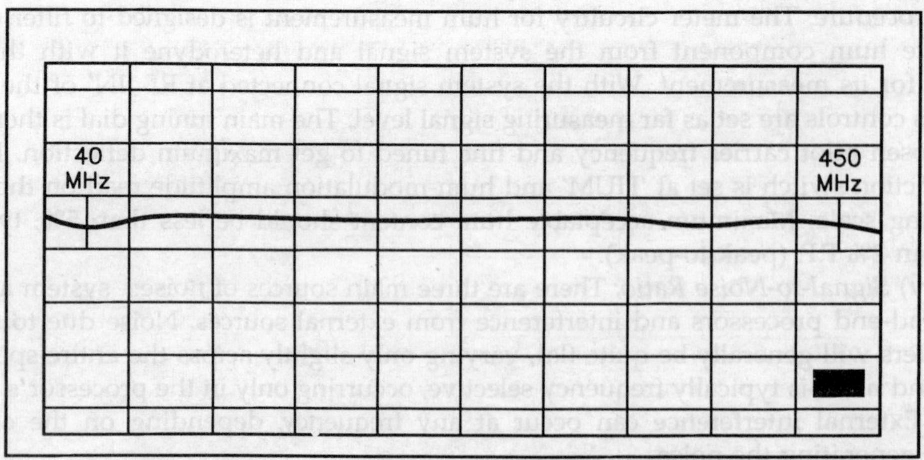

Fig. 18.3. Ideal frequency response of trunk line amplifiers.

Cable TV Sweep Systems: There are three types of sweep systems available for system sweeping. These are: (1) high level sweep, (2) low level sweep, and (3) no level sweep. However, all sweep systems measure amplitude and frequency response of the CATV plant and are essentially a combination of a signal generator called transmitter and a receiver along with a spectrum analyzer for display of these measurements. The transmitter generates necessary signals for various tests in a particular sequence and sends these to the receiver which is designed to look at the (check the) cable system sequentially to record and display the results. Before operating any sweep equipment, it is necessary to read its instruction manual and work with the system engineer or trained technician to learn how to properly operate it.

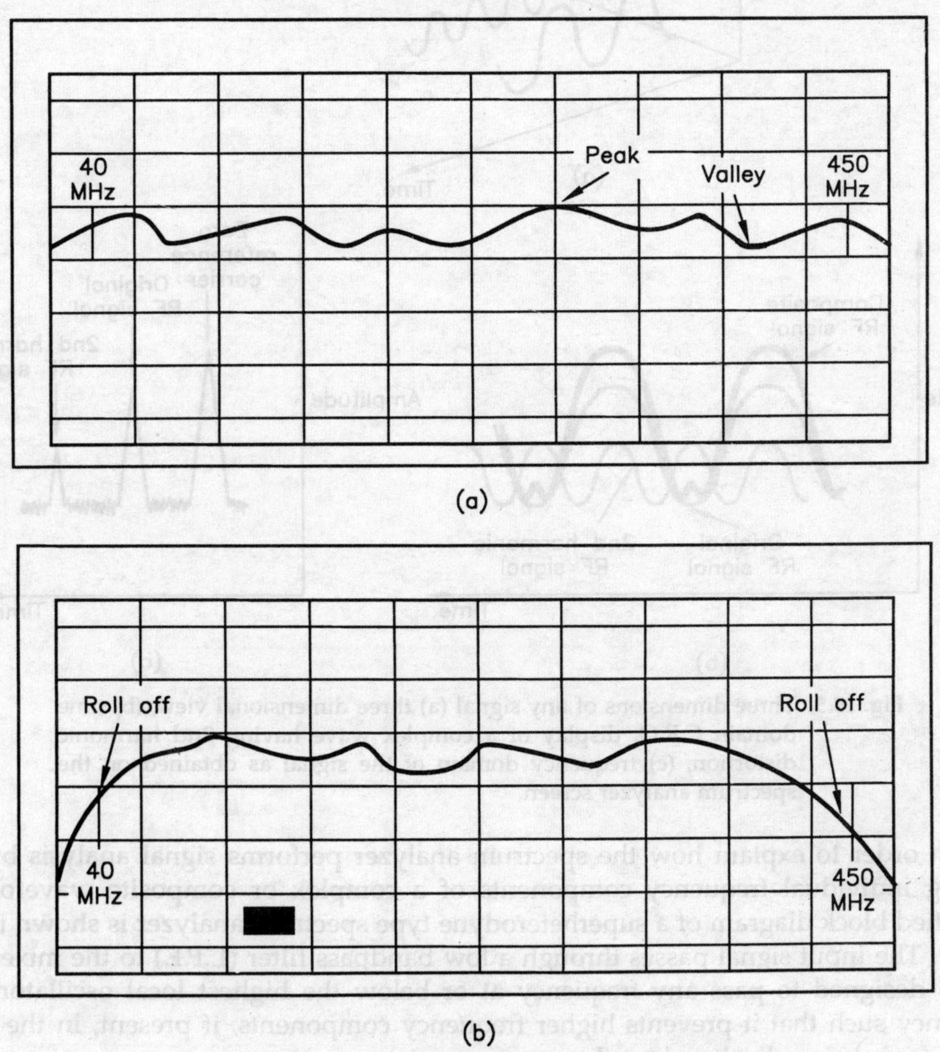

Fig. 18.4 (a) Distortion due to peaks and valleys.
(b) Poor roll-off frequency response.

Spectrum Analyzer: Since the results are finally displayed on the spectrum analyzer screen, it is necessary to know, how it functions to do so. Any RF signal or for that matter any signal is three-dimensional *i.e.* it consists of amplitude, frequency and time as illustrated in Fig. 18.5(a). An oscilloscope (CRO) looks at the time domain *i.e.* amplitude versus time [Fig. 18.5(b)] of the signal while a spectrum analyzer displays the frequency domain aspect *i.e.* amplitude and frequency as shown in Fig. 18.5(c).

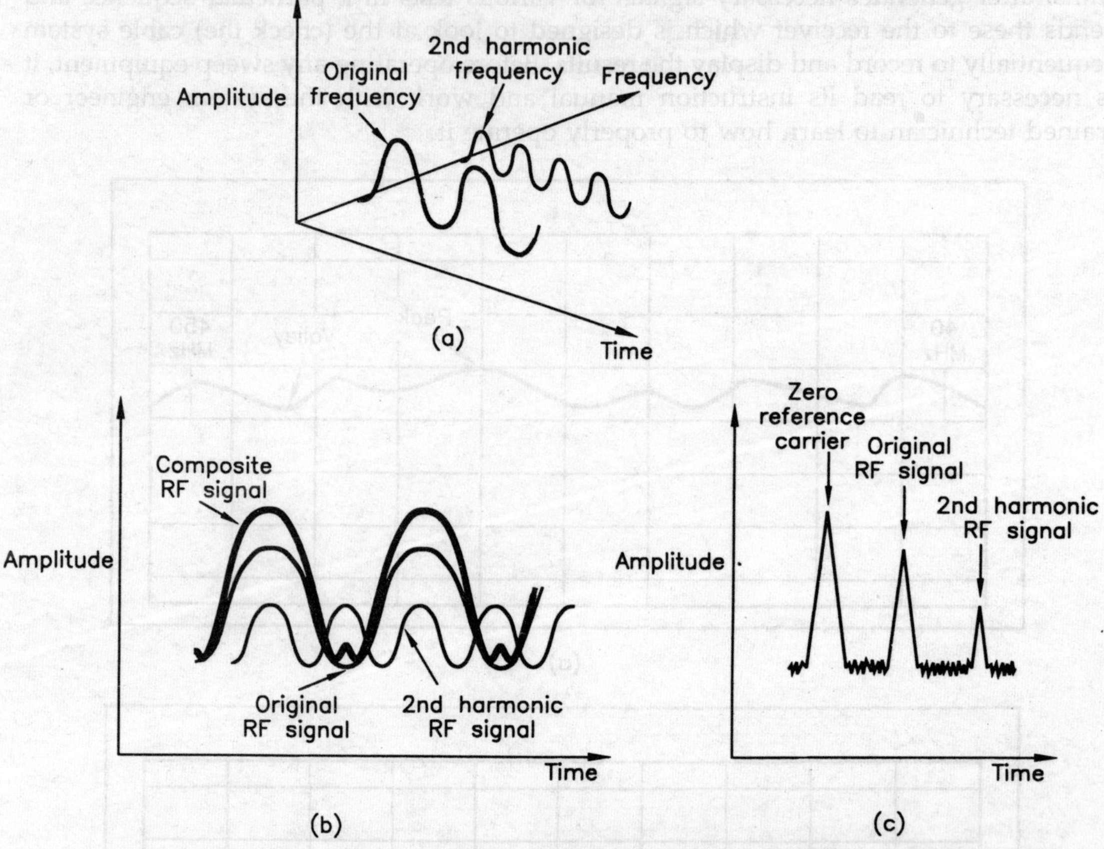

Fig. 18.5. Three dimensions of any signal (a) three dimensional view, (b) time domain C.R.O. display of a complex wave having 2nd harmonic distortion, (c) frequency domain of the signal as obtained on the spectrum analyzer screen.

In order to explain how the spectrum analyzer performs signal analysis by displaying individual frequency components of a complex or composite waveform, a simplified block diagram of a superheterodyne type spectrum analyzer is shown in Fig. 18.6(a). The input signal passes through a low bandpass filter (L.P.F.) to the mixer. The LPF is designed to pass any frequency at or below the highest local oscillator (LO) frequency such that it prevents higher frequency components, if present, in the input signal, from being displayed on the screen.

The upper part of the block diagram is seen to be essentially that of superheterodyne receiver, except that the LO is electronically swept back and forth between two frequency limits at a linear rate. At the same time, the illuminated spot on the face of cathode-ray tube (C.R.T.) is caused to move horizontally across the face of tube in synchronism such that deflection is a function of the frequency of local oscillator. The sweep voltage waveform is a ramp (sawtooth type) in nature to cause the spot to travel horizontally at a linear rate. The flyback time of the sweep (ramp) is essentially zero, meaning that the spot returns to the starting point in almost zero time duration.

As the amplitude of ramp rises, it causes the frequency of LO to rise linearly from minimum to some maximum value. The process repeats with each cycle of waveform with the result that the difference product at the output of mixer corresponds at each

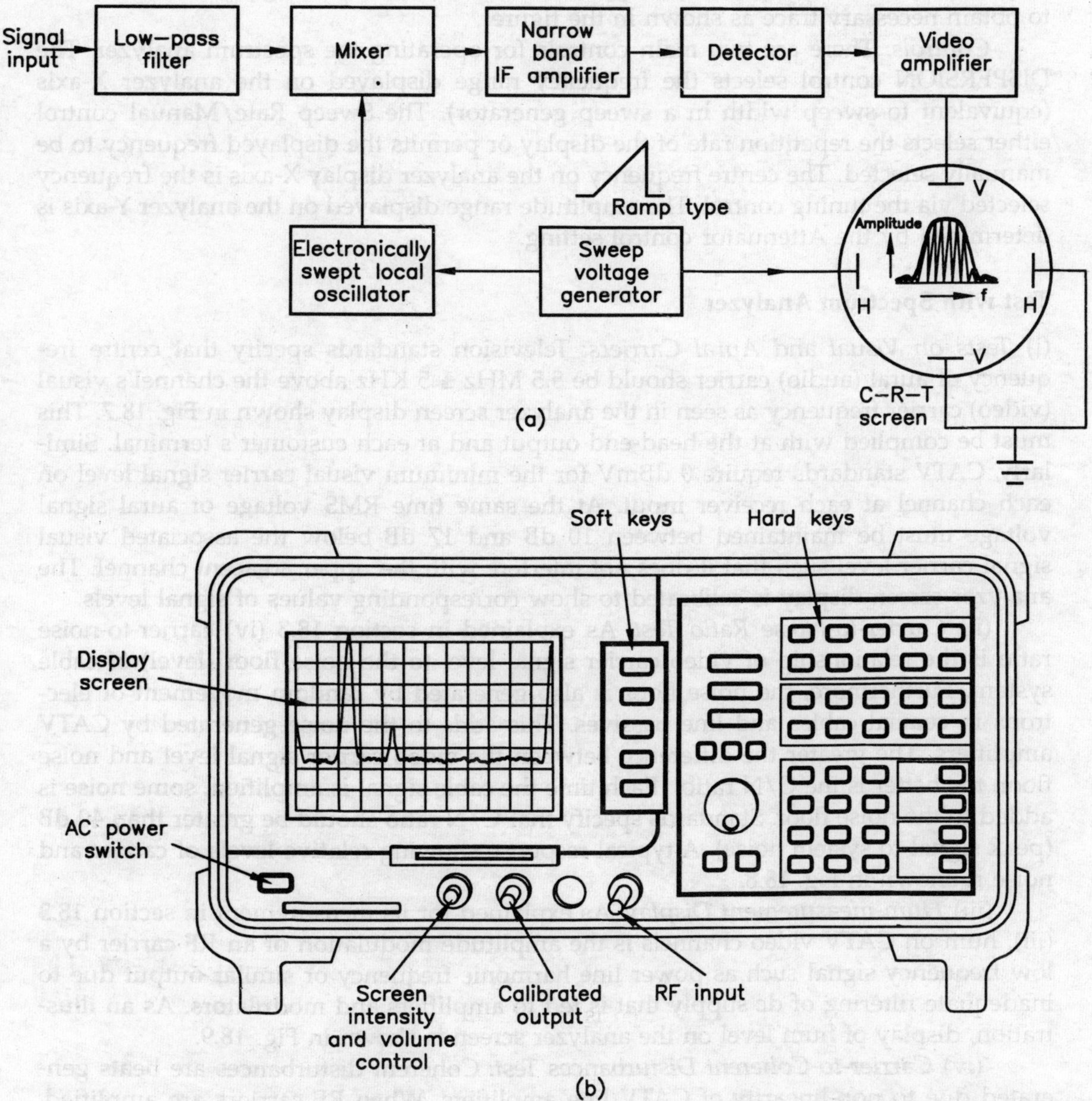

Fig. 18.6. (a) Block diagram of a superheterodyne type of spectrum analyzer, (b) Front panel view of a typical spectrum analyzer.

instant to a different frequency out of the wave-band of input cable signal. This means that the IF amplifier output at any instant carries information about the amplitude of each frequency component present in the input signal. The IF frequency amplifier is designed to have a very narrow band of I KHz and if possible, down to 100 Hz in order to provide highest resolution of the waveform under investigation. The amplifier output is detected by a diode type envelope detector. The detected output represents cable signal amplitude on a continuous basis in synchronism with the horizontal trace. It is amplified by a video amplifier and applied across vertical deflecting plates of the C.R.T. to obtain necessary trace as shown in the figure.

Controls: There are two main controls for operating the spectrum analyzer. The DISPERSION control selects the frequency range displayed on the analyzer X-axis (equivalent to sweep width in a sweep generator). The Sweep Rate/Manual control either selects the repetition rate of the display or permits the displayed frequency to be manually selected. The centre frequency on the analyzer display X-axis is the frequency selected via the tuning control. The amplitude range displayed on the analyzer Y-axis is determined by the Attenuator control setting.

Test with Spectrum Analyzer

(i) *Tests on Visual and Aural Carriers:* Television standards specify that centre frequency of aural (audio) carrier should be 5.5 MHz ± 5 KHz above the channel's visual (video) carrier frequency as seen in the analyzer screen display shown in Fig. 18.7. This must be complied with at the head-end output and at each customer's terminal. Similarly, CATV standards require 0 dBmV for the minimum visual carrier signal level on each channel at each receiver input. At the same time RMS voltage of aural signal voltage must be maintained between 10 dB and 17 dB below the associated visual signal carrier level such that it does not interfere with the upper adjacent channel. The analyzer screen display is calibrated to show corresponding values of signal levels.

(ii) *Carrier-to-Noise Ratio Test:* As explained in section 18.3 (iv) carrier-to-noise ratio is the relationship of video carrier signal level to the noise floor (level) of cable system. Furthermore, the noise floor is also generated by random movement of electrons in coaxial cables and line passives. This adds to the noise generated by CATV amplifiers. The greater the difference between the video carrier signal level and noise floor, the better is the C/N ratio. Each time the cable signal is amplified, some noise is added to the noise floor. Standards specify that C/N ratio should be greater than 40 dB (peak signal to system noise). A typical response showing relative levels of carrier and noise is shown in Fig. 18.8.

(iii) *Hum-measurement Display:* As explained for its measurement in section 18.3 (iii), hum on CATV video channels is the amplitude modulation of an RF carrier by a low frequency signal such as power line harmonic frequency or similar output due to inadequate filtering of dc supply that is fed to amplifiers and modulators. As an illustration, display of hum level on the analyzer screen is shown in Fig. 18.9.

(iv) *Carrier-to-Coherent Disturbances Test:* Coherent disturbances are beats generated due to non-linearity of CATV line amplifiers. When RF carriers are amplified, some unwanted sum and difference frequency products are generated due to mixing action amongst them. Similarly, video and audio carriers mixing gives rise to their

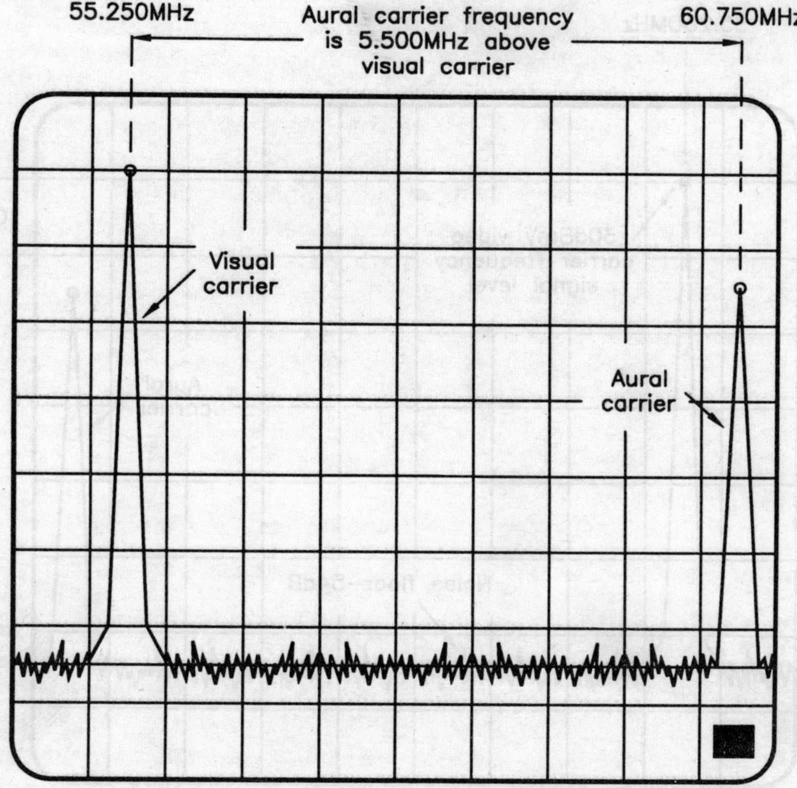

Fig. 18.7. Display on spectrum analyzer screen of aural and visual carrier levels and spacing between them.

harmonics and new frequency components called Beats. Such beat outputs can cause distortion in the reproduced picture. Standards specify that the ratio of video carrier signal level to the RMS amplitude of coherent disturbances that cause beat formations should not be less than about 45 dB. The possible location of such beats is shown in Fig. 18.10.

18.5 Signal Level Meters

A popular signal level meter is DEVISER-1870. Its basic circuitry is essentially the same as that of SAM-1 as described in section 18.2. As such, the functions of various keys provided on its front panel perform in a similar way to carry out various measurement. The names of various keys and parts located on the front panel are: Level Indicator Meter LCD display, Multifunction knob, RF/AC. input connector, Attenuator selector key, Selection reducing key, volume control second function key, Power Off key, Power 'on' key, Charger port and Charger LED indicator. As obvious, the measurement procedures and techniques are nearly the same at that of SAM-1. The manual that is supplied with the meter gives necessary details for all the available measurement.

Leader (953): Another wide band signal level meter preferred by big cable companies is Leader (model 953) manufactured by Leader Electronics Corp., JAPAN and

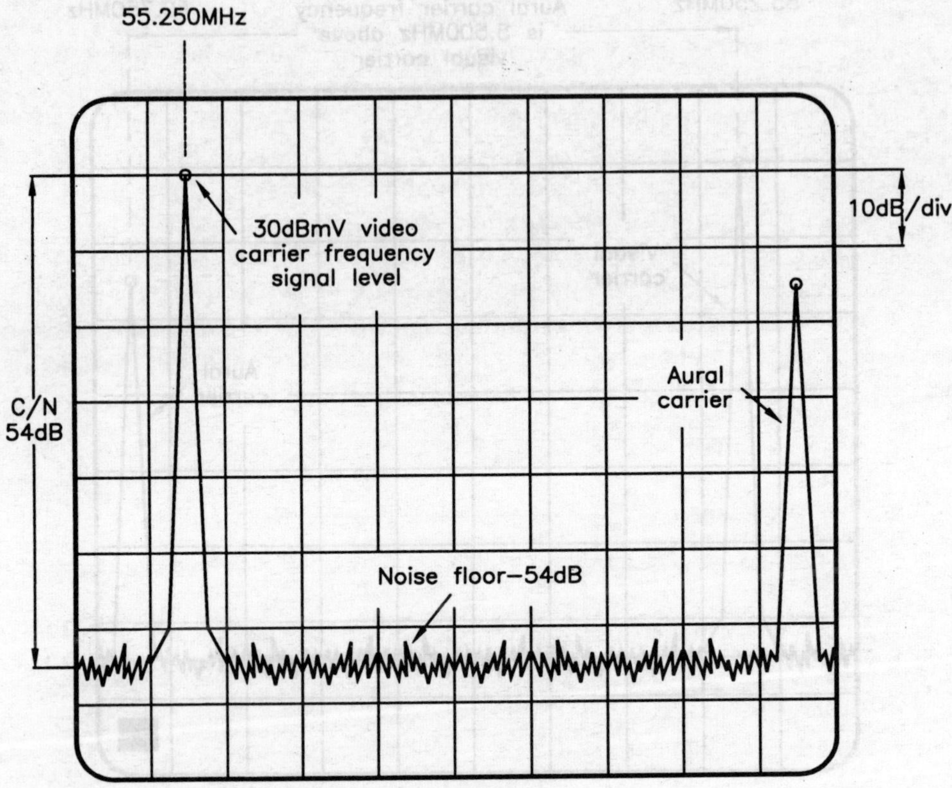

Fig. 18.8. Spectrum analyzer display of the unmodulated video and audio carrier signals and noise floor.

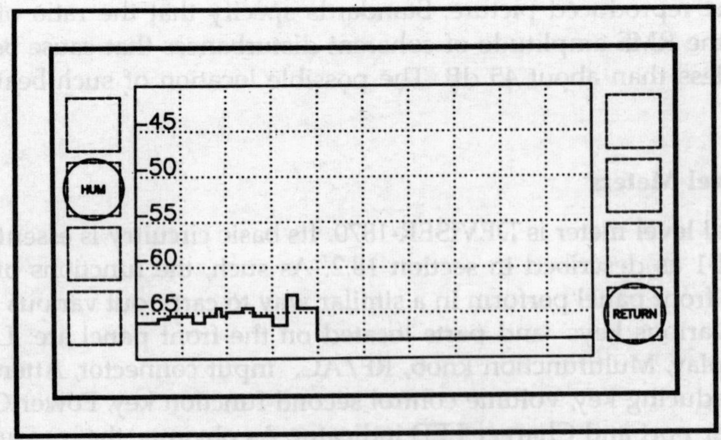

Fig. 18.9. Hum-measurement display on the spectrum analyzer screen.

conforms to PAL colour 625 line TV standards as adopted by India and many Asian countries. Its range is up to 1 GHz and covers VHF, UHF, and CATV channels including upstream lines. The LCD display spectrum analysis function easily detects interface components. Up to 128 channel levels can be displayed on the LCD with ± 2dB accuracy or better. This versatile model is ideal for periodic CATV measurements using data logger function. It can also be used for C/N ratio, hum, and stability measurements.

In the spectrum measurement mode, frequency synthesis ensures highly accurate frequency setting. It has built in speaker for sound monitor buzz check and beeper. Remote Control facility is also available via RS 232C interface.

The measurement techniques and procedures are nearly the same as explained in section 18.4. While the accompanying manual gives necessary instructions for all the measurements, it is advisable to ask the supplier to demonstrate meter's use for different functions.

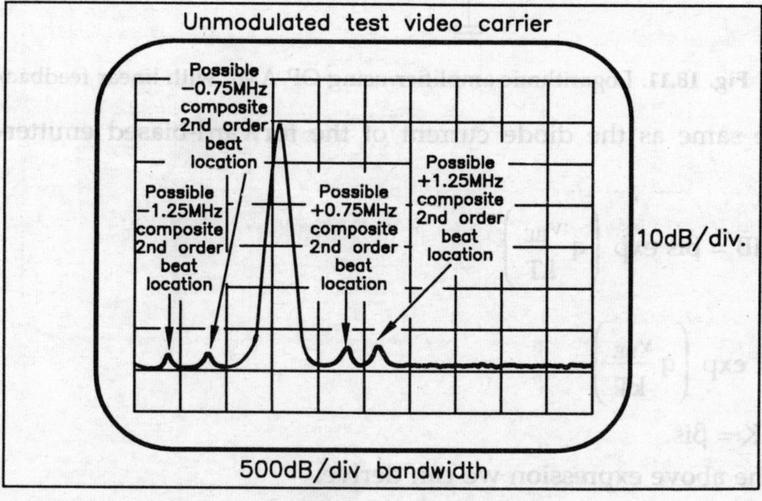

Fig. 18.10. Location of possible 2nd order beats as seen on the analyzer screen.

18.6 Decibel Meter

The term logarithmic is applied to an electronic voltmeter when the current or voltage produced in the indicating instrument by an applied voltage is proportional to the logarithmic of the applied voltage. Such a characteristic leads to a linear decibel scale for the indicating instruments and finds numerous applications in electronics. The reading on the meter scale is calibrated in decibels and hence, the instrument is called a dB voltmeter or simply dB meter.

While earlier dB meters employed several stages of variable - mu vacuum tubes to obtain logarithmic characteristics, the present-day meters use a diode in the feedback loop of an operational amplifier (OP-AMP). In practical meter circuits, a grounded base transistor is preferred for realising a log-amplifier because the exponential relation between current and voltage extends over a much wider voltage range. A typical circuit is shown in Fig. 18.11.

Since v_1 is effectively equal to zero, and the transistor base is grounded, $v_{CB} = 0$.

The collector current $i_c = \beta iB$ where the base current is given by $i_B = i_s \exp\left(q\frac{v_{BE}}{kT}\right)$

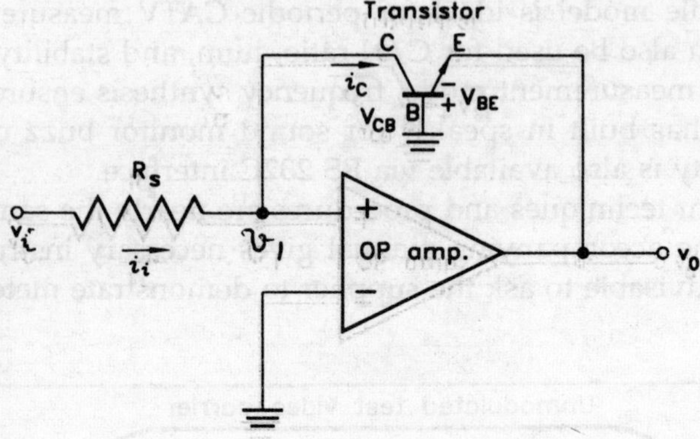

Fig. 18.11. Logarithmic amplifier using OP-AMP with linear feedback.

which is the same as the diode current of the forward-biased emitter-base junction. Hence

$$i_c = \beta iB = \beta is \exp\left(q\frac{v_{BE}}{kT}\right)$$

$$= K \exp\left(q\frac{v_{BE}}{kT}\right)$$

where K = βis.

From the above expression we can derive:

$$v_{BE} = \frac{kT}{q}\ln\left(\frac{ic}{K}\right)$$

Since emitter of the transistor is connected to output of OP-AMP, $v_0 = v_{EB} = -v_{BE}$

Also $i_c = i_i = v_i / R_S$ [see Fig. 18.11]

Hence $v_0 = -\frac{kT}{q}\ln\left(\frac{v_i}{KR_S}\right)$, meaning v_o is proportional to log of v_i. Thus with v_i · as the voltage to be measured, the output voltage being proportional to log v_i is greatly compressed and this enables the desired decibel (dB) scale. Based on this, the functional circuit of dB meter can be represented by the block diagram shown in Fig. 18.12(a). A typical analog dB meter is shown in Fig. 18.12(b) and Fig. 18.12(c) shows the front panel view of a digital version of such a dB signal level meter.

Decibel Table: The table given on page 291 when kept handy can be used for quickly computing dB values for corresponding powers and voltage ratios.

* The basics of OPAMP are described in section 14.10.

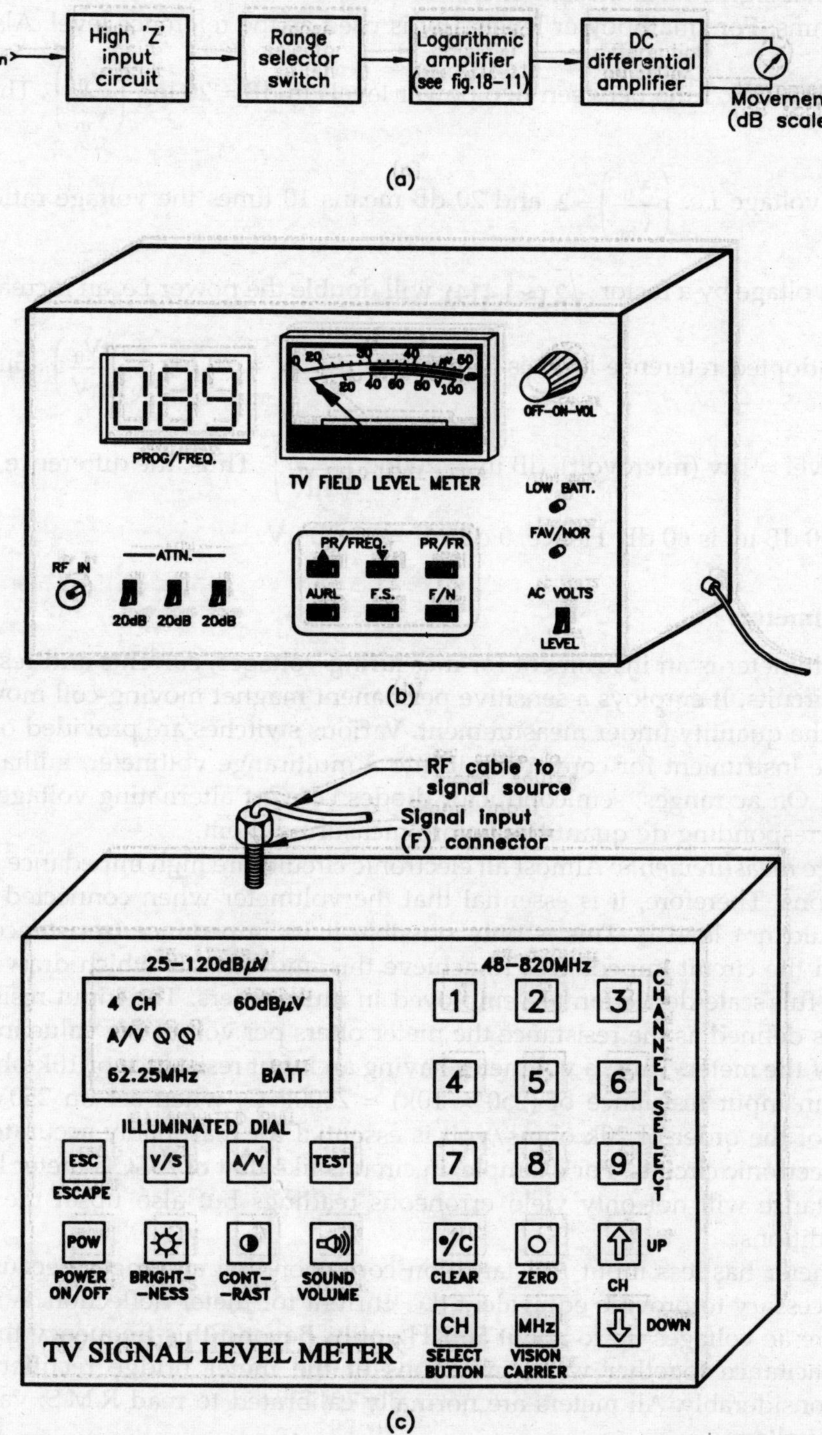

Fig. 18.12. (a) Simplified block diagram of dB meter, (b) A typical dB meter (analog), (c) Digital signal level dB meter.

Note that signal level measurements in cable TV are made with reference to 1 mV across 75 ohms. For small power levels 1μv is used as the reference level. Also remember that logarithmic ratio between two power levels in dB = 20 log $\left(\frac{V_0}{V_{in}}\right)$. Thus, 6 dB is double the voltage *i.e.* $\left(\frac{V_0}{V_{in}}\right) = 2$ and 20 dB means 10 times the voltage ratio. Besides, raising the voltage by a factor $\sqrt{2}$ (≈ 1.414) will double the power *i.e.* an increase of 3 dB. When the adopted reference level is 1 mV then dBmV = 20 log $\left(\frac{mV_0}{1mV}\right)$. Similarly for reference level = 1μv (microvolt), dB μv = 20 log $\left(\frac{\mu v_0}{1\mu v}\right)$. Thus, the difference between 0 dB mv and 0 dB μv is 60 dB. Hence, 0 dBmV = 60 dB μV.

18.7 Multimeters

A basic multimeter is an instrument for measuring voltages, currents and resistances in electronic circuits. It employs a sensitive permanent magnet moving-coil movement for indicating the quantity under measurement. Various switches are provided on the front panel of the instrument for converting it into a multirange voltmeter, milliammeter or ohm-meter. On ac ranges, semiconductor diodes convert alternating voltages and currents to corresponding dc quantities before measuring them.

Voltage measurements: Almost all electronic circuits are high impedance low power configurations. Therefore, it is essential that the voltmeter when connected across the circuit should not load it. This is only possible if its impedance (resistance) is much higher than the circuit impedance. To achieve this, movements which draw only 50 to 100 μA for full scale deflection are employed in multimeters. The input resistance of a voltmeter is defined as the resistance the meter offers per volt of the value indicated by the range of the meter. Thus, a voltmeter having an input resistance of 10k-ohm per volt will have an input resistance of (250 × 10k) = 2500k Ω, when set on 250V range. A sensitivity of the order of 20k ohms/volt is essential for reasonably accurate measurements in electronic circuits. For example, in circuits like that of AGC a meter having low input resistance will not only yield erroneous readings but also upset the prevailing circuit conditions.

The meter has less input resistance on corresponding ac ranges because more ac input is necessary to provide equivalent d.c. current for meter deflection. A multimeter can measure ac voltages up to about 50 KHz only. Beyond this frequency the lead and circuit capacitance together with limitations of the meter bridge rectifier affect the accuracy considerably. All meters are normally calibrated to read R.M.S. values of the applied ac voltage.

Current Measurement: The same movement is used for voltage and current

dBmV/Microvolt chart

0 dBmV = 1.0 mV across 75 ohms

dBmV	μV	dBmV	μV	dBmV	μV	dBmV	μV
–60	1.00	–25	56.23	+10	3.162	+45	177.800
–59	1.12	–24	63.10	+11	3.548	+46	199.500
–58	1.26	–23	70.79	+12	3.981	+47	223.900
–57	1.41	–22	79.43	+13	4.467	+48	251.200
–56	1.58	–21	89.13	+14	5.012	+49	281.800
–55	1.78	–20	100.0	+15	5.623	+50	316.200
–54	1.99	–19	112.2	+16	6.310	+51	354.800
–53	2.24	–18	125.9	+17	7.079	+52	398.100
–52	2.51	–17	141.3	+18	7.943	+53	446.700
–51	2.82	–16	158.5	+19	8.913	+54	501.200
–50	3.16	–15	177.8	+20	10.000	+55	562.300
–49	3.55	–14	199.5	+21	11.220	+56	631.000
–48	3.98	–13	223.9	+22	12.590	+57	707.900
–47	4.47	–12	251.2	+23	14.130	+58	793.300
–46	5.01	–11	281.8	+24	15.850	+59	891.300
–45	5.62	–10	316.2	+25	17.780	60	1,000.000
–44	6.31	– 9	354.8	+26	19.950	+61	1,122.000
–43	7.08	– 8	398.1	+27	22.390	+62	1,259.000
–42	7.94	– 7	446.7	+28	25.120	+63	1,413.000
–41	8.91	–6	501.2	+29	28.180	+64	1,585.000
–40	10.00	–5	562.3	+30	31.620	+65	1,788.000
–39	11.22	–4	631.0	+31	35.480	+66	1,995.000
–38	12.59	–3	707.9	+32	39.810	+67	2,239.000
–37	14.13	–2	794.3	+33	44.670	+68	2,512.000
–36	15.85	–1	891.3	+34	50.120	+69	2,818,000
–35	17.78	0	1.000	+35	56.230	+70	3,162.000
–34	19.95	+1	1.122	+36	63.100	+71	3,548.000
–33	22.39	+2	1.259	+37	70.790	+72	3,981.000
–32	25.12	+3	1.413	+38	79.430	+73	4,467.000
–31	28.18	+4	1.585	+39	89.130	+74	5,012.000
–30	31.62	+5	1.778	+40	100.000	+75	5,623.000
–29	35.48	+6	1.995	+41	112.200	+76	6,310.000
–28	39.81	+7	2.239	+42	125.900	+77	7,079.000
–27	44.67	+8	2.512	+43	141.300	+78	7,943.000
–26	50.12	+9	2.818	+44	158.500	+79	8,913.000
						+80	10,000.000

Note: In the above drawn chart the dot (.) in μV values needs clarification. 0 dBmV corresponds to 1.000 meaning 1000 μV i.e. 1 mV and 20 dBmV corresponds to 10.000 meaning 10000 μV i.e. 10 mV and so on for other values.

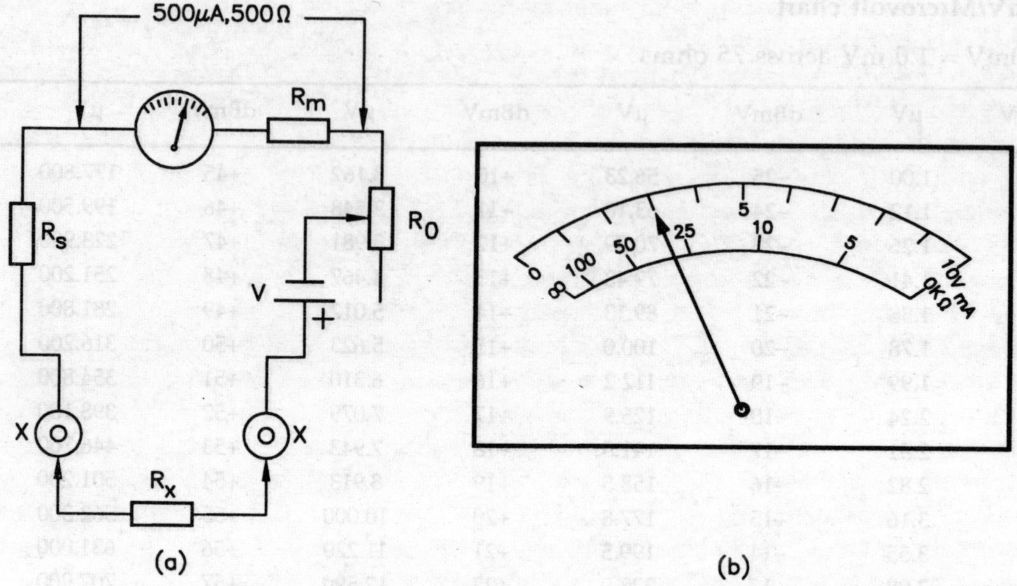

Fig. 18.13. Ohm-meter section of a multimeter (a) basic circuit, (b) scale.

measurements. Since the movement can carry only a very small current, shunts are provided to extend its range. The values of shunt resistance are so chosen that for the current indicated on the range switch, the meter draws full-scale deflection current. For ac current measurements, a bridge rectifier converts ac to dc before applying it to the same circuit configuration.

Resistance Measurement: The same movement (meter) is used for measuring resistance. For medium value resistances found in radio and television circuits, the test piece (R_x) is connected in series with the meter as shown in Fig. 18.13. A battery is required to supply current to energize the movement. The variable resistor R_o works as zero-adjust. It compensates for any change in battery voltage 'V' with ageing. The series resistor R_s limits the current to full scale deflection when test terminals X-X are shorted. If the external terminals are shorted to simulate ZERO ohm, the current in the circuit is maximum. If necessary, zero-adjust can be varied to obtain maximum deflection on the scale. Similarly, when test terminals 'X-X' are left open, an infinite resistance is simulated. The current through the meter is then zero and the pointer does not deflect at all. Thus, there is full scale deflection for zero-resistance and no deflection for infinite resistance. This explains why zero of the ohm-meter scale is at the right end and infinite (∞) at the left as shown in Fig. 18.13 (b). Since the circuit current is proportional to $1/R_x$, where R_x is the resistance to be measured, the ohms scale is highly non-linear as shown on the scale [Fig. 18.13b]. On account of excessive scale crowding near the zero current value, higher value resistances cannot be accurately measured unless a separate range is provided. This requires higher battery voltage in the meter.

A word of caution is necessary. An ohm-meter is never applied to an energized circuit because this may send excessive current through the movement and damage it. Another point to be noted is that, while making resistance measurements in a circuit, it should be ensured that there is no parallel branch across the component that is being measured. When in doubt, one terminal of the resistance to be measured should be disconnected from rest of the circuit.

Solid State Multimeters: Vacuum-tube voltmeters (VTVMs) that were in use for accurate measurement of voltage and current for over six decades have since been replaced with solid state multimeters. These are transistorized or IC version of the VTVM with the advantage of portability as no ac supply is required and the battery cells needed for its operation can be kept inside the instruments.

As shown in Fig. 18.14, a solid-state multimeter is basically a dc voltmeter. The range switch selects desired range and thereby attenuates the input voltage to a level that can be accommodated by the dc amplifier. The input stage of the amplifier consists of a Field-Effect Transistor (FET) because of its very high input impedance. The two transistors Q_1 or Q_2 that form the dc amplifier are so connected that any change in the FET drain current caused by the input voltage affects their biasing in such a way that the emitter current of Q_1 that flows into the meter movement varies linearity with variations in voltage under measurement. Thus deflection of meter movement is directly proportional to the input voltage. The scale is accordingly calibrated to indicate different voltage levels. As in other multimeters, with additional circuitry the meter can be used to measure ac voltage, current and resistance values.

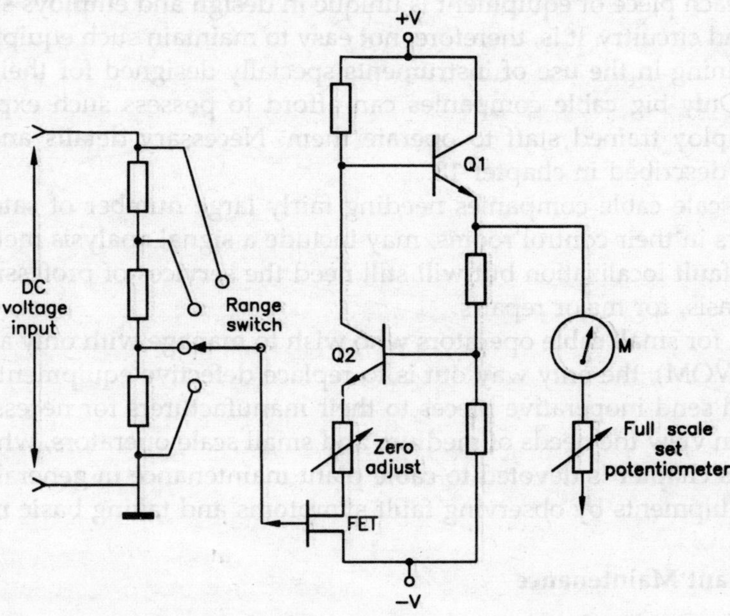

Fig. 18.14. Basic circuit of a solid state dc voltmeter.

Digital Multimeters: A digital multimeter (DMM) displays voltage, current or resistance measurements, as discrete numerals instead of a pointer deflection on a continuous scale as in an analog meter. Numerical readout is advantageous because it reduces parallax, human reading and interpolation errors. Features such as autoranging and automatic polarity selection further reduce measurement errors and any possible instrument damage through overload or reversed polarity.

19

CABLE SYSTEM MAINTENANCE AND FAULT ISOLATION

Introduction

Any TV cable system comprises of three main sections, each employing sophisticated equipments to receive, process and distribute television signals. Their operation encompasses a very wide frequency range that extends from microwaves to audio frequencies. As such, each piece of equipment is unique in design and employs special devices, components and circuitry. It is, therefore, not easy to maintain such equipments without specialised training in the use of instruments specially designed for their performance verifications. Only big cable companies can afford to possess such expensive instruments and employ trained staff to operate them. Necessary details and use of such instruments is described in chapter 18.

Medium scale cable companies needing fairly large number of satellite receivers and modulators in their control rooms, may include a signal analysis meter in the tools kit for ease of fault localization but will still need the services of professionals, may be on part-time basis, for major repairs.

However, for small cable operators who wish to manage with only a dB meter and a multimeter (VOM), the only way out is to replace defective equipments with known good units and send inoperative pieces to their manufacturers for necessary repairs.

Keeping in view the needs of medium and small scale operators, whose number is quite large, this chapter is devoted to cable plant maintenance in general and isolation of defective equipments by observing fault symptoms and taking basic measurements.

19.1 Cable Plant Maintenance

While proper installation, as detailed in chapter 17, is essential for trouble free operation of cable plants, their regular maintenance is equally important to prevent sudden failure of any section of the cable system. For this, the following precautions and procedures are recommended:

1. Front End Section

Maintenance needs of dish antennas, LNBCs and special cables that connect them are as under:

(i) *Antennas:* The antenna performance may be subject to seasonal problems. In high altitude areas in winter time, formation of ice and snow on the reflector can cause signal reduction. If the deposit is not too thick, a broom can be used to remove frozen precipitation. Feed horn should also be cleared of any deposit taking care that its window is not damaged. For excessive ice deposits anti-freeze or de-icing sprays can be used.

Another problem in certain locations is loss of signal due to new growth on nearby young trees partially blocking the line of sight to the satellite. Because of the involved time element, this problem is difficult to detect but it should be an early suspect during spring months.

At least every six months, all structural fastners of the antenna mount should be checked to make sure that these are in place and not obviously loosened. The foundation and mount should also be checked for any structural failure and if necessary suitably repaired. Also, the elevation and azimuth adjustment mechanisms should be greased occasionally to ensure their smooth and easy movement.

Whenever, any dish is readjusted to receive signals from another satellite, the azimuth—elevation mounting bolts should be tightened firmly after setting it at the desired look angle. Furthermore, antennas installed for receiving local terrestrial broadcasts, should be checked regularly for correct orientation, specially after strong winds and storms, during summer and monsoon periods.

(ii) *LNBCs:* Low noise block converters, if installed correctly, do not need any maintenance except that these are kept well protected with rain covers to ensure that moisture does not enter the assembly. The active devices used in the LNBC circuitry are highly voltage sensitive and therefore regular check of dc supply to converter units must be made to keep it at the set value, usually + 15 volts. The dc supply to bullet amplifiers should also be checked regularly.

(iii) *Cables:* The cable lengths used in the antenna unit must stay moisture free. Should moisture become evident at any time, the entire cable length should be replaced with a known equivalent length of same quality.

2. *Head-End Equipments*

Satellite receivers, modulators, signal processors, VCRs and combiner units that are installed in the control room are designed to operate trouble free continuously. Even then, it is desirable to check periodically the presence and quality of programmes on all the cable channels. This will enable detection of any deterioration in picture and or sound quality and enable immediate remedial action and thus prevent failure of any channel.

A constant check on the functioning of mains voltage stabilizer is also necessary to maintain constant power supply to the equipments. In case an Inverter Unit is installed as an alternative source of ac supply, its functioning and the condition of batteries should be regularly checked to ensure smooth changeover when mains supply fails.

3. *Cable Network*

Cables are spread in the open over a large area and therefore need maximum attention against dislocations caused by power lines and telephone maintenance staff, construction workers, thefts and strong winds. As stated in chapter 17 safer routes should be chosen while laying cables and firm connections made at all distribution points. In any case, routine maintenance should include the following checks:

(i) The cable plant technician should take at least one round daily of the network area to see that there are no dislocations or breaks in cable lengths and at distribution points on different routes.

(ii) Voltage supply to trunk and line amplifiers should be checked often to see that it is at the required value.

(iii) RF signal levels should be measured at all tap-off points periodically and compared with corresponding readings noted at the time of commissioning the cable station.

(iv) If on providing additional connections on a particular route to new subscribers, the signal level falls, an additional amplifier should be provided or gain of the existing amplifier increased if there is provision to do so, to restore signal amplitude to its original level.

19.2 Fault Tracing at Cable Stations

From dish antenna down to subscriber, the link is by coaxial cables. It is, therefore, necessary to use good quality cables and weather-proof connections to prevent any discontinuity in signal flow across the cable system. That is why, both ends of cable lengths that connect two equipments are carefully soldered to connector pairs to prevent any open circuit or short of signal with cable sheath which is grounded. As such, faults in connecting cable lengths are rare, more so in the control room which is well protected.

Therefore, while describing fault finding procedure at the cable station set-up, cable lengths will be the last suspect. Fault finding procedure described next is based on observations made at the signal combiner unit output on the monitor (TV Receiver) screen.

Condition	Corrective Action
1. No output from combiner *i.e.* Monitor Receiver Screen is blank for all cable channels.	Since all the cable channels inclusive of VCR and local broadcasts cannot fail at the same time, the fault is in the combiner unit. It can be due to a break in the combiner windings, interconnections or failure of amplifier if the combiner is of ACTIVE type. The failure can also be due to a fault in the distribution amplifier if this forms part of the combiner unit. Use common fault finding techniques to locate the fault and rectify it. If you do not succeed, replace the combiner with a known good unit and send the defective piece to its manufacturer for repairs.

Condition	Corrective Action
2. No signal (pictures) from all satellite channels received through one particular dish antenna set-up.	Failure of satellite transponders is rare. Therefore, if it is transmission time for all the missing channels, the fault is in the Front-End *i.e.* antenna section. To locate the fault proceed as under and stop where normal reception gets restored: (i) Check dc supply to LNBC and bullet amplifier if in use. If not present, measure it at the satellite receiver from where it is received. If necessary, locate fault in the path between receiver's dc output point and LNBC and rectify it. If dc supply is not present at the satellite receiver itself, troubleshoot power supply section of the receiver to restore it. (ii) Check orientation of the dish to make sure that it has not moved away from the set look angle. Such a fault is common in areas where strong winds blow. If necessary, do the needful to obtain normal operation on all the channels. (iii) If the problem still persists, check feed unit to the LNBC to make sure that is in the correct polarization position. Also check for any obstruction on the face of feed unit. Sometimes birds tend to make their nests at the mouth of feed horns. If found so, clear the mess and carefully clean the area. (iv) Failure of bullet amplifier or an open connection at its input or output ends may be causing open circuit in the signal path. This can happen since bullet amplifiers are often left suspended without proper support along the path of RF cable. If such a fault is not there, by-pass the bullet amplifier with a small length of RF cable making proper connections at both the ends. If on doing so, normal operation is restored, replace the bullet amplifier. (v) If necessary, next check the Power Divider by replacing it with a known good piece. If the fault gets cleared the power divider is obviously defective.

Condition	Corrective Action
	(vi) Also check RF cable connections at the LNBC output socket and if found loose tighten them. It may also be noted that the LNBC connector is sealed with water sealing compound to prevent seaping-in of water during rains. Any crack or leakage in the seal will cause water to reach the LNBC unit and thus damage it. In fact, this is a common cause that makes LNBCs defective. Therefore, ensure by regular inspection that sealing of connector is proper and where necessary do the needful.
	(vii) Next in tracing the fault, check continuity with a multimeter (VOM) set as 'OHM-METER' of the RF cable that connects LNBC output to the bullet amplifier and, or power divider after disconnecting it at both ends. In case of open circuit, replace the defective length of RF cable.
	(viii) If trouble still persists, as a last resort, replace the LNBC unit. This should enable reception from all the channels.
3. No signal (reception) from any one channel out of several received through a common dish antenna.	(i) Make sure from the programme chart that this is the transmission time for the missing channel.
	(ii) If the pilot (indicator) light of the associated satellite receiver is glowing and there is no deflection on the tuning meter, try feeding incoming IF signal to the receiver from another port of the power divider. If this enables reception, locate fault in power divider and replace it, if necessary.
	(iii) If deflection on the tuning meter is low than normal, try retuning the receiver. If this does not help, use another satellite receiver to make sure that the existing one is good or defective.
	(iv) In case the receiver is not defective, try changing the associated modulator unit. If found inoperative replace it.
4. No reception from one local TV broadcast.	(i) Check continuity of the twin-lead cable that connects the antenna to signal processor/

Condition	Corrective Action
	RF amplifier. Also make sure that orientation of antenna and connections of cable to it are correct. Where necessary, do the needful. (ii) If reception is not restored, connect the receive end of the twin-lead cable to the input of a TV receiver. If programme is received, the signal processor/RF amplifier is defective. Service or replace it. (iii) If the channel programme is still not available at the output of combiner, try replacing the modulator of this channel. (iv) Make sure that it is transmission time of this channel and the local transmitter has not become inoperative, though this is rare.
5. No output from VCR channel.	(i) Feed a TV receiver from the modulated RF output socket of the VCR. If normal reception is obtained, the fault is in the associated signal processor or modulator. Proceed step by step and replace the defective unit. (ii) If still there is no reception on the TV receiver, proceed to locate fault in the VCR and if necessary, send it for servicing.
6. Weak colours on pictures of one particular satellite channel.	(i) Retune corresponding satellite receiver. If the problem persists try increasing video level of the associated modulator. If trouble still persists, change the modulator. (ii) If the channel happens to be an edge channel of the transponder, try re-setting the look angle accurately.
7. Weak sound on one particular satellite channel.	(i) Check and if necessary, readjust the sound subcarrier frequency and its bandwidth on the satellite receiver. (ii) Re-adjust sound carrier frequency on the associated modulator. (iii) As a lost measure, try replacing the satellite receiver and modulator one after the other.

Condition	Corrective Action
8. Weak picture and sound from one channel.	(i) Retune the satellite receiver both for video and sound at the given carrier frequencies. (ii) If the tuning meter still shows low signal level, tune this channel on another receiver. If normal output is obtained, replace the existing receiver. (iii) If the receiver checks normal, replace associated modulator. But before doing so, check dc supply to the existing modulator.
9. Hum in output of any one channel.	(i) Check dc supply to the modulator for effective filtering. This can be easily done by connecting a 16/32 μF electrolytic capacitor across the supply points. If hum disappears, connect the capacitor permanently across the supply point, and remove the faulty capacitor. (ii) Check if the sheath of input cable to the modulator is properly earthed. (iii) If hum still persists, replace the modulator.
10. Intermittent reception on all channels coming from one dish antenna unit.	(i) Check for loose cable connections in the signal path from LNBC onwards and tighten if found loose on any link cable. (ii) Check step-by-step for a faulty cable length and replace it if necessary. (iii) Replace LNBC if the fault still persists. (iv) Check for loose nuts and bolts on the dish unit as wind pressure may be forcing the dish antenna to shake causing intermittent reception.
11. Black and white spikes on the picture and noise in sound output on one particular channel.	(i) Check tuning of this channel's receiver and if necessary, fine tune it. (ii) In case the fault persists, try replacing the satellite receiver.

Condition	Corrective Action
12. Spikes on picture and noise in sound output on all channels received from one particular satellite.	This can be due to weak signal reception from the satellite. To rectify the fault proceed as follows: (i) Check for setting of correct look angle of the receiving dish antenna. (ii) Look for any obstruction in the signal path to the dish antenna, LNBC and feed horn. (iii) Fine tune the associated satellite receivers. (iv) If the trouble still persists it can be due to insufficient dish diameter chosen for the location of cable station.

19.3 Fault Finding in Cable Network

The procedure described below for locating fault(s) in cable network is based on the assumption that programmes on all the cable channels are available at the output of combining network at the control room.

1. *Signal not present on all cable routes*

Distribution amplifier that amplifies combiner network output signal is inoperative. Check its power supply, input-output connections and if necessary, replace it.

2. *Signal not available to customers on one distribution route*

(i) Check for open circuit at the splitter that separates the route from the main cable.

(ii) Check operation of first trunk amplifier on the distribution route.

(iii) Check continuity of the cable that connects splitter to the lst trunk amplifier.

3. *Signal not available to subscribers after a particular line amplifier*

(i) Check operation of this line amplifier and its input-output connections.

(ii) Check for any break in ac supply to the line amplifier.

4. *No Signal to only one Customer*

(i) Check connection at the input of TV receiver.

(ii) Check connections and network continuity at the tap-off unit.

(iii) Look for any open circuit in the cable that connects the customer TV from the tap-off point.

5. ***Weak signal to all TV sets connected on one route***

(i) Check gain of the trunk amplifier, if a spare is available, try replacing it with the already connected amplifier.

(ii) Check for loose or bad connections at input-output points of the amplifier.

(iii) With an ohmmeter, check for partial short in the trunk line cable.

6. ***Weak signal to any one customer's TV set***

(i) Check if the TV receiver operates normal with input taken from the roof-top antenna.

(ii) Check tap-off network unit and if necessary replace it.

(iii) Check for partial short in the connecting cable.

7. ***Ghost images on receivers on a particular route***

(i) Line terminating resistance may be open or disconnected. Do the needful.

(ii) Impedance mis-match at the output of line amplifier. Restore correct impedance match.

(iii) Impedance mismatch of tap-off network with line impedance. If so, do the needful.

8. Bars on pictures of some channels

(i) Check channel tunings on the TV receiver. Accurate fine tuning may eliminate the bars.

(ii) Line cable having poor shielding can pick-up signal(s) of local broadcast(s) and the resulting beat can cause bars. Keep the CATV cable away from the local broadcast receiving antenna if located close by.

9. ***Intermittent reception on one or more receivers***

(i) Proceed step by step to trace a loose connection on the line or lines that form part of the cable network where reception is intermittent.

(ii) Check operation of the line amplifier(s).

(iii) Check performance of TV receiver(s).

10. ***Interference on pictures of some customer receivers***

As explained in section 17.11, there is nothing much that can be done to eliminate interference from ignition systems of passing vehicles, welding sets and local broadcasts but interference originating in the customer's house can perhaps be reduced by connecting spark quenching networks across brushes of household pump motors, electronic fan regulators or defective tube-light starters. Radiations from such appliances being strong are picked up directly by the TV receiver.

11. ***Overlapping of pictures of two different channels***

(i) If the unwanted channel happens to be the adjacent channel, try reducing its signal strength if found quite high.

(ii) If the interfering channel happens to be a few channels away, then the overlapping could be due to spurious signal being generated by the modulator of the unwanted channel. Identify the modulator and service/replace it.

12. ***Deterioration in picture quality of higher channels***

This could be due to poor frequency response of distribution and trunk amplifiers towards higher end of their bandwidths. The picture quality will also suffer because of losses over lengthy distribution cables which increase with rise in frequency. Adjustment of the slope control of amplifiers can be tried to improve the picture quality.

13. ***Netting on the picture on any route of the distribution network***

This could be due to unwanted oscillations in a particular amplifier on the route. Identify this and service/replace it.

19.4 Servicing Of CATV Equipments

Servicing of CATV equipments, as already explained is quite involved and needs full familiarity with voltages and their waveshapes that are obtained at various points in the circuit of the equipment when it is operating normally. It is also important to be fully conversant with the functions and the effect of manipulation of all controls and adjustments provided in the equipment.

Equipment trouble shooting is a skill in itself and does not come automatically by remembering circuits but needs long experience and there are no short-cuts to it. Nevertheless, a technician with basic knowledge of circuit blocks in any equipment can locate simple faults and thus restore the unit to normal operation. The first step towards finding the cause of failure of any equipment is usual inspection of components, transistors and ICs mounted on various printed circuit boards and modules on its chassis. Such an inspection sometimes leads quickly to the defective device or part of the unit. For example, burnt or charred resistors can often be spotted by visual observations. Similarly, overheated transformers, oil or wax filled capacitors and coils can be located by a peculiar smell caused by overheating and shorting. In addition, a dis-soldered wire of a coil or component, broken lead of a transistor and other similar faults can be found out by a preliminary visual inspection.

When any such fault is located, it is good practice to immediately ensure with an ohmmeter (multimeter set as ohmmeter) that there is no partial or full short in any section of the power supply. If a faulty capacitor is found to be the cause of excessive current drain, it should be replaced by a new one having the same capacity and power (voltage) rating. Similarly, broken leads, if any, should be carefully soldered at their correct locations.

To ensure that the short has been removed, another quick check of the dc voltage supply should be carried out before reconnecting the equipment to the mains supply or restoring dc supply to it. A short in some section of the circuit (load) receiving dc supply, will cause all voltages to read low and often cause the fuse to blow out or the circuit breaker, if provided, to trip. Excessive current drain due to shorts in the load can be traced by opening connections at different branching points of dc supply line and monitoring dc voltages with a voltmeter (multimeter set as voltmeter). When there is a partial short, the trouble is most likely at the point of lowest voltage. Then an ohmmeter can be used to locate the defective component in the supply line.

Even otherwise, absence of any short in the dc supply line is indicated by a momentary deflection of the ohmmeter needle that is caused by a surge of charging current of filter capacitors when the meter is connected across any positive point and chassis ground.

In case the equipment does not perform after restoring normal dc supply or when no fault is found in the power supply section and no visible indication is available, a voltmeter set at low range can be used to measure voltages between emitter and base of various transistors to see if these are forward biased. If such a test indicates that any transistor is inoperative, the same should be replaced to put the circuit to normal operation. If still the equipment fails to operate, it should be sent to the manufacturer for thorough checking and servicing.

Module Isolation Method: In Western countries where sub-assemblies (modules) and PCBs of various equipments are easily available, the servicing technique is directed towards isolation of the defective module and then replacing it. As an illustration, the module isolation guide for a particular satellite receiver is as under:

Condition	Corrective Action
1. Power indicator not illuminated.	(i) Check ac input fuse, line cord, line voltage, and external circuit breaker.
	(ii) After the items in step (i), have been checked carefully, check plus and minus 15 V dc with VOM at test points on the rear panel of the receiver.
	(a) If dc voltages are not present, check for a power supply short. If no short is found, replace the power supply module.
2. VOM indicates High or Low dc Supply.	If dc supply is more or less by 10% of the normal (15 V) value, replace the power supply module.
3. Video output distorted.	(i) Check level and modulation settings of external equipments on output side of receiver.
	(ii) Check that other video (if unused) is open or terminated in ≤ 75 ohms.
	(iii) Check composite base band (BB) output for:

Condition	Corrective Action
	(a) If C/N (carrier/noise) reading is 10-20 dB, replace video demodulator and recheck video output. (b) If C/N reading is less than 10dB, check IF output level of the LNBC. If low than normal, LNBC failure is probable. If original level is normal, receiver's IF amplifier failure is probable.
4. Video output not present but audio output present.	(i) Replace video clamp (dc restorer) board and re-check video output. If video out put still not present, re-insert original video clamp and replace video demodulator module.
5. Neither video or audio is present.	(i) Check C/N value. If C/N is 10 to 20 dB, check external equipment at the receiver's output. (ii) If C/N output is less than 10 dB, check output of IF amplifier. If should be on 70 MHz at –5 dBm ± 0.5 dB.
6. Audio not acceptable, video acceptable.	(i) Replace audio demodulator. (ii) Recheck audio output. If output is still unacceptable, reinsert original audio demodulator module and replace video demodulator module.
7. Option (second) audio unacceptable, video normal, (first) audio acceptable.	Replace optional (2nd) audio printed-wiring board.

The troubleshooting procedure given above is for a particular make and model of satellite receiver. The servicing engineer can find similar guide charts in the service manuals of the receivers to be repaired. The same is true about other major equipment like LNBCs, modulators, signal processors and VCRs.

19.5 Customer Complaints

Most complaints that can come from a customer (subscriber) have already been dealt with in previous sections under different heads. Additional complaints which are specific to each customer are dealt with in this section.

It is quite possible that reception problem(s) that the subscriber experiences pertain to faults in his/her television receiver. As such, the technician should first check the performance of the TV receiver and if faults like excessive rolling of picture, insufficient width or height of picture, distorted sound etc., are noticed, the customer should be advised to get the receiver serviced. Other faults which can arise because of cable connection, their remedies are as follows:

1. *Over loaded (distorted) pictures on all channels:* It could be due to excessive signal amplitude at the input of TV receiver. While an input signal level at 60 to 70 dB μV is considered satisfactory, it may be higher for a TV receiver of high sensitivity. The technician should measure the input signal level and insert an attenuator, if necessary, to reduce its magnitude to obtain pictures of normal resolution.

2. *Weak pictures on all channels:* This should need increasing the cable signal amplitude being fed to the TV receiver. Necessary adjustment can be made at the TAP-OFF point or by adjusting gain of the line amplifier.

3. *Picture not steady on one particular channel:* This could be misturning on the tuner. It may be checked and fine tuned to obtained normal reception.

4. *Poor or no reception on most channels with Table Top Converter in use:* Table Top Converter should be by-passed to see if some channels are received normally. If so, the customer should be advised to get the top-converter serviced.

20

ELEMENTS OF DIGITAL COMMUNICATION

Introduction

Digitisation of signal path has gained importance due to many advantages of the digital system. As the digital signal is in a 1 or 0 i.e. high or low pulse amplitude form, there is no degradation of video and audio components due to noise or distortion, while processing it in the studio chain, in VCRs or in receivers. In addition, digital bit stream can be stored as long as necessary to intorduce signal correction and for conversion to other television standards. These and other merits of digital processing have resulted in rapid transition from analog to digital form of transmission and reception of television and other communication systems.

Television signal distribution through cables is getting replaced by Direct-To-Home (DTH) also called Direct-Satellite-Broadcast (DSB) system where satellite signals are directly received and decoded at the subscriber's place. Also, Conditional Access System (CAS) has now been introduced by which a customer can opt for the channels of his choice and pay for these only.

With rapid advances in TV technology, now we have High Definition Television (HDTV) where a higher line scanning rate and an aspect ratio of 16:9 is used to enhance the quality of reception with large size pictures.

The price paid for change over to digital technology is that transmission channels need large channel bandwidths. However, to circumvent this, data-compression techniques have been developed to reduce data bit rate and thus conserve bandwidth.

While all aspects of digital transmission and reception are described in subsequent chapters, this chapter explains elements of digital processing, transmission and reception mostly for readers not fully conversant with digital technology.

20.1 Digital Data Processing

Electronic signals may be analog or digital. An analog signal, like the output of a microphone, can have any amplitude (within a range) for different frequency components. In contrast, in a digital system, information is represented in discrete or digitized form rather than continuous as in analog. The most common is the binary form, which means that for any signal only two discrete states are possible, which are normally denoted as 0 and 1. In digital circuits, the '0' state is usually represented by 0 volts or ground and the '1' state is usually represented by the supply voltage for digital circuitry (commonly called VDD). The usage of only two discrete signal levels results in a very significant noise immunity advantage for digital circuits. If the noise level at the input of a digital circuit is smaller than a specified critical magnitude (called the noise margin), the noise signal is sharply attenuated between the input and output, while the desired logic signals are transmitted at full amplitude (0 or VDD volts) without error. This is the cause of increased robustness of digital circuits over analog. Digital signals are generated and manipulated using high-speed transistor switches. Based upon only two signal levels, digital systems represent information in the binary system.

Binary Numbers: The decimal system expresses numbers in powers of 10 (*i.e.,* to the base 10). For example:

$$134 = 1 \times 10^2 + 3 \times 10^1 + 4 \times 10^0 = 100 + 30 + 4$$

The same approach is used in the binary system. However, here the base is 2 (powers of 2 are used rather than powers of 10 as in decimal system) and only two coefficients (0 and 1) are allowed (which correspond to 0 volts and VDD volts in actual digital circuits). As an example, decimal number 19 is expressed in the binary system as 10011 and read as one, zero, zero, one, one (and not ten thousand and eleven). This representation is correct because

$$10011 = 1 \times 2^4 + 0 \times 2^3 + 0 \times 2^2 + 1 \times 2^1 + 1 \times 2^0 = 16 + 0 + 0 + 2 + 1$$

Similarly, it can be shown that decimal number 25 is equal to 11001 in binary form and will look in signal form as shown in Fig. 20.1(a). As obvious, the binary equivalents will become bigger and bigger as the number (which could be representing analog signal amplitude) grows. For example, the decimal number 10,000 in the binary form is 10011100010000. While it is too big, digital processing of such numbers does not present any serious problem. This is because of the ease with which a very large number of logic elements can be designed and manufactured in a tiny integrated circuit (IC) and the very high speed and reliability with which digital electronic circuits can operate to carry out desired operations.

Data Communication Code: For transmitting or processing data in digital form, it is necessary to have a code to represent various numbers or letters. A commonly used code is ASCII (pronounced ASKEY). It is a 7-bit binary code which can represent 128 different numbers, letters, punctuation marks and other symbols. For example, the letter 'J' is represented in this binary code as 1001010 and will look in signal form as shown in Fig. 20.1(b). A single piece of binary information is called a "bit" and a 8-bit sequence is called a "byte". Please note that there is an 8-bit ASCII code also.

Binary Arithmetic: Arithmetic operations on binary numbers are done exactly in the same way as on decimal numbers. For example, in decimal system, if we add 6 to 5, we get a sum of 1 and a carry of 1. Similarly, in binary system, 1+1 results in a sum of 0 and a carry of 1 since here the base is 2 rather than 10. In the same way, for subtraction, 1–0 = 1 (difference=1, borrow=0) and 00 = 0 (difference = 0, borrow = 0) and so on. Multiplication and division is also carried on similar lines as in the decimal system. However, it may be noted that multiplication circuits are basically implemented using adder circuits. To multiply 22 by 6, 22 is added 6 times. Similarly, division is carried out by repeated subtractions. Since the time taken by each operation is so small, repeated additions and subtractions are an effective way of implementing multiplication and division.

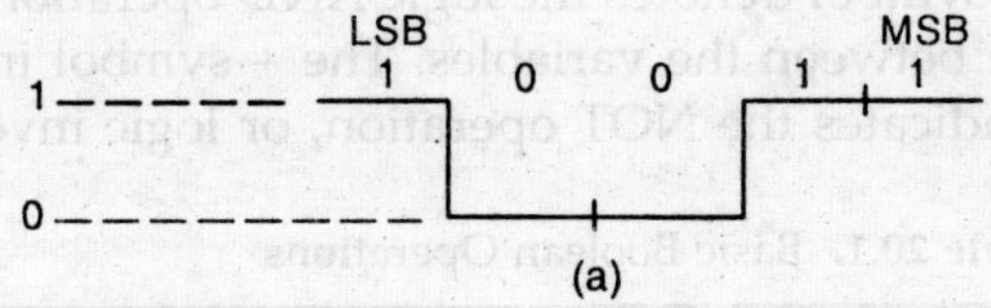

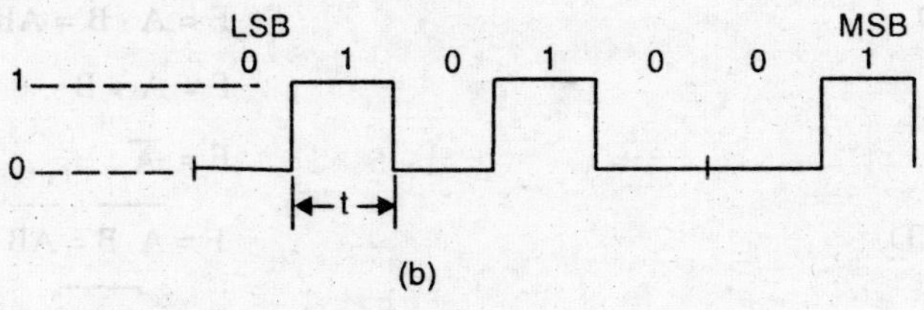

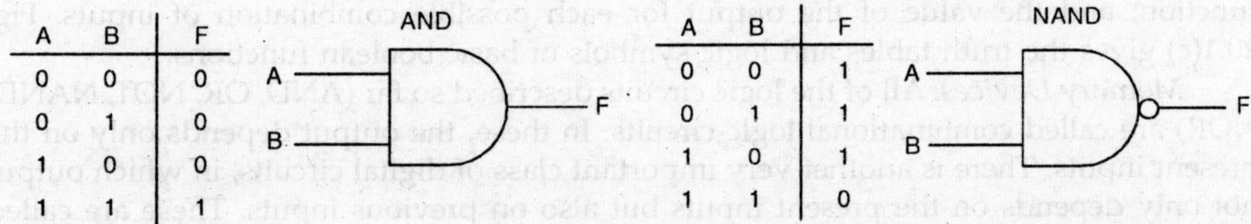

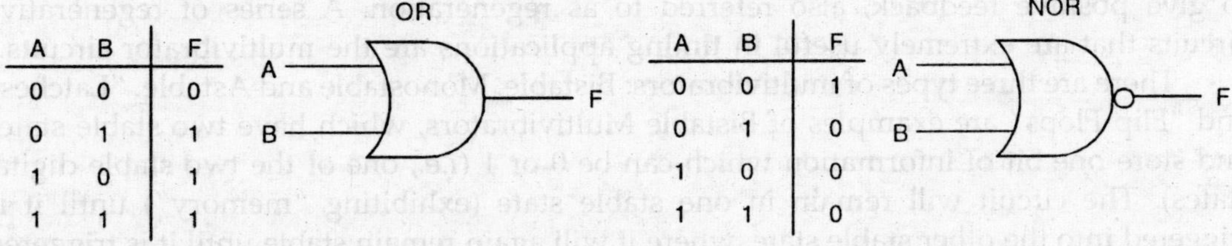

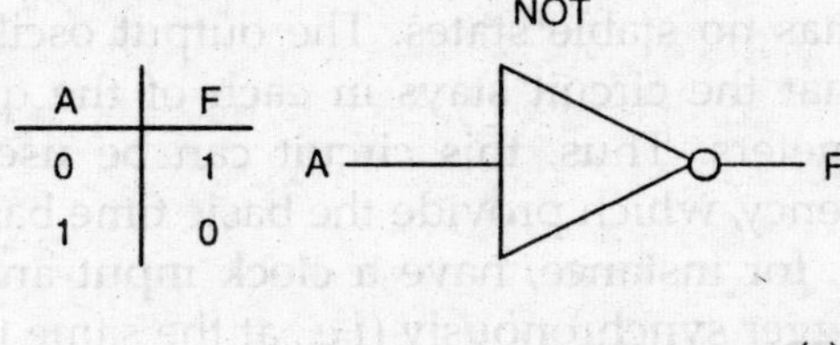

Fig. 20.1. (a) Decimal number 25 in binary code (11001), (b) Letter J in binary form (1001010), (c) Truth Tables of basic boolean functions.

Binary Algebra and Basic Logic Functions: A special algebra, which is applicable to the binary system is refereed to as Boolean Algebra. The basic boolean operations are included in Table 20.1. The '•' symbol denotes the logic AND operation, but please note that it is usually omitted from between the variables. The + symbol indicates a logical OR. A bar over the variable indicates the NOT operation, or logic inversion.

Table 20.1. Basic Boolean Operations

Operation	Boolean Expression
AND	$F = A \cdot B = AB$
OR	$F = A + B$
NOT	$F = \overline{A}$
NAND	$F = \overline{A \cdot B} = \overline{AB}$
NOR	$F = \overline{A + B}$

A truth table is a systematic listing of all possible inputs to a circuit or logic function, and the value of the output for each possible combination of inputs. Fig. 20.1(c) gives the truth tables and logic symbols of basic boolean functions.

Memory Devices: All of the logic circuits described so far (AND, OR, NOT, NAND, NOR) are called combinational logic circuits. In these, the output depends only on the present inputs. There is another very important class of digital circuits, in which output not only depends on the present inputs but also on previous inputs. These are called Sequential Circuits which have "memory" and are able to "store" information. A characteristic of sequential circuits is that one or more outputs are internally fed back to inputs to give positive feedback, also referred to as regeneration. A series of regenerative circuits that are extremely useful in timing applications are the multivibrator circuits.

There are three types of multivibrators: Bistable, Monostable and Astable. "Latches" and "Flip-Flops" are examples of Bistable Multivibrators, which have two stable states and store one bit of information which can be 0 or 1 (*i.e.,* one of the two stable digital states). The circuit will remain in one stable state (exhibiting "memory") until it is triggered into the other stable state, where it will again remain stable until it is triggered back to the original stable state. The monostable multivibrator has only one stable state. When triggered, it temporarily goes into the other state and then returns to the stable state on its own. The astable multivibrator has no stable states. The output oscillates between two quasi-stable states. The time that the circuit stays in each of the quasi-stable states is determined by circuit parameters. Thus, this circuit can be used to generate "Clock" pulses of the required frequency, which provide the basic time base or synchronization in digital circuits. Flip-flops, for instance, have a clock input and all flip-flops that are driven by the same clock trigger synchronously (*i.e.,* at the same time) to sample respective new inputs and to go to the respective next states. The clock frequency is a very important parameter of digital circuits and is indicative of the speed at which operations take place.

Registers: A register is an array of flip-flops all triggered by a common clock. Registers are implemented to store information. For example, an 8-bit register consists of eight flip-flops, and can store 8-bit long pieces of information like an "instruction" in a microprocessor or other pieces of 8-bit data. In addition to storing information, a class of registers (shift registers) can shift data left or right as well.

Counters: Flip-flops can also be connected to function as counters to count random events or to divide a clock frequency or to measure a parameter such as time. A "divide by n" counter produces one output for "n" input clock pulses. Counters that are used to directly display information for human use usually need to do so in the decimal system. However, digital circuits work using the binary system. Therefore, binary numbers must be coded to decimal form and the counters that are designed to do so are called decade counters.

20.2 A/D and D/A Converters

Due to the very significant advantages of digital circuits over analog, more and more circuits are being implemented using digital techniques. While it is very advantageous to do signal processing in digital form, the real world is analog. Therefore, analog signals must first be converted to digital form before processing can be done, and the digital results must be converted back to analog for human consumption. As an example, a singer sings in front of a microphone to generate analog output, which must be converted to digital form before it can be further processed to generate a music Compact Disc (CD). Similarly, when a CD is played, the digital output after it is retrieved and processed, must be converted to analog form before it can be amplified and fed to a speaker so that humans can enjoy crystal clear sound. It may be noted that the increased sound clarity is a direct consequence of keeping most of the processing digital (thus noise free) after the initial A/D and before the final D/A conversion.

D/A Converter: D/A conversion involves translating digital information into equivalent analog information. This amounts to changing the "n" digital voltage levels into one equivalent analog voltage. This procedure is illustrated with the help of a basic D/A circuit shown in Fig. 20.2(a). In actual practical circuits, the logic switches shown as simple SPDT switches would be implemented using transistors. Assuming that at any instant the digital input is 0011, *i.e.*, $(0 \times 2^3 + 0 \times 2^2 + 1 \times 2^1 + 1 \times 2^0 = 3)$ and knowing that the least significant bit (LSB) is the right most bit, on application of 0011 binary input to the D/A converter, the switches will toggle to contacts 1 or 0 as shown in the figure. Thus 40 ohms and 80 ohms resistors are both connected to 15V supply while the 10 ohm and 20 ohm resistors are grounded. As such the upper leg(1s) of the voltage divider has a value equal to $\frac{40 \times 80}{40 + 80} = \frac{80}{3}$ ohms while the value of lower leg (0s) is $\frac{10 \times 20}{10 + 20} = \frac{20}{3}$ ohms. The division ratio of the divider is thus $\frac{20}{20 + 80} = \frac{1}{5}$ and thus the output voltage will be 1/5 of 15V *i.e.* 3 volts and this is the required value as shown above. The input digital data in binary form changes at a fast rate thus enabling almost a continuous analog signal at the output. In television receivers which use digital circuitry to process incoming signals, D/A converters are used to convert digital video and audio outputs in analog form to drive picture tube guns and audio amplifiers respectively.

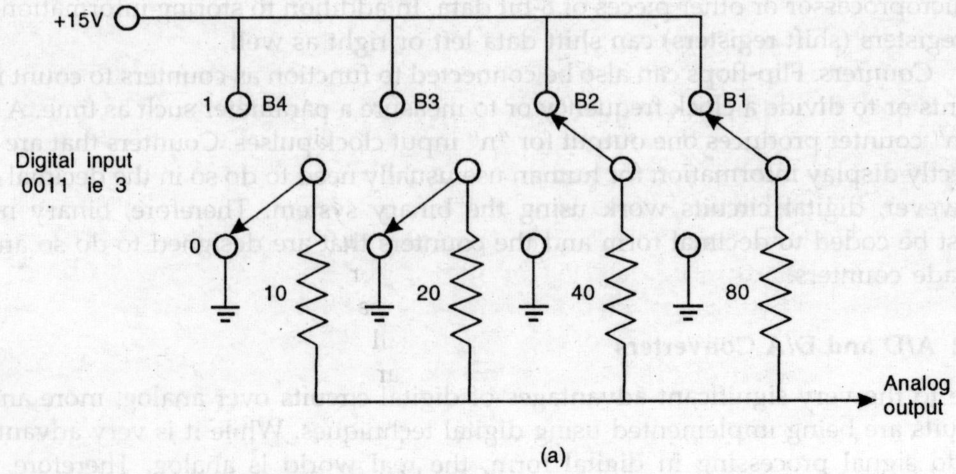

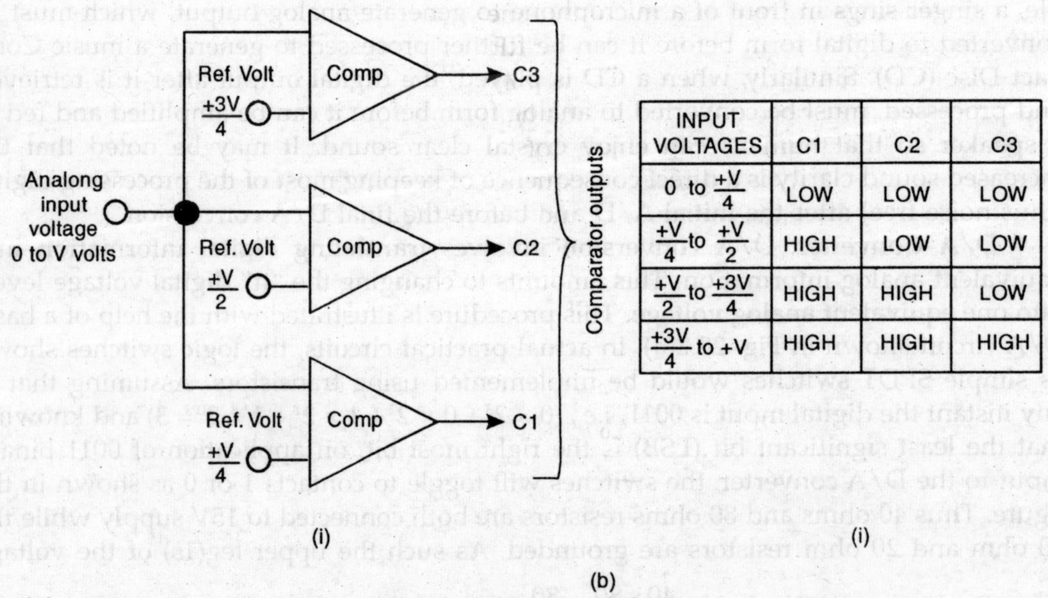

INPUT VOLTAGES	C1	C2	C3
0 to $\frac{+V}{4}$	LOW	LOW	LOW
$\frac{+V}{4}$ to $\frac{+V}{2}$	HIGH	LOW	LOW
$\frac{+V}{2}$ to $\frac{+3V}{4}$	HIGH	HIGH	LOW
$\frac{+3V}{4}$ to +V	HIGH	HIGH	HIGH

Fig. 20.2. (a) Basic circuit of a D/A converter,
(b) Simultaneous A/D conversion (i) basic logic circuit,
(ii) comparator outputs for input voltages ranges.

A/D Converter: Analog to digital conversion is somewhat more complex. As such, a number of different methods have been developed, the simplest of which is the simultaneous method as illustrated in Fig. 20.2(b). This is based on the use of a number of comparator circuits which are basically difference amplifiers. The figure shows three comparator circuits though in practice the number is usually quite large. The analog signal to be digitized serves as one of the inputs to each comparator. The second input is a standard reference voltage. The reference voltages used in this illustration are +V/4, +V/2 and +3V/4 and are obtained from potentiometer networks of the type shown in Fig. 20.2(a). The system is then capable of accepting an analog input voltage between 0 and V volts.

If the analog input signal exceeds the reference voltage to any comparator, the comparator turns "ON", *i.e.*, its output goes high (1). Similarly, if the input signal is less than the reference voltage, the comparator turns OFF, *i.e.*, its output goes low (0). Thus, depending on the instantaneous level of the input, analog signal comparator outputs will switch as shown in the chart. Examination of Fig. 20.2(b) shows that there are four voltage ranges that can be detected by the converter. Four ranges can be effectively discerned (seen) by 2 digital bits. The three comparator output can then be fed into a coding network to provide two bits which will be equivalent to the input analog voltage. The bits out of the coding network can then be entered into a register for storage where necessary.

20.3 Digital Computer Organisation and Signal Processing

A digital computer consists of several devices linked together to form an operational unit. All the electrical, mechanical and magnetic parts that become part of the computing system are together called HARDWARE. All other units, like keyboard, monitor and printer are referred to as peripherals. Another necessary but invisible element of any computer is the PROGRAM with the help of which the processor unit operates the computer. A program consists of a series of instructions designed to guide the computer step by step to perform the desired operation. On a broader scale, all the instructions that combine to make the computer function including the programs provided by the manufacturer are referred to as the SOFTWARE.

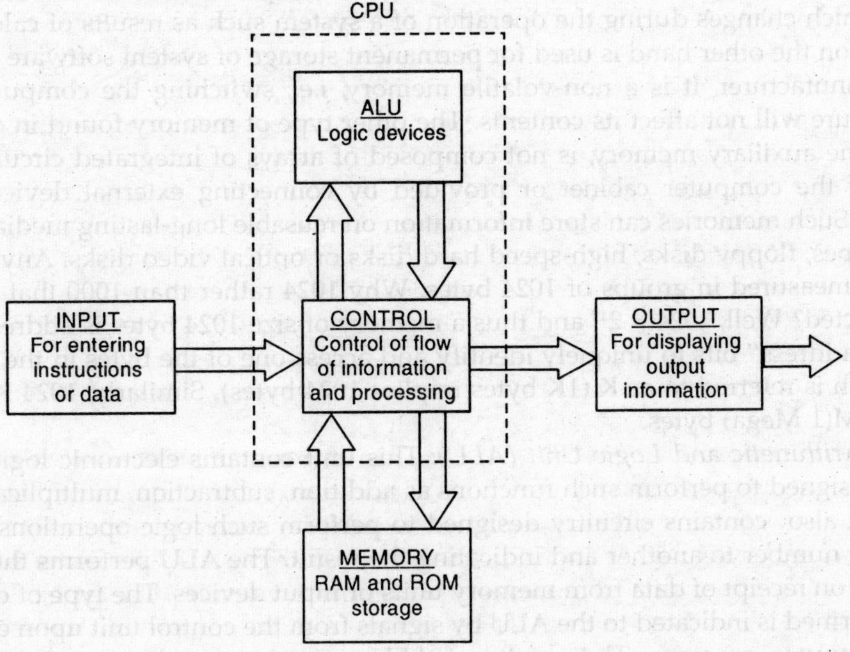

Fig. 20.3. Block diagram of a digital computer system.

All computer systems have five essential blocks: Input, Output, Memory, Arithmetic Logic Unit (ALU) and the Control Unit as shown in Fig. 20.3. Each block is now briefly explained.

(i) *Input:* Each computer has one or more devices which allow inputs to be fed into the system. The most common is the keyboard, like that of a typewriter, which besides usual keys has special function keys that pertain directly to the operation of the computer. Along with the keyboard is an assortment of special pointing devices such as mouse, trackball, light pen, graphics tablet or joystick for communicating with the computer for specific applications. The most commonly used is the mouse, which when operated enables the movement of cursor on the monitor screen. The cursor is usually a small blinking symbol which indicates where the computer's attention is currently focussed within a certain program. The mouse, when moved over the surface of a desktop causes the corresponding movement of cursor on the monitor's screen.

(ii) *Output:* The output unit can take several forms. The most common are the monitor and printer on which results are displayed or printed. While general purpose monitors use a cathode ray tube (CRT) for the screen, the laptop and pocket computers have monitors made of liquid crystal display (LCD) screens which makes them portable. When results are required in printed form, a printer is used. When outputs are required in analog form, as for initiating various functions of a TV receiver, D/A converters are used to first convert the output in analog form.

(iii) *Memory:* Most computer systems have two types of memory (storage) areas—main and auxiliary. The main memories found inside the computer consist of arrays of integrated circuits capable of storing millions of bits of information. These are the Random Access Memory (RAM) and the Read Only Memory (ROM). RAM is capable of receiving and storing data which can be changed as required. It is used to store information which changes during the operation of a system such as results of calculations. The ROM on the other hand is used for permanent storage of system software provided by the manufacturer. It is a non-volatile memory, *i.e.*, switching the computer off or power failure will not affect its contents. The other type of memory found in computer systems, the auxiliary memory, is not composed of arrays of integrated circuits. It can be part of the computer cabinet or provided by connecting external devices to the computer. Such memories can store information on reusable long-lasting media, such as cassette tapes, floppy disks, high-speed hard disks or optical video disks. Any memory is usually measured in groups of 1024 bytes. Why 1024 rather than 1000 that you may have expected? Well, $1024 = 2^{10}$ and thus a memory of size 1024 bytes is addressable by 10 input "address" bits to uniquely identify and access one of the bytes in the group of 1024, which is referred to as K (1K bytes implies 1024 bytes). Similarly, 1024 K bytes is same as 1M(1 Mega) bytes.

(iv) *Arithmetic and Logic Unit (ALU):* This unit contains electronic logic circuits that are designed to perform such functions as addition, subtraction, multiplication and division. It also contains circuitry designed to perform such logic operations as comparing one number to another and indicating the result. The ALU performs the desired operations on receipt of data from memory units or input devices. The type of operation to be performed is indicated to the ALU by signals from the control unit upon execution of the computer program. The results of ALU operation can be transferred to the memory unit or output devices.

(v) *Control Unit:* The flow of instructions, data and results is governed by the control unit. As such, in accordance with the software program, it enables the desired sequence of arithmetic or logic operations on retrieving instructions from the memory. The control unit also provides the timing signals which regulate the speed of all operations.

Central Processing Unit (CPU): The ALU and the control unit together form what is referred to as the Central Processing Unit of a computer as indicated by dotted chain lines on the figure. The CPU is really the brain of a computer as it is the CPU which controls all other units and directs all operations as required by the software or program.

All inputs to a computer must be in binary form, and that applies to programs as well. However, coding large programs in binary format is very tedious and error prone. Hence, high level languages like C, FORTRAN and COBOL have been developed which allow the users to write programs more easily which are then automatically compiled or translated into computer executable binary format.

Microprocessor: Very Large Scale Integration (VLSI) has made it possible to integrate the CPU on a single chip which is called Microprocessor Unit (MPU) or µP.

Microcomputer (µC): When in addition to µP, memories and input/output interfaces are also integrated, the resulting single chip device is referred to as a microcomputer. For example, special µCs have been developed for use in TV technology for digital signal processing, channel selection and remote control of receiver functions.

Data Transmission: As discussed earlier, any information to be processed or transmitted is first converted to a binary representation and stored in computing equipment. For processing within computers, most of the time data transfer from one circuit to another (e.g., from one register to another) happens in parallel format. This means that if a data (8 bits) or a word of data (2 bytes, or 16 bits) needs to be transferred, there are as many interface lines as the number of bits, enabling simultaneous transfer of all the bits (the entire byte or word). While such parallel interfaces enable fast processing of data within computers, such interfaces are not economically feasible for long distance transmission from one computer to another. This is because such an approach would require eight separate communication channels if all eight bits of a byte were to be transferred in parallel. Long distance communication channels, instead handle one bit a time. Therefore on the sending end data must first be converted from parallel to serial using a shift register, transferred one bit at a time, and then on the receiving end again converted from serial to parallel (again using a shift register) before further processing.

20.4 Binary Transmission by Frequency Shift Keying

An important concept in data communication and analog communication for that matter is that the amount of information that can be transmitted in a given time depends on the bandwidth of the communication channel. This means that if the signal to be transmitted occupies large bandwidth, it would need longer time for transmission. Since binary signals consist of rectangular pulses, which are made up of a fundamental sinewave plus many higher harmonics, the channel bandwidth must be large enough to pass all the harmonics to preserve the binary pulse wave shape. Voice grade telephone lines, for example have an upper cut-off frequency of about 3400 Hz and if used for binary data transmission as such would result in severe distortion because higher harmonics will not get transmitted. However, the telephone network is a vast world-wide communication system that is available and thus must be somehow used to transmit digital data. This is in fact achieved by converting data bits into low frequency AC form that lies within the telephone-line channel bandwidth of about 300 to 3000 Hz before transmission. Obviously, a reverse procedure is required at the receiving end to recover the original binary bit data.

The process of converting binary signals to AC form is called Frequency Shift Keying (FSK). In its simplest form, two sine wave frequencies are used to represent logic 0 and logic 1. Logic 0 (usually called space) is allotted a frequency of 1070 Hz and logic 1 (called mark) is allotted a frequency of 1270 Hz. The two frequencies are transmitted as shown in Fig. 20.4(i), to create equivalent sine wave data. To permit simultaneous transmit and receive modes, another set of frequencies are allotted with "space" at 2025 Hz and "mark" at 2225 Hz. The circuits used to affect necessary conversions from binary to AC form and vice versa are called FSK modulators and demodulators respectively.

FSK, as described above, enables data transmission speed of about 300 baud *i.e.*, 300 bits per second, which is very small for most applications. Higher keying speeds up to 1200 bauds are obtained by using binary phase shift keying (BPSK) where as shown in Fig. 20.4(ii) to transmit logic 0s, the carrier signal (sine wave) is transmitted with one phase (say 0°) and for logic 1s the phase is shifted by 180°. Still higher baud rates are obtained with what is known as Quadrature PS Keying (QPSK) modulation. In QPSK, four reversals are obtained by using push-pull type balanced modulators (180° phase shift) fed with a carrier that is phase shifted by 45° before applying to the two modulators to cause another phase shift, thus making four phase shifts in one cycle of the carrier. For still higher baud rates, another trick is used where besides QPSK, alternate amplitude modulation is introduced to create eight reversals. Such a version is called

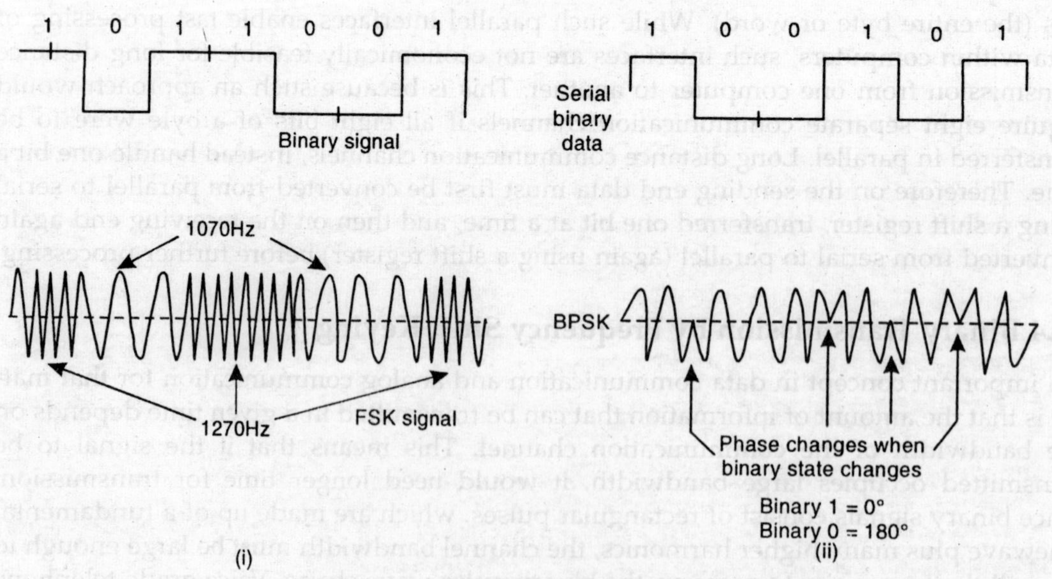

Fig. 20.4. (i) Frequency Shift Keying,
(ii) Binary Phase Shift Keying.

8-QAM. A constraint with such FSK circuits is that a phase reference of the carrier is needed for demodulation at the receiving end for correct recovery of the transmitted data. This problem is solved by the addition of previous bit before the present bit. This is achieved by delaying the previous bit by one clock pulse, whereby, at the receiving end the demodulator compares the phase of the present bit with that of the previous bit to correctly recover all the incoming bits.

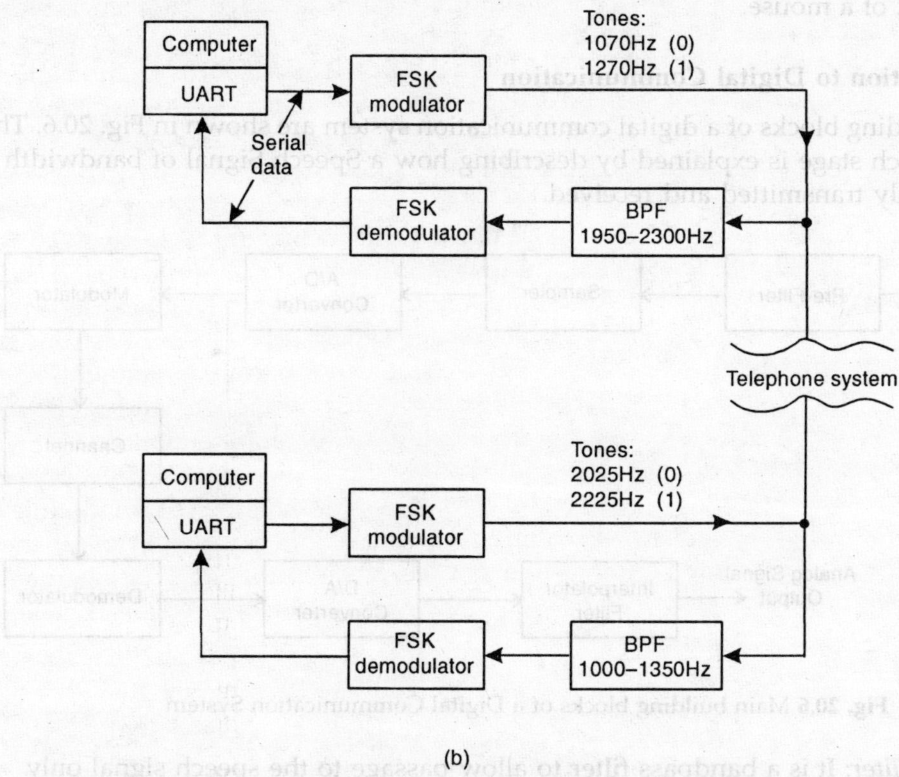

Fig. 20.5. Block diagram of a MODEM employing FSK modulation.

FSK MODEM: A modem is a device containing both a modulator and a demodulator, associated bandpass filters and a device called the Universal Asynchronous Receiver Transmitter (UART), which performs parallel to serial and serial to parallel transformations of outgoing and incoming data needed for long distance communications as explained earlier.

As an illustration of how data transmission takes place through a MODEM, the block schematic of a simple FSK MODEM is shown in Fig. 20.5. Both sending and receiving ends contain FSK MODEMS so that data transmission is possible both ways. For separating the two paths, bandpass filters are provided at either end. Similarly, UARTs are needed at both ends. The type of MODEM will vary depending upon the type of underlying technology (*i.e.*, FSL, BPSK, QPSK, 8-QAM, etc).

While we have described above some basic approaches of implementing MODEMS to illustrate the concept of digital data transmission over regular phone lines, there have been tremendous advances in technology since the above techniques were developed. Today's state of the art MODEMS employ sophisticated Digital Signal Processors (DSPs) to implement advanced algorithms and as a result, baud rates of up to 56K are now common. Still higher baud rates are beginning to get implemented using Digital Subscriber Line Technology. Such high baud rates are further fueling the internet revolution where very large amounts of data must be transmitted or received upon the click of a mouse.

20.5 Introduction to Digital Communication

The basic building blocks of a digital communication system are shown in Fig. 20.6. The function of each stage is explained by describing how a Speech Signal of bandwidth 4 KHz is digitally transmitted and received.

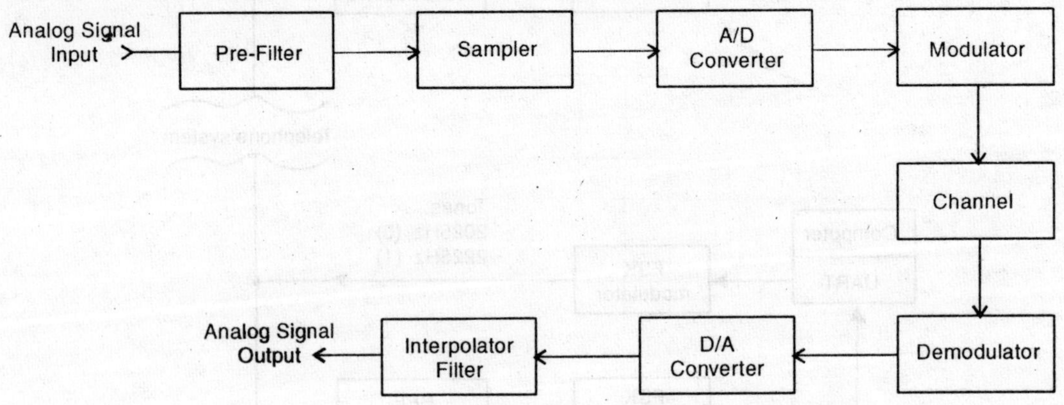

Fig. 20.6 Main building blocks of a Digital Communication System

(1) *Pre-Filter:* It is a bandpass filter to allow passage to the speech signal only.

(2) *Sampler:* In digital transmission, it is not wise to attempt to reproduce accurately the amplitudes of modulating signal because of certain uncertainty introduced by any noise present in the modulating signal. Therefore, the signal is quantised and sampled in such a way that if the amplitude of modulating signal lies anywhere within a specified range, a single and fixed amplitude signal gets transmitted. This is illustrated in Fig. 20.7 where the original signal, its sampling levels and shape on reception are shown separately. However, in our example of speech signal transmission, only four levels of sampling are considered for ease of explaining the sequence of operations.

Sampling Rate: According to NYQUIST theorem, the sampling rate should be at least two times the highest modulating frequency for obtaining reasonably accurate output at the receiving end. Thus, a sampling rate of 8000 (2 × 4000) per second will do for the speech signal of 4 KHz.

(3) *A/D Converter:* The sampling pulses in the sampler result in output voltages proportional to each quantised level. The stream of output signals is fed to an A/D converter to obtain output in the binary form. Since there are four sampling levels, a 2 bit ($2^2 = 4$) A/D converter would be necessary.

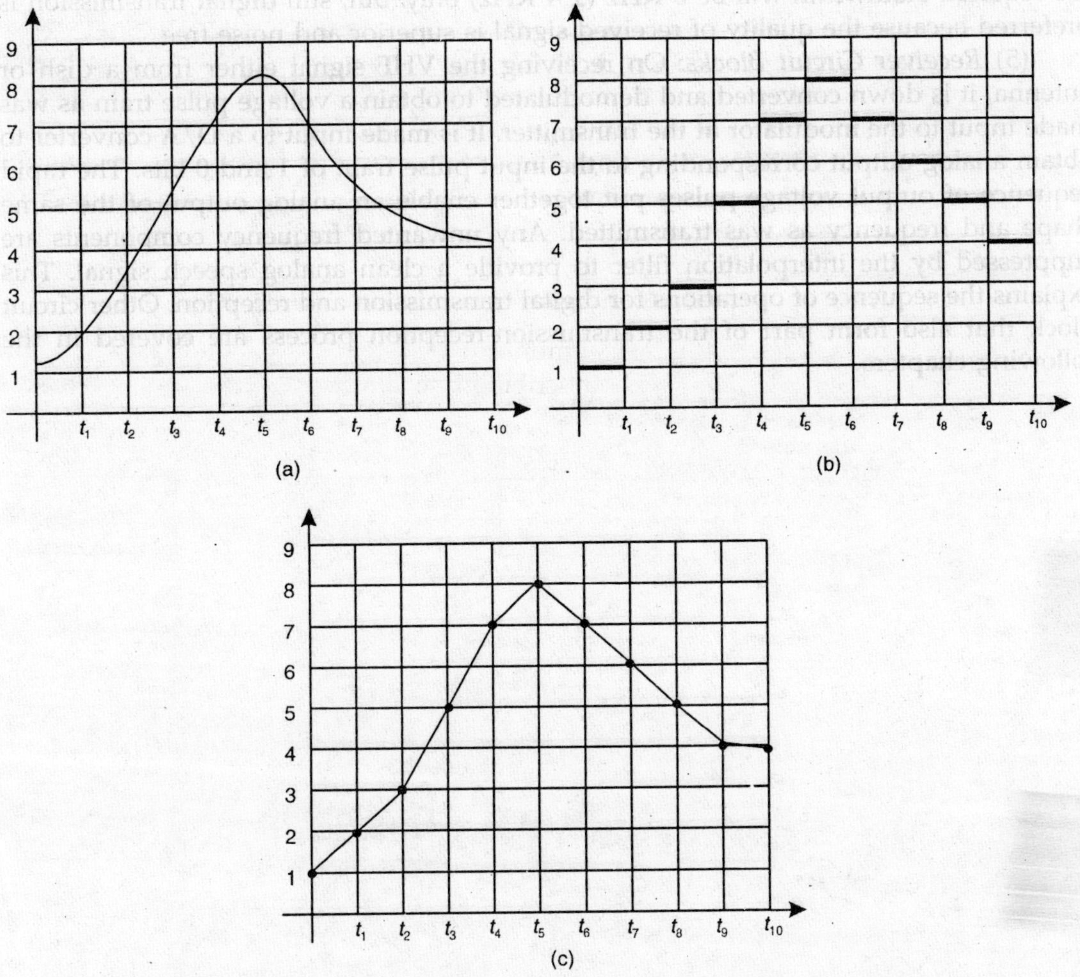

Fig. 20.7 Sampling of analog signals (a) original signal (b) its sampling levels and (c) its shape on reception.

Thus each sampling level will get represented by a 2 bit binary code out of 00, 01, 10 and 11 depending on its voltage amplitude. The result will thus be a 2 × 8000 = 16000 bits output per second. Each bit is converted into a voltage pulse before modulation. Assuming each '1' bit causes a positive voltage and each '0' a negative voltage swing, there will be 16000 pulses each of duration 1/16000 second.

(4) *Modulation:* The output data, that is now in the form of voltage pulses is FM modulated on a suitable sine wave carrier and then up converted to the channel transmission frequency. It will be in the UHF band for terrestrial broadcast and in C-BAND for satellite transmission.

Bandwidth Requirement: It can be shown that if pulses occur in a random manner and the probability of occurrence of 1s and 0s is same, a channel bandwidth equal to 32 KHz (± 16 KHz) will be necessary for proper transmission of the speech signal. However, if the same speech signal is transmitted in an analog format using AM modulation, the required bandwidth will be 8 KHz (± 4 KHz) only. But, still digital transmission is preferred because the quality of received signal is superior and noise free.

(5) *Receiver Circuit Blocks:* On receiving the VHF signal either from a dish or antenna, it is down converted and demodulated to obtain a voltage pulse train as was made input to the modulator at the transmitter. It is made input to a D/A converter to obtain analog output corresponding to the input pulse train of 1 and 0 bits. The rapid sequence of output voltage pulses put together enable an analog output of the same shape and frequency as was transmitted. Any unwanted frequency components are suppressed by the interpolation filter to provide a clean analog speech signal. This explains the sequence of operations for digital transmission and reception. Other circuit block that also form part of the transmission-reception process are covered in the following chapters.

21
SIGNAL ENCODING AND COMPRESSION

Introduction

In the last decade or so a transformation has taken place in the way we communicate and is still underway. The transformation includes the ever growing Internet, the explosive expansion of mobile phone services and increasing importance of video and audio communication. Data compression is one of the enabling techniques of each these aspects of multimedia revolution. It would not have been possible to put images, let alone video and audio on web sites in the absence of data compression methods commonly called ALGORITHMS. The advent of digital TV became possible only when data compression techniques got developed. Data Compression in effect means reducing the number of bits (1, 0) required to represent images, sounds or say a television sequence. It is thus an art or science of representing information in a compact form. Compression techniques vary with their application and are very complex developed after lot of research effort.

In this chapter, first the bandwidth requirements for video and audio transmission are deduced followed by means of encoding them to obtain their values in the form of binary data, which is then compressed before modulation. Later sections of the chapter are devoted to data compression methods necessary for television signals and video-audio recording on compact discs.

21.1 Digital Television Signal Bandwidth

Digital data is obtained on quantisation and sampling the analog signal. It has been practically verified that sampling the generated analog picture signal at 256 levels is sufficient for the reproduced picture to be almost undetectable from the transmitted picture. To achieve this, the video signal is sampled at a very fast rate and then converted to binary code by an A/D converter. For 256 levels an 8 bit converter would be necessary since $2^8 = 256$.

The binary signal is square wave in nature and to retain its shape during transmission, it is necessary that the sampling rate is at least two to three times the base band of video signal which is 5 MHz for the 625/50 PAL television system. It is desirable to choose the sampling rate to be a multiple of colour subcarrier frequency (fsc) which is 4.43361875 MHz in the PAL system. This ensures desired co-relation between scanning rates and other operations for obtaining high quality pictures. In practice, a sampling rate equal to 4 × fsc is used for encoding the luminance (Y) signal. Similarly, a sampling rate equal to 2 × fsc is used for encoding chroma i.e. colour-difference signals U & V because their base band is almost half that of the Y signal.

Thus, with an 8 bit A/D converter and the chosen sampling rates, bit rate data for the transmission of Y signal will be 8 × 4 × fsc ≈ 8 × 4 × 4.34 = 142 Mega bits per second (142 MB/S). Similarly, for the two chroma signals put together, the bit rate will be 2 (8 × 2 × fsc) = 142 MB/S). Thus, a binary data equal to 284 MB/S (142 + 142) will need to be transmitted for Colour TV signals.

For such a high bit rate, a channel bandwidth of our 200 MHz would be necessary. Since it is not possible to allow such a large channel width in the present satellite transmission system, the bit rate is compressed by a factor of about 50:1 before modulation and transmission. For the TV sound signal 128 levels of sampling are sufficient for acceptable sound reproduction. To know how quantised signal looks, a cycle of audio signal is shown in Fig. 21.1, where deliberately 16 sampling levels have been chosen to show clearly the nature of sampled signal which as seen is staircase in nature. For 16 sampling levels a 4 bit A/D converter would be necessary ($2^4 = 16$). The resulting binary code is labelled in the figure for all the levels. Obviously for the TV audio signal which is sampled at 128 different levels, the staircase levels will be too small and close to each other. With a sampling rate equal to 600 KHz which is nearly four times the audio signal

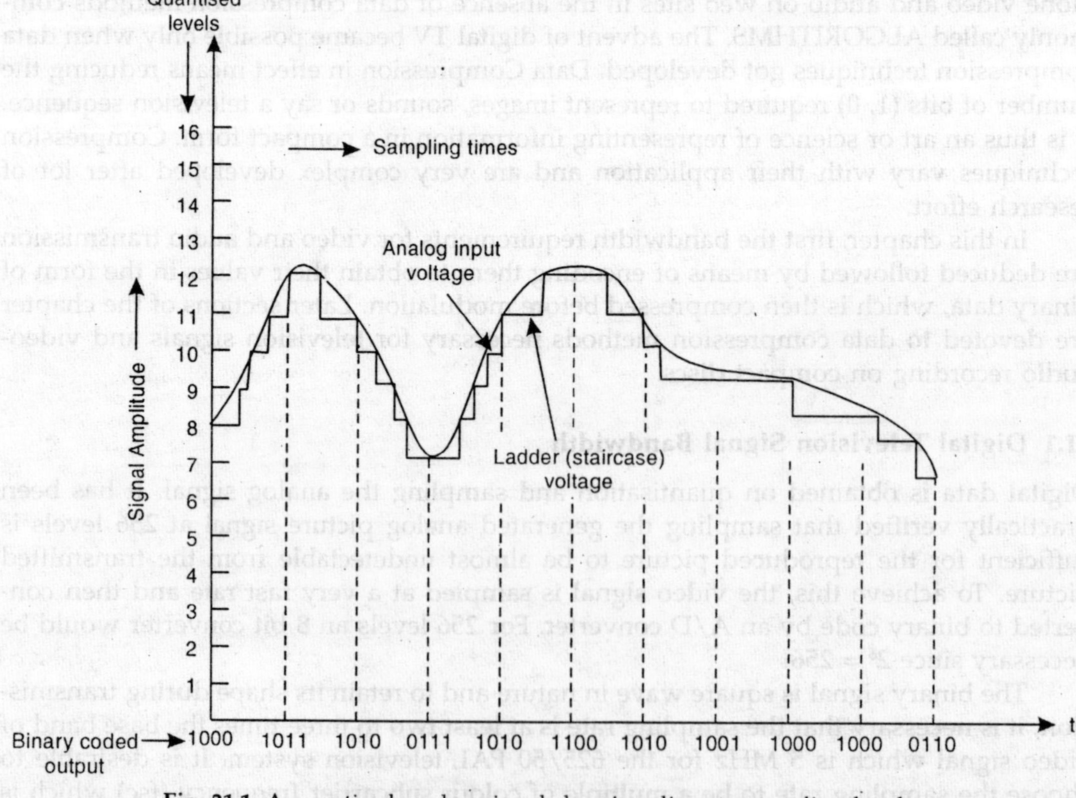

Fig. 21.1. A quantised analog signal showing discrete sampling levels.

Note: In the figure steps are shown quite exaggerated for clarity.

base band, the bit rate will come to 7 × 600 = 42 KB/S, because a 7 bit A/D converter will be needed ($2^7 = 128$). The bit rate will be little more than double for the stereo sound. Though this is not too high a bit rate, still it is compressed like the luminance (Y) and chroma signals before transmission.

21.2 Video and Audio Signal Encoding

The encoding of video and audio signals proceeds as shown in Fig. 21.2. The video signal amplitude variations for one line period (64 micro second) are pictured in Fig. 21.3. Because of the 8 bit converter each of the 256 levels will have a distinct 8 bit code. For example, the 8th level will be represented as (00001000) and 19th level as (00010011). In the same way black level amplitude of the video signal which lies at the 64th level will have a code (01000001) and peak white at level 204 a code (11001100) as labelled on the figure.

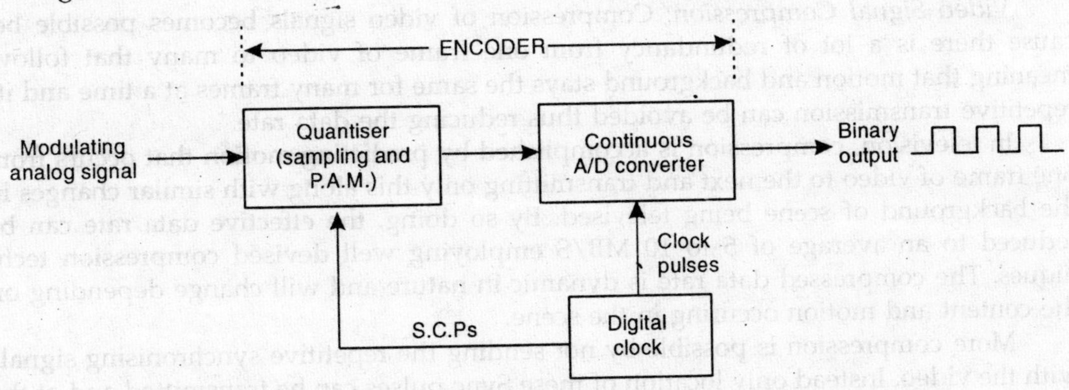

Fig. 21.2. Block schematic of a PCAM encoder for video and audio signals.

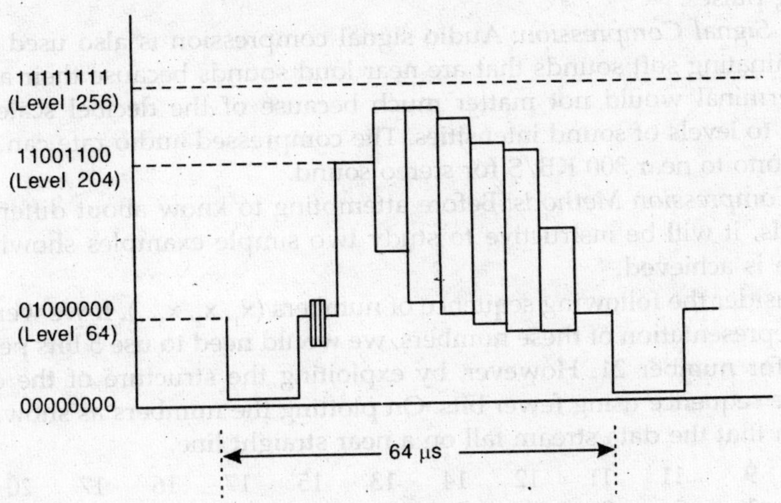

Fig. 21.3. Video signal quantisation levels.

For encoding, Pulse Amplitude Modulation (PAM) is preferred where the period of sampling clock pulses remains constant while their amplitude follows the level of modulating signal. As shown in Fig. 21.2 the quantised output from the modulator is input to the clock controlled A/D converter whose output corresponds to the original (video or audio) amplitudes but in the binary form. The complete encoding configuration is termed as ENCODER as labelled in the figure.

21.3 Data Compression Techniques

Data compression process is similar to the animation of a cartoon, with only the character's movement being redrawn, thus conserving bandwidth and allowing more programming per transponder. Digital compression is a mathematical process (algorithm) in which the data is combined using space or common information so that less data transmission is needed to process or pass-along the information.

Video Signal Compression: Compression of video signals becomes possible because there is a lot of redundancy from one frame of video to many that follow, meaning that motion and background stays the same for many frames at a time and its repetitive transmission can be avoided thus reducing the data rate.

In television, compression is accomplished by predicting motion that occurs from one frame of video to the next and transmitting only this along with similar changes in the background of scene being televised. By so doing, the effective data rate can be reduced to an average of 5 to 10 MB/S employing well devised compression techniques. The compressed data rate is dynamic in nature and will change depending on the content and motion occuring in the scene.

More compression is possible by not sending the repetitive synchronising signals with the video. Instead only location of these Sync pulses can be transmitted and at the receiving end locally generated pulses can be inserted for necessary synchronisation. Similarly, saving in data is possible by removal and reinsertion of vertical and horizontal blanking pulses.

Audio Signal Compression: Audio signal compression is also used to reduce bit rate by eliminating soft sounds that are near loud sounds because their absence at the receiving terminal would not matter much because of the decibel scale response of human ears to levels of sound intensities. The compressed audio rate can vary from 50 KB/S for mono to near 300 KB/S for stereo sound.

Data Compression Methods: Before attempting to know about different compression methods, it will be instructive to study two simple examples showing how data compression is achieved.

(1) Consider the following sequence of numbers (x_1, x_2, x_3...). If we were to transmit the binary representation of these numbers, we would need to use 5 bits per sample like (1 0 1 0 1) for number 21. However, by exploiting the structure of the data, we can represent the sequence using fewer bits. On plotting the numbers as shown in Fig. 21.4 (a), it is seen that the data stream fall on a near straight line.

Data	9	11	11	12	14	13	15	17	16	17	20	21
S. No.	1	2	3	4	5	6	7	8	9	10	11	12

A model for the data can then be an equation—

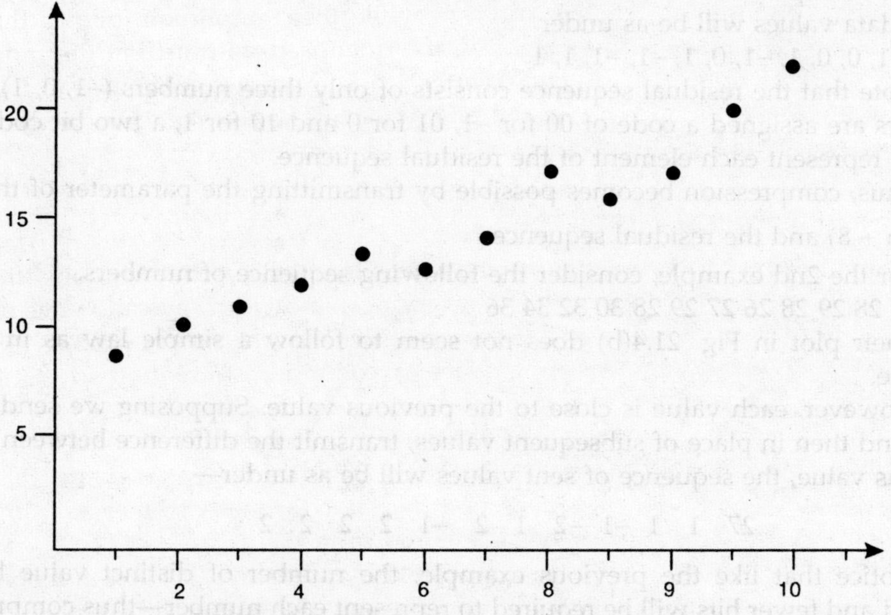

Fig. 21.4(a) A plot of sequence of data values

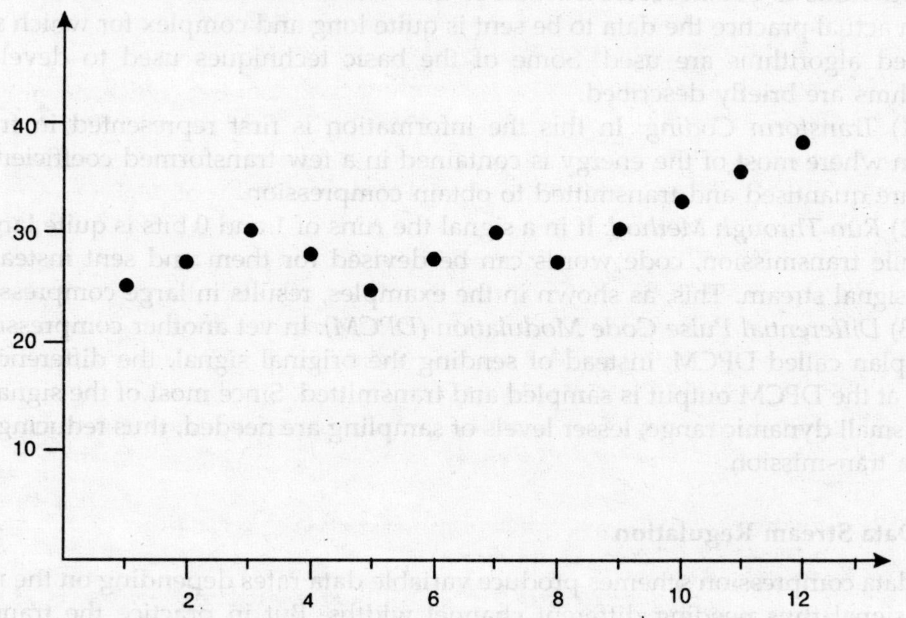

Fig. 21.4(b). Data values plotted against the sequence.

$\hat{X}_n = n + 8$, where n is the serial number of the data to be transmitted. The difference between data and the model an be expressed as:–

$$e_n = x_n - \hat{x}_n$$

For example with n = 1, the difference will be 9 – (1 + 8) = 0. Thus the difference for all the data values will be as under

0, 1, 0, 0, 1, –1, 0, 1, –1, –1, 1, 1

Note that the residual sequence consists of only three numbers (–1, 0, 1). If these numbers are assigned a code of 00 for –1, 01 for 0 and 10 for 1, a two bit code can be used to represent each element of the residual sequence.

Thus, compression becomes possible by transmitting the parameter of the model $\hat{X}_n = (n + 8)$ and the residual sequence.

For the 2nd example, consider the following sequence of numbers.

27 28 29 28 26 27 29 28 30 32 34 36

Their plot in Fig. 21.4(b) does not seem to follow a simple law as in the first example.

However, each value is close to the previous value. Supposing we send the first value and then in place of subsequent values, transmit the difference between this and previous value, the sequence of sent values will be as under—

27 1 1 –1 –2 1 2 –1 2 2 2 2

Notice that like the previous example, the number of distinct value has been reduced and fewer bits will be required to represent each number—thus compression is achieved. At the receiving end, the decoder adds each received value to the previous decoded value to obtain reconstruction of the actual value.

In actual practice the data to be sent is quite long and complex for which specially designed algorithms are used. Some of the basic techniques used to develop such algorithms are briefly described.

(1) *Transform Coding:* In this the information is first represented in frequency domain where most of the energy is contained in a few transformed coefficients. Only these are quantised and transmitted to obtain compression.

(2) *Run Through Method:* If in a signal the runs of 1 and 0 bits is quite large as for Facsimile transmission, code words can be devised for them and sent instead of the entire signal stream. This, as shown in the examples, results in large compression.

(3) *Differential Pulse Code Modulation (DPCM):* In yet another compression algorithm plan called DPCM, instead of sending the original signal, the difference as obtained at the DPCM output is sampled and transmitted. Since most of the signal energy is in a small dynamic range, lesser levels of sampling are needed, thus reducing the rate of data transmission.

21.4 Data Stream Regulation

Most data compression schemes produce variable data rates depending on the nature of input signal thus needing different channel widths. But in practice the transmission bandwidth is fixed for a given application requiring a uniform data rate of transmission. To overcome this problem a BUFFER is provided at the output of compression algorithm (system) as shown in Fig. 21.5. It regulates the data rate by a feedback mechanism. As the data rate tends to increase above the normal value, the buffer senses it and sends a signal to the data compression system which then switches to another algorithm to increase compression. Thus, the output data rate remains within the prescribed data run.

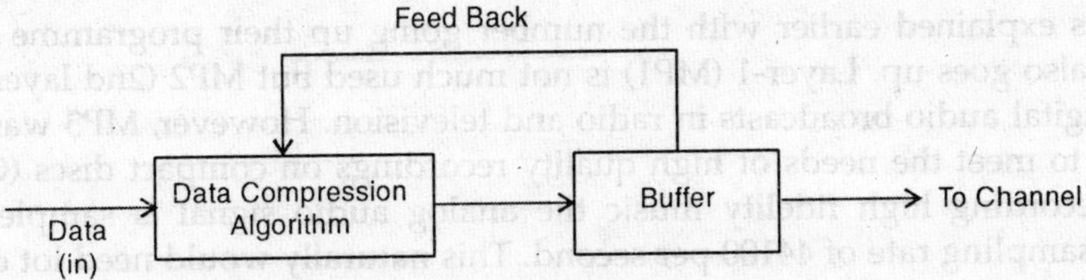

Fig. 21.5. Buffer to regulate flow of data stream.

21.5 Video Compression Formats

MPEG is a acronym for MOTION PICTURES EXPERT GROUP which developed compression systems both for video and audio signals. There has been some confusion regarding numbering of MPEG codes. For one, the word layer probably made sense to the developers because Codes MPEG-1 and MPEG-2 are called layered in the sense that the higher numbered code was built upon the previous one. But to users the naming appears strange-perhaps levels would have been better. MPEG codes are for video signal compression and are numbered as MPEG-1, MPEG-2 etc. The corresponding codec developed for audio compression are called MPEG audio layers 1, 2, or 3. For example 'MPEG-2, layer 3 audio' is actually MPEG-2 video compression codec and MP3 the audio compression codec.

MPEG FORMATS. Initially MPEG-1 was developed for video compression but was soon replaced by MPEG-2 which is now a standard format for satellite broadcasts, DVD movies and High Definition Television (HDTV). Later development efforts resulted in MPEG-3, 4, 5, 6 and now MPEG-7 is being considered. The newer versions are based on detailed study of the response of human brain to different inputs as received via the eye and associated optic nerve. However, the gains have not been enough to justify their superiority over MPEG-2 which then continues to be the standard format for video signal compressions.

"MPEG-2 ENCODER" It analyzes each frame and decides how to encode it. As such it eliminates redundant or irrelevant data and extrapolates information from other frames to reduce the overall size of the file (sequence). The rate of compression depends on the nature of programme. If the encoder is converting a newscast, it can use a lot more predicted frames because most of the scene stays the same from one frame to the next. For other programmes, such as an action movie or music videos where things change very quickly from one frame to the next, the encoder has to create more inter-frames. Thus news-casts or similar low action programmes generally get compressed to a much smaller size than others like action movies or fast games like hockey. The reverse happens at the receiving end where a decoder decompresses the data to its original form before feeding to the D/A converter to obtain original video signal.

At the broadcasting centers the received video binary data is compressed with MPEG-2 algorithms to around 5 to 10 MB/S before transmission. This is the crucial step that has made DTH satellite service a success.

21.6 Audio Compression Formats

Under MPEG-1, three audio coding methods were developed. These are called layers 1,

2 and 3. As explained earlier with the number going up their programme and implementation also goes up. Layer-1 (MP1) is not much used but MP2 (2nd layer) is widely used for digital audio broadcasts in radio and television. However, MP3 was especially developed to meet the needs of high quality recordings on compact discs (CDs).

For recording high fidelity music the analog audio signal is sampled at 65536 levels at a sampling rate of 44100 per second. This naturally would need lot of compression. To demonstrate this let us calculate the data that will get stored on a CD for recording a three minutes stereo song. Since a 16 bit A/D converter would be needed ($2^{16} = 65536$), and stereo would mean two channels, the total data generated will be

$$44100 \times 16 \times 2 \times 3 \times 60 = 254016000 \text{ bits}$$

$$\approx 32 \text{ million bytes (one byte = 8 bits)}$$

Such a huge storage on a CD is quite large for one song especially when a 56 K modem will take close to two hours to download it from the Internet. Obviously compression of this data is a must and MP3 format does so by a factor of 10 to 14 without noticeably effecting the quality of sound. Thus, with MP-3 algorithm, a 32 megabite song on a CD compresses down to about 3 megabites. This enables downloading the song from Internet in minutes rather than hours. A CD can store up to 74 minutes of music and thus has the capacity to store more than 50 songs in compressed state.

MP3 ALGORITHMS: In order to develop such a high rate of compression a technique called "PERCEPTUAL NOISE SHAPING" is used to develop compression algorithms. It is "perceptual" partly because MP3 format uses intricate characteristics of the human ear besides its decibel response to sound intensities. These are:

(i) There are certain sounds that the human ear hears much better than others.

(ii) There are certain sounds that the human ear cannot listen.

(iii) If there are two sounds playing simultaneously, we hear the louder one but cannot hear the softer one.

These characteristics are illustrated in Fig. 21.6 where it is shown that if there is a loud sound in one band the compression algorithm can ignore all of the other bands.

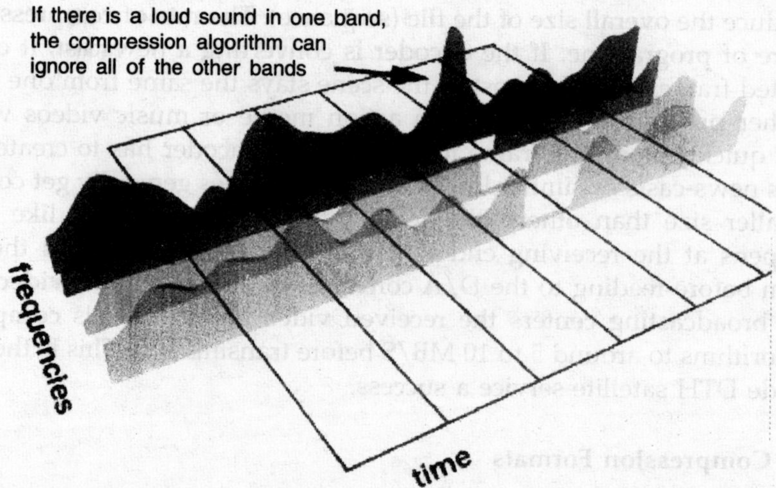

Fig. 21.6. Illustration of how MP3 compression works on Sound signals.

Using the factors listed above, certain parts of the song can be eliminated without significantly affecting the quality of song for the listner. This enables a reduction in data rate by a factor close to ten. Thus it is seen that MP3 is nothing magical, it is simply a file format that compresses music into a smaller size making easier to move around on the internet and store, if necessary. To further improve speech transmission on the Internet a code called Advanced Audio Codec AAC-3 has been developed. It is also referred to as MP4 Codec. Development work continues and we can expect better video and audio algorithms in years to come.

22

DIGITAL SATELLITE TRANSMISSION AND RECEPTION

Introduction

As in other areas of electronics, digital technology has made its impact on television transmission and reception enabling almost complete transition from analog to digital form of signal processing. This has resulted in much improved pictures and sound output quality. However, digitization of video signals in particular, needs high speed A/D converters and means of high bit rate of transmission. But with rapid strides in VLSI technology, high speed A/D and D/A converter are now available. The problem of high bit rate of transmission that needs large channel width has also been solved by signal compression techniques as explained in the previous chapter.

The change over from analog to digital mode of transmission and reception has not in any way affected satellite communication system. The up-link and down-link frequencies are the same in the C-band of frequency spectrum enabling a total channel width of 500 MHz to provide 12 channels of width equal to 36 MHz + 4 MHz as guard band. Frequency 'REUSE' effectively doubles the available number of channels. There are as many transponders in the satellite to carry signals of various types of communication.

Both for International and National broadcasts, channels are hired by prospective users for airing their programmes. All the channel programmes are sent to the broadcasting stations where these are duly processed before up-linking. For all this the reader is advised to revise chapters 12 to 15, if necessary.

22.1 Digital Signal Processing

The various stages of digital signal processing before packetising and multiplexing are—Signal Encoding, Data compression and Encryption. Each stage is separately explained.

(1) *Signal Encoding:* The process of signal encoding of both video and audio signals is fully explained in chapter-21. As stated there, the quantization levels and sampling rates are so chosen that the reproduced pictures are almost undetectable from those transmitted and sound output is also of high quality. The A/D converters deliver corresponding outputs in the binary form as high and low square wave pulses.

(2) *Data Compression:* The need for data compression as explained in chapter-21, is to accommodate the entire data within a limited channel width. From economics point of view the data should be compressed till significant digital artifacts (drawbacks) appear. For this, MPEG-2 plan is used for video signal compression and MP-2 for audio signal compression. With these formats, programmes containing fast moving objects can be compressed 3 or 4 to a transponder. Similarly, programmes containing mostly still images can be accommodated to about 5 to 6 in a transponder. Movies can be compressed up to 7 or 8 in one transponder by cutting down repetitive data which can be added at the receiving end.

(3) *Data Encryption:* The purpose of data encryption by scrambling the video pulse train is to restrict access of Pay Channels to those subscribers only who opt and pay for their viewing. In the digital broadcast system, the video signal is encrypted before up-linking by inserting special keys (data) in the video pulse train to disturb its sequence to prevent normal reproduction of pictures. The keys for de-encryption are also transmitted with the channel data to enable authorised customers to descramble i.e. restore normal sequence of video signal to enable watching of the selected channel. More details on data encryption are given in the next chapter.

22.2 Data Packetising and Multiplexing

The entire information to be transmitted is digital in nature and is sent in the form of data packets which also have signals to register identity to each packet thus enabling their easy separation at the receiving end.

Data Packets: The five types of data packets are: (a) Video that contains visual information after due encoding, compression and if necessary encryption, (b) Audio that contains aural information, (c) Conditional Access (CA) that is addressed to individual receivers, (d) PC Compatible which provides additional useful information to the customer, and (e) Program Guide that keeps the customer informed about channel numbers and current program details. The various basic blocks together with packet listing (see Fig. 22.1) is referred to as UP-LINK set-up. As necessary, for transmission, the data packets are time multiplexed into serial data and sent to the transmitting site for further processing.

Clock System: As shown in Fig. 22.1 all the data processing blocks are fed with clock pulses for initiation, rigorous time duration and sequencing control. For this, timed clock pulses are accurately generated by an electronic digital clock system which is driven by a crystal controlled oscillator, the frequency of which is multiplied and divided as necessary for specific time durations. Such time durations are in the form of pulse trains to be fed to allotted processing blocks. Some output pulses are delayed by storing them in registers to be retrieved at the exact time a sequence is to be encoded in the data packet. Thus, the clock system is actually a complex digital network controlled by a crystal oscillator operating in the MHz frequency range. By common usage, such a complex time reference digital circuitry is simply referred to as "CLOCK" or "CLOCK SYSTEM".

Multiplexing: The data stored in packets is in parallel form and needs to be converted to a serial stream before transmission. For this, the packets are time-multiplexed where data of each packet is transferred to the data stream line in a sequencial manner.

The sequencing mechanism is electronic in nature with gating circuits which open and close as timed by the multiplexing and sequencing control system.

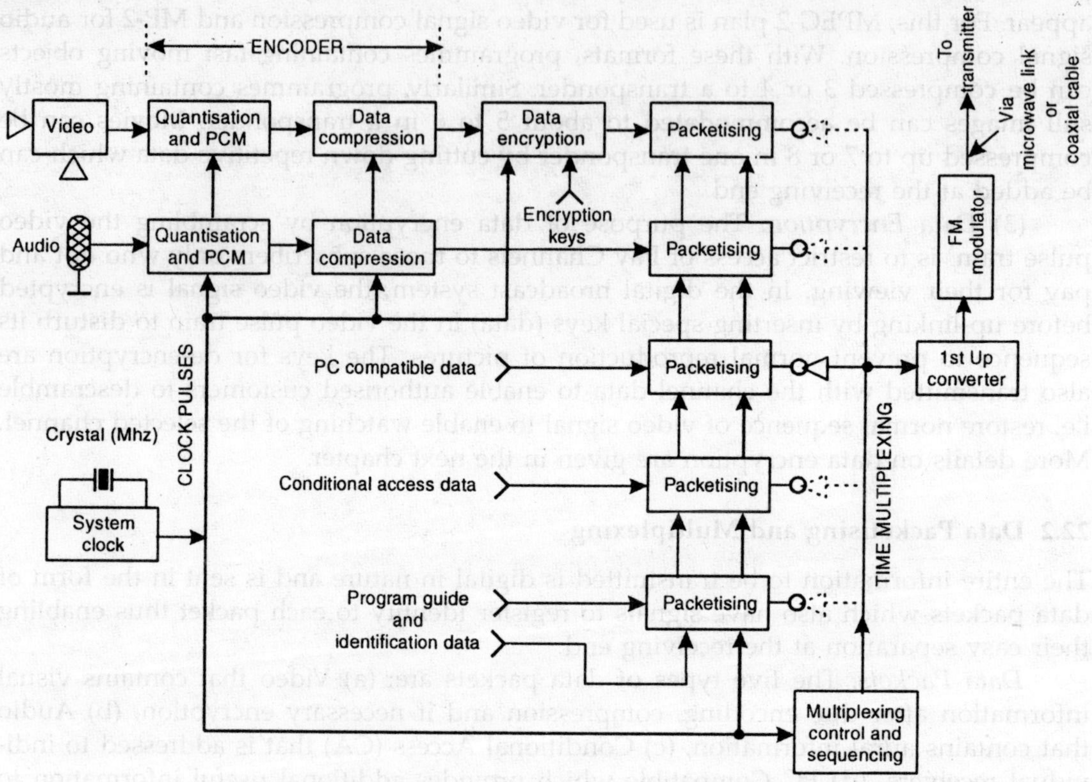

Fig. 22.1. Simplified UP-Link set-up for digital transmission. Multiplexing shown by a moving switch is actually electronic in nature.

22.3 Modulation and Transmission

The data stream as obtained on multiplexing is up-converted and FM modulated on the allotted channel carrier. In earlier days broadcasting stations were located in remote country locations because of their enormous dish antennas and other heavy equipments. Now these stations are much less complex and dish antennas are smaller. Thus many up-link stations are now located on top of tall buildings or in locations where the programme actually originates or is processed. However, when there are small distances between the processing location and up-linking site, a microwave link is used for the transport of data instead of providing a coaxial cable link between them. The advantage of microwave link is that parabolic reflector dish antennas can be used to obtain high gains with a very narrow beam of propagation wave. It may be necessary to up-convert the signal with a suitable carrier before transmitting it via the microwave link.

Transmission: The signal on reception at the transmitting location is up-converted to take it to the allotted GHz range before further amplification and finally connecting it to the transmitting dish antenna. A typical transmitter set-up close to the antenna location is shown in Fig. 22.2. The amplification is done by power amplifiers. The initial stage is an intermediate power amplifier (IPA).

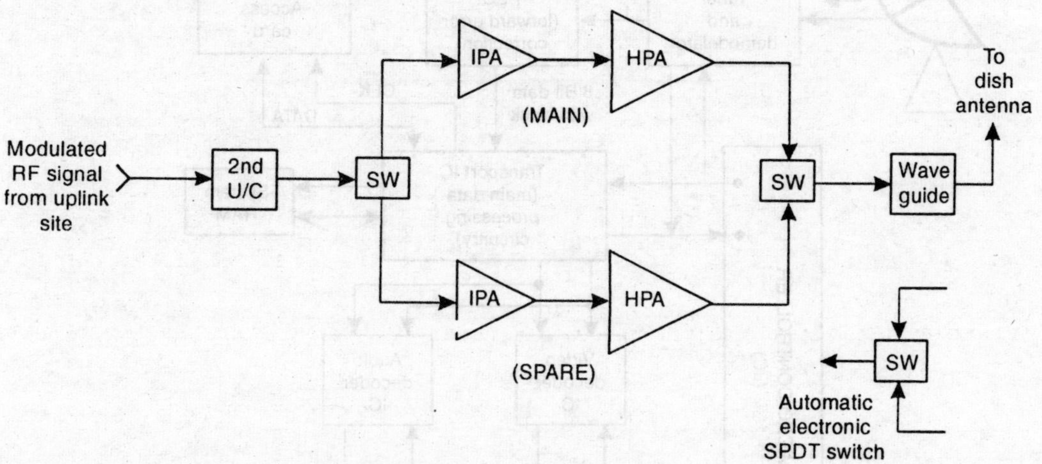

Fig. 22.2. A typical transmitter block schematic at the up-linking broadcast site.

This provides sufficient drive to the final high power amplifier (HPA). Notice in the figure that redundant amplifiers are used. The main IPA and HPA are used until a failure occurs. The switches (SW) automatically disconnect the defective main amplifier path and connect the signal to a spare unit to ensure continuous operation. The amplified signal is sent to the antenna via a wave-guide for radiation towards the satellite.

22.4 Reception and Decoding

The radiated microwave signal on reception by the dish antenna located at the cable station is converted by the LNBC (Low Noise Block Converter) to RF Output at 950 to 1450 MHz. It is then processed by the receiver-decoder to obtain composite video and audio outputs. These are then amplitude modulated in the usual way on assigned cable channel carriers for distribution on the cable network. A simplified functional block schematic of the receiver-decoder is shown in Fig. 22.3.

Its various sections are:

Tuner-Demodulator: The tuner is controlled by the micro-comtroller on receipt of specific code pulses for each channel from the remote control unit. The separated channel signal is demodulated by the demodulator to obtain original data stream. This is made input to the "Forward Error Correction" (FEC) section.

Forward Error Correction: This section re-assembles the data and corrects it if any error is detected. The error corrected data is made input to the Transport IC via an 8-bit parallel interface.

Transport IC: The transport IC is the heart of receiver data processing circuitry. It functions under the control of Micro-Controller unit and isolates video, audio and associated data packets. The separated video and audio payloads are sent to respective video and audio decoders.

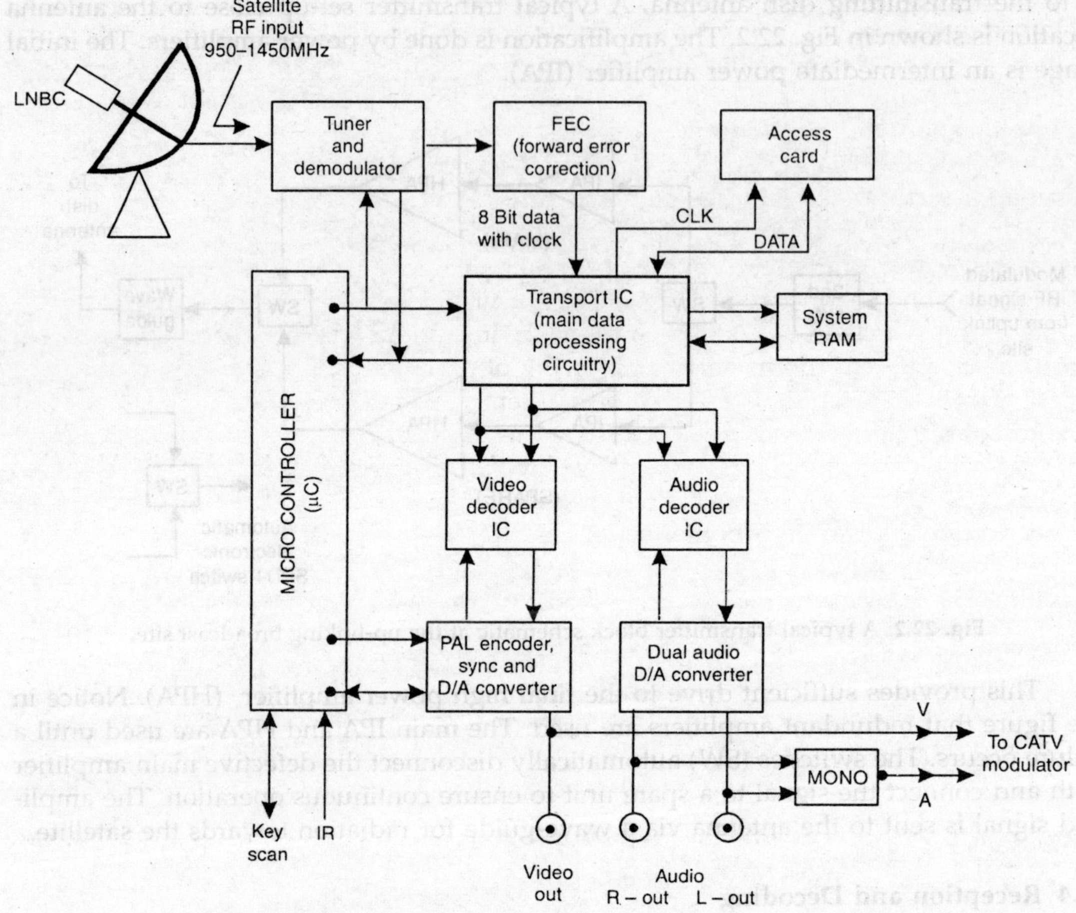

Fig. 22.3. A simplified block schematic of a digital broadcast receiver.

Video and Audio Decoders: The two ICs separately decode the compressed video and audio data and send these to corresponding PAL video and audio encoders.

Video Encoders: The video encoder converts the digital video information into PAL analog video in the desired format for modulation (AM) on any one CATV channel.

Audio Encoder: This IC is a processor cum D/A converter to obtain audio (mono or stereo) signal in analog form. In case of direct reception system, the video and audio signals will become input to the TV receiver. However, if the reception is via a cable station, the two signals feed into corresponding modulators set at carriers of the assigned cable channel.

22.5 Cable TV Signal Distribution

Cable TV signal distribution system is essentially a coaxial cable network to feed multiplexed channel signals from the output (combiner) of the cable station to its subscribers. One millivolt (0 dBmV) is the minimum signal input voltage for receivers having input impedance equal to 75 ohms and 2 mV (6 dBmV) for 300 ohms input impedance receivers. This enables noise free reception. In practice signal voltage amplitude is kept somewhat higher to ensure good picture reception and sufficient sound output. Chapter 15 is devoted to all aspects of cable signal reception, modulation amplification and distribution.

While DTH i.e. Direct to Home satellite broadcast system which delivers much superior pictures and sound quality has come as a better alternative to cable TV system but it will take time for its wide acceptance as cable TV is going to stay for many reasons. Cable operators provide 50 to 100 channels to the customer with choice to select any one channel instantly with the television receiver tuner. It is possible to view one channel and record another on the VCR. In addition, more cable connections can be taken for different rooms enabling viewing of different channels on separate receivers.

For DTH reception, besides a dish antenna, an Integrated Receiver Decoder (IRD) commonly referred to as Set Top Box is necessary. It is expensive and may prohibit many viewers from immediate change over to direct reception of TV programmes. Also, with DTH, only one channel can be tuned in and watched at a time. For access to any additional channel, another set of costly equipment will be necessary. Many cable operators are offering digital signal via special coaxial cables to stay in competition with DTH service. Chapter 24 is devoted to all aspects of DTH Satellite system.

23
TELEVISION CONDITIONAL ACCESS SYSTEM (CAS)

Introduction

The television conditional access system is a technology for signal distribution without having to communicate with any external agency. It is utilized for digital satellite broadcasting. The CAS limits programme viewing to channels opted by individual subscribers and blocks access to the remaining channels. For this, video signals of all Pay Channels are scrambled before transmission to prevent unauthorized reception. At the subscriber's premises for receiving any channel, its data is descrambled by a coded signal that is transmitted with the video signal.

There is no blocking of free-to-air channels but for special programmes such as new film releases, special sports events, movies without commercial interruptions and the like, there are additional charges. The accounting is done at the billing center, located far away, on receipt of information from the subscriber's receiver-decoder. How all this happens, is the subject matter of this chapter.

23.1 System Overview

In the digital broadcast system, the video signal is encrypted before uplinking by inserting special keys (data) in the video pulses train to disturb its sequence and content with the aim to prevent normal reproduction of pictures on the receiver screen. The keys for de-encryption are also transmitted with the channel data to enable the authorized (paying) customer to de-scramble i.e. restore normal sequence of the video signal.

In the receiver, this is achieved by inserting in the Integrated Receiver Decoder (IRD) a card called 'SMART CARD' supplied along with the decoder. It has the correct de-encryption algorithm and security keys which are different for each customer to prevent its use on another receiver. The insertion of 'SMART CARD' also activates coded signal which goes to the billing center over normal telephone lines via a MODEM which is a part of the receiver decoder. In case the telephone line is busy or out of order, the signal remains stored in the receiver and sent when line becomes available.

Requirements to be met for CAS: The scrambling and de-scrambling processes must not alter perceptibly the quality of received pictures, sound and other signals. The system must also provide security against unauthorized access to the service. Special locking signals are inserted in the data stream to prevent any breach of security.

Different 'CAS' Plans: The two CAS schemes in use are:

(1) *Single Conditional Access System:* The term 'SIMULCRYPT' has been coined for the system which defines a common scrambling algorithm. By using it in conjunction with the standard MPEG data, transport and selection mechanisms, it is possible to incorporate in Digital Video Broadcasting (DVB) transmission the means to carry multiple messages which enable control of the same scrambled broadcast but are generated by a number of different CA systems. This Simulcrypt technique allows both the delivery of one programme to a number of different decoder populations that contain different CA systems and also for the transition between CA systems in any decoder population.

(2) *Common Interface:* The common interface or 'MULTICRYPT' operates at the MPEG transport stream level and although specifically intended for conditional access, it can prove more useful in more generic applications such as a worldwide box, which has support for multiple TV standards, languages and different scrambling algorithms. The CA function of decryption and de-scrambling takes place in a small plug-in module similar to size and appearance to a PCMCIA Card as popularly used with laptop PCs. The CA module may operate in conjunction with a smart card through an associated smart card reader.

23.2 Conditional Access Operation

The manner in which CAS operates both at sending and receiving ends is detailed in Fig. 23.1. Conditional Access requires that the information must be scrambled before it is broadcast. This process is under the control of a scrambling sequence obtained from a pseudo-random generator.

The de-scrambling process at the receiving end requires the same sequence to recover the original signal. To provide this sequence and to ensure synchronism between the sending and receiving process, the starting condition of the pseudo-random bit stream generator is controlled by an initialization word as shown in the figure. The initialization word has two components, the control word and the initialization modifier.

The control word is the basic element of security. Its value is chosen arbitrarily and it may be changed during the service operation to enhance security. The control word is communicated to the receiver along with video signal broadcasts. The second component, the initialization modifier is used to impose sufficiently short scrambling sequences to provide security, while avoiding need for too frequent calculations of the control word. It is also broadcast as part of the video signal.

23.3 Functional Sequence of 'CAS' System

The functional block schematic of the complete digital CAS system which complies with the DVB Simulcrypt recommendations is shown in Fig. 23.2.

Subscribers Authorization System (SAS): The SAS is in charge of delivering entitlements to the subscriber smart card. It receives commands from the Subscriber Management System (SMS) also called billing system. The SAS output are ciphered (coded) Entitlement Management Messages (EMMS) ready to be transmitted to the subscriber. The SAS can be connected to several SMSs.

EMM Ciphering Unit: The EMM ciphering unit is connected to the SAS and is in charge of ciphering the messages generated by the SAS.

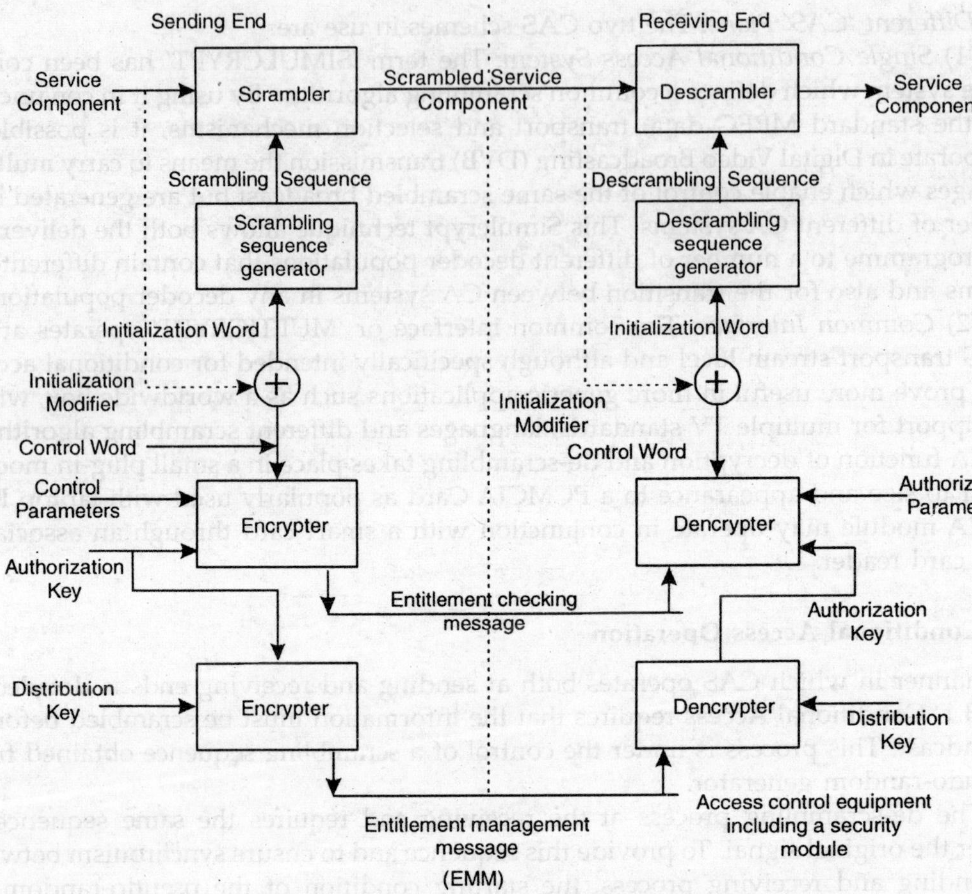

Sending End
Receiving End
Service Component
Scrambler
Scrambled Service Component
Descrambler
Service Component
Scrambling Sequence
Descrambling Sequence
Scrambling sequence generator
Descrambling sequence generator
Initialization Word
Initialization Word
Initialization Modifier
Initialization Modifier
Control Word
Control Word
Control Parameters
Encrypter
Dencrypter
Authorization Parameters
Authorization Key
Entitlement checking message
Authorization Key
Distribution Key
Encrypter
Dencrypter
Distribution Key
Access control equipment including a security module
Entitlement management message (EMM)

Fig. 23.1. Detailed structure of the process of conditional Access System.

EMM ciphering unit
ECM ciphering unit
SMS
SAS
Multiplexer
Satellite
or
Terrestrial
or
Cable
Return Channel
PSTN, Cable Modem
TV
Set Top Box

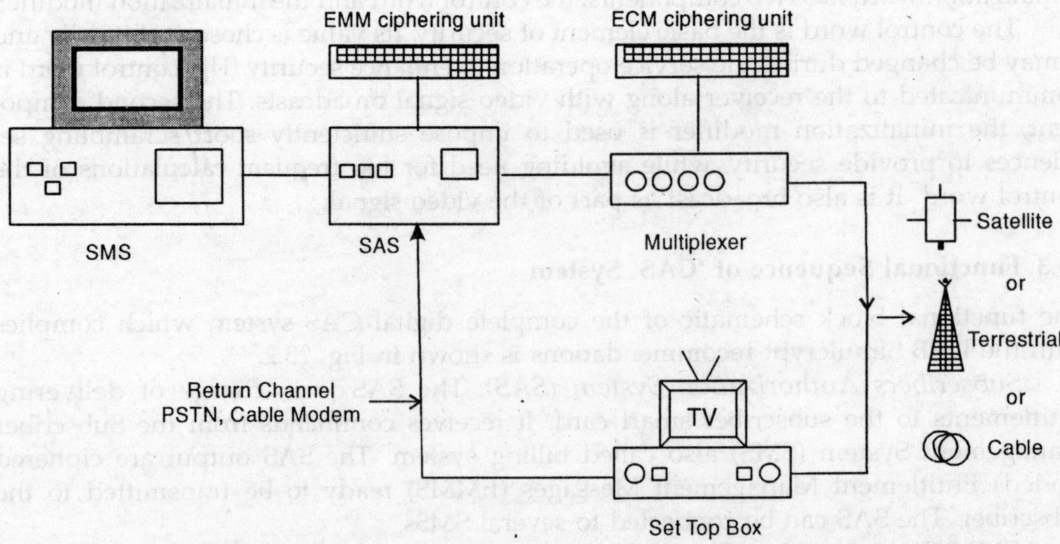

Fig. 23.2. Functional Block Schematic of the Simulcrypt CAS system.

ECM Ciphering Unit: The Entitlement Control Message (ECM) unit is a private conditional access information that specifies control word and possibly other stream—specific scrambling and/or control parameters. ECM's ciphering unit i.e. code generation unit is connected to the multiplexer/scrambler and sends ECMs when requested by the scrambler to do so. The multiplexer output is uplinked to the satellite or sent over cable or broadcast terrestrically depending on the chosen mode of transmission.

Subscriber Smart Card: The Subscriber Smart Card on insertion in the Set-Top Box receives EMMs from the SAS for entitlements modification. It is also able to decipher (decode) the ECM to let the subscriber watch the programme.

23.4 CAS Implementation in Cable System

The Conditional Access System for subscribers who receive programme via cables is somewhat simpler than for direct reception (DTH). Where CAS has not been introduced, each subscriber pays a fixed monthly charge for all the channels that the cable operator provides. In case of default, the subscriber's cable line is disconnected and restored only when due payment is received. Similarly, if the cable station owner (Franchisee) fails to pay in time to the programme provider, for any particular channel, its reception at the cable station is blocked by sending a specific coded key via satellite transmission which prevents functioning of the corresponding receiver decoder. The channel is restored when the bill is cleared.

However, when CAS is introduced in the cable system of signal distribution, a set-top-box for each subscriber become necessary. The block schematic of such a delivery system is drawn in Fig. 23.3. The digital satellite transmission is received at the cable station and the Integrated Receiver Decoder on processing enables video and audio outputs. From the HEADEND, management software is encoded in the video signal. The encoding information also includes encryption and scrambling codecs. The now available data stream and audio are modulated on a pre-selected carrier and sent to subscribers over the cable network.

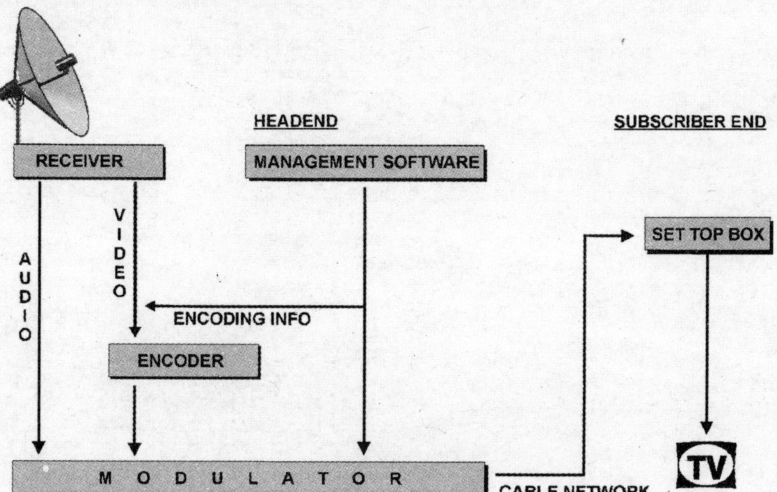

Fig. 23.3. Block Schematic of CAS system for cable network Subscribers (Courtesy FCH Software India Ltd)

When the signal arrives at the set-top-box, it is in scrambled form. Any hacker/ pirate attempting to gain access to the signal will receive the scrambled signal and hence cannot use it. The set-top-box contains a card reader into which the Smart Card is plugged in. The Smart Card contains the subscriber's own private key. The box uses this private key and the public key embedded in the transmitted signal to decode the wanted channel and sends it as pure video-audio steam to the television set.

Normally with fixed number of channels opted by individual subscriber there should be no billing problem as there will be a fixed monthly charge. However, when any subscriber wishes to see any other channel, the set top box informs the accounting service who then authorizes the Head End to send keys to the set-top-box for de-scrambling the asked-for channel. In due course of time, when DTH service commences the cable network will become redundant for families who will change over to direct reception of digital satellite transmission. Of course, they will then come under the rigorous regime of Conditional Access System.

24

DIRECT-TO-HOME (DTH) SATELLITE TELEVISION

Introduction

The Direct-to-Home or Direct Satellite Broadcast (DSB) system enables viewers to access directly many channels of high quality digital video programming over a vast area from one or more high powered KU-Band satellites. There is no need for complex cable networks for signal distribution. For billing purposes, the receiver connects through a modem to the customers' telephone line for communicating channel viewing details to the service computer. The DTH system otherwise functions in the same way for signal encoding, compression packetising and up-linking as explained in Chapter-22 for conventional satellite transmission and reception.

However, many other features of DTH make it superior over conventional satellite broadcasting. The received pictures are of high definition because of the use of high power satellites and transmission in the GHz range. In addition, to prevent unauthorised reception, a special encryption technique called "Conditional Access System" is used. In fact, it was especially developed for DTH service. CAS is fully explained in the previous chapter.

Another important distinguishing feature of DTH is the use of "Multiplexed Analog Component" (MAC) encoding where luminance and chrominance signals are time-division multiplexed to prevent any cross luminance and cross colour effects on the picture in areas of high colour saturation. MAC encoding and other aspects of DTH system are detailed in this chapter.

24.1 MAC Encoding of Television Signals

In the conventional PAL/NTSC systems the luminance and chrominance information is transmitted simultaneously for each line in interleaved form. For sound information, there is an additional carrier and there are two separate carriers in the case of stereo sound. This creates problems of interference beats called cross luminance and cross colour effects associated with areas of high colour saturation. This problem had to be avoided in the satellite standards for DSB transmission wherein bandwidth is much less restricted than in terrestrial broadcasts in VHF/UHF bands.

With this in view, a new encoding system called 'Multiplexed Analog Components' (MAC) was proposed. In this, i.e. MAC encoding, luminance and chrominance signals are time division multiplexed i.e. sent in time sequence as separate components in the time of a normal horizontal line and not simultaneously as in the conventional compatible PAL/NTSC systems.

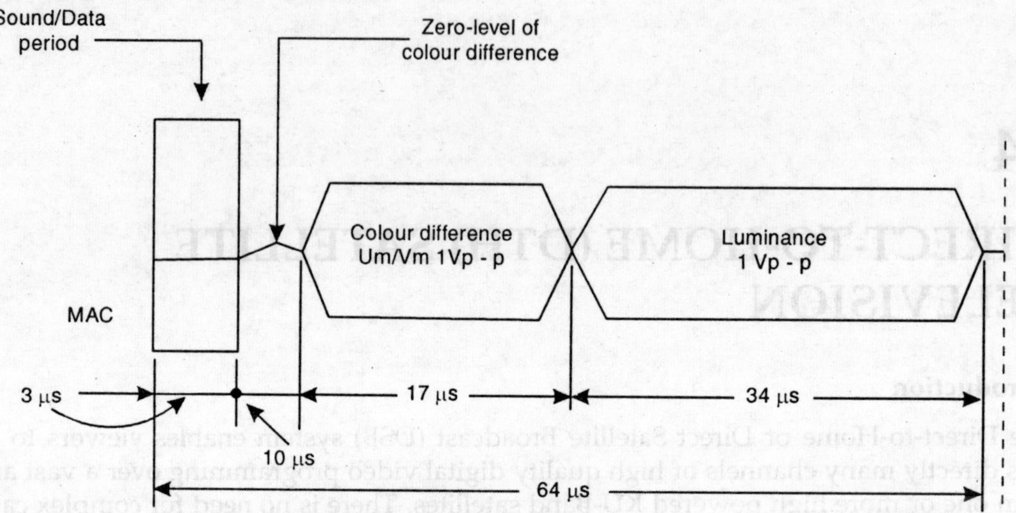

Fig. 24.1. MAC encoding format.

Compression of TV Signal: In MAC colour and brightness informations no longer appear simultaneously in each line, but are transmitted sequentially during the same line period, therefore, the two must be time compressed. The luminance and chrominance signals, U and V are individually time compressed so as to accommodate both in the 64µS line scanning period. Digital technology is cleverly employed here. The luminance and chrominance signals are separately digitized and buffer stored. The stored line data is then read at a faster bit rate so as to compress it into a smaller time interval than the line scan period to leave about 10 µS for digital sync and sound data. The stored data thus read out at a faster rate is in a compressed form and is transmitted separately in sequence—the colour signal U/V first for 17µS followed by luminance for 34µS as shown in Fig. 24.1.

Several versions of MAC format have been developed, in particular the B-MAC, C-MAC and D-MAC systems. The main difference among these relates to the way in which the digital sound and data information are multiplexed with the MAC vision signal.

24.2 Programme Material

The programme material as provided by its owner producer must be of digital quality and thus compatible with the excellence of digital processing, transmission and reception. For this, digitally controlled cameras are used. The signal output from such cameras can be stored in digital memory for any period of time and read and written at different speeds. It can also be stretched, delayed, compressed and read in a form different from the one in which it was written. All this has made possible creation of special effects by which signals of scenes taken at different times and locations can be mixed as desired. This is done with the help of a control computer which is programmed accordingly for a particular presentation. It thus becomes possible for the producer to call pre-recorded background scenes of jungles, rivers, fields, flowers, buildings, etc.,and to mix any one or more with the programme to add elegance to the whole show. To a viewer, while watching such a presentation, it appears that the actors, singers or dancers were taken to all these spots for recording the programme while actually they do not move out of the studio. Similarly, electronic Character-Generators, which are also digitally operated enable imposition of titles, lettering, etc., that appear in virtually every television programme. To enable comparable audio quality, digital audio recorders are used where the coded signals can be stored and processed as necessary before transmission.

The matter does not end with televising scenes and recording accompanying sounds, because lot of editing is necessary before the programme material becomes ready for transmission. Computer assisted editing systems that are now available offer the ultimate in sophisticated video tape editing operation. Such a system combines a data processing computer with television production equipment enabling the operator to easily accomplish a wide range of complex editing in a relatively short period of time. Similarly, digitally recorded audio signals can be edited with ease. Furthermore, for any direct broadcast system, additional information received from the producer includes Conditional Access Data, PC compatible data and Program Guide menu. This comes via computers where it is stored earlier. The video, audio and additional information data on due processing is put into separate data packets prior to uplinking. The video cameras, microphones and computers shown in Fig. 24.2 are thus only symbolic representations of various inputs because, as explained above, lot of production effort goes into bringing each input to the form in which it is transmitted and later presented to the viewer.

24.3 Data processing and Packetising

The assumed data processing configuration shown in Fig. 24.2 was earlier for one transponder. Now with data compression techniques and digital processing it is possible to broadcast four to six channels from one transponder. And, in addition to TV channels, music video and music only channels are also broadcast. The music only channels can be played through the usual stereo system or even through the TV receiver. In Fig. 24.2 two TV video channels with associated audio for stereo or bilingual and additional data input is shown to indicate the nature of programme information that gets transmitted through one channel.

(1) *Data Encoding and Compression:* The quantisation of analog data and its conversion to digital form by PCM is shown in Fig. 21.1 and explained in section 21.2. For compression, while the MPEG-2 pattern is followed but data is transmitted in accordance with MAC encoding plan as explained in the previous section. As shown in Fig. 24.2 clock pulses are fed to both the PCM encoder and compression blocks to control data transfer rate and maintain synchronism between all the processing units.

(2) *Data Encryption:* As explained in the previous chapter, video data is encrypted at the up-link site to prevent unauthorised reception of restricted TV channels. A secure encryption formula known as Conditional Access System (CAS) is used to modify the encoded video information. The keys (codec) for encryption are obtained from

Conditional Access Data that is available in a computer memory and transmitted through one of the packets.

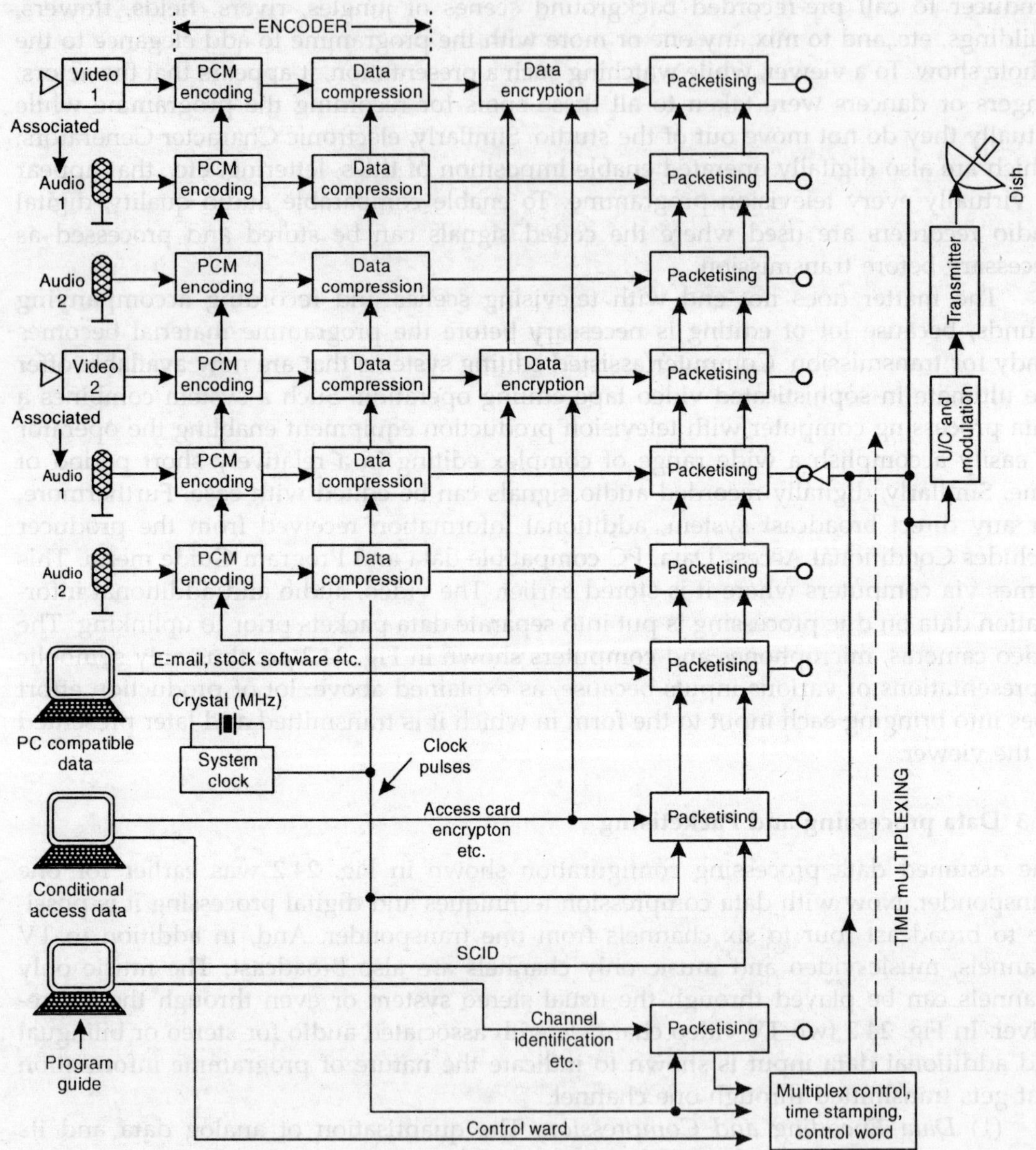

Fig. 24.2. A representative data processing set-up for one television channel.

Decoding keys to be used at the receiver are also available at the Conditional Access Data source and are added to the serial data. The Smart Card which is supplied along with the receiver is used which on insertion in the decoder identifies de-encryption keys which allow the receiver to decode video information to its normal format.

(3) *Data Packets:* The entire information to be transmitted is digital in nature like that in a computer. It is sent in the form of data packets which also contain signals to register identity of each packet thus enabling their easy separation at the receiving end. The five types of data packets used in the DTH system are:

(i) *Video:* This packet contains, visual information as its payload after due encoding, compression and encryption.

(ii) *Audio:* In this packet the useful payload is the aural component of the scene as obtained from the microphone after necessary encoding.

(iii) *Conditional Access Data (CA):* This packet contains coded information that is addressed to individual receiver such as activation data, customer's E-mail and channels the receiver is authorised to decode.

(iv) *PC Compatible:* The PC compatible serial data packet can obtain any form of data the program provider wants to transmit such as stock reporting, and any additional software.

(v) *Programme Guide:* It maps channel numbers to transponders in the satellite and contains SCIDs (more on this later). The guide packet also gives the customer current TV and other programme listing information.

Data Packet Composition: Each data packet is around 150 bytes long depending on the TV system and compression technique used as shown in Fig. 24.3.

(a) *First Two Bytes contain SCID information and flags:* The SCID (Service Channel's Identification) is a unique 12 bit number which identifies the packet's data channel. The flags (keys) that contain the remaining 4 bits, are for telling whether or not the packet is encrypted and in what form.

(b) *Third Byte has two components each of four bits:* The first is for indicating packet type and when combined with SCID code, determines how the packet is to be used. The second 4-bit part constitutes a continuity counter which increments once for each packet type and SCID.

(c) *Next packet of nearly 130 bytes* is the payload data which is the actual useful information to be transmitted. The nature of payload varies as the video and audio signals vary from instant to instant.

(d) *Last 17 Bytes contain Forward Error Correction (FEC)* data that enables verification of the correctness of transmitted information at the receiving end. Errors that occur during transmission get corrected at the receiving end on comparing the arriving data with the FEC data available with it. The constituents of each data packet as detailed above can vary depending on the type of programming. Also as explained earlier, the number of total bytes in a packet can vary somewhat depending on the television system in use.

(4) *Signal Multiplexing into Serial Data:* The time multiplexing of data from all the packets into a serial stream is done through the multiplexing control unit (See Fig. 24.2) which is designed to establish contact with the output point of each packet sequentially on a continuous basis. The data stream thus obtained is frequency modulated and up-converted in frequency before further modulation and transmission.

24.4 Signal Uplinking and Satellites

Up-linking: The up-conversion of the final signal to the assigned GHz range is similar in all satellite transmission systems as detailed in the 22nd chapter. However, there can be some variation depending on system design and power of the transmitter.

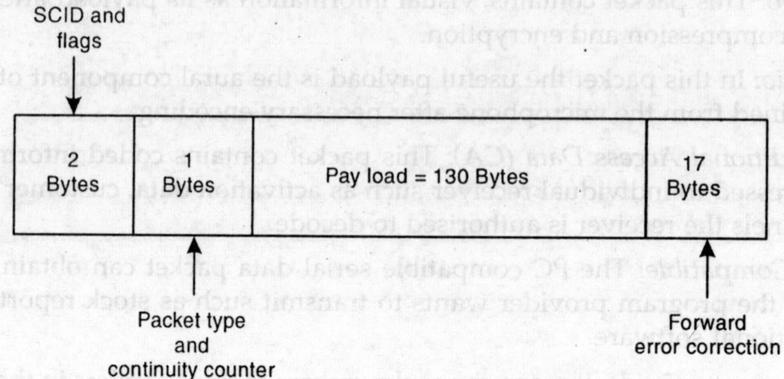

Fig. 24.3. Various constituents of a data packet.

As explained in Chapter 4 (section 4.3), special tubes have been developed for amplifying signals at microwave frequencies. For Direct Broadcast Systems, high power transmitters are preferred where Magnetrons and Travelling Wave tubes are often employed.

Satellites: For digital transmission, Ku-band frequencies are preferred because these are not prone to interference from, ground point-to-point communications and also need much smaller diameter dish antennas. Ku-band satellites may be classified into two groups-low and medium power. These transmit in the 11.7 to 12.2 GHz range. But the new high powered Ku-band satellites are assigned 12.2 to 12.7 GHz band for efficient operation of DTH transmissions. In some regions where DTH is yet to be introduced, digital transmission is taking place in C-band and signals are received with the same large diameter dish.

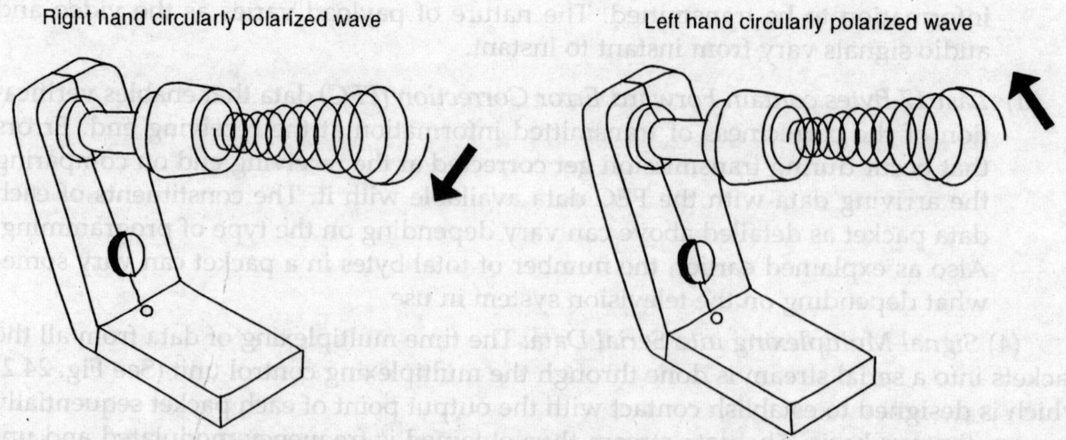

Fig. 24.4. Right hand and left-hand circular polarization.

Polarization: Unlike C-band satellites, which employ horizontal and vertical polarization, the DSB (DTH) transmitters are designed for circular polarization. With this, the microwave energy travels in a spiral-like pattern as pictured in Fig. 24.4. If the number of channels to be provided is large—say more than 50, two high powered satellites are used with a spacing of only 0.5 degree from each other thus enabling the use of one dish antenna at a fixed orientation. However, to prevent mutual interference, one satellite uses Right Hand Circular (RHC) polarization and the other is designed for Left Hand Circular (LHC) polarization. Modern satellites have a life expectancy of about 12 years.

The LNBC: The LNBC converts the 12.2 to 12.7 GHz down link signal from the satellite to the 950-1450 MHz range to be fed to the receiver tuner. Two types of LNBCs are available-single and dual output. The single output version has only one RF connector while the dual output type can be used to feed a second receiver to tune and receive another channel or to connect to some other signal distribution arrangement.

24.5 DTH Satellite Receiver

The DTH television receiver is a complex digital signal processor employing specially produced VLSI chips and micro-processors. The binary data received from the satellite is decoded and digitally processed. No analog signals are to be found in any section of the receiver except in the video encoder and audio digital-to-analog converter (DAC) as indicated by dotted chain lines in Fig. 24.5. The main manufacturers of DTH receivers are SONY, JVC, HITACHI and TOSHIBA of Japan, RCA of USA and ITT of Germany. While the basic building blocks of all designs are essentially the same, each has unique features that suit different TV standards. A simplified block schematic of a typical DTH satellite receiver is shown in Fig. 24.5 and each circuit block is briefly described.

Tuner-Demodulator: On down converting the satellite signal to 950-1450 MHz by the LNBC it is routed to the receiver input socket by a short length of coaxial cable and fed to the receiver tuner. The turner isolates the required channel and the demodulator converts the modulated data to a digital data stream. The channel selection is controlled by the micro-controller on receipt of coded input from the remote unit via IR (infra-red) radiation. It is different for each channel.

Forward Error Correction (FEC): The data is encoded at the transmitter site by a process that enables the decoder to reassemble the data, verify it and correct errors that may have occurred during transmission. This process is called "Forward Error Correction". As shown in the figure, the data output on FEC is made input to the transport IC via a 8-bit parallel interface.

Transport IC: The transport IC is the main computing unit of the receiver. It has its own RAM and its functions are controlled by the micro-controller. The data on receipt from FEC section is processed by the transport IC and sent to respective Video and audio decoders. The micro-controller communicates with video and audio decoders through this IC. Furthermore, the 'Smart Card' interface is also processed through it.

Smart Card: The Smart Card also called Access Card receives encrypted keys for decoding a scrambled channel from the transport IC. The keys i.e. data for descrambling is also transmitted with the channel data to enable the customer to descramble i.e. restore normal sequence of video data without having to communicate with any external agency. This is achieved by inserting the Smart Card into a slot provided for it on the front panel of the receiver. This operation also activates a coded signal which goes to the billing centre over normal telephone line via the Modem. Necessary billing for other services availed by the customer are also accounted-for through the access (smart) card.

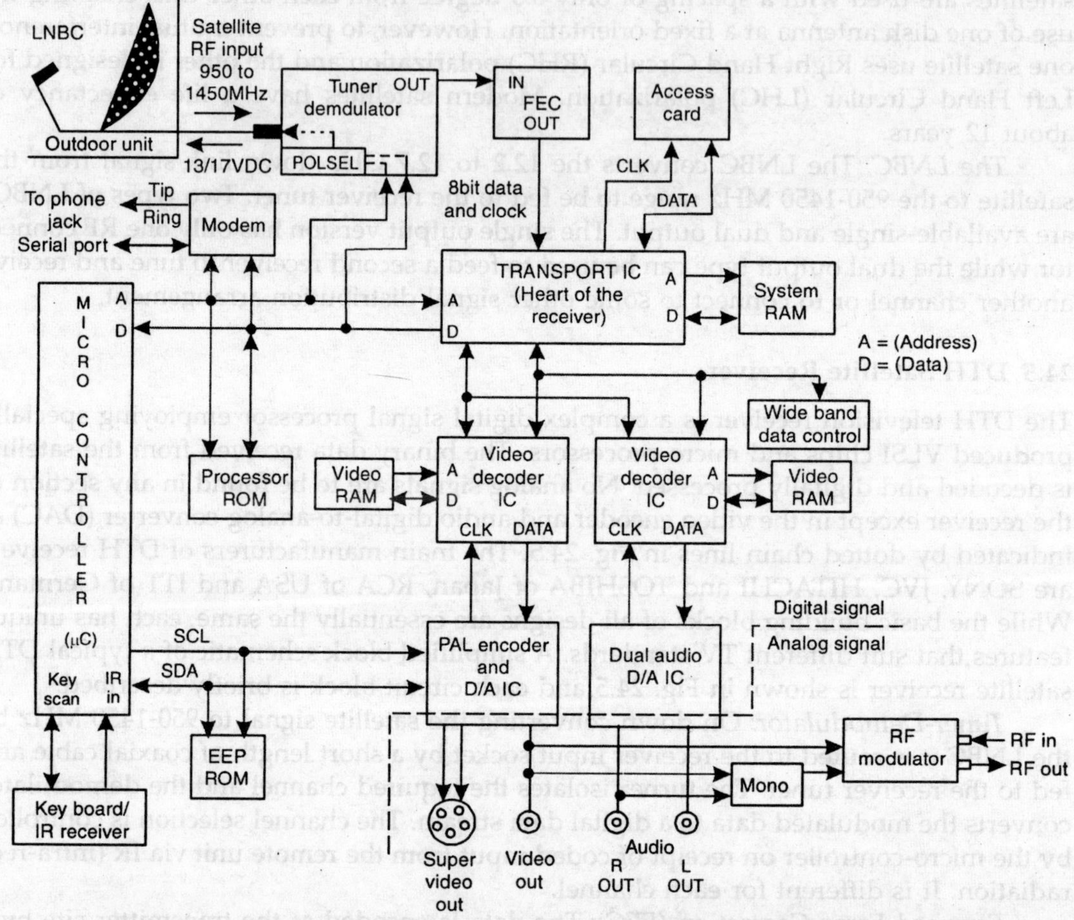

Fig. 24.5. Block schematic of a DTH satellite receiver.

When the Smart Card is activated in the receiver for the first time, the serial number of the receiver is encoded (entered) on it. This prevents the card from activating any other receiver except the one in which it is initially authorized. Also, the receiver does not function with Smart Card removed from it.

Video Decoder IC: This IC on interfacing with its own RAM isolates the video payload data from the packet, decompresses it into its normal data stream and feeds it to the encoder and D/A IC.

PAL Encoder and D/A IC: Here the received digital video data is converted into analog form and encoded to obtain standard composite PAL video output that can be fed directly to the TV receiver. It enables another output called S-Video (Super-Video) that can provide best possible picture quality but is of use only if the available TV receiver has S-Video capability.

Audio Decoder IC: The function of this IC is similar to Video decoder IC but it deals with the associated audio signal. The isolated and decompressed signal is fed to the dual Audio and D/A IC.

Dual Audio and D/A IC: This IC is designed to separate left and right audio channel digital data and to convert it back into analog stereo output labelled as R-out and L-out in the figure. The output can also be obtained as two independent bilingual audio outputs if transmitted this way.

Micro-controller (μC): The micro-controller or processor IC receives, decodes, controls and initiates dedicated functions on receipt of binary command signals via IR receiver and keyscan inputs. Its software is contained in the Processor ROM which is also interfaced with the Modem. As shown in figure (Fig. 24.5) the μC also controls other digital devices (Ics) of the receiver. It is also programmed to turn ON, the green (ON/OFF) LED located on the front panel of the receiver.

Modem: The modem connects to the customer's phone line. It calls the programme provider's computer and transmits the subscriber's purchase for billing purposes. When the modem first attempts to dial, it sends its first number as touch-tone. If the dial tone continues after the first number, the modem switches to pulse dialing and redials the number. If the dial tone stops after the first number, the modem continues to dial the rest of the number. The modem also automatically releases the phone line if the customer picks-up another phone on the same extension.

24.6 DTH Receiving Equipment

Fig. 24.6 shows the complete DTH receiving equipment. Its three essential components are the receiving dish, receiver-decoder unit and the interactive remote control.

Receiver Front Panel: The front panel of a typical DTH receiver is shown in Fig. 24.7(a). The different operation controls and buttons provided on it are:

The ON/OFF button turns the receiver ON and OFF. When the receiver is turned OFF, a flashing light close to the label 'Message' indicates that necessary information has been sent to the customer service centre.

TV/DTH or TV/DSB button switches between the satellite signal and the terrestrial antenna (or cable) when the receiver is turned 'ON'.

Four cornered arrows glow to point to different items in the menu system and programme guide and also for channel surfing while viewing video. On pressing the MENU button, a list of available programmes appear on the TV screen.

The Select/Display provision is for selecting items out of the program guide and menu charts.

Access (Smart) Card Slot accepts the Access Card. It is a Credit Card sized device which on plugging in the slots allows the decoder to receive authorised programming.

Receiver Back Panel: A view of the DTH receiver back panel and connection locations are shown in Fig. 24.7(b). Its connections are both inputs and outputs.

Receiver Inputs (i) Phone Jack: This connects the phone line to the DTH Receiver, (ii) Satellite In—This connects the 950-1450 MHz input from the LNBC, (iii) IN From Antenna—It is used to connect the normal terrestrial reception antenna and cable service to the receiver.

Receiver Outputs (i) Out to TV: This is the RF out jack (either on channel 3 or 4) of the receiver. This RF can be one of two signals—one, the RF signal applied to the

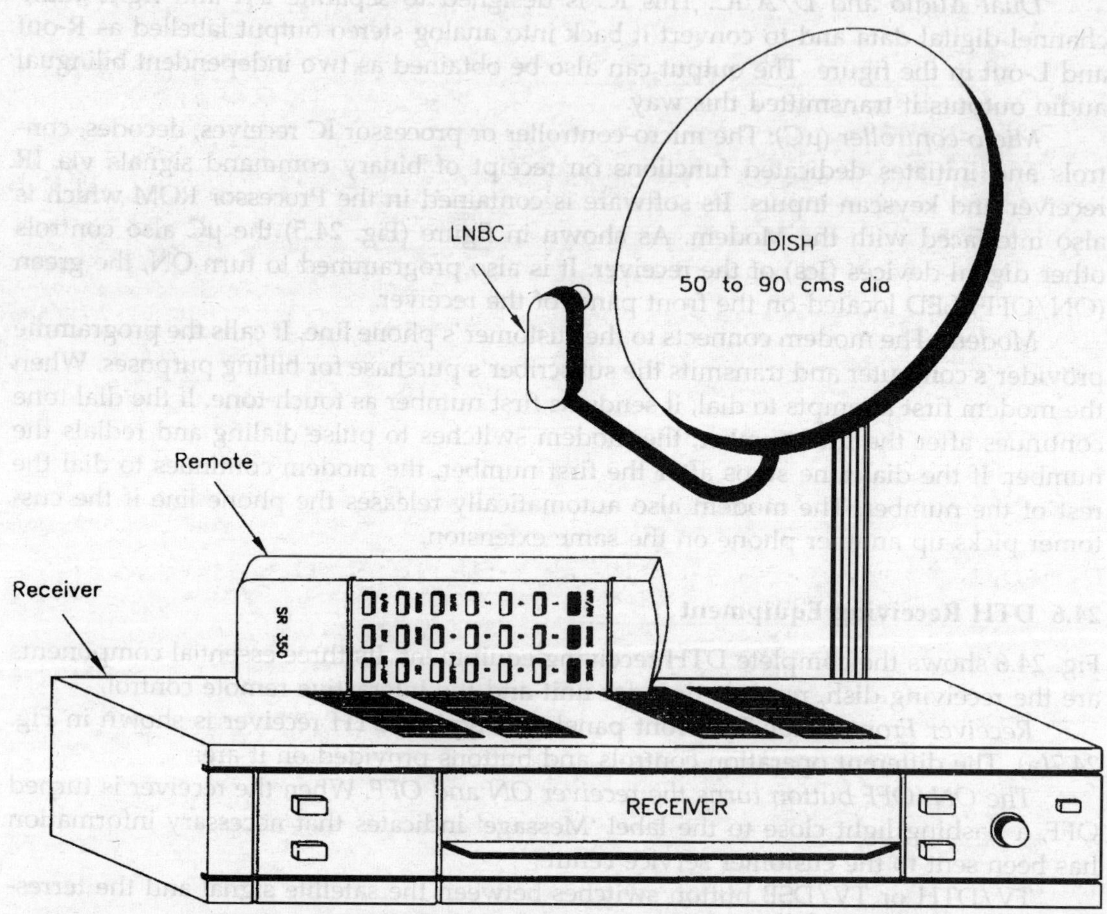

Fig. 24.6. The three units of the receiving DTH satellite system—The Dish, Receiver-Decoder and Remote Controller.

connector marked "In From Ant" and the other 'output of the receiver'. The TV/DTH (DSB) button on the front panel of the receiver determines which signal is present on the jack, (ii) S. Video—This connector connects Super Video to the compatible television set or VCR. (iii) Video—This connector outputs video signal from the DTH receiver to a monitor type of TV or VCR. (iv) Audio (Left) and Audio (Right). These are left and right channel audio signals from the receiver. In a monitor type of receiver, these are connected to audio inputs. These inputs can also be connected to audio input of stereo amplifier. As shown in Fig. 24.5, mono input is also available, (v) Wide Band Data. This connector (also shown in Fig. 24.5) is designed to be used in conjunction with technologies such as High Definition TV, Interactive TV, etc.

Remote Control of Receiver: The remote control pictured in Fig. 24.6 functions in a similar manner as for TV/VCR receivers. Its functions include DTH/VCR, ON/OFF, ALT (alternatively) input, guide, arrows, clear, TV/DTH receiver, channel, volume, mute, menu and many more. These are fully listed and explained in the manual that is supplied with the receiver. The term Menu is rather loosely used and includes information on programme guide, coming attractions. Mailbox, options in Menu, alternative audio, help menu, etc., which appear on the TV receiver screen as on computer monitors.

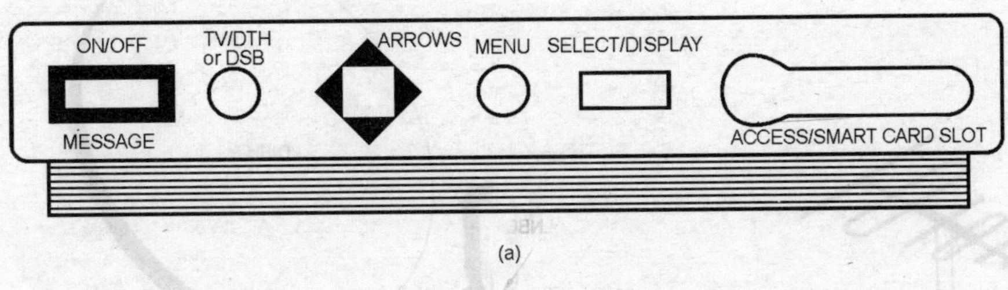

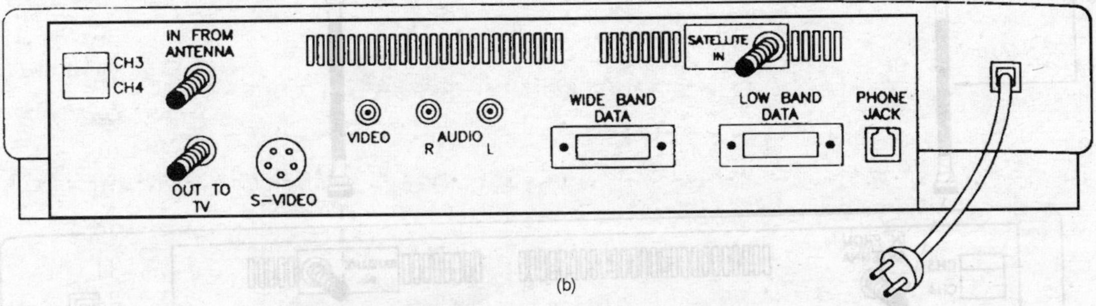

Fig. 24.7. (a) DTH receiver front panel buttons/controls.
(b) DTH receiver back panel connections.

System Connections: Fig. 24.8 shows basic connections from a satellite DTH receiver to a standard TV receiver. The TV/DSB button on the front panel of the receiver switches the RF signal applied to the TV receiver between the antenna or cable signal and the signal from the receiver. If the TV receiver is capable of receiving video and audio signals directly from the receiver, video and audio cables need to be added. To watch satellite signals, monitor inputs to the TV receiver should be selected. To connect a stereo amplifier to such a system, make Audio L, Audio R connections to the music system.

Pointing the Dish: The dish must be accurately pointed towards the satellite using both elevation and azimuth adjustments. Unlike analog transmissions, the alignment for Digital System is critical for satisfactory performance. When the dish is pointed directly at the satellite, a strong signal is delivered to the receiver. Any dropouts in the signal would be small and get corrected in the receiver. If the dish is not positioned properly, the signal will be weak and the number of dropouts then increase. Depending on the size and number of these dropouts, the receiver may not be able to correct for these and the picture will be blank or freeze frame until good signal is once again received. This would become worse on cloudy or windy days.

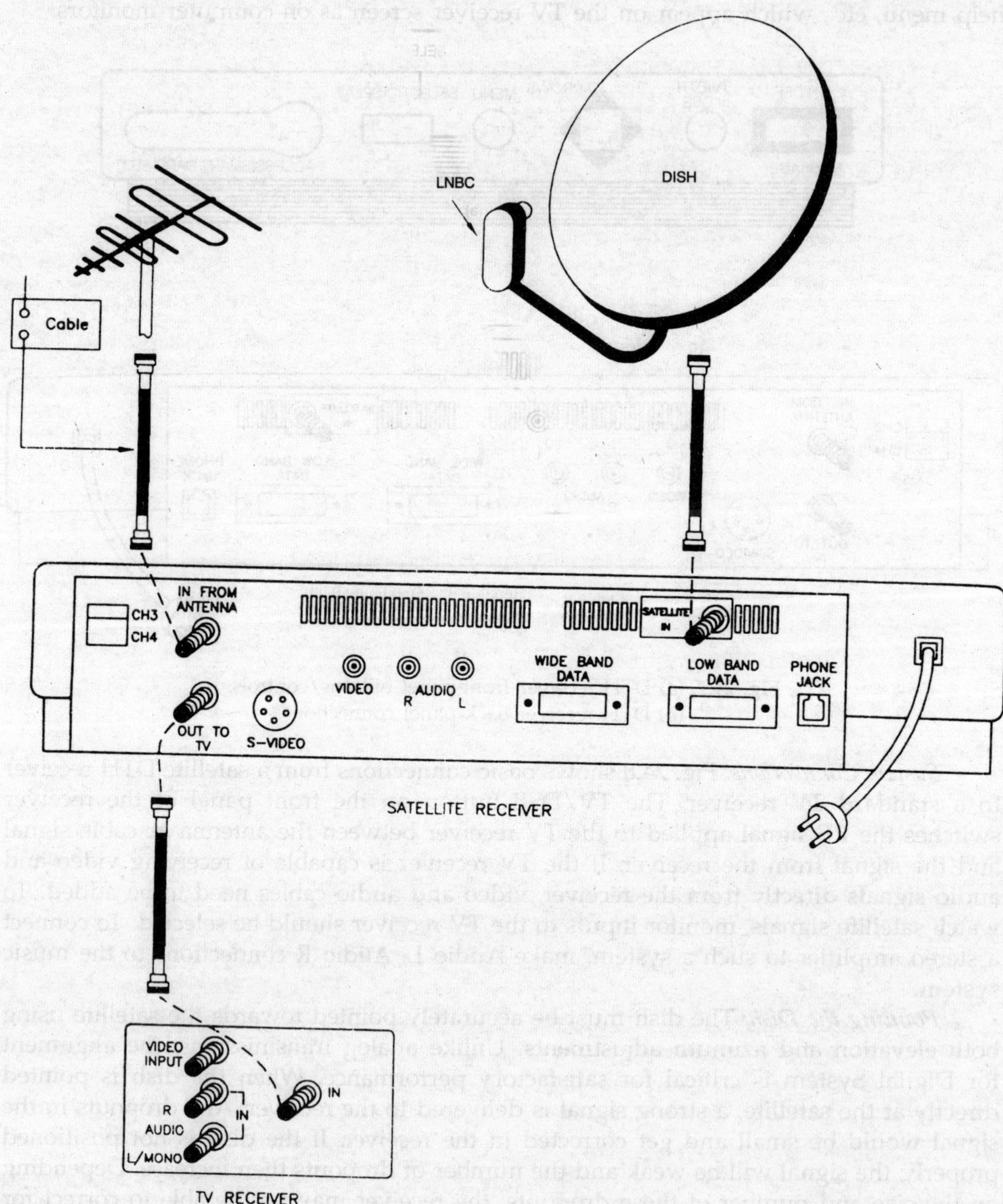

Fig. 24.8. Basic connections from the rear panel of a DTH Satellite receiver to a standard TV receiver.

24.7 Consumer Guide to DTH Satellite System

While DTH satellite broadcast system is fully explained in the previous sections, its salient features are enumerated next for the benefit of those planning to get connected to this service.

(1) Direct-To-Home system is designed to deliver TV and other programmes directly from the broadcaster to the consumer. The received pictures are crystal clear and the sound better than now available out of TV sets.

(2) In addition to TV and music video, music only programmes are also broadcast which can be played on the stereo music system or even though the TV receiver.

(3) Pay-Per-View is made possible via connection to the customer's telephone line. The phone serves as the two-way interactive part of the pay-per-view letting the programmer know that the customer has ordered a pay event.

(4) Just with a keystroke on the keypad on the remote, latest block-buster movies and other important events can be received on the TV screen via pay-per-view service.

(5) With the remote control in hand, it is possible to directly select the desired channel, check E-mail sent by the programmer, turn volume up and down, mute the sound, return instantly to the previous channel and even set-up favourite programmes that can be called later with a keystroke. With advanced remotes, it is possible even to operate one's TV, VCR and other units such as audio/video receivers. Another very useful feature of the remote is that besides enabling the viewers to surf channels from the comfort of one's chair, on-screen programme guide can also be accessed.

(6) DTH is compatible with all TV sets currently in use. All that is required is a physical connection from the DTH receiver-decoder to the TV set and also if desired to the VCR. Even the audio portion of the received signal can be fed to a stereo receiver or processor. Finally, if the TV receiver has Super(s) video capability, the DTH receiver has S-video output that can be availed of through a separate cable connection between the two units.

(7) Local terrestrial broadcasts are the best source for local news events and weather reports. These can also be viewed on the same TV set by connecting the local RF signal to a separate input socket on the DTH receiver.

(8) A single receiver-decoder can decode a single DTH channel which can then be routed to several TV sets and or VCRs throughout the house. However, a separate receiver-decoder unit is needed for each TV or VCR if it is desired to receive different satellite channels simultaneously. Therefore, to watch two different satellite channels at the same time or to tape (record) one channel while viewing another, it is necessary to have two receiver-decoder units.

(9) DTH receive equipment (dish, receiver and remote) is expensive. Therefore, unless the number of available DTH channels is more than 10, the investment is perhaps not justified.

(10) If the DTH channels are broadcast from more than one distant satellite, separate dish antennas will be needed.

25

HIGH DEFINITION TELEVISION (HDTV)

Introduction

Even with the improvements implemented in the DSB (DTH) systems, the functional capabilities of existing TV systems are inadequate to fully utilize the psychological characteristics of the human vision system. Researchers have, therefore, been working on a new television system now called 'High Definition Television'. It aims at (i) improvement in both vertical and horizontal resolution of the reproduced picture by approximately 2:1 over existing standards, (ii) much improved colour reproduction, (iii) higher aspect ratio of at least 5:3, and (iv) stereophonic sound. Their implementation results in a picture quality as clear as obtained from 35 mm cine films and sound as good as from digital audio discs.

The worldwide pioneers in HDTV are Japan and America. The standards adopted include 1125 scanning lines per frame, 60 fields per second, 2:1 interlace scan and an aspect ratio of 16:9. The 1125 lines were chosen to get an approximate doubling of vertical resolution and to allow for 9/5 and 15/7 down conversion to Europe's 625 line PAL standards and America's 525 line NTSC standard respectively. However, European countries have developed their own HDTV called "EUREKA" to be compatible with CCIR standards but with line numbers equal to 1250/1251 and an aspect ratio of 16:9. The bandwidth required to transmit one TV channel on compression is 10 MHz, meaning one transponder will carry lesser such channels. The real advantage of HDTV is that large size pictures can be seen with clarity because of the greatly improved resolution. For this, CRT industry has pushed the picture tube screen to a diagonal of 40 inches. However, screen size is no more a problem because *projection TV systems are now available where the screen, made of cloth, is a separate unit on which the picture is projected. At present HDTV transmission is available on a few channels because of the problems of implementation to be compatible with the existing receivers and multiple TV systems. All aspects of HDTV are briefly described in this chapter.

25.1 HDTV Parameters

HDTV is a high resolution digital television system with Dolby Digital *surround sound which delivers excellent pictures and high quality sound output. High resolution

---

* Projection television sets and Surround sound formats are described in the next chapter.

pictures are obtained because of the very high scanning rate as compared to 625 lines in PAL and 525 in the NTSC system. The pictures as received with 4:3 and 16:9 aspect rates are shown in Fig. 25.1.

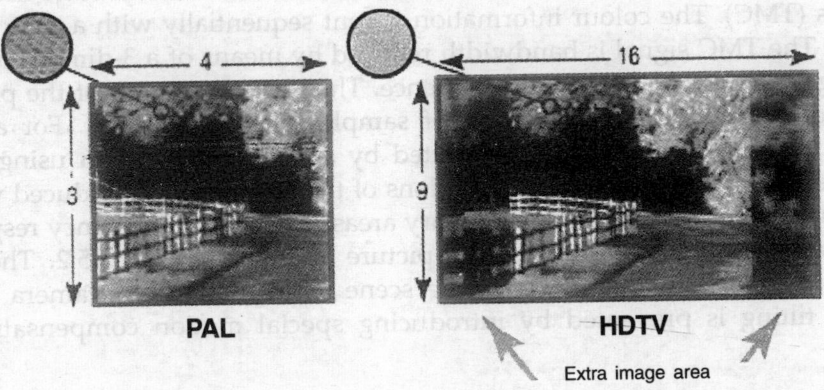

Fig. 25.1. Reproduced pictures with 4:3 and 16:9 aspect ratios.

In USA at present two transmitting formats are in use—One with progressive scanning and 16:9 aspect ratio and the other with interlaced scanning and 4:3 aspect ratio to avoid inconvenience to viewers during the transition period. Similarly, because of the distinctive features of HDTV, new production and transmission equipments are necessary at the transmitting stations and broadcasters are taking time to implement such a change over.

25.2 HDTV Signal Compression and Encoding

Digital TV in general relies on MPEG-2 compression and encoding scheme to fit its images into a reasonable bandwidth. As explained in an earlier chapter, its software records just enough of the picture without making it look as if something is missing. The MPEG-2 encoding reduces the amount of data by about 50:1 to squeeze picture details and stereo sound into the same channel width of 7 MHz for PAL, 6 MHz for NTSC and 8 MHz for the British TV system. For satellite transmission of HDTV, a channel width between 10 to 12 MHz is provided.

Improved Colour Picture: The conventional colour TV systems in use, introduce colour errors and other impairments in the displayed picture largely because of the assumptions about colour primaries which are not quite correct.

Such impairments were considered acceptable in the present day systems alongwith other limitations because they go unnoticed due to smaller picture screen. When HDTV was introduced, due attention was paid to improve colour rendition by generating more accurate primaries. For this, EUREKA evolved new compression and colour reproduction methods. However, Japan found a new approach called MUSE to counter colour picture deficiencies and this found wide acceptance all over the world.

Muse Encoding: Muse stands for Multiple-Sub-Nyquist Sampling Encoding and is an HDTV bandwidth compression scheme developed by NHK of Japan. It uses the fundamental concepts of performance exchange in the Spatic-temporal (transitory transformation) domain along with motion compression to reduce transmission bandwidth down to about 10 MHz. The processed HDTV signal can then be transmitted using a single satellite channel.

In MUSE, the luminance and colour information are sent by Time-Multiplexed Components (TMC). The colour information is sent sequentially with a time compression of four. The TMC signal is bandwidth reduced by means of a 3-dimensional offset subsampling pattern over a four-field sequence. The stationary areas of the picture are reconstructed by temporal interpolation of samples from four fields. For a moving picture area the final picture is reconstructed by spatial interpolation using samples from a single field. Hence, the moving portions of the picture are reproduced with one-quarter the spatial resolution of the stationary areas. The spatial frequency response for both stationary and moving areas of the picture is shown in Fig. 25.2. The lack of resolution during movements of the entire scene as in the case of camera panning, zooming or tilting is prevented by introducing special motion compensation techniques.

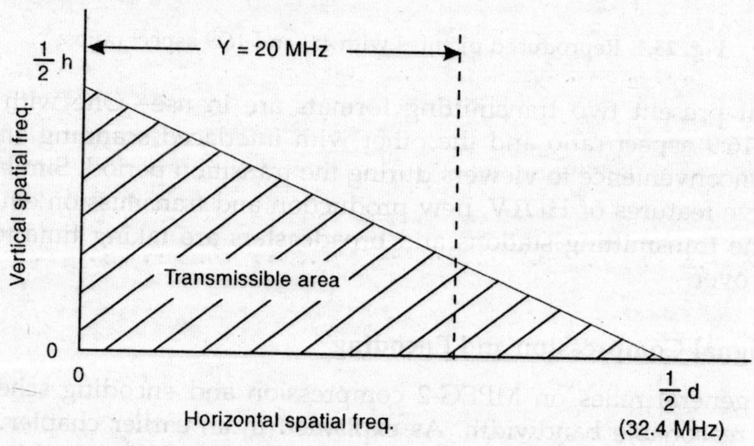

(a) For stationary portion of the picture (Temporal Interpolation)

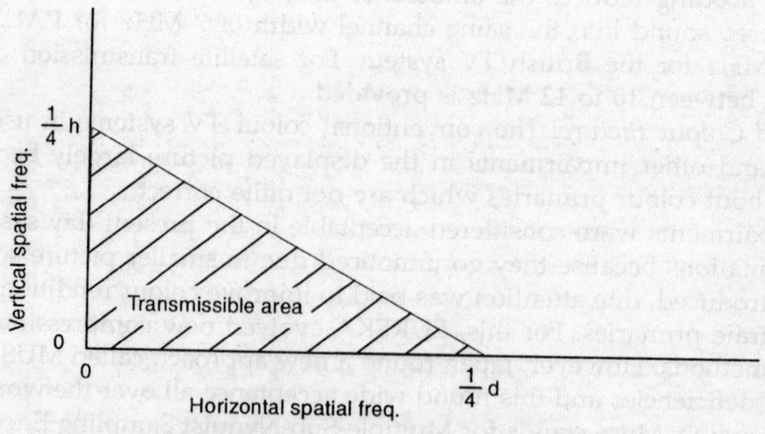

(b) For moving portion of the picture (Spatial Interpolation)

Fig. 25.2. Transmissible spatial frequency area of the MUSE System (Courtesy IEEE)

A vector representing motion of the scene is calculated for each field at the encoder. This signal is multiplexed in the vertical blanking interval and transmitted to the receiver. In the decoder, the read-out addresses of pixels (picture elements) from previous fields are shifted according to the information provided by the motion vector so that the data can be processed in the still-picture mode. These two modes of interpolation, interframe processing for stationary pictures and intra-field averaging for moving portions of the picture, are switched by detecting the moving areas at the decoder. Audio transmission is by 4-Phase DPSK which is multiplexed with the processed video signal in the vertical blanking interval after frequency modulation of the transmission carrier by the video signal.

25.3 HDTV Transmission and Equipments

Satellite transmission of HDTV signals is carried out in the KU-Band frequency spectrum of 11 to 13 GHz because of the following advantages:

(i) availability of a potentially low cost of frequency spectrum.

(ii) C-Band and KU-Band signals do not interfere with each other allowing a single satellite to relay both types of transmission.

(iii) more channel bandwidth that is needed for HDTV transmission can be easily provided in the KU-Band.

(iv) antenna have higher efficiency in high band frequencies.

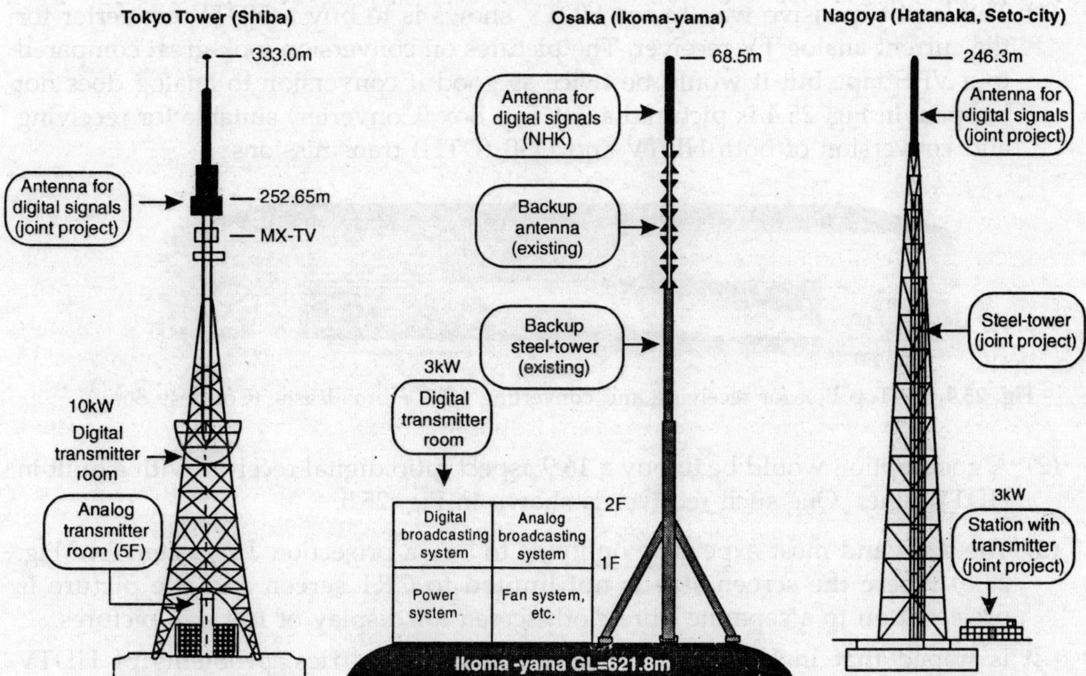

Fig. 25.3. Terrestrial transmission towers located in Tokyo, Okasa and Nagoya (Courtesy MPHPT of Japan)

Transmission Equipments: Because of different standards for HDTV, broadcasters need to buy more equipments such as cameras, remote broadcast units, sound equipments and additional components and cables for control rooms. In addition, the width to height ratio used in theatrical movies, typically 1.85:1 and 2.35:1 do not match with the 16:9 (1.78:1) aspect ratio of HDTV and therefore broadcasters must either pan and scan the images down to 4:3 eliminating part of every scan or present the full picture only on middle part of the screen with black bars above and below. However, 16:9 screen panning and scanning of theatrical movies does not remove much from the original pictures. As of now, because of the above listed constraints, HDTV transmission by satellite is limited to a few channels. Instead, terrestrial transmission is more common around big cities where more people can afford to buy expensive converters for viewing on existing analog TV receivers. Families having HDTV receivers can, however, receive programmes directly. Terrestrial transmission is carried out in the upper UHF channels and received through multi-element YAGI type antenna. It is more common in Japan. The picture shown in Fig. 25.3 is of three different transmission towers installed in Tokyo, Nagoya and Osaka.

25.4 HDTV Reception Options

In areas where HDTV transmission is available, most households are getting hooked up to it because of its merits such as—wider images, 5.1 channel Dolby Digital CD quality surround sound and provision to either send data directly to a screen or to a PC as a download. Three different ways of receiving HDTV broadcasts are:

(1) The least expensive way to see HDTV shows is to buy a HDTV converter for the current analog TV receiver. The pictures on conversion look great compared to a VHS tape but it would be twice as good if conversion to analog does not happen. In Fig. 25.4 is pictured a Set-Top Box (Converter) suitable for receiving and conversion of both HDTV and DSB (DTH) transmissions.

Fig. 25.4. Set-Top Box for receiving and converting HDTV broadcasts. (Courtesy Sony)

(2) A good option would be to buy a 16:9 aspect ratio digital receiver with a built in HDTV tuner. One such receiver is shown in Fig. 25.5.

(3) The third and most expensive option is to buy a projection TV system (see Fig. 25.6) where the screen size is not limited to CRT screen and the picture is projected on to a separate fibre cloth screen for display of full size pictures.

It is hoped that in USA and other developed countries, problems of HDTV transmission and reception will soon be overcome and most families will have 16:9 aspect ratio large screen receivers. However, many countries including India will have to wait for viewing HDTV shows.

Fig. 25.5. A 42" Plasma HDTV receiver with built-in tuner (Courtesy Sony)

Fig. 25.6. Front projection TV system where the screen is a separate unit (Courtesy SHARP)

26

TELEVISION HOME ENTERTAINMENT THEATRES

Introduction

Nearly two decades back only a small percentage of household had cable television and VCR was a luxury item. People visited Theaters to enjoy movies with pictures on large screens and stereo sound from a network of speakers. However, with rapid advances over the years in satellite communication, the world of home entertainment has changed radically. Now most families have colour TV receivers and get atleast 50 channels of cable television showing news, movies and other entertaining programmes. This gradually lead to the idea of setting a entertainment theatre at home designed to create the experience of watching movies on big screens with surround sound effects. For this, the two essential requirements are a television set with a bigger screen and a multi-speaker stereo sound system. Other components that add to the usefulness of a home theatre are a VCR and a DVD player. This chapter is devoted to explaining such equipments and suggestions on obtaining and installing them at home.

26.1 Stereo Sound

Our ears can detect the direction of incoming sounds because of relative time lag of striking of sound waves on the drum of each ear. In order to simulate such characteristics with speakers, it is necessary to have two independent channels for pick-up and processing of sounds and separate speakers for hearing them. For this, two microphones are placed, one on either side of the centre of sound pick-up area to take the place of right and left ears. Stereo effect arises due to the difference in lengths of sound path lines and hence the time delay with respect to each other in striking the diaphragms of the two microphones. This is illustrated in Fig. 26.1.

In practice more than one microphone is used to represent the range of each ear to obtain realistic sound pick-up. At the receiving end, far away or in the same hall, each channel output on due processing is fed to separate speakers located on either side of the listener who then gets the feel of stereo sound i.e. as if coming from different locations. The two sound channels are referred to as left (L) and right (R) channels for convenience of their identification, processing and reproduction.

Stereo Sound in Television: For stereo sound transmission in television, the left and right channel signals are accommodated in a different way than in FM stereo broadcasts. The IF spectrum of the TV signal is modified to have a first sound IF carrier at 5.5 MHz and a second IF carrier at 5.7421875 MHz from the vision (picture) IF carrier in the PAL system. It may be noted that the two sound carriers are separated by 15.5 × f_H, the horizontal scanning frequency to minimise any interference (5.5 MHz × 15.5 × 0.15625 MHz = 5.7421875 MHz). The first sound IF is frequency modulated with (L + R) information so that mono TV receivers can still reproduce mono sound when stereo is being transmitted. Similarly, the second sound IF carrier is FM modulated with (2 × R) information. To identify whether the transmitted signal is mono, stereo or dual sound, the 2nd IF carrier also carries an identification signal which is a 54.6875 KHz pilot carrier amplitude modulated with 117.5 Hz for stereo or 274.1 Hz for dual transmission. The frequency spectrum of the intercarrier IF signal is shown if Fig. 26.2 (a) and the audio spectrum after filtering and demodulation in Fig. 26.2(b). The stereo information can, therefore, be simply dematrixed by doubling the amplitude of 1st audio signal and subtracting the demodulated second audio signal from it i.e. 2 (L + R) −2R = 2L. After the identifications signal is demodulated, it is used to activate logic circuits to route the required information to the sound output, under control of the viewer.

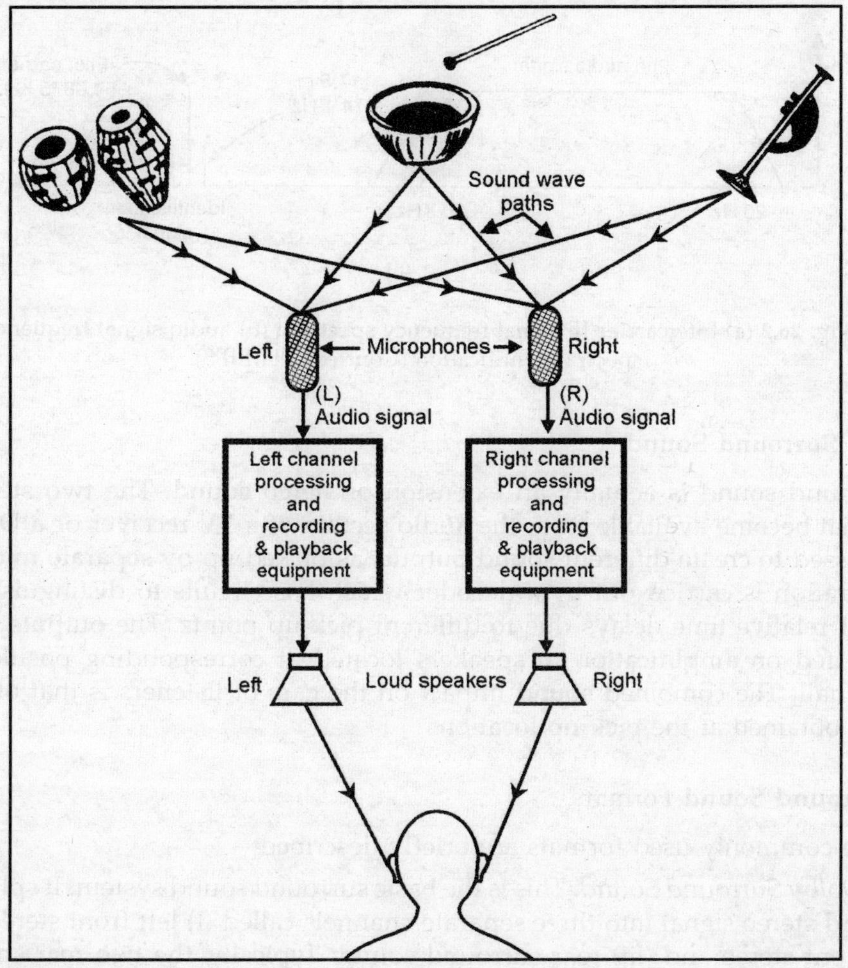

Fig. 26.1. Sound waves pick-up by the left and right Channel microphones.

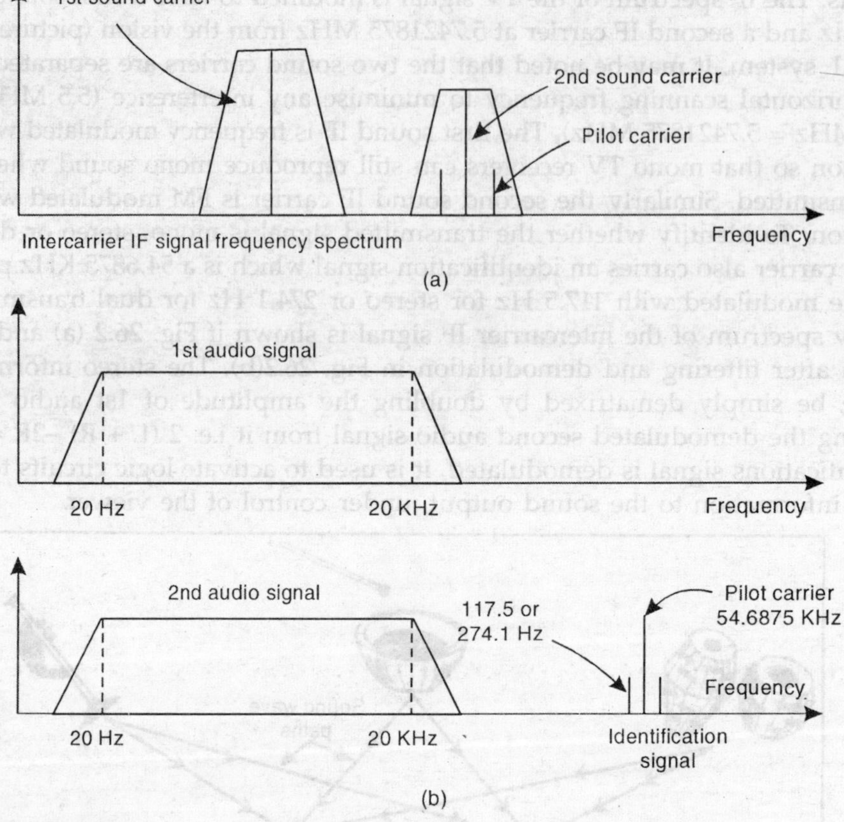

Fig. 26.2 (a) Intercarrier IF signal frequency spectrum (b) audio signal frequency spectra identification (Courtesy PHILIPS)

26.2 The Surround Sound

The surround sound is actually an extension of stereo sound. The two stereo sound signals that become available from the audio section of a TV receiver or a DVD player are processed to create different sound outputs as picked up by separate microphones. The separation is carried out by a decoder which has circuits to distinguish between sounds of relative time delays due to different pick up points. The outputs thus separated are fed on amplification to speakers located at corresponding positions in the listening hall. The combined sound impact on the ears of listeners is that of surround sound as obtained at the pick up locations.

26.3 Surround Sound Formats

The three commonly used formats are briefly described—

(1) *Dolby Surround Sound:* This is the basic surround sound system. It splits a standard stereo signal into three separate channels called (i) left front stereo, (ii) right front stereo and (iii) rear surround sound. Typically, the two rear speakers are

connected to play the same rear surround channel. Dolby surround can be installed in small halls without much professional assistance.

(2) *Dolby Prologic:* As illustrated in Fig. 26.3, it is a four channel five speaker format. Separate channels power the central front speaker, the left front speaker and the right front speaker. The two rear speakers play the same (fourth) channel. Prologic gives rich stereo sound in front and good overall response. This format is used for stereo sound recording at television broadcast stations.

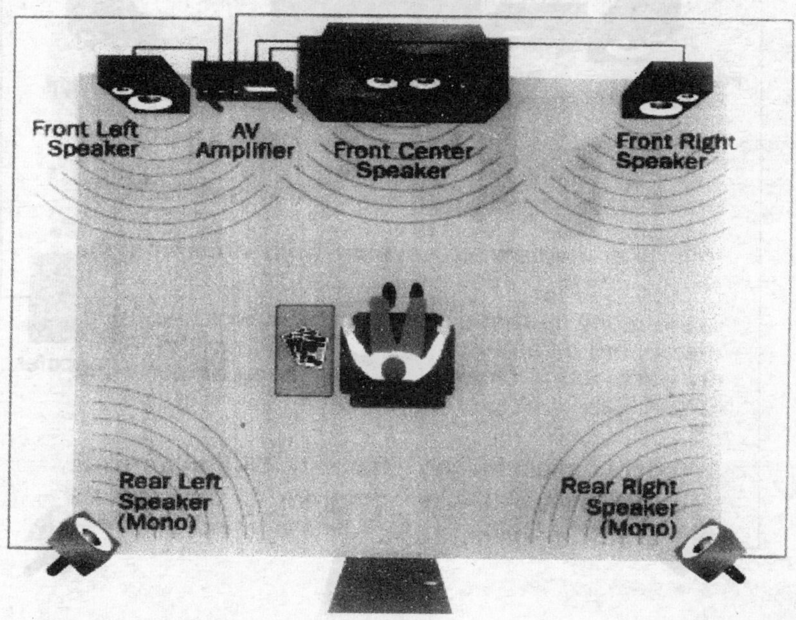

Fig. 26.3. Dolby Prologic installation—2 × front stereo speakers, 2 × rear Mono speakers and 1 × Central speaker.

(3) *Dolby Digital:* The main difference between prologic and dolby digital is that dolby digital has two separate rear speaker channels as well as subwoofer channel (see Fig. 26.4). Dolby digital is also called 5.1 format indicating five ordinary channels and one "effects" channel. The two rear speakers are positioned to the right and left of the listener rather than behind him to give a more precise surround sound listening. The subwoofer channel carries low frequency sounds to give 'BASS' boost and create rumbling effect for certain special sounds such as an explosion or a passing train. Another, significant difference between dolby prologic and dolby digital is that the latter is processed and transmitted in digital form. As such, dolby digital gives a richer and clearer sound output. This is now a standard format for HDTV transmissions.

26.4 Surround Sound Stereo Receiver

This is an important component of any home theatre. It receives signals from various inputs like VCR, DVD player and satellite receiver-decoder. There is provision to choose any one of the inputs and its delivery on processing to outputs units. On receipt, the signals are interpolated and sent to corresponding output devices—television receiver or sound system. The stereo signals are decoded as explained in section 26-2 and on amplification sent to different loudspeakers. Such a surround sound stereo receiver is pictured in Fig. 26.5.

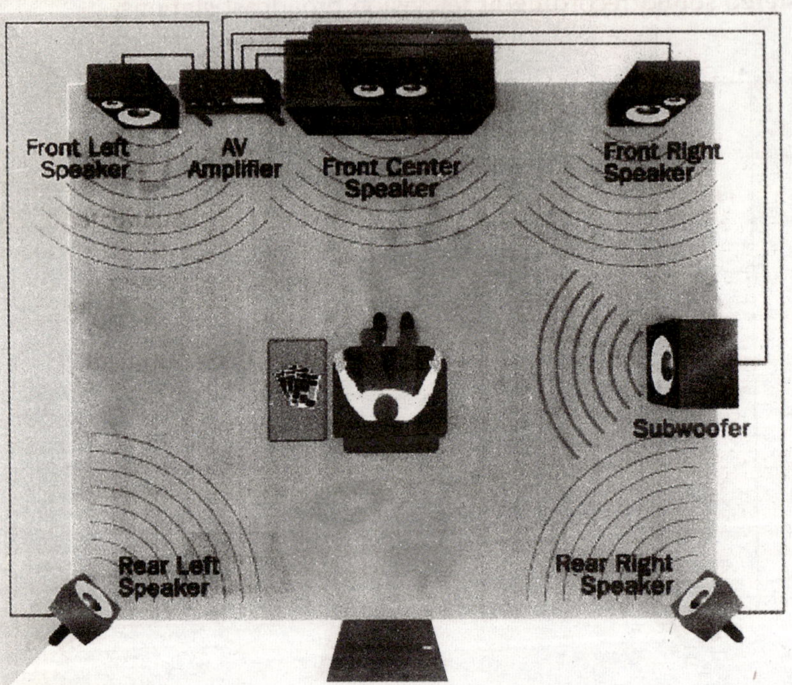

Fig. 26.4. Dolby Digital installation—2 × front stereo speakers, 2 × rear stereo speakers, 1 × Central speaker and 1 × subwoofer.

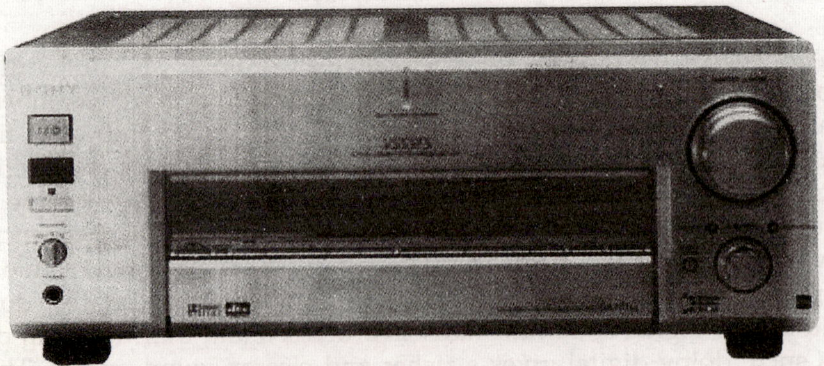

Fig. 26.5. A surround sound stereo receiver (Courtesy Sony)

Loud Speakers: Another very important component of the home theatre system is the speakers. Even with top-of-the line DVD player and audio/video receivers, the sound quality will be poor if quality speakers are not installed. The characteristics that need to be considered are audio power handling capacity, frequency range and response matching with each other to avoid any imbalance sound effect. To make things easier, many manufacturers offer home theatre speaker packages. In many countries there is certification system by which the quality and performance of loudspeakers is guaranteed. Infact the supplier company undertakes to install speakers to the satisfaction of customers.

26.5 Large Screen Televisions

The biggest variable in the home theatre system is the television receiver. The two main types in use are—(1) DIRECT VIEW and (2) PROJECTION SCREEN.

Standard Direct View Television: This is the most commonly used TV receiver. It has a Cathode-ray tube (CRT) with three in-built guns which on receiving different (R, G, B) video signals paint a colour picture on the screen. While such receivers produce very good pictures but because of limitations in glass tube technology, the screen size is limited to around 40 inches measured diagonally. A 27 inch or preferably a 32 inch direct view television receiver is quite suitable for a medium size home theatre.

The general rule for television size is that the screen size should be one-third the distance from the position of the viewer. If the screen size is bigger or the viewer sits closer, scanning lines that produce the picture become noticeable showing picture of poor quality. It may also be noticed that a television with darker screen enables better pictures because there will be a strong contrast between light and dark meaning black will actually appear black rather than grey. Furthermore, the now available flat screen receivers produce better pictures because there is no distortion due to curvature of the screen more so at the edges where it is maximum.

Projection Television: With the introduction of Direct Satellite and HDTV broadcasts, large screens are desirable for watching movies. For this, projection TV sets are now available. In these, the screen size constraint is not there because the cathode-ray tube is not used for display—instead a separate projection screen is provided. The tuner-decoder of such receivers is a separate unit which produces video signal for the picture projection and audio signal for the surround sound receiver.

(i) *Rear Projection Television:* Rear Projection Televisions actually use three different cathode-ray tubes which on splitting the video signal produce red, green and blue images. Their combined effect depending on the strength of each colour light produces the entire visual spectrum. For this, the three CRTs project on to a mirror which bounces the full colour image on the screen located on the upper half of the television box. One such 53 inch wide screen TV is shown in Fig. 26.6.

(ii) *Front Projection Television System:* Front Projection Television, though works almost the same way as a rear projection TV, but is more like a film projector. In this, the three CRT projectors are combined in one unit and the picture is focussed on a separate glare resistant fabric screen located at a distance from the projector. The advantage of this arrangement is that large screen sizes become possible. Screen sizes up to 120″ are available but a 90″ screen is quite suitable for medium size theatre halls.

Fig. 26.6. A 53″ SONY wide screen rear projection television. (Courtesy SONY)

The front projection unit is mounted at a height from the floor or hung from the ceiling in front of the screen. It receives video signal from a separate tuner-decoder located close by. Such TV units are designed to show both 4:3 and 16:9 aspect ratio pictures. A high end digital front projection unit is shown in Fig. 26.7.

Fig. 26.7. A high end digital front projection TV system produced by SHARP.

(iii) *Liquid Crystal Display Television:* In this television there is only one projector and lens assembly. Thus configuring of different colours is not necessary. The video picture is generated on a LCD screen and so has limited brightness. Otherwise LCD sets are very light because they do not have CRTs or projection devices and can be easily mounted on the wall.

(iv) *DLP Projection System:* While front projectors are rated as the best choice, but another option is now available. It is based on the Texas Instruments Digital Light Processing technology (DLP). DLP projectors generate pictures by manipulating a panel of thousands of tiny sensitive mirrors hinged on a semiconductor chip called the Digital Micro-mirror Device or DMD chip. A high-power lamp shines light through a rotating colour wheel which rapidly changes the colour from red to green to blue. The coloured light hits the semiconductor chip on which tiny mirrors are hinged.

Based on the information encoded in the video signal, the semiconductor DMD chip tilts some of the tiny mirrors to reflect the coloured light out of the projector towards the screen and tilts some mirrors such that they do not reflect any light. Thus collectively the tiny dots of reflected light form a monochromatic image. It is claimed that DLP projectors produce excellent and clear pictures.

The Screenplay 7205 front projection television of Infocus Corporation of Oregon (USA) uses DLP technology. Its video quality is not only unsurpassed but unmatched. Though very expensive, Screenplay 7205 offers impeachable home theatre images. Its versions for PAL and SECAM systems are also available with offices at Amsterdam in Europe and Singapore in Asia.

26.6 Satellite OFF-AIR Tuner and Receiver

Instead of installing separate tuner-receiver units for different incoming OFF-AIR signals a single unit is preferred. SONY, PHILIPS and many companies in USA manufacture such units. One such is DISH 811 by DISH NETWORK Corporation. It is a high definition satellite receiver with integrated off-air tuner for reception of high definition (HD) and standard definition (SD) programming. It can also receive local digital and analog broadcasts. Both HD and SD content can be received on a HDTV using HD video connections. As necessary, SD content is up converted. The tuner takes input from a dish antenna and through small multi element YAGI antennas for local DSB and HDTV terrestrial broadcasts. All options can be selected with a Remote Control. The DISH 811 High Definition receiver is shown in Fig. 26.8.

26.7 VCRs and DVD Players

Video Cassette Recorders: When VCRs first came in the scene, they completely revolutionised the home entertainment industry. These days, however, VCRs are outdone mostly by DVD players because of their relatively higher horizontal resolution. It may be noted that the earlier VCRs had mono sound—meaning they feed a single audio track along with the video picture. If used in an advanced home theatre system, having stereo sound input, sound will come only out of the front speakers, or the same sound track out of multiple speakers.

While VCRs do not give best quality pictures they are certainly an important component of the home theatre partly because there are so many classic movies available only on VHS video and need a VCR to view them. A VCR is also useful because of its own tuner and timer which allows the user to record on one channel while viewing another and the facility to see recorded programmes later, after days or even weeks. Despite these merits, it appears that even the HI-FI VCRs will get replaced by digital video formats like DVD but for the immediate future VCRs are going to be used as home entertainment equipment.

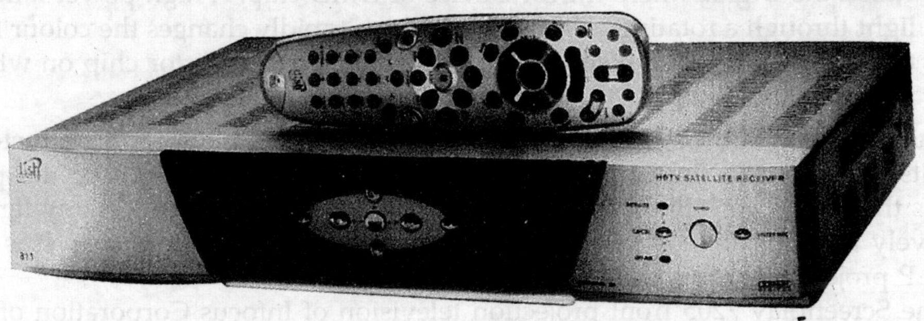

Fig. 26.8. DISH-811 High Definition Tuner-receiver.
(Courtesy Dish NETWORK Corporation of USA)

DVD Players: DVDs (digital versalite discs) offer excellent pictures and clear rich sounds. Most DVDs are formatted for one or more surround sound formats. The main features of present day DVD players are:

(i) fully compatible with DVD, SVCD, CVD, MP3 and DVCD.
(ii) compatible with full DVD playback functions.
(iii) advanced 10 bit/27 MHz video D/A converter with over 500 lines of horizontal resolution.
(iv) both normal and S-video outputs.
(v) Dolby digital 2 channel audio output.
(vi) 24 bit/96 KHz audio D/A converter.
(vii) NTSC/PAL TV system compatible.

All these qualities make the DVD player as an ideal equipment for home entertainment theatres. However, present DVDs lack one most useful capability of a VCR that they cannot record video but can only play it. Also, lack of timer facility is another drawback. But it will soon be a thing of the past. Digital Video Recorders (DVDs) have already hit the market where lot of programming can be recorded to be viewed later.

26.8 Baic Home Theatres

In united States, Europe, Japan and some other countries many signal options are available besides VCRs and DVD. Cable TV is now having competition from DSB transmission, HDTV terrestrial broadcasts, local digital broadcasts and digital cable. Analog Cable, digital cable and DSB essentially carry the same programmes. Most families prefer the existing analog cable, DSB which is digital and HDTV service. The equipments are accordingly procured and installed.

In India and many other countries, the only external signal source available is the satellite TV transmission. It is made available to subscribers on cables in analog format. Therefore, most families do not go beyond buying a colour receiver, a VCR and a DVD player. Improved sound quality becomes possible because all present day direct view TV receivers have two/three speakers located on either side of the screen. The mid-range speaker reproduces almost the entire audio range while the tweeter is to boost high frequencies and woofer to enable enhanced 'BASS' in the sound output. The combined impact on the listener is close to that of surround sound. External speakers cannot be used because this would need an external audio amplifier.

26.9 Economical Home Theatres

Many enthusiasts and music lovers wish to install a regular home theatre to see movies on relatively large screen and enjoy surround sound music. For them a large screen direct view TV receiver and a medium power HI-FI component system as offered by both SONY and PHILIPS is most suitable. The system marketed by SONY has a powerful audio amplifier to drive five loudspeakers that come with it to enable stereo surround sound output. In addition, the system also has a DVD player, an audio cassette player and a AM-FM radio receiver. A VCR can also be made part of the music system. The selection of any programme item is both direct and through a Remote Control unit.

While all assistance is given by the suppliers for installing the system, the positioning of speakers is illustrated in Fig. 26.9. The front speakers are at the level of TV receiver and at an angle of 45 degree from the listening position.

Similarly, the centre speaker is put at almost the same height as the front speaker. The surround speakers face each other at about 60 to 90 cm above the listening position.

For families who have spacious accommodation and expense is not much of a constraint, can set up an elaborate entertainment theatre on the lines detailed in the next section. However, it would be necessary to keep provision for installing additional equipments when DTH and HDTV services become available.

26.10 Top of the Line Home Theatre

In countries where all the OFF-AIR transmissions are available many families have and more are setting up their own entertainment theatres. The choice of equipments depends on several factors but a top of the line theatre should have a Dolby Digital surround sound speaker system, preferably a DLP based front projection system, HDTV receiver and multi-system tuner-decoder besides a DVD player, a digital recorder and a HI-FI VCR. The merits of these components are detailed in previous sections of this chapter. For a new installation, a home theatre expert, if hired, will prove very useful.

In general, the home theatre should be like a movie theatre that is air conditioned, enclosed, rectangular, spacious, well furnished and carpeted but with less outside light coming in. Another important aspect to be considered is the lighting in the hall. While

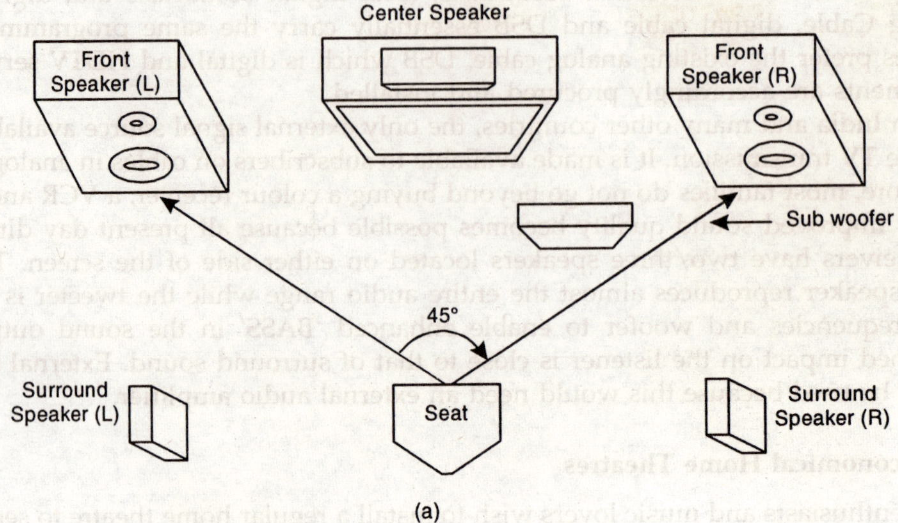

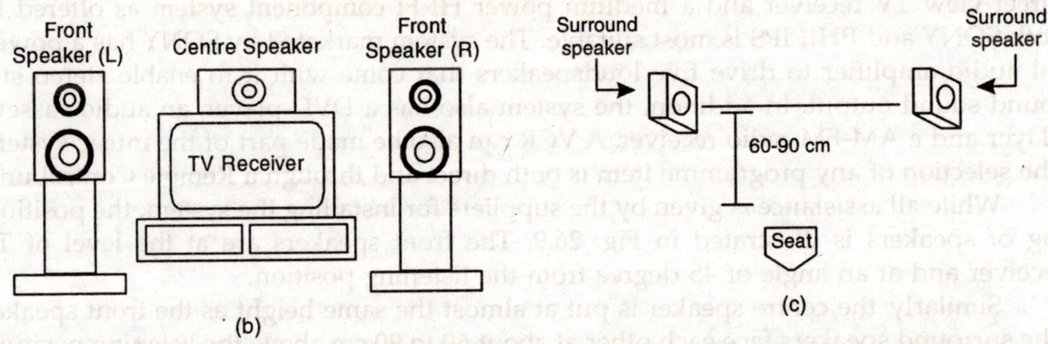

Fig. 26.9. Positioning of speakers. Position of (a) front speakers, (b) centre speaker and (c) surround speakers.

it is important that it should not be brightly lit yet not dark either because the contrast of light from the screen may strain the eyes of viewers.

Ideally, the hall should have soft ambient lighting connected to an automatic dimmer via ON-OFF switch of either the audio or video receiver. With this, when the movie projection commences, the lights will go down automatically to the preset level. The theatre hall may also be configured with curtains operated by a remote control.

Some families provide-for in the same hall or in another room additional equipments together with a receiving dish for direct reception of other countries broadcasts. This is common with NRIs in USA for viewing Indian TV programmes.

Two photographs are printed in Fig. 26.10 to show the set up in a sophisticated home entertainment theatre hall.

Fig. 26.10 (a). The picture shows the DLP projector mounted close to the ceiling. From there it projects pictures on the cloth screen.

Fig. 26.10 (b). The picture shows cloth screen and four speakers. The cabinet below the screen is for satellite receiver, DVD player, Audio/Video processor and seven channel power amplifier.

GLOSSARY

ACCESS CARD: It is like a 'Credit Card' provided along with the Direct Satellite Service (DSS) receiver. It identifies the user to the service or program provider without which the receiver does not function optimally.

ADDRESS: The unique identifier of a terminal on a network.

A/D CONVERTER: Circuitry devised to convert analog signals to digital form.

ADJACENT CHANNEL: An adjacent channel is one which is immediately next to another channel. For example, channels 3 and 4 and channels 8 and 9 are adjacent channels. But channels 4 and 5 are not adjacent as these have other frequency allocations (FM) between them.

ALTERNATE AUDIO: Alternate audio means option to choose among two or more available audios like Hindi, Urdu or English.

ANALOG: A signal that conveys information by means of changes in signal intensity and frequency like the output signal from a microphone.

ALU: An Arithmetic Logic Unit that contains electronic logic circuits designed to perform addition, subtraction, multiplication and division in a computer.

ANTENNA POSITIONING SYSTEM (APS): A mechanical electronic system used for pointing the dish-antenna towards the wanted satellite.

APERTURE: A cross-sectional area of the antenna actually exposed to the satellite signal.

APOGEE: The point in the elliptical satellite orbit that is farthest from the surface of earth.

ATTENUATOR: A passive electronic devise to reduce signal strength by a specific value (usually expressed in dB).

AUTOMATIC FREQUENCY CONTROL (AFC): A circuit that locks receiving unit on to the frequency chosen and prevents drifting from that frequency.

AUTOMATIC GAIN CONTROL (AGC): A circuit that boosts or reduces gain to compensate for temporary gain shifts in the system.

AZIMUTH ANGLE: The angle of rotation (horizontal) that a ground based parabolic antenna must be rotated through to point to a specific satellite in a geosynchronous orbit. The Azimuth angle for any particular satellite can be determined for any point on the surface of Earth given the latitude and longitude of that point. It is defined with respect to due North for convenience.

BALUN (MATCHING TRANSFORMER): A passive electro-magnetic device that matches the impedance of one component, transmission line or circuit to that of another to prevent loss of signal strength and other unwanted characteristics. In antenna systems, baluns typically are used to match 75 ohm cable to the 300 ohm output of an antenna or the 300 ohm input of a TV receiver.

BASEBAND: The basic direct 7 MHz output signal from a television camera, satellite television receiver or videotape recorder. Baseband signals can be viewed only on studio monitors. To display the baseband signal on a conventional TV receiver, a frequency-translator is required to convert the baseband signal to one of the VHF or UHF television channels that the TV set can be tuned to receive.

BEAM-WIDTH: The angle formed by two compass directions that outline the boundaries of the area from which the front of an antenna can intercept signals and deliver these to the output at relatively equal values. Generally, the narrower the beamwidth of an antenna, the greater its directivity and gain.

BIRD: Nickname for a satellite.

BLOCK DOWN-CONVERTER: A frequency band converter that changes higher frequency bands to lower frequency bands. A low-noise block converter changes the higher frequency satellite transmission band (C-band or Ku-band) at its input to a lower intermediate band (IF) at its output.

C-BAND: The 3.7 to 4.2 GHz band of frequencies, which is the dominant mode of satellite broadcast to domestic dish antennas.

CARRIER-TO-NOISE RATIO (C/N): C/N ratio pertains to difference in amplitude between the desired RF carrier and the noise present in that portion of the spectrum occupied by the desired signal. It is expressed in decibels to denote the ratio or relationship between the two. Usually measured in dB at the LNBC output.

CIRCULAR POLARIZATION: A mode of transmission in which signals are downlinked in a rotating corkscrew pattern. A satellite's transmission capacity can be doubled by using both right hand (RH) and left hand (LH) circular polarizations.

COAXIAL CABLE: A type of round transmission line composed of a central conductor wire surrounded in turn by an insulating (dielectric) cover and a metallic shielding material that typically is braided and acts as the second conductor. Coaxial Cables (COAX) used as domestic (home) transmission lines have an impedance of 75 ohms.

CO-LOCATION: The placing of several satellites near each other in orbits. Co-location allows a single fixed receiving antenna to receiver signals from all such satellites without having to move the dish for tracking them.

COMMUNICATION SATELLITE: An artificial satellite usually placed in geostationary orbit to relay radio transmissions.

CONDITIONAL ACCESS: It is a system designed to limit programme viewing to only those TV channels opted and paid for by individual subscribers.

CROSS-MODULATION: A condition in which a strong signal will overload an amplifier and cause interference on other channels. It amounts to one type called third order distortion which occurs in all amplifiers as the signal level is increased.

COUNTERS: Digital circuit's connected to count random events or to keep track of operations in digital circuits and instrumentation.

CROSS-TALK: The unwanted leakage of signal between supposedly independent channels.

dBI: Antenna gain in decibels, relative to an isotropic (omnidirectional) antenna. It is the true term to express gain of a satellite antenna which is normally around 40 dBI.

DATA COMPRESSION: In order to reduce data transfer rate to a possible level, the information is compressed by eliminating redundancy that is present in the signal.

DATA PACKETS: In a direct-satellite system a data packet is an assemblance of useful data like video and other coded signals necessary to identify various inputs from each other for specific processing at the receiving end.

dBM: The ratio in dB of the power of a signal as compared with a one-milliwatt reference power.

dBMV: The ratio in dB of the power of a signal compared to one millivolt level in a 75 ohm system.

dBW: The ratio of power to one watt expressed in decibels.

DECIBEL (dB): A measurement term that describes the strength (level) of a signal in logarithmic relation to a reference level. For example, when the signal is expressed in micro-volts, an increase of signal by 6 dB means that the signal strength has been doubled.

DECODER: A device that reconstructs an encrypted signal so that it can be clearly received.

DIGITAL: Conversion of information into bits (0, 1) of data for transmission through wire or over air.

DIRECT BROADCAST SYSTEM (DBS): Refers to service that uses satellites to broadcast multiple channels of television programming direct to small dish antennas.

DOWN LINK: Term used to describe the retransmitting of signals from a satellite back to earth.

D/A CONVERTER: Circuitry devised to convert digital signals to analog version.

ECM: The entitlement control message is a private conditional access information that specifies control word in CAS.

EMM: These are coded entitlement management messages from SAS ready to be transmitted to the subscriber.

ENCODER: A device used to electronically alter a signal so that it can only be viewed on a receiver equipped with a special decoder.

ENCRYPTION: Altering a programme for the purpose of controlling access. In television a programme is encrypted with a scrambler at the source of transmission. A descrambler is required to decrypt the programme.

EQUALIZATION: The technique of compensating for difference in attenuation of a signal at different frequencies.

F-CONNECTOR: A small metallic male type connecting device with internal threads which is attached at the end of a coaxial cable to secure and electrically connect the coaxial cable to a female F-fitting. The internal threads of the male connector screw on to the external threads of the female connector. Such 'F' connector sets are used to connect signals from one equipment device to another without causing any signal leakage.

FEED-HORN: A device that gathers microwave signals reflected from the surface of a dish and feeds these to the LNBC.

FEED-LINE: A special coaxial cable (transmission line) that carries signals from a dish antenna to the satellite receiver.

FOCAL LENGTH: Distance from the feed horn to the centre of antenna reflector.

FOCAL POINT: The area towards which the primary reflector directs and concentrates the received signals.

FOOT PRINT: The geographic area towards which a satellite directs its signal.

FORWARD ERROR CORRECTION (FEC): Forward Error Correction relates to data used in digital signals which is sent to enable correction of the data being transmitted.

FREQUENCY-AGILE: The ability of a satellite TV receiver to select or tune all channels (transponders) from a satellite. Receivers not frequency-agile are dedicated to a single channel as used at cable stations. Frequency-agility can be of continuously variable tuning type or discrete-step type *i.e.* channel selection tuning. Similarly, there are frequency agile modulators as compared to fixed frequency carrier modulators.

FREQUENCY SHIFT KEYING (FSK): A technique in which binary levels are expressed with a shift in frequency.

GAIN OVER NOISE TEMPERATURE (G/T): The ratio of antenna gain (G) to the antenna system noise, a figure of merit - the higher the better.

GEOSYNCHRONOUS ORBIT (GSO): An orbit around the earth with an average distance from the centre of earth of about 36000 kms., in which a satellite would have a period equal to the rotation period of the earth. Most communication satellites are in geosynchronous orbit.

HEAD-END: The master distribution centre of a CATV system in which the incoming television signals from space and distant broadcasting stations are received, amplified, and remodulated on to television channels for transmission down the CATV coaxial cables.

HIGH DEFINITION TELEVISION (HDTV): A television system of higher picture quality than the standard PAL/NTSC/SECAM TV systems. This usually involves more television lines as for PAL 1250 lines rather than 625 and a different aspect ratio 9 : 5 rather than 4 : 3.

INSERTION LOSS: Signal lost when a piece of equipment is inserted in the line, commonly referred to as 'feed through' loss.

INTERACTIVE TV: In this system of TV, the viewer has a choice to determine which aspect of the scene should appear on the TV screen.

KELVIN (K°): The temperature measurement scale used in scientific studies. Zero degree Kelvin represents absolute zero and corresponds to –237.5°. Thermal noise characteristics of an LNBC are measured in degree Kelvin, with respect to room temperature (see noise temperature).

LATITUDE: An angular measurement of a point on the earth above or below the equator. The equator represents zero degree, the north pole + 90° and the south pole –90°.

LONGITUDE: An angular measurement of a point on the surface of earth in relation to the meridian of Greenwich (LONDON). The earth is divided into 360° of longitude, beginning at the Greenwich mean. As one travels west around the globe, the longitude increases.

LOOK ANGLE: The angle of antenna dish above the local horizontal when the dish is looking at a satellite.

LOW NOISE AMPLIFIER (LNA): A preamplifier used in the feed-horn of the satellite antenna to amplify weak satellite signals.

LOW NOISE BLOCK DOWN-CONVERTER (LNB/LNBC): An assembly unit that converts the incoming microwave signals to a lower (IF) frequency and amplifies them.

MOTION PICTURE EXPERT GROUP (MPEG): A group of experts to decide video and audio compression of data to be transmitted.

MEMORY: The data storage area in computers is called Memory.

MMDS: Multipoint Microwave Distribution System.

MULTICRYPT: This is another CAS plan that operates at the MPEG transport stream level.

MULTIPLEXER (MUX): It provides for a method to accommodate more than one signal within a specified bandwidth, which could be shared by time, frequency or a combination of both.

MUSE: Multiple-sub-nyquist sampling encoding is a HDTV bandwidth compression scheme developed by NKH of Japan.

NOISE: Any unwanted contribution to a signal. It may be natural or interference from other signals.

NOISE FIGURE: Generally, a numerical rating that indicates how much electromagnetic "noise" there will be at the output of a circuit or system compared to the noise at the input. If the strength of noise is too close that of the picture producing signal, noise produced specks called 'snow' will be seen on the TV receiver screen. The lower the noise figure of a circuit, system or component, the lower the output noise level will be compared to the output signal level and in turn, lower the probability of snow effect on the reproduced picture.

NOISE TEMPERATURE: The noise temperature of a low-noise amplifier is expressed in degrees Kelvin with respect to room temperature.

OFF-THE-AIR: The reception of a TV signal that has been broadcast over the air without the assistance of microwave relay.

OFFSET FEED: A specified feed offset from the actual dish antenna focus point.

OMNI DIRECTIONAL ANTENNA: An antenna capable of receiving signals from all compass directions equally well. Such an antenna is non-directional.

PAY-PER-VIEW (PPV): A system where a CATV customer can order individual programmes such as films or sporting events either immediate or for some specific time later.

PERIGEE: The point in an elliptical satellite orbit that is closest to the surface of Earth.

PLUMB LINE: A plumb line is a weight attached at one-end of a string. The weighted string provides a reference line perpendicular to ground.

PROJECTION TELEVISION: It is a TV viewing system where the picture is viewed on a separate cloth screen instead of the CRT screen.

QUADRATURE PHASE SHIFT KEYING (QPSK): A form of phase modulation where there are four binary phase levels.

QUANTISATION: Sampling of analog signals (video or audio) at discrete amplitude levels.

REGISTERS: Memory devices used in digital systems to store data.

RAM: Random-Access Memory. It receives and stores digital data which can be changed as required.

ROM: Read-Only Memory. It is used for permanent storage of system software provided by the manufacturer.

REMOTE CONTROL: It is used to control various functions of the receiver from a distance as for TV sets and VCRs.

SAMPLER: In digital transmission the entire modulating analog signal is not transmitted. Instead signal voltages at specified levels are transmitted. This is called sampling of the signal and the circuit block for doing so is called Sampler.

SATELLITE: A sophisticated electronic communication relay station orbiting 36000 km (≈22300 miles) above the equator moving in fixed orbit at the same speed and direction as the Earth at about 7000 mph (≈11260 kmsph) east to west.

SATELLITE RECEIVER: A digital or wideband FM receiver operating in the microwave range, converting the incoming C-band or Ku-band RF signals to a standard baseband video signal.

SAS: This is subscribers authorization system incharge of delivering entitlements to subscriber smart card.

SCRAMBLING: Any system intended to render a received message unintelligible or inviewable without authorization. Also called encryption and encoding.

SIGNAL TO NOISE RATIO (S/N): Signal to noise ratio is a comparison of signal to noise level.

SIMULCRYPT: This is one of the conditional access (CAS) plans that is commonly used.

SNOW: A form of noise picked-up by a television receiver caused by a weak signal. Snow is characterized by alternate dark and white dots randomly.

SMS: Subscribers Management System is the billing system in CAS.

SMART CARD: See Access Card.

STEREO SOUND: Stereo sound output from an audio system enables sense of direction from which sound is coming.

SUBCARRIER: A second signal modulated onto a main signal to carry additional information. A carrier of a lower frequency modulating a main carrier with a higher frequency.

SURROUND SOUND: The surround sound is an extention of stereo sound which enables detection of sounds originating from different locations.

TAP-OFF: A device that when installed in a feeder line will allow a specific amount of energy to be removed from the through line to be fed to a TV receiver.

TERRESTRIAL TELEVISION: Ordinary VHF and UHF television transmission limited to an effective range of about 150 km.

TILT CONTROL: A device on a distribution amplifier that adjusts the amplification to compensate for the attenuation slope of cable loss.

TRANSPORT IC: This is the main signal processing IC in a receiver-decoder of the DSS system.

THREE DIMENSIONS (3-D) TV: In this system of television along with length and breadth the depth of any scene is also depicted on the screen.

TRANSPONDER: A combination receiver, transmitter and antenna package on a satellite. The signal is received from the up-link station and downlinked back to earth on a different frequency. Most C-band satellites have 24 transponders with 12 vertical and 12 horizontal polarization. Most Ku-band satellites have 16 transponders. Some satellites have a mix of C-band and Ku-band transponders.

TRAP: A frequency sensitive device that is used to attenuate specific signals that cause interference.

TUNER: That portion of a receiver, that can variably select, under user control, a desired signal from a group of signals in a frequency band.

UPLINK: A signal from an earth transmit station to a satellite or an antenna facility that transmits signals to a satellite.

VERTICAL POLARIZATION: Vertical polarization along with horizontal polarization enables doubling the useful bandwidth on adjacent frequencies and provides for 30 dB isolation from each other.

VIDEO COMPRESSION: A digital technique to compress several video channels into the bandwidth normally required by one channel.

VOLTAGE STANDING WAVE RATIO (VSWR): A measurement of the difference between the minimum and maximum voltage along the transmission line. The ideal VSWR is 1 : 0. As the SVWR increases so does the possibility of ghosting of pictures.

WAVEGUIDE: A metal tube used to guide or transmit microwave signals between two given points.

WEATHER BOOT: A rubber like covering that is used to protect outdoor electrical connections from rain, ice etc. (also called rain cover).

X-BAND: Used loosely to refer to satellites operating in the 8/7GHz range.

REVIEW QUESTIONS

Chapter 1. Elements of Cable Television

01. Explain how the launch of communication satellites enabled the expansion of cable television in a big way, thus making MATV unnecessary and earlier CATV system redundant.

02. State the three main sections in which a modern cable TV system can be divided and explain briefly the function of each section.

03. Why is it necessary to down convert microwave signals received from satellites to a relatively low frequency range before processing? Explain with a block diagram how is this done with an LNB converter to obtain an IF output at 950 to 1450 MHz.

04. Draw simple diagram of a cable station and fully label it to show how the IF signal obtained at the output of LNBC is divided to feed many satellite receivers which are tuned to select different TV channels. Also indicate how the detected video and audio outputs from the receiver are fed to corresponding modulators to obtain RF output on assigned channel carriers. Explain what you understand by the term 'baseband signal'.

05. Explain with a block diagram how the modulator outputs are combined and all the cable channel signals thus obtained distributed to customers over the network. In particular, explain the functions of splitters and line terminating resistors.

06. What is a 'SET-TOP' converter and explain why its use is necessary with earlier TV receivers, the tuners of which are not designed to be 'Cable-Ready'.

Chapter 2. RF Waves-Generation and Propagation

01. Describe with necessary diagrams the radiation mechanism of electro-magnetic (E-M) waves from transmitting antennas.

02. Define polarization of E-M waves. Explain with sketches how the orientation of electric and magnetic fields with respect to earth determines vertical and horizontal polarizations.

03. Explain fully why the transmitting and receiving antennas must be set for the same polarization for optimum transmission and reception.

04. Define field intensity of transmitted radio waves and explain why the signal strength of such waves falls-off sharply due to spreading losses even at not too long distances from the transmitting antenna.

05. Name and indicate by drawing radiation patterns, the three distinct methods by which wave propagation takes place through space and thus bring out the limitations of each type. Mention the suitability of each type for specific frequency bands.

06. Explain why space-wave propagation is used for terrestrial transmission of television signals. What steps are taken to increase the range of line-of-sight communication enabling reception at distant locations.

07. Explain briefly the factors which determine the frequency band for satellite communication. Why is the frequency band from 1 GHz to 10 GHz often referred to as the 'Atmospheric-Window'?

Chapter 3. Antennas and Transmission lines

01. Describe the basic principle of a receiving antenna. Show that for maximum signal pick-up, the length of half-wave dipole should equal $\lambda/2$, where λ is the wavelength of the signal frequency to be received.

02. Define antenna directivity, antenna gain and antenna impedance, as applied to receiving antennas. Explain why it is not possible to use the same antenna for both i.e. lower and upper VHF channels.

03. Describe with suitable sketches constructional details of a 300 ohm twin-lead balanced transmission line and a coaxial cable. What is the essential difference between a balanced and an unbalanced line? Why coaxial cables are preferred for laying distribution network lines in cable TV systems.

04. Draw electrical representation of a high frequency transmission line in terms of lumped parameters and explain the significance of each parameter. Explain why at low and medium frequencies, the only effective parameter is resistance of conductors that constitute the line.

05. Explain why is it necessary, that irrespective of the length of connecting cable, an impedance match between the source and load (receiver) must be obtained for maximum transfer of power.

06. Describe fully, what is meant by characteristic impedance (Zo) of a transmission line or that of coaxial cable. Explain how it is determined electrically and also from physical dimensions.

07. Explain why distribution cables in CATV must be terminated in their characteristic impedance, which for most cables is 75 ohms. A coaxial cable has a shield braid with an inside diameter of 0.2 and its central conductor diameter is 0.057, both expressed in the same units. Show that the cable's Zo $\approx$ 75 ohms.

Chapter 4. Microwave Techniques and Antennas

01. What range of frequencies come under microwaves? What is the main benefit of the use of such extremely high frequencies? Mention some major applications of microwaves.

02. Describe main reasons which make it difficult to work with microwave signals as compared to relatively low frequency signals. Explain the solutions devised to overcome these constraints.

03. For what reasons electric signals are reflected back to source from load side on a transmission line and hence, define Standing Wave Ratio (SWR). Under what conditions the SWR can become either one or infinity. Illustrate this with corresponding wave shapes of wave travel on the line.

04. Explain how at microwave frequencies, short lengths of transmission lines can be used as inductors and capacitors in place of very small values of L and C, which cannot be physically realized.

05. Describe how one-quarter and one-half wavelength transmission lines can be used as reactive or resonant circuits in microwave circuitry. Explain by sketching waveshapes, why an open $\lambda/4$ line and a shorted $\lambda/2$ line behave as a series resonant circuit at the corresponding microwave frequency.

06. What is a quarter-wave matching stub and explain how it can be used for impedance matching-say between an antenna and a transmission line. Similarly, explain how a balun or balancing transformer can be used to match a 300 ohm transmission line coming from an antenna to the 75 ohm input impedance of a television receiver.

07. Describe with suitable sketches, constructional details of a strip-line and a microstrip which as printed-circuit board implementations of transmission lines are widely used to create resonant circuits and filters in microwave circuits.

08. What is a Wave-Guide and explain how it enables transmission of microwave energy, without any appreciable loss, from one point to another, separated by a short distance from each other.

09. Explain by showing associated electric and magnetic fields, the mechanism of wave travel in a wave-guide. Describe how is the microwave signal injected at the sending end of a wave-guide and collected at its receiving end. What is the significance of cut-off frequency of a wave-guide?

10. Describe briefly the mechanism of operation of a (i) cavity resonator, (ii) a schotthy diode and a (iii) GaAs FET and their applications in microwave circuits.

11. Explain why antennas designed for transmission and reception of VHF and UHF signals cannot be used at microwave frequencies. Describe how a HORN along with a parabolic reflector is successfully used as an antenna at microwave frequencies.

12. Explain what is the focal point 'F' of a parabolic reflector and how it enables concentration of energy into a narrow beam and thus enables gain. Describe with suitable sketches the technique of sending and receiving microwave signals with parabolic reflector antennas.

Chapter 5. Performance Characteristics of Amplifiers

01. Define voltage gain and power gain of an amplifier. Explain how it is convenient to express difference in RF signal levels in decibels in a cable TV system. Derive expressions for power and voltage gains in decibels.

02. Explain to justify that the same numerical dB value does not mean equal voltage and power gains. Show that 40 dB corresponds to a voltage gain of 100 and also for a power gain of 1000.

03. Explain why the reference level for measurements in cable systems has been chosen as 1 μV across 75 ohms. Also explain why is it convenient to use 1 μv as reference for measurements of very weak signals as received from satellites. How is the loss in voltage and power relations expressed in dBs.

04. Describe briefly the techniques used to extend the bandwidth of an RC coupled amplifier to meet video signal requirements. Explain why phase-shift correction is essential in video amplifiers but not so in audio amplifiers.

05. Define harmonic or non-linear distortion. Reason out with a circuit diagram, how it gets greatly reduced in push-pull amplifier configurations.

06. Why gain stability of amplifiers is not very high? Explain how with negative feedback a high degree of gain stability becomes possible. Derive necessary expression to prove that gain stability is achieved at the cost of reduction in gain. How is this loss made up, if necessary?

Chapter 6. Resonant Load RF Amplifiers

01. Compare series and parallel resonant circuits with respect to (a) resonant frequency (b) impedance at resonance, (c) bandwidth and (d) applications.

02. Define quality factor 'Q' as applied to tuned circuits and explain with response waveforms how it affects frequency discriminating capability of tuned RF amplifiers.

03. Enumerate distinguishing features of VHF, UHF and super UHF amplifiers as used in cable and satellite circuit applications. Explain why it is necessary to use superior grade active devices (transistors, tubes) and circuit components in very high frequency amplifiers.

04. Explain how thermal noise affects the sensitivity of UHF and super UHF amplifiers. In this context, define 'NOISE FACTOR' and explain how it determine the usefulness of such amplifiers at very low signal levels.

05. Define 'NOISE TEMPERATURE' and develop its relation with noise factor. Explain why special RF amplifiers commonly called 'Low Noise Amplifiers' i.e. having very low noise temperature value, are used at the first stage of equipments meant for receiving and processing very weak signals coming from satellites at SHF frequencies.

Chapter 7. Sinusoidal Wave Generation and Frequency Control

01. Explain why positive and negative feedback is necessary in an amplifier to pro-

vide oscillations. State the condition that must be met to obtain sustained amplitude oscillations. What determines frequency of oscillations and how oscillations start at this frequency without any external starting voltage source.

02. Explain why L.C. oscillators are prone to drift from the set frequency. How is this drawback overcome in crystal oscillators?

03. Explain how a varactor diode is used to change the oscillator frequency in modern TV tuners to select different channels. Why such an oscillator is called a Voltage Controlled Oscillator (VCO)? Describe the method by which the Automatic Frequency Tuning (AFT) circuit associated with the tuner keeps the oscillator frequency stable at the set value.

04. Explain with a block diagram, how the Phase Locked Loop (PLL) technique is used to obtain several exact low frequency outputs with the accuracy of a crystal oscillator from only one high frequency oscillator.

05. What is a Frequency Synthesizer? Draw block diagram of such a synthesizer as employed in modern TV tuners and explain how the desired channel frequency is chosen and kept constant at this value.

06. It is very difficult to generate frequencies of the order of GHz as needed in oscillators used in satellite transmission and reception systems. Explain how frequency multipliers with the use of varactor diodes enable generation and selection of super high frequencies with input from relatively low frequency crystal oscillators.

Chapter 8. Waveform Modulation and Demodulation

01. Why is it necessary to modulate TV picture and sound signals before transmission? How is frequency modulation different from amplitude modulation?

02. Explain how sidebands are generated in amplitude modulation (AM). Show that in this system, the required channel width for transmission of any signal without distortion is 2fm centred on fc where fm is the highest modulating frequency and fc the carrier frequency.

03. Explain with a circuit diagram how the amplitude modulated signal is detected by a diode envelope detector to obtain the modulating signal. Why is the associated R-C filter modified when the modulating signal frequency is high as that of video signals.

04. State the salient features of frequency modulation (FM). Why is it preferred for TV sound signal transmission over AM in terrestrial broadcasts? Write CARSON'S rule for finding channel bandwidth with FM and thus justify the allocation of 150 KHz bandwidth for sound signal transmission in the 625 line CCIR television system.

05. Explain how with FM modulation, the use of a 'PRE-EMPHASIS' circuit at the transmitting end helps to eliminate noise interference. Explain why a corresponding 'DE-EMPHASIS' circuit becomes necessary after the FM detector at the receiving end.

06. Explain briefly how a slope detector enables FM detection as used in FM radio and TV receiver circuits. What is the main merit of phase-locked loop method of FM detection as used in modern receivers.

Chapter 9. Frequency Translation and Filters

01. What is frequency translation, also known as mixing or heterodyning? Draw basic circuit of a frequency converter and explain how by operating the transistor in its non-linear region, new frequencies get generated.
02. In communication receivers, the incoming high frequency AM or FM signals are down converted at the tuner stage to a low centre frequency, usually called intermediate (IF) frequency. What are the merits of doing so?
03. Explain why a diode mixer instead of the usual transistor type converter is used at UHF frequencies as in satellite receivers. Draw basic circuit of a diode mixer and explain its operation.
04. Differentiate between an attenuator, a pad and an equalizer as used in communication networks. State few applications of such networks as found in trunk amplifiers and cable distribution networks.
05. Give specific characteristics and applications of Low-Pass, High-Pass and Band-Pass filters. Explain how a SAW-FILTER designed for use in TV receivers serves the purpose of several discrete filter sections in the IF sub-system of a TV receiver.
06. Explain why physical sections of filter networks are not realizable at microwave frequencies. Describe how micro-strip lines serves as filters in equipments that carry and process microwave signals.

Chapter 10. Voltage Regulation and Cable Powering

01. What is cable powering as used in CATV systems? Explain why it is necessary to regulate both ac and dc supplies before feeding to cable station equipments and trunk/line amplifiers.
02. Describe how Switched Mode Power Supply (SMPS) is considered superior to that obtained through a Series Voltage Regulator. Explain why rigorous shielding of SMPS units is necessary to prevent interference in nearby equipments carrying RF cable signals.
03. Explain with a block diagram how a ferromagnetic transformer along with a suitable reactor unit enables regulated ac power supply, usually at 60V, to power supply units of trunk/line amplifiers from the 220V, 50 Hz ac mains supply.
04. Explain with the help of a block diagram, how a Power Insertor functions to feed 60V ac supply to trunk/line amplifiers through the same cable that carries RF signals in the cable distribution network.
05. Explain with a block diagram how by the Direct Trunk Amplifier Powering method, ac supply is fed to power units of various amplifiers.

06. Describe briefly the alternative AC powering methods used in small cable TV stations for feeding ac/dc supply to trunk and line amplifiers.

Chapter 11. Terrestrial Television

Part–A, Television Transmission

01. Why is it necessary to restore to scanning in TV transmission and reception and how is it carried out at a very fast rate? Explain how interlaced scanning reduces flicker and conserves bandwidth.

02. Why three pick-up tubes (vidicons) are provided in a colour camera? Define luminance (Y-signal) and explain why is it necessary to transmit it along with corresponding chroma (colour) signal during colour transmission.

03. Describe briefly the following tonal and colour characteristics of television pictures—(i) Brightness, (ii) Contrast, (iii) Luminance, (iv) Hue, (v) Saturation and (vi) Chroma.

04. Why is compatibility essential for the co-existence of monochrome and colour TV systems. Explain how 'C' signal is interleaved in the 'Y' signal spectrum to conserve bandwidth and thus keep channel bandwidth at 7 MHz.

05. Explain how the colour-difference signals, (R-Y) and (B-Y) are generated, encoded and then combined with 'Y' signal before transmission. Why (G-Y) signal is not transmitted?

06. Sketch PAL colour TV channel bandwidth details and label on it VHF channel 3 (54 to 61 MHz) frequency components to show that these add upto 7 MHz.

Part–B, Television Reception

01. Explain why is it necessary to prevent any deviation (drift) of IF frequency in a colour receiver. Describe how the AFT circuit keeps it fixed at the chosen value.

02. Describe how AGC voltage is developed in a TV receiver and applied to IF subsystem to keep its output amplitude almost constant. Why is a voltage delay circuit inserted in the path of AGC to be applied to RF amplifier in the tuner?

03. Explain with block diagram of corresponding sections of a colour receiver, how contrast and brightness controls operate in the 'Y' signal path and saturation and colour-killer circuits function through chroma amplifier of the decoder.

04. Explain the delay line technique used in PAL colour receivers to separate U and V colour phasors. Illustrate your answer with input colour signals of two consecutive lines.

05. Explain how the colour-burst signal transmitted after each scanning line is used at the receiver to keep the colour reference oscillator frequency stable at 4.43361875 MHz.

06. Explain with the help of a block diagram how the colour difference signals (R-Y),

(B-Y) and (G-Y) are obtained on demodulation of U and V signals. Why are these de-weighted before feeding to R, G and B matrixing circuits?

07. Show various sections of raster circuitry of a TV receiver and explain how outputs of vertical and horizontal deflection oscillators are modified to obtain linear deflection of the electron beam on the picture tube screen.

08. Justify the need for applying very high dc voltage at the accelerating anodes of television picture tubes. Explain how EHT voltage of the order of 25 KV is developed to be applied at the final anode of colour picture tube.

09. Explain with a sketch how the three colour beams in a P.I.L. picture tube on passing through the aperture grille (mask) land on corresponding colour phosphor stripes coated on the inner side of face plate. Describe how the absence of one or more colour beams and/or variation in their intensities result in the formation of different colours, as seen by the eye on the receiver screen.

10. Which functions of the colour receiver are usually controlled with a remote control unit? Explain briefly how a different command signal for each control is generated in the transmitter of the remote control unit. How are the generated pulse trains transmitted to the receive senser located on the front penal of TV receiver.

Chapter 12. Satellite Communications

01. Explain how communication satellites orbiting around the earth enable transmission of television, telephone and similar data signals over very long distances.

02. Explain why communication satellites are set to orbit at a distance close to 35887 Km in the sky from the surface of earth. Why are up-link and down-link frequencies chosen to be in the GHz range and kept fairly apart from each other? What limits the life of a satellite once put in orbit?

03. What is a transponder as used in satellites? Explain with a block diagram how a transponder on receipt of up-link transmission, amplifies and down converts it before feeding it to the down-link transmission antenna. Why television picture signals are frequency modulated for transmission via satellites instead of amplitude modulation as done in terrestrial broadcasts?

04. How many channels are accommodated in the 500 Mhz width of the C-band satellite communication system? What is the actual and effective bandwidth of each channel. Draw a chart showing centre frequencies of these channels.

05. Draw block schematic of a 12 channel transponder architecture and explain how each up-link channel is separated, amplified, frequency converted and again amplified before combining it with other similar outputs for down-link transmission.

06. Explain how by employing 'V' and 'H' polarization of E-M waves, two identical sets of 12 up-link channels becomes possible within the same 500 Mhz transmit bandwidth. Draw block diagram of such a 24 channel transponder arrangement and label it suitably.

07. In the frequency reuse method of obtaining 24 channels within the same 500 MHz

transmit-receive band-width, the centre frequencies of channels having identical numerical values are offset by 20 MHz between the two i.e. 'V' and 'H' polarizations and by 40 MHz between those having same polarizations. Based on this, draw charts to indicate relative locations of all the 24 channels for both up-link and down-link transmissions. Also indicate channel numbers and centre-frequency of each channel on the chart.

08. Both at send and receive earth stations, frequency conversion is done in two stages to overcome image signal interference. Usually, the first conversion contains local-oscillator (L.O) frequency control to select different channels. For this, frequency-synthesis technique is used to obtain exact and stable L.O. frequencies. To illustrate this, draw a 12 channel down converter arrangement employing frequency synthesizers.

09. Draw block diagram of a TV 'RECEIVE ONLY' (TV RO) earth station receiver and explain how the converted down-link IF signal of a TV channel is processed to obtain both video and audio signals.

10. Draw block schematic of a terrestrial transmitter and explain how the video and audio signals as obtained from the earth station receiver are processed and then combined before feeding to the transmitting antenna.

Chapter 13. Cable Television Channels

01. Explain why is it desirable to restrict CATV channel frequencies in the lower region of VHF bands. While allotting lower frequency CATV channels, what steps are taken to minimize adjacent channel interference?

02. Make a chart to show how the 46 cable TV channels marked for the 625 line CCIR B&W and PAL P&G systems are distributed in the vacant bands after TV channel 4, in mid-band, super-band and hyper-band. Also explain why the allotted channel space is raised to 8 MHz in the hyper-band.

03. Explain why the TV channels of bands VHF-I and VHF-III not used for terrestrial transmissions, are the first to be allotted in any CATV set-up for showing popular TV channels. What is the main demerit of using cable channels located in super and hyper bands?

04. Explain in which un-used band, additional cable channels are located when it becomes necessary to extend total cable channels from 46 to 50.

Chapter 14. Cable Signal Processing

01. Define angles of latitude and longitude which are used to locate any place or point on the surface of earth. Explain how these angles become reference points for the acquisition of satellites. Why is the angle of longitude (E or W) enough to tell the location of a communication satellite parked on the equatorial orbit?

02. Define and explain clearly with illustrative diagrams, the significance of Azimuth and Elevation angles which enable correct orientation of dish antennas for acquiring desired satellites. Why is the angle of elevation large for locations of cable stations close to the equator?

03. What are the main parts of a dish antenna unit? Show the location of feed horn together with LNBC on a dish and explain how the horn is orientated to collect maximum possible signal from the dish. How is the collected signal in the form of E-M wave intensity converted to proportionate voltage before feeding it to the low-noise amplifier in the LNBC?

04. Explain with a suitable diagram the 'OFFSET' technique of locating two feed horns on the same dish for acquiring two satellites if available within a reasonable arc area of the dish. Also explain with a separate diagram, how an orthogonal mode transducer (OMT) is used to receive both 'V' and 'H' polarized signals from the same satellite without any mix-up between them.

05. Draw block diagram of a typical C-band LNBC unit and explain how the 'RF-IN' from the dish antenna is amplified, down-converted and further processed to obtain IF output at 950 to 1450 MHz. How is the dc supply fed to the LNBC unit? Why in some signal reception set ups, a bullet amplifier is provided in the coaxial cable length that carries IF signal to the satellite receiver in the control room?

06. What is the purpose of a power divider (splitter)? Sketch a 5 port power divider to show how the input power gets equally divided. Explain how dc supply to the LNBC coming from the satellite receiver is kept away from the splitter unit.

07. Draw simplified block diagram of a manually operated satellite receiver, label it fully and briefly explain the functions of all the modular sections.

08. Draw detailed block schematic of the tuner of a manually operated satellite receiver and explain how channel selection takes place. Also explain how a sample of the IF output is used to drive the signal level meter.

09. Explain how in a remote controlled satellite receiver, the µP which is the control processing unit (CPU) of the receiver, on receipt of different command signals from the hand-held remote control unit, initiates the execution of different functions. In particular, explain with the help of relevant receiver blocks, how the two audio outputs are obtained for stereophonic sound reproduction.

10. Draw complete block diagram of a fixed frequency channel modulator and explain how the video and audio inputs from the satellite receiver are processed, combined to form a composite signal and then up-converted at the IF section before modulation on the allotted channel carrier. Also explain, how any channel carrier on generation is maintained at its fixed RF frequency.

11. How is an agile modulator different from a fixed frequency modulator? Draw basic block diagram of an agile channel modulator and explain the need for two IF sections in this modulator.

12. Describe essential features of a Character Generator, commonly used at TV and CATV stations. Explain how various letters, graphics and related data are generated and later combined in the modulator for display on the receiver screen along with the main programme.

13. What is the difference between a Signal Processor and a modulator? Explain how

the signal of a local broadcast on receiving through a conventional antenna is demodulated and transferred to the cable channel carrier assigned to it.

14. What is the function of a combining unit at the cable TV station? Why it has to be a linear mixer and not of heterodyne type? While a resistive summing network is suitable for smaller CATV units, transformer—coupled signal combiners are preferred for stations offering large number of channels. Draw block schematic of a section of such a combiner and explain how outputs from smaller units are combined before amplification and distribution on the cable network.

15. A summer employing an operational amplifier (OP-AMP) enables mixing and amplification of RF signals obtained from the outputs of modulators. Explain with a block diagram how the OP-AMP performs the two functions. Why such a combiner is called an Active Combiner?

Chapter 15. Cable Signal Distribution

01. Draw flow chart to show how the cable signal is distributed over the network to subscribers. Label each part of the network and briefly explain its function.

02. Distribution cables cause considerable signal loss, more so, for channels located in the high frequency domain. Account for such losses and explain what steps are taken to reduce them so as to keep the signal level nearby same for all the channels.

03. Cable network amplifiers tend to introduce many types of distortions in cable signals. State the causes of such distortions and explain the methods used to keep these as low as possible.

04. State essential performance characteristics of trunk and line amplifiers and explain with a block diagram how the AGC and ASC controls operate to keep overall gain of these amplifiers nearly the same over the entire frequency range of cable signals.

05. Draw functional block diagram of a push-pull trunk amplifier and explain how the AGC and ASC controls located on the control board are used to obtain uniform gain and frequency response over its entire bandwidth.

06. Draw simplified block diagram of a power doubling trunk amplifier and explain how is this different from the push-pull configuration. Give details of the housings used for these amplifiers and the need for ports on them. How is the signal level measured through these ports?

07. What is the purpose of line splitters that are provided along distribution lines of CATV networks. Draw circuit details of two-way splitters and explain how the signal power gets equally divided between the two branches yet maintaining the desired 75 ohm impedance between them. Describe how the splitter network is modified to enable bypass flow of ac power supply to successive amplifiers.

08. How is a directional coupler different from a line splitter? Draw the circuit of a two-way hybrid directional coupler and explain briefly how it operates. On which section of cable lines are line taps located and for what purpose?

09. Set-Top cable signal converters are usually of 'UP-DOWN' conversion type. Ex-

plain why is it necessary to first shift the incoming channel signals to a band in the UHF region and later down convert these to a common TV channel domain, often channel 3 or 4, for actual display on the receiver screen. Draw block diagram of such a converter and explain its operation.

10. For precision tuning in the VCO of up-converter in the SET-TOP converter, a PLL frequency synthesizer is used. Draw block diagram of such a converter to explain how the frequency selection for each channel is obtained. Also explain how the remote control communicates with the frequency synthesizer unit.

Chapter 16. Cable Television Systems

01. Keeping in view the availability of satellites of interest in your land location, suggest a 6 channel CATV system for a small locality. Illustrate your answer by drawing the proposed plan of the head-end equipment.

02. Assuming that the number of subscribers is sufficiently large, suggest and draw head-end layouts for both 11 channel and 16 channel cable systems, preferably using a different combining unit. Indicate and justify the chosen cable channel allocations.

03. Describe how you will proceed to design a 32 channel cable TV system with provision to extend it to 50 channels. Justify the proposed number of dish antennas to be installed based on the choice of channels to be shown from different satellites. Also indicate channel allocations in the VHF and higher bands.

04. Explain how an orthogonal mode transducer (OMT) is installed along with the LNBC to avail of both 'V' and 'H' polarized channels coming from the same satellite. Illustrate your answer by drawing a complete layout meant to receive 8 channels with 4 from each polarized mode.

05. How is the direct reception system (DRS) different from the conventional CATV set up? Why is a special satellite receiver recommended for exclusive use in individual homes? Draw block schematic of a DRS showing two possible ways of connecting a VCR for recording incoming channel programmes when desired.

Chapter 17. Cable System Planning and Installation

01. Describe briefly the factors which should be taken into account before venturing to install a CATV centre in any locality. State the conditions which must be met while choosing sites for control room and dish antenna farm for the proposed cable plant.

02. What factors need consideration while deciding on the number, size and type of front-end section equipments and accessories for the proposed cable station? Also give main specifications of equipments, namely—dish antenna, LNBC, bullet amplifier and power dividers, not forgetting that these must conform to the TV standards adopted by the user country.

03. Why is it desirable to decide on signal distribution cable routes in such a way that load on each trunk route is nearly the same? On knowing cable-run length for

different routes, state the factors which need consideration while deciding on the number of trunk and line amplifiers on each route. Give main specifications of a power-pass distribution amplifier.

04. In order to keep the cable TV plant cost low, narrate the sequence in which equipments should be ordered and with this aim in view, state the preparations that should be made beforehand at the cable station and dish antenna farm sites. Why is it advisable to buy all the front-end and head-end equipments from a single and reliable manufacturer/supplier?

05. Explain how would you proceed to select cable types and lengths for each section of the proposed cable system. What precautions are necessary while laying cables on different signal distribution routes? Describe with a sketch how a loop is formed to prevent any excessive pull on the joint made between ends of two cables.

06. Describe the type of foundations recommended for dish antennas of different diameters. How is the dish mount fixed firmly on the foundation? State briefly the mechanisms provided on the mount for setting the dish at the required look angle.

07. Describe with necessary sketches, the provisions made on the dish mount for setting big diameter dishes at the desired Azimuth (AZ) and Elevation (EL) angles. How are these operated to obtain the exact look angle?

08. Describe with the help of a sketch, how the LNBC and associated feed units are mounted above the dish on a suitable support. Also explain how the feed unit is set for the required polarization.

09. Describe the procedure to be followed for installing folded dipole (YAGI) antennas at the antenna site for receiving local broadcasts. What precautions you will take while setting the antenna to ensure maximum possible signal pick-up without any ghost images on the receiver screen? Why is it advisable to obtain local broadcast programmes via the satellite if available?

10. Enumerate and discuss all the preparations that are necessary before proceeding to set dish antennas at the required look angles.

11. Explain with a sketch how a magnetic compass can be used for setting any antenna dish at the required Azimuth (AZ) angle. Give step-by-step procedure to be followed while doing so. Illustrate the dish location for an AZ angle $\approx \angle 193°$.

12. Describe how an angle meter can be used for setting the antenna dish at a particular elevation angle. Illustrate this with sketches showing how the angle meter is first set at the wanted angle say $\approx \angle 67°$ and then used to raise the dish at this elevation angle. Also illustrate how a 'D' angle protector can instead be used for the same purpose.

13. State the procedure and equipment needed for tuning and thus checking availability of all the wanted channels from different dish antennas. Give block diagrams to show equipment hook-ups as done at the site of dish antenna farm. The three possible settings will depend on the available equipments. These are: (i) the available

colour receiver does not have 'V' and 'A' input sockets and hence a VCR is needed in between the satellite receiver and colour TV receiver. (ii) the colour receiver has 'V' and 'A' input sockets. (iii) the satellite receiver has an in-built modulator for enabling composite modulated output on the carrier of a TV channel, usually 2/4.

14. Assuming that the equipments to be used for tuning-in various satellite channels are, a satellite receiver, a VCR and a colour receiver, describe how on interconnections will you set various controls on these equipments for receiving a particular channel. On connecting power supply to the equipments state how will you manipulate these controls to obtain picture and sound of the channel selected at the satellite receiver.

15. While tuning a particular channel, if no picture appears on the TV receiver screen, explain how will you proceed to locate the faulty setting or connection and if necessary setting of corresponding dish at the correct look angle and the connection of LNBC to the satellite receiver.

16. If the channels to be received from a satellite are both in vertical and horizontal modes, explain with a block diagram, how will you connect the two satellite receivers to the common colour receiver through a two-way switch and proceed to tune-in both 'V' and 'H' mode channels without any mutual interference.

17. On checking the availability of all the wanted TV channels at the antenna farm site and connecting the LNBC outputs of all the dishes to the chosen satellite receivers in the control room, explain how would you proceed to set various equipments located there to obtain RF outputs at the combining network on the assigned cable channel carrier frequencies.

18. On commissioning the cable plant at the control room, explain how would you proceed to check the availability of all the channels with sufficient and almost same signal strength on all the trunk routes and at tap-off points.

19. Before making cable connection to the customer's receiver, why is it necessary to first check that the signal level at this point is around 55 to 65 dB (μV). If necessary, how will you alter the signal level.

20. Explain briefly how a 'SET-TOP' converter operates to enable all the available cable channels even on old versions of TV receivers which do not have 'CABLE-READY' tuners. Such converters are mostly of remote control type. State novel features that are commonly provided via remote control units for easy operation and quick access to any wanted channel.

Chapter 18. Cable Television Performance Measurements

01. Explain how a signal analysis meter (SAM), also called signal level meter (SLM) enables signal level measurements at a specific frequency out of a complex RF signal. Draw block schematic of such a meter and explain how it functions to provide signal level measurement at any desired RF frequency.

02. Describe how will you proceed to measure with a SAM, the magnitudes of video

carrier and audio carrier of a particular TV channel. What is the desirable difference in carrier levels of video and audio in any channel?

03. Explain what you understand by (i) hum modulation and (ii) signal to noise ratio. How these get added to the channel signal? Describe briefly the procedure for measuring these with a signal level meter.

04. What is the main difference between an oscilloscope and a spectrum analyzer. Draw block diagram of a superheterodyne type of spectrum analyzer and explain how it functions to display frequency components of a complex waveform. Draw amplitude waveforms that will appear on the spectrum analyzer screen on measuring video and audio carrier levels and showing frequency spacing between them.

05. How is a decibel meter different from a conventional voltmeter and why is it preferred for signal level measurements in cable systems. Deduce, with the help of a functional diagram, how an operational amplifier (OP-AMP) with linear feedback applied, produces an output voltage that is proportional to the logarithmic of applied input voltage.

06. Explain how the same meter movement in a multimeter is used to measure voltage, current and resistance in electrical circuits. How is a digital multimeter different from an analog type meter and why is it preferred for measurements in electronic circuits.

Chapter 19. Cable System Maintenance and Fault Isolation

01. What periodic checks are necessary on dish antennas, LNBCs and associated cable connectors in the Front-End section for trouble free operation of a cable TV system.

02. Give a list of checks which should be regularly carried out on equipments located in the control room to ensure good signal quality and uninterrupted reception of all the channel programmes.

03. Signal distribution cables lay spread in the open over the entire area of cable network. State regular checks that are necessary to prevent any disconnection in these cables. Also describe performance checks that are needed on trunk/line amplifiers and associated passive networks to maintain good quality service.

04. Assuming connecting cables on dish antenna and control room equipments to be fault free, account for the following discrepancies noticed on the monitor receiver connected at the output of combining network at the cable station:

   (i) monitor receiver screen remains blank for all the cable channels (no signal output condition).

   (ii) no signal output from one particular channel out of several being received from the same dish antenna unit.

   (iii) no output from VCR/CD player channels.

05. Describe the checks necessary for locating the following faults as observed on the monitor receiver connected across output point of the combining network:

(i) weak colours on pictures of one particular channel.

(ii) weak sound output from one channel as compared to other channels.

(iii) hum noise in the output of one particular channel.

06. Describe the procedure to be followed for locating the following faults:

(i) intermittent reception on all channels being received from one dish antenna unit.

(ii) weak pictures and sound from one particular channel.

(iii) black and white spikes on the picture and noise in sound output on one channel out of the many being received from a particular dish antenna.

07. Assuming normal reception on all the channels at the output of combining network, describe the procedure to be followed for locating the following faults occuring in the signal distribution network:

(i) signal not available to customers on one distribution route.

(ii) no signal to one particular customer.

(iii) weak signal delivery to all the TV sets fed from one particular cable route.

08. How would you proceed to isolate and correct the following drawbacks in reception on some TV sets:

(i) ghost images on receivers fed from a particular cable sub-route.

(ii) appearance of bars on pictures of some channels.

(iii) intermittent reception on one or more receivers.

09. Reason out the conditions which can cause the following faults on some customer's receivers:

(i) overlapping of pictures of two different channels.

(ii) deterioration in picture quality of channels located in super and hyper bands.

(iii) netting on pictures of receivers connected on one route of the distribution network.

10. For restoring normal operation of a faulty equipment at the cable station, make a list of check you will carry out by inspection, continuity tests and dc voltage measurements before deciding to send it for servicing to its manufacturer or supplier.

11. In modern equipment, modular technique is used in their production. While servicing one such equipment—say a satellite receiver, suggest which module(s) should be replaced for the following faulty conditions:

(i) video output is distorted.

(ii) video output not present but audio is available.

(iii) power indicator is not illuminated.

12. The complaints listed below pertain to individual subscribers. Explain how you will proceed to restore normal reception.

(i) overloaded (distorted) pictures on all channels.

(ii) weak pictures on all channels.

(iii) picture rolls and is of insufficient height.

(iv) picture not steady on a particular channel.

(v) on reception with 'SET-TOP' converter is use.

Chapter 20. Elements of Digital Communication

01. What are the main advantages of signal transmission in digital form over that in analog form? In decimal system, the number 25 can be expressed as $2\times10^1 + 5\times 10^0$, show that the same number when expressed in binary form, where the base is 2 instead of 10, is written as 11001.

02. Since the real world is analog, data in this form is first converted to binary form for processing in digital equipments, like computers and results are converted back to analog form for human consumption. As necessary special methods have been evolved for analog to digital (A/D) and digital to analog (D/A) conversions. Illustrate with basic circuits the techniques used for such conversions.

03. Draw elementary block diagram of a digital computer and with this as the base, describe what you understand by the terms:

(i) hardware, (ii) software, (iii) program, (iv) random access memory (RAM), (v) read only memory (ROM), (vi) central processing unit (CPU), (vii) microprocessor (μP) and (viii) microcomputer (μC).

04. Signal processing in digital form is mostly done in parallel format which needs as many interface lines as the number of bits in the data, which is usually quite large. However, for long distance transmission on conventional electrical lines, the data is first converted to serial form and then sent bit-by-bit one after the other. Explain why it becomes necessary to do so. Also name the equipment that is used for parallel to series and series to parallel transformations.

05. Explain why long distance telephone lines are not suitable for data transmission in digital form. Describe how the problem is solved by converting binary data to low frequency ac form before transmission. How the identity of 0s and 1s is retained on their conversion to sinusoidal form?

06. Describe with suitable waveforms how the digital data is converted to ac form by Phase Shift Keying (PSK) method. While doing so, why separate frequencies are allotted for 0s and 1s. Also explain how the transmission speed on long distance lines is increased by restoring to quadrature phase shift keying (QPSK) and if necessary to 8-QAM for still higher speeds.

07. For digital data transmission over telephone links, modems have been devised for necessary conversion and transformation of data at either end of the line. Draw

block diagram of such a modem employing PSK modulation and explain how by the use of B-P filters two way communication becomes possible.

08. Draw basic block diagram of a digital transmission and reception system. Explain function of each block assuming that a speech signal of 5KHz width is to be transmitted to another station.

Chapter 21. Signal Encoding and Compression

01. Why is video signal quantised at 256 levels and sampled at a rate equal to 4 × f_{sc} for conversion to binary code? Show that with these rates a binary data equal to 284 MB/S becomes necessary for transmitting colour TV signals.

02. Show that for digital transmission of television audio, a data rate close to 42 KB/S becomes necessary. What will be the bit rate for stereo audio transmission?

03. The first stage for digital satellite transmission of TV signals and associated data is encoding by which analog video and audio signals are first quantised and then converted to digital form. Draw block diagram of a pulse amplitude modulation (PAM) type of encoder and explain how video and audio signals are first quantised and then converted to digital (binary) form by continuous A/D conversion.

04. Why is it necessary to compress digital video and audio signals before transmission? What are the factors that make this possible?

05. Describe briefly the techniques used to compress video and audio signals. Why is data stream regulation necessary and how is it achieved?

06. Describe the development of video and audio signal compression formats and justify the use of MPEG-2 format for video and MP-2 for audio signal compressions.

07. Why very high quantisation levels and sampling rates are used for recording music on compact discs (CDs)? Explain how Perceptual Noise shaping is used to develop MP3 compression algorithms for audio signals to enable a compression ratio of 10 to 14.

Chapter 22. Digital Satellite Transmission and Reception

01. On obtaining binary data from the A/D converter, it is further processed in three stages before modulation and transmission—the first stage is data compression, which usually forms part of the encoder unit, the second stage is encryption and the third packetising. Draw block diagram of a UP-LINK set up which includes these stages and explain fully the need and method of affecting such operations. What is the need of clock pulses in the UP-LINK section?

02. Draw block diagram of a transmission set-up that follows up-linking unit and explain how the modulated signal on receipt from the up-linking site is first frequency converted to the assigned GHz range and then amplified before feeding to the dish antenna for radiation. Explain what provision is usually made at the transmitter to ensure uninterrupted transmission.

03. Draw functional block diagram of a digital broadcast receiver and explain how it decodes the down linked TV signal, fed to it via LNBC, to obtain video and audio outputs. What is the role of micro-controller (μC) in the functioning of digital receiver?

Chapter 23. Television Conditional Access System (CAS)

01. What is conditional access system and why is it employed in TV transmission? Give an overview of this system, and state requirements to be met for CAS implementation.

02. Describe briefly the two main schemes of conditional access system and describe operation of CAS with a suitable block diagram.

03. Draw functional block schematic of the Simulcrypt CAS and explain how the subscriber is authorised to receive pay channels. What is the procedure for billing if additional channels are to be received?

Chapter 24. Direct-To-Home (DTH) Satellite Television

01. State distinguishing features of Direct-To-Home or Digital Satellite Broadcast System and explain how in the absence of two-way cable network system, the receiver communicates with the service computer through a modem for billing purposes and to have access to additional services provided by the owner producer.

02. Explain how the TV program material provided by its producer is processed and audited at the studio before sending it to the up-link site. What is the purpose of conditional access data, P.C. compatible data and program guide that is provided as additional material by the owner producer?

03. Draw block schematic of a representative up-linking set-up designed for a D.T.H. broadcasting system. In the diagram include at least two TV channels with stereo sound pick up provision. Explain fully how the camera and microphone outputs are encoded, compressed, encrypted where necessary and combined to form packet units on including other services provided by the producer as digital data via computers. Also explain how all the packetised outputs are sampled sequentially and the serial output thus obtained processed for modulation and onward transmission to the transmitting site. Why are KU band frequencies preferred for digital satellite broadcasting?

04. Draw block schematic of a DTH satellite receiver and explain how the signal picked-up by the dish antenna is processed through the tuner, Frequency Error Correction (FEC) unit, transport IC, video and audio decoders and finally by the PAL encoder and associated dual audio D/A converter to enable analog video and audio signals. Also explain, how the desired TV channel is selected through the micro-controller on receipt of coded input from the hand-held remote control unit. What is the purpose of Access Card (Smart Card) that is supplied with the DTH receiver?

05. Sketch front and back panels of a typical DTH receiver and label on these all the

provided controls and connecting sockets. Explain briefly the need and purpose of all such provisions. In particular explain the use of 'S' video 'out' and wide and low band data outputs.

06. Describe all relevant features of the DTH TV system for the benefit of subscribers who intend to get connected to this service to fully understand the merits and demerits of doing so.

Chapter 25. High Definition Television (HDTV)

01. State parameters of HDTV and explain how is it different from conventional satellite TV broadcasts.

02. Explain how 'MUSE' encoding of colour TV signals removes impairments that otherwise become noticeable on large size HDTV pictures.

03. Why is HDTV transmission carried out in the KU band? What additional equipments become necessary for its transmission? Why is terrestrial transmission of HDTV more common at present?

04. What are the reception options for subscribers who wish to have access to HDTV transmission? Why is it expensive to do so?

Chapter 26. Television Home Entertainment Theatres

01. Explain fully how stereo sound is created. How does it lead to surround sound audio system?

02. Describe briefly the three main surround sound formats. Why is Dolby Digital format superior from Dolby Prologic?

03. What are the functions of a surround sound stereo receiver? Why is the choice of loudspeakers critical for high quality sound reproduction?

04. Describe distinguishing features of the Direct-View Projection Screen television units. Why are projection TV systems more suitable for receiving HDTV transmission and viewing movies?

05. Explain briefly how DLP projection television operates to display pictures. Why is this considered superior to other TV receivers?

06. What are the distinguishing features of VCRs and DVD players? Will DVD players eventually replace VCRs?

07. What are the Home Theatre options for subscribers for whom the only available OFF-AIR transmission is TV programmes either through satellite or terrestrial broadcasts?

08. Describe the requirements of a TOP OF THE LINE Home Theatre set up and explain how is it possible to receive different transmissions through a common tuner equipment.